ADVANCES IN CONDENSED MATTER AND MATERIALS RESEARCH

ADVANCES IN CONDENSED MATTER AND MATERIALS RESEARCH

VOLUME 10

ADVANCES IN CONDENSED MATTER AND MATERIALS RESEARCH

Additional books in this series can be found on Nova's website
under the Series tab.

Additional E-books in this series can be found on Nova's website
under the E-books tab.

ADVANCES IN CONDENSED MATTER AND MATERIALS RESEARCH

ADVANCES IN CONDENSED MATTER AND MATERIALS RESEARCH

VOLUME 10

HANS GEELVINCK
AND
SJAAK REYNST
EDITORS

Nova Science Publishers, Inc.
New York

For permission to use material from this book please contact us:
Telephone 631-231-7269; Fax 631-231-8175
Web Site: http://www.novapublishers.com

Library of Congress Cataloging-in-Publication Data

ISSN: 2159-1709

ISBN 978-1-61209-533-2

Published by Nova Science Publishers, Inc. † New York

CONTENTS

PREFACE

Condensed matter physics is the field of physics that deals with the macroscopic and microscopic physical properties of matter. This new book discusses topics such as the field electron emission theory; organic-inorganic hybrid proton exchange membrane electrolytes for medium temperature non-humidified H2/O2 fuel cells; electronic band structure calculations based on empirical psuedopotential formalism; Fractional Hall effect; topological insulators; rate-dependent hysteresis in piezoelectric actuators and an itinerant-electron metamagnetic system with two-band Eigenvalue spectrum.

Chapter 1 - The field electron emission theory has been successfully applied to various vacuum electron sources and devices. However, recent rapid progress of nano technology provides many new choices of the emitter materials, such as nano wire and nano tubes. The field emission experiments indicate some novel phenomena in field emission of carbon nanotubes, which inspires us to reconsider the fundamental concepts and assumptions in the conventional field emission theory. The authors will review the basic field emission theory and its new development in nanoscale material emitters recent years, including the introduction to the basic conceptual modification and generalization for nanoscale emitters, the new field emission theories for the carbon-nanotube, Luttinger liquid and nanowire, and the general empirical theory. The authors also briefly introduce the computer simulation for carbon nanotubes and some novel results in the carbon nanotube field emission. These new developments in field emission theory provide a new physical perspective and understanding of the nanoscale material field emission, and a calculation framework to investigate the field emission phenomena of the nanoscale-materials field emission. The authors also give an outlook in understanding of the nanoscale-material field emission.

Chapter 2 - Anhydrous proton conducting inorganic-organic hybrid membranes were fabricated by sol-gel process with tetraethoxysilane/ tetramethoxysilane/ methyltri-ethoxysilane/trimethylphosphate and 1-ethyl-3 methylimidazolium tetrafluoroborate [$EMIMBF_4$] ionic liquid as precursors. These hybrid membranes were studied with respect to their structural, thermal, conductivity, and hydrogen permeability properties. The Fourier transform infrared spectroscopy (FT-IR) and ^{31}P, ^{29}Si, ^{1}H, ^{13}C, and ^{19}F Nuclear magnetic resonance (NMR) measurements have shown good chemical stability, and complexation of $PO(OCH_3)_3$ with [$EMIMBF_4$] ionic liquid in the studied hybrid membranes. For all the prepared membranes, average pore sizes and specific surface areas were investigated by the Brunauer–Emmett–Teller (BET) method. The average pore size was increased proportionally with the ionic liquid weight percent ratio in the host phosphosilicate matrix within the range 1 - 9 nm, respectively. Thermogravimetric analysis (TGA) and differential thermal analysis (DTA) measurements confirmed that the hybrid membranes were thermally stable up to 290 ^{0}C. Thermal stability of the hybrid membranes was significantly enhanced due to the presence of inorganic SiO_2 framework and high stability of [BF_4] anion. The effect of [$EMIMBF_4$] ionic liquid addition on the microstructure of the membranes was studied by Scanning electron microscopy (SEM) and Energy dispersive X-ray analysis (EDX) micrographs and no phase separation at the surfaces of the prepared membranes was observed and also homogeneous distribution of all the elements was confirmed. Conductivity of all the prepared membranes was measured within temperature range -25 to 155 ^{0}C, and high conductivity of 1 x 10^{-2} S/cm was obtained for 40 wt% [$EMIMBF_4$] doped 30TMOS-30TEOS-30MTEOS-10PO(OCH_3)$_3$ (mol%) hybrid membrane, at 155 ^{0}C under anhydrous conditions. The hydrogen permeability was found to decrease from 1.41 x 10^{-11} to 0.1 x 10^{-12} mol/cm.s.Pa for 40 wt% [$EMIMBF_4$] doped hybrid membrane as the temperature increases from 20 to 150 ^{0}C. For 40 wt% [$EMIMBF_4$] doped hybrid membrane, the membrane electrode assemblies were prepared and a maximum power density value of 11 mW/cm^2 at 26 mA/ cm^2 as well as a current density of 40.57 mA/ cm^2 were achieved at 150 ^{0}C under non-humidified conditions during utilization in a H_2/O_2 fuel cell.

Chapter 3 - Electronic band structures for zinc blende (GaAs, AlAs) and diamond (Ge, Si) nanostructure semiconductors were performed. The calculations were based on the empirical pseudopotential method (EPM), which takes into account the effects of both hydrostatic pressure and temperature. The results showed an interesting behavior, that the pseudopotential form factors associated with the reciprocal lattice vectors of $|\vec{G'} - \vec{G}|^2 = 11$ are more affected by both pressure and temperature than the other values of the form factors. Therefore, the effects of pressure and temperature must be considered when dealing with electronic band structure calculations based on empirical pseudopotential formalism.

Chapter 4 - In this chapter, the phenomenon of rate-dependent hysteresis in piezoelectric actuators is described based on an experimental result. Then, a brief analysis on this phenomenon is presented. After that, an expanded input space based modeling approach previously proposed by the authors is presented. On this constructed expanded input space, the multi-valued mapping of rate-independent hysteresis can be transformed to a one-to-one mapping. Thus, any traditional modeling methods can be applied to modeling of the behavior of rate-independent hysteresis on this expanded input space. Then, a model with parallel structure is proposed to approximate both hysteretic behavior and rate-dependent feature of the piezoelectric actuators. In the parallel model, the rate-independent hysteretic performance is approximated by a neural network based on the expanded input space method, while the rate-dependent behavior of the hysteresis is approximated by a network with the sum of weighted first-order difference operators. For the compensation of the effect of hysteresis, the corresponding inverse model for rate-dependent hysteresis is obtained. Furthermore, a compensator is constructed to compensate for the effect of the rate-dependent hysteresis in piezoelectric actuators. Moreover, the experimental results of applying the proposed method to a piezoelectric actuator are presented.

Chapter 5 - The process of light interaction with condensed matter is an important problem of the modern solid state physics and quantum electronics. Holographic recording in wide-gap photorefractive materials is of particular interest from the scientific and practical point of view. Such a recording includes the stage of space charge formation, which is associated with the spatial redistribution of photoinduced carriers. A number of new methods studying the dynamics of space charge and photoconductivity in widegap semiconductors were proposed during recent years.

The approach based on non-steady-state photocurrent investigation is considered to be very attractive and powerful technique for characterization of wide-gap semiconductors, molecular crystals, amorphous materials, photorefractive polymers. The development of the new experimental and theoretical approaches for investigation of wide-gap semiconductors and nanostructured materials including photorefractive sillenites grown in the air and argon atmosphere, molecular crystal SnS_2, boron nitride crystal and nanosized GaN and Se within porous matrices is the main goal of this work.

Chapter 6 - Liquid crystals are signature systems of soft condensed matter media. Their unique features make them very susceptible to the action of external agents. The authors focus here in the study of their hydrodynamical and optical properties so the authors first review the formalism to describe these properties. Nematic liquid crystals are fluids that exhibit long-range orientational order over distances many times larger than the dimensions of the molecules of which they are composed. Their intrinsic anisotropy is responsible for considering liquid crystals as interesting fluids for electrorheological and optical applications. They offer obvious advantages over the more conventional fluids such as avoiding the problems associated with the settling of the dispersed phase; these complications are nonexistent for liquid crystals since they are homogeneous phases. The authors study various viscometric properties of a nematic submitted to distinct external imposed fields and confined under different geometries including the planar and cylindrical ones. The authors show the existence of an electrorheological effect for which the effective viscosity increased various magnitude orders its value upon the application of an electric field. The authors discuss also the appearance of various non newtonian and directional behavior as a function of an external applied electric field. The authors develop an asymptotic geometrical formalism to describe the propagation or light in liquid crystals. The authors use this to study the propagation of monochromatic rays in nematic and smectic bent core liquid crystals confined in planar cells, cylindrical pores and droplets. The authors show how the photo refractive effect can be controlled by externally imposed agents like electric fields and imposed flows. Various phenomena related like beam steering, dispersion, total internal reflection and scattering of light are also discussed.

Chapter 7 - Wide bandgap semiconductors such as SiC, GaN, and diamond have a potential for use in high power and high frequency devices, while narrow bandgap semiconductors such as the GaSb family have a potential for near- and mid-infrared laser diodes and photo-detectors for detecting CO_2, CH_4, NO_x, and SO_x. In these next-generation semiconductors, it is essential to precisely determine the densities and energy levels of dopants (donors or acceptors) as well as unintentionally-introduced impurities and defects, which affect the majority-carrier concentrations in semiconductors. The authors have developed a graphical peak analysis method called Free Carrier Concentration Spectroscopy (FCCS), which can accurately determine them using the temperature dependence of the majority-carrier concentration without any assumptions regarding dopant species, impurities, and defects. The authors have determined the densities and energy levels in undoped, N-doped or Al-doped SiC. Moreover, the dependence of the energy level of each dopant species on dopant density has been obtained. From the temperature dependence of the majority-carrier concentration in SiC irradiated by high-energy electrons, the dependence of the density of each dopant or defect on fluence has been determined. Because wide bandgap semiconductors have a low dielectric constant and a hole effective mass heavier than an electron effective mass, the energy levels of acceptor species are expected to be deep. When x is larger, on the other hand, the donor level of Te in $Al_xGa_{1-x}Sb$ is changed from shallow to deep, just like $Al_xGa_{1-x}As$. In these semiconductors, the density of dopants determined from the temperature dependence of the majority-carrier concentration is much higher than the concentration of dopant species determined by secondary ion mass spectroscopy. When the energy level of dopant species is deep, the excited states of the dopant should affect the majoritycarrier concentration. Therefore, the authors have introduced the occupation function of the dopant with deep energy level, which includes the influence of its excited states, instead of the Fermi-Dirac distribution function. Using the occupation function the authors have proposed, they have investigated semiconductors with deep level dopants by FCCS using the temperature dependence of the majority-carrier concentration.

Recently, high-resistivity or semi-insulating semiconductors have been required to use as substrates of GaN field-effect transistors and as active layers of X-ray detectors capable of operating at room temperature. Because these defects affect the majority-carrier concentrations in high-resistivity or semi-insulating semiconductors and degrade the device performance, it is required to accurately determine the densities and energy levels of defects with deep energy levels. Transient capacitance methods such as Deep Level Transient Spectroscopy cannot apply higher-resistivity semiconductors. Therefore, the authors have been developing a graphical peak analysis method using an isothermal transient current in the diode, called Discharge Current Transient Spectroscopy (DCTS). Using DCTS, the authors have determined the densities, emission rates, and energy levels of deep defects in semi-insulating SiC and thin insulators. Moreover, the density-of-states in high-resistivity amorphous semiconductors are investigated.

Chapter 8 - At very low temperatures, a two-dimensional electron system in a perpendicular magnetic field exhibits remarkable quantum phenomena where the fractional quantum Hall effect (FQHE) stands out as one of the most important discoveries in condensed matter physics for the last decades. FQHE represents a unique example of a novel collective quantum liquid state of matter that originates from strong electronic correlations only. A perpendicular magnetic field leads to the creation of massively degenerate discrete quantum states known as Landau levels. The energy quantization associated with such states is the foundation of many new experimental and theoretical advances in condensed matter physics. The filling factor, defined as the ratio of the number of electrons to the degeneracy (number of available states) of each Landau level, represents an important characteristic parameter of the system. Typical liquid FQHE states are more pronounced in the extreme quantum limit of very high perpendicular magnetic field where the lowest Landau level is fractionally filled with electrons. Stabilization of these novel electronic phases happens at special filling factors that generally have odd denominators. Among them, the most robust FQHE states correspond to filling factors 1◁3 and 1◁5 and are well described by Laughlin's theory in terms of trial wave functions. Differently from odd-denominator-filled states in the lowest Landau level, even-denominator-filled states with filling factor 1◁2, 1◁4 and 1◁6 do not show typical FQHE features and behave as isotropic compressible metallic Fermi liquid states. The composite fermion theory for the FQHE shed light on the Fermi-liquid nature of such even-denominator-filled states. On the other hand,Wigner crystallization occurs when the filling factor becomes around or less than 1◁7. In this work, the authors attempt to give a brief overview of the basic theory of the fractional quantum Hall effect. To this effect and for brevity of treatment, the authors focus their attention only on states on the lowest Landau level and try to explain in a simple manner some of the key ideas and models used to study such an intriguing phenomenon.

Chapter 9 – The authors introduce new methods for investigating propagation of electrons in a multi band system with spin orbit interaction which are time reversal invariant . Therefore Kramer's theorem imposes constraints which give rise to non-trivial Berry connections. As a result chiral zero modes appear at the interface between two domains characterized by parameters which are above or below some critical values. The variation of the parameters is due to disorder, geometry or topological disorder such as dislocations and disclinations. The mechanism might explain the high conductivity coming from the bulk of the topological insulators. The authors introduced the method of curved geometry and study the effect of dislocations on the Topological Insulators. As a demonstration of the emergent Majorana Fermions the authors consider a P-wave wire coupled to two metallic rings.

In: Advances in Condensed Matter ... Volume 10
Editors: H. Geelvinck and S. Reynst
ISBN: 978-1-61209-533-2

Chapter 1

FIELD ELECTRON EMISSION THEORY AND ITS NEW DEVELOPMENT

Shi-Dong Liang*
State Key Laboratory of Optoelectronic Material and Technology and
School of Physics and Engineering, Sun Yat-Sen University,
Guangzhou, People's Republic of China

ABSTRACT

The field electron emission theory has been successfully applied to various vacuum electron sources and devices. However, recent rapid progress of nano technology provides many new choices of the emitter materials, such as nano wire and nano tubes. The field emission experiments indicate some novel phenomena in field emission of carbon nanotubes, which inspires us to reconsider the fundamental concepts and assumptions in the conventional field emission theory. We will review the basic field emission theory and its new development in nanoscale material emitters recent years, including the introduction to the basic conceptual modification and generalization for nanoscale emitters, the new field emission theories for the carbon-nanotube, Luttinger liquid and nanowire, and the general empirical theory. We also briefly introduce the computer simulation for carbon nanotubes and some novel results in the carbon nanotube field emission. These new developments in field emission theory provide a new physical perspective and understanding of the nanoscale material field emission, and a calculation framework to investigate the field emission phenomena of the nanoscale-materials field emission. We also give an outlook in understanding of the nanoscale-material field emission.

* E-mail: stslsd@mail.sysu.edu.cn.

I. INTRODUCTION

Electron sources are widely used nowadays in modern society and play a central role in information display. The conventional electron sources are implemented based on mostly thermoelectron emission called cathode ray tubes. Recent years, more and more instruments requires source with brightness and monochromatic emission. Cold electron sources based on field emission promise a potential to realize high monochromatic electron beam, which is becoming growing attractive. Although some field emission devices have been commercially available, such as Mo tips, researchers are actively seeking for alternative materials.[Heer, Rinzler] The rapid advance of the nano-technology allows us to successfully synthesis many low-dimensional and nano scale materials (NSM), such as carbon nanotubes (CN), GaN nanotubes and ZnO nanowires. These new quasi-one-dimensional materials not only provide us many choices for the field emitter but also bring us many new problems and challenges.[Falvo, Bonard] The experimental studies have demonstrated that CNs exhibit excellent field-emission performance, such as low turn-on field, high current density and high coherent energy spectrum.[Bonard] Some interesting phenomena, for example the emission current saturation in strong electric fields and both the single- and multi- peaks of the energy spectra, were observed, which deviates the conventional Fowler-Nordheim (FN) theory. [Bonard] Thus, the CN field emission brings us some fundamental issues. How do we understand theoretically these properties of CN field emission? Is there any new field emission mechanism for nanoscale materials? What is the basic physics of the NSM field emission?

Recent years, there have been a lot of effort focused on the field emission properties of CNs, many new theories emerge, such as the Dirac electron theory and Luttinger liquid theory to the basic physics of NSM, and the computer simulation to the current-voltage characteristic,[Bonard] the energy band effect in the emission current and the emission energy spectra,[Liang 03, 06] the cap effect,[Peng, Kim, Mayer] the defect effect,[Lu, Kim] and the field penetration effect.[Peng] The computer simulation methods cover the tight-binding approach, [Liang 03-06] the first-principle calculation,[Zheng,Peng,Lu] and the scattering theory.[Mayer, Adessi] These works investigate the field emission properties of CNs from different points and levels, and give some meaningful hints for understanding the novel properties of the CN field emission..

In this chapter, we will review the theoretical progress on the field emission in recent years. We will first introduce briefly the basic concepts of the field emission and the conventional field emission theory in the section II. After reviewing the experimental observations and new phenomena in the field emission of CNs, we will give the theoretical issues and the physical considerations on the NSM field emission in the section III. In the section IV, we will review the new development of field emission theory, including the Dirac electron theory of CN field emission, Luttinger liquid theory of field emission, and the general empirical theory. In the section V, we will present the computer simulation for the CN field emission and their results. The spin polarized field emission will be presented in the section VI. We give a discussion and perspective in the section VII. Finally, we will a conclusion.

II. Basic Concepts of Field Emission and Fowler-Nordheim Theory

2.1. Basic Concepts and Aims of Field Electron Emission

The study of the electron emission from metals under an applied electric field has a long history, which may trace back to about seventy decades ago. Physically, electrons emitted from metals have two mechanisms. One is that electrons tunnel through the vacuum potential barrier under a high applied field, which is called as field emission, the other is that electrons overcome the vacuum potential barrier coming out from metals in high temperature, which is called as the thermoinic emission. In practice, the basic aims of the field emission include (1) high emission current density, (2) high coherent electron spectra, (3) low driving field, (4) long life time, (5) stability and controllability. The key point to solve these problems involves how to select excellent materials and how to design optimal emitter. Besides the experimental investigation theoretical study can clarify the physical mechanism of electron emission, providing a guideline to seek the crucial physical factors to optimize the field emission performance.

The successful model of field emission was proposed by Fowler-Nordheim, [Fowler] in which the emitter is modeled by the free electron gas and the vacuum potential is assumed by the triangle potential. The I-V characteristic and the emission energy distribution can be obtained analytically in some approximations.[Forbes99] In order to give a complete description of the field emission theory, we will give a briefly introduction of FN theory in the following subsection.

2.2. Fowler-Nordheim Field Emission Theory

A complete field emission system includes three basic parts, the substrate, the emitter (cathode) and anode that offers an applied electric field. Physically, electrons in conduction bands of emitters tunnel through the vacuum potential barrier under an applied electric field. The key physical process happens at the tip of the emitter. The standard FN theory replies on a series of assumptions and approximations [Forbes99]:

(i) the effect of the substrate in field emission is so small that its effect can be neglected because the main physical process of field emission happens at the tip of the emitter;
(ii) the electron tunneling happens only along one direction because the tip of emitter is usually very sharp and electrons tunnel through the vacuum potential from the thinnest barrier of the potential barrier;
(iii) the emitter has a parabolic free-electron band;
(iv) electrons obey Fermi-Dirac statistics;
(v) the emitter has a smooth surface;
(vi) the work function is independent of external field;

(vii) the exchange and correlation interaction between the emitted electrons and the surface can be represented by the image potential;

(viii) the tunneling probability of emitted electrons may be evaluated using WKB approximation.

In general, above assumptions are physically reasonable and approximately valid for metallic emitters. Based on these assumptions, the emission current density can be written as [Forbes99]

$$j_d = \rho v f D \tag{2.1}$$

where ρ is the charge density; v is the group velocity of electron and f is the occupation probability for states. D is the tunneling probability through the vacuum potential. For the free-electron model, the kinetic energy can be written as

$$E = \frac{\hbar^2}{2m}(k_n^2 + k_p^2), \tag{2.2}$$

where k_n is the wave vector along the direction of the electron emission, and k_p is the wave vector perpendicular to the direction of the electron emission. Thus, the group velocity is written as

$$v = \frac{1}{\hbar}\frac{\partial E}{\partial k_n} \tag{2.3}$$

The emission current density can be expressed as [Forbes99]

$$j = z\int\int f(\varepsilon)D(\varepsilon, E_p)dE_p d\varepsilon, \tag{2.4}$$

where $z = \frac{em}{2\pi^2\hbar^3}$ is a constant; $\varepsilon = E - E_F$ is the total energy E relative to Fermi energy and $E_p = \frac{\hbar^2}{2m}k_p^2$ is the electronic kinetic energy perpendicular to the emitting direction. The tunneling probability can be written in terms of the Gamow exponent [Forbes99]

$$D(\varepsilon, E_p) = \exp(-G(h)), \tag{2.5}$$

where $h = \varphi - \varepsilon + E_p$ is the height of the vacuum potential barrier; φ is the local work function of the emitter. Using WKB method, G(h) can be obtained from,

$$G(h) = g\int_0^{x_c}\sqrt{V(x)}dx \quad , \tag{2.6}$$

where $g = \frac{2\sqrt{2m}}{\hbar} \approx 10.246(eV)^{-1/2}(nm)^{-1}$; V(x) is the vacuum potential barrier, and x_c satisfies $V(x_c) = 0$. The triangular barrier is a simplest form for an analytic result, namely,

$$V(x) = h - eFx \quad . \tag{2.7}$$

Thus, $x_c = \frac{h}{eF}$, Inserting the Eq.(2.7) into Eq.(2.6) and integrating Eq.(2.6), the G(h) can be evaluated

$$G(h) = \frac{bh^{3/2}}{F}, \tag{2.8}$$

where $b = \frac{2}{3e}g$. Making the Taylor-expansion of G(h) at $h = \varphi$ to the first order, and $(\frac{\partial G}{\partial h})_\varphi = \frac{3}{2F}b\varphi^{1/2} \equiv d_F^{-1}$, the tunneling probability becomes [Forbes99]

$$D(\varepsilon, E_p) = \exp(-b\frac{\varphi^{3/2}}{F})\exp(\frac{\varepsilon - E_p}{d_F}) \quad , \tag{2.9}$$

Putting Eq.(2.9) into Eq.(2.4) and considering $E_F << d_F$, the emission current at zero temperature can be obtained

$$j_{FN} = a\frac{F^2}{\varphi}\exp(-b\frac{\varphi^{3/2}}{F}) \quad , \tag{2.10}$$

where $a = \frac{e^3}{4\pi^2\hbar} \approx 1.54\mu AeV/V^2$, and $b = \frac{4\sqrt{2m}}{3e\hbar} \approx 6.83(eV)^{-3/2}V/nm$. The Eq. (2.10) is a well-known FN equation. [Fowler] For finite temperature, considering the Fermi-Dirac function and using the Beta function theory, the emission current can be obtained [Murphy, Forbes99]

$$j_{FN,T} = \Theta(T)j_{FN}(F,\varphi) \quad , \tag{2.11}$$

where

$$\Theta(T) = \frac{\pi k_B T}{d_F} \sin(\frac{\pi k_B T}{d_F}) \tag{2.12}$$

is a correction factor for the effect of temperature. At room temperature, typical values $\Theta(T) \approx 1.02 \sim 1.1$.

For a general barrier form, the Gamow exponent factor may be modified by [Forbes99]

$$G(h) = \frac{bvh^{3/2}}{F}, \tag{2.13}$$

where v is a correction factor. Similarly, the tunneling probability can be obtained [Forbes99]

$$D(\varepsilon, E_p) = \exp(bv(\varphi)\frac{\varphi^{3/2}}{F})\exp(\frac{\varepsilon - E_p}{d_F(\varphi)}), \tag{2.14}$$

where $d_F(\varphi) = \frac{d}{\tau(\varphi)}$, where τ is a correction factor defined by [Forbes]

$$\tau(h) = v + \frac{2}{3} h \frac{\partial v}{\partial h} \tag{2.15}$$

The generalized FN equation for the general barrier can be written [Forbes99]

$$j_{FN,G} = a\frac{F^2}{\varphi\tau(\varphi)}\exp(-bv(\varphi)\frac{\varphi^{3/2}}{F}) \tag{2.16}$$

For the standard Schottky-Nordheim barrier that consider the effect of the image potential, $V(x) = h - eFx - e^2 / 4x$, $v(\varphi)$ and $\tau(\varphi)$ are reduced to the well-known special elliptic function [Forbes99, Gadzuk].

The FN equations (2.10) and (2.16) give the characteristic of the field emission from metals. This characteristic of the current-voltage (I-V) can be identified by

$$\ln(\frac{j_{FN}}{F^2}) = C - B\frac{\varphi^{3/2}}{F} \tag{2.17}$$

where $C = \ln(a/\varphi)$ ($C = \ln(a/\varphi\tau(\varphi))$) and $B = b$ ($B = bv(\varphi)$) are independent of the applied electric field F, respectively for (2.10) and (2.17). This relationship between the

$\ln(j_{FN}/F^2)$ and $1/F$ is linear, which is called as FN plot and give a critical standard of the I-V characteristic for FN theory. The slope of the FN plot is related to the work function,

$$S_{FN} = \frac{d\ln(j_{FN}/F^2)}{d(F^{-1})} = -B\varphi^{3/2} \tag{2.18}$$

In practice, one can obtain the work function from the experimental value of the slope of the FN plot.

2.3. Richardson-Schottky Thermionic Emission Theory

In high temperature and weak applied fields, electrons are excited to high energy states in the conduction band. Most of emitting electrons come out from the top of the potential barrier, namely, electronic thermoinic emission. Using high temperature approximation for the Fermi-Dirac function with the tedium derivation, the thermionic emission current can be obtained [Richardson, Schottky, Murphy]

$$J_{MS} = AT^2\exp[-\frac{\varphi-(e^3F)^{1/2}}{k_BT}], \tag{2.19}$$

where $A = \frac{emk_B^2}{2\pi^2\hbar^3}$. This is the well-known Richardson-Schottky (RS) formula. The Eq. (2.19) gives the characteristic of thermoinic emission. Similarly to field emission, the Eq. (2.19) can be rewritten as $\ln(J_{MS}/T^2) = \ln A - [\varphi-(e^3F)^{1/2}]/k_BT$. This equation shows a linear relationship between $\ln(J_{MS}/T^2)$ and $1/T$, which gives the typical characteristic of the thermoinic emission. Although the FN and RS formalisms are obtained based on a series assumptions and approximations, it works well for the conventional metallic emitters. However, some novel phenomena are observed in recent experiments from nano-scale materials, which deviates the FN and RS theories and lead to new developments of field emission theory.

III. New Phenomena and Challenging Problems for Nano-Scale Materials

3.1. New Phenomena in Carbon Nanotube Field Emission

As a typical nano-scale material carbon nanotubes (CN) has been attracted much effort in field emission. It has been found experimentally that CN has excellent field emission performance, such as low turn-on field, high current density and high coherent energy spectra.

However, many novel properties of field emission have been observed experimentally. (1) The I-V characteristic of the CN field emission violates FN-type behavior in high current density. The slope of the FN plot changes typically 10%~30% increasing or decreasing depending on samples, without discontinuities or instabilities in the I-V characteristics up to 1μA.[Bonard] (2) Most of emitters, including the single multi-wall carbon nanotubes (MWCN) and nanotube films, show a strong saturation of the emission current around 1μA.[Bonard] The maximum current from one tube can reach 0.1~0.2mA, which is higher than that of conventional metallic emitters. (3) The turn-on fields for various CN emitter are in the range of 0.9~5.6V/μm, which is lower than that of most of conventional emitters, such as the diamond emitter.[Bonard] Actually, the work function of CN is estimated experimentally around 5eV,[Bonard] which is higher than most of conventional emitters. (4) The energy distribution of the emission electron is much narrower than that of conventional emitters. The full width at half maximum is typically in 0.1~0.3eV at room temperature. [Bonard] (5) The multi-peak energy distribution of emission electrons is observed experimentally on one single tube.[Bonard] These characteristics of the emission energy spectrum strongly suggest that the electrons are not emitted from a metallic continuum, but from several energy bands. These novel phenomena cannot be understood by FN theory, and lead to some fundamental questions. Is there new field-emission mechanism for CN? What do physical factors govern the nonlinear behavior of FN plot? Why do CNs with higher work function shows lower turn-on field? Do the multi-peak energy spectra stem from the cap of the tube, layer-layer interaction or intrinsic property in SWCN?

There have been some explanations to these novel phenomena based on FN theory. The emission current saturation in high field region was explained as the space charge effect [Xu] and the band bending.[Zheng] In order to explain phenomenally why high work function corresponding to low turn-on field, one may assume that the local field at the tip of tube is enhanced due to the sharp metallic tip, but the field enhancement factor has to be estimated 1000~3000, which seems to be physically unseasonable. Actually, these explanations cannot still clarify above puzzles.

3.2. Challenging Problems and Physical Thought

We think that two main reasons block us to clarify these field-emission phenomena of CN. One comes from experiments. Most of emitters in experiments are composed of various CN, such as the CN film mixed by SWCN and MWCN, or capped and open CNs, as well as some impurities and defects. For such dirty emitters many factors affect the emission current, such as the layer-layer and tube-tube interactions, and the localized states from impurities or defects. We cannot clarify which factor dominating the field-emission behavior and the emission mechanism. The other reason stems from the theoretical method. Most of the theoretical analysis are still based on the FN theory, which validity relies on a number of physical assumptions and approximations.[Forbes99] The main assumption is that the emitter has a parabolic free-electron energy band, which is continuous states near the Fermi energy. This assumption is no longer valid for CN because the energy band of CN is linear near Fermi energy and discrete due to the quantum confinement of electrons in the transversal direction. For such nanoscale materials (NSM), the energy band structure, impurities and defects should play an important role in field emission. Thus, what do basic physical elements dominate the intrinsic characteristic of CN field emission?. To clarify these novel characteristics of CN field emission we should go beyond FN formalism. Actually CN as a typical quasi-one-dimensional material includes the basic physical properties of nano-scale materials. It is necessary to reconsider the basic physics of the field emission from nano-scale materials.

VI. New Development of Computer Simulation of Field Emission

4.1. General Theoretical Strategy on Field Emission of Nano-Scale Materials

The novel phenomena observed in the CN field emission inspire us to reconsider the validity of FN theory for NSM. The novel phenomena in CN field emission originate mainly from the intrinsic properties of CN. The key point is to clarify basic physical differences between NSM and conventional bulk emitters. In principle, at least there are three basic physical factors of NSM that are different from the conventional bulk emitters. The first is that the energy band of NSM is not parabolic and discrete due to quantum confinement. Especially there are only a few states near Fermi energy even for metallic CN. The second is that the localized state or surface states could play some roles for NSM field emission, even dominate the field emission behavior. The third is that the field enhancement factor and the image potential will be not well-defined for NSM. The challenging problem is which factor dominates the behavior of the NSM field emission. It is difficult to find out this answer from experimental investigations.

In general, the theoretical methods are developed mainly along two ways. One is to set up a theoretical model to capture the key physics of NSM field emission. The other is to simulate the NSM field emission based on the first-principle idea. These two methods have their own advantages and flaws. They compensate each other to understand all physics of NSM field emission. The key point is to take the crucial physical factors of NSM into account for models and simulations. Actually, the basic physical process of NSM field emission is same to the field emission of conventional metals. The electrons in the conduction and valence bands move along the tube or quasi-one-dimensional emitter in an applied electric field and tunnel through the vacuum potential. The assumptions (i), (ii), (iv), (vi), and (viii) in the FN theory may be still valid. The energy band structure (iii) of NSM is no longer parabolic free-electron band, namely the energy dispersion will be not of the form in Eq. (2.2). The exchange and correlation interaction (vii) between the emitted electrons and the surface cannot be described by the image potential for NSM because the tip surface of NSM is usually dirty in nano-scale space, such as the tips of CNs being open or capped. The different energy bands between NSM and conventional metals could lead to different electron emission mechanism. The emitted electrons go out from a continuous conduction band for the conventional metals, while electrons go out from the discrete bands or localized states for NSM.

From the theoretical point of views we can set up a model of NSM and take each factor into account separately to understand the basic physical effect of this factor in the field emission behavior. On the other band, one can also put all physical factors together to build in a computational framework from the first-principle calculation.

4.2. Basic Concepts of Carbon Nanotubes

As a typical example of NSM, CN is a quasi-one-dimensional system. Before we present our theory of CN field emission we first introduce the basic physical properties of CN. The CN forms either the single-wall CN (SWCN) or multi-wall CN (MWCN). Each SWCN is identified by the chiral index (n,m). The tube ending may be capped or opened. The diameter of the SWCN can be written as $d = L/\pi$, where $L = a\sqrt{n^2 + m^2 + nm}$ is the circumference of the tube, and a is the lattice constant of the hexagon. The SWCN may be regarded as rolling up of the two-dimensional graphite. In the tight-binding approximation, the energy dispersion of the SWCN can be obtained [Satio98]

$$\varepsilon_q(k) = \pm t\sqrt{1 + 4\cos(\frac{\sqrt{3}k_x a}{2})\cos(\frac{k_y a}{2}) + 4\cos^2(\frac{k_y a}{2})} \tag{4.1}$$

where t is the nearest-neighbor hopping amplitude. The wave vector can be expressed as

$$\begin{pmatrix} k_x \\ k_y \end{pmatrix} = \begin{pmatrix} \frac{a}{2L}k(m-n) + \frac{\sqrt{3}\pi a}{L^2}q(n+m) \\ \frac{\sqrt{3}a}{2L}k(m+n) + \frac{\pi a}{L^2}q(n-m) \end{pmatrix}, \tag{4.2}$$

where k and q are quantum numbers, labeling the longitudinal wave vector and transversal modes of CN, respectively. The k values are within $-\pi/T_c < k < \pi/T_c$ and $q = 0,1,2..N-1$ where $T_c = \sqrt{3}L/d_R$ is the length of the translational vector of CN, and $N = 2L^2/a^2 d_R$ is the hexagon number per unit cell of CN, where d_R is the greatest common divisor of 2n+m and 2m+n. From the Eqs. (4.1) and (4.2), we can find that the energy band is gapless when the chiral index satisfies $\mathrm{mod}(n-m,3) = 0$.[Satio98] This kind of SWCN is metallic. The other SWCN is semiconducting. The detail structure of the energy band depends on the chirality of SWCN. In fact, the mean diameter of SWCN has been observed experimentally about 1.5nm and the length of SWCN can reach to several microns. It should be noted that the valence and conduction bands are discrete due to the quantum confinement for the nano-scale diameter of CN. There are only a few states near Fermi energy even for metallic tubes. This characteristic of the energy band structure is quite different from the conventional metals.

For the capped SWCN, the translational symmetry of SWCN is broken. There could be some localized states at the cap of SWCN, which could play some role in the field emission.

MWCN is composed of a series of coaxial SWCNs, which may be classified into three kinds based on their lattice symmetries, the commensurate, quasi-commensurate, and incommensurate MWCN. [Liang 05, Satio 00] The experimental observation and theoretical studies indicate that most of MWCN are metallic due to the interlayer coupling.[Liang 05, Kociak] The typical diameter of MWCN has 8~15nm.[Bonard]

4.3. Theory of Carbon Nanotube Field Emission

Since the energy band structure of NSM plays a crucial role in field emission we consider a simplified field emission model to study the energy band effect in CN field emission. Our theory starts from three basic assumptions: (1) the CN energy dispersion relation is linear with discrete transverse modes, which is the characteristic of low energy of CN; (2) the main features of the vacuum potential barrier of field emission can be described approximately by a triangular-shape barrier, which is still valid for the basic field emission properties; (3) the defect and impurity effects at the tip of the emitter are negligible because our attention is focused on the energy–band effect. Actually, the effects of defects and impurities and the modified triangle-potential barrier can be considered separately from the theoretical point of views such that we can understand the basic physical properties of different physical factors. The emission current is in general given by [Gadzuk, Flip, Liang 03,08];

$$j_{CN} = \sum_{q} \int_{BZ} N[\varepsilon_q(k),T] D[\varepsilon_q(\mathrm{k}),F] \frac{dk}{\pi}, \tag{4.3}$$

where $\varepsilon_q(k)$ is the energy band of CN measured to the Fermi energy of CN (see Eq. (4.1)). The $N[\varepsilon_q(k),T] = ev_g f(\varepsilon_q(k)) > 0$ is the so-called supply function, where $f(\varepsilon_q(k))$ is the Fermi-Dirac function and $v_g = \frac{1}{\hbar}\frac{\partial \varepsilon_q(k)}{\partial k}$ is the group velocity of electrons. The $D[\varepsilon_q(k),F]$ is the transmission coefficient of emitted electrons through the vacuum potential barrier. We use the triangular-shape vacuum barrier $U(x,F) = \varphi - \varepsilon_q(k) - eFx$, where φ is the work function and F is the local electric field. The transmission coefficient can be written as [Murphy, Liang08]

$$D(\varepsilon_q(k),F) \approx \begin{cases} \exp[-b_0 + c_0 \varepsilon_q(k)] & for\, \varepsilon_q(k) < \varphi \\ 1 & for\, \varepsilon_q(k) \geq \varphi \end{cases} \tag{4.4}$$

where $b_0 = (4/3)\sqrt{2m/\hbar^2}\varphi^{3/2}/eF$; $c_0 = 2\sqrt{2m/\hbar^2}\varphi^{1/2}/eF$. For SWCN, the energy-dispersion relation can be obtained from that of graphene. In the low-energy approximation, the energy dispersion relation of graphene can be obtained [Mintmire] $E(k) = \pm\sqrt{3}at\,|\mathbf{k} - \mathbf{k}_F|/2$, where a is the lattice constant of the hexagon and t is the electron hopping amplitude. The energy-dispersion relation is linear and radial symmetric around the point K. This is a typical characteristic of Dirac electrons [Bockrath, Liang08,10]. These states near the Fermi energy play the main role in field emission. It is noted that one of the discrete lines pass the K point in the first Brillouin zone for the metallic SWCNs, and the discrete lines do not passes the K point for the semiconducting SWCNs [Satio98]. Using the Murphy and Good's approach, [Murphy] we can obtain [Liang08]

$$j_{CN} = \alpha \frac{F}{\varphi^{1/2}} e^{-b\phi^{3/2}/F} \begin{cases} \coth(\gamma\varphi^{1/2}/Fd) & for \quad metallic \quad tubes \\ \dfrac{\cosh(\gamma\varphi^{1/2}/3Fd)}{\sinh(\gamma\varphi^{1/2}/Fd)} & for\ semiconducting\ tubes \end{cases}. \quad (4.5)$$

where $\alpha = e^2/2\pi\sqrt{2m}$; $b = 4\sqrt{2m}/3e\hbar$; $\gamma = at\sqrt{6m}/\hbar e$; the d is the diameter of CN. This field-emission equation (4.5) gives a general formula of field emission currents for the metallic and semiconducting SWCNs, respectively, which is a generalized FN formula of the emission current. The FN-like plot of the generalized field emission current may be written as

$$\ln\frac{j_{CN}}{F} = \ln\frac{\alpha}{\varphi^{1/2}} - \frac{b\varphi^{3/2}}{F} + \begin{cases} \ln\coth(\gamma\varphi^{1/2}/Fd) & for \quad metallic \quad tubes \\ \ln\dfrac{\cosh(\gamma\varphi^{1/2}/3Fd)}{\sinh(\gamma\varphi^{1/2}/Fd)} & for\ semiconducting\ tubes \end{cases} \quad (4.6)$$

It can be seen that the FN-like plot of CN field emission is quite different from the FN plot. The last term in Eq. (4.6) modifies the slope of the FN-like plot. To compare this formula in Eq. (4.5) with the FN-type formula in Eq. (2.10), we introduce a parameter $g_D = \gamma\varphi^{1/2}/Fd$ to distinguish different parameter regions. Let us consider two special cases, large-diameter tubes and high fields $g_D << 1$ and small-diameter tubes and low fields $g_D >> 1$, the emission current in Eq.(4.5) reduces [Liang08] to:

$$j_{CN,m} = \begin{cases} \alpha(\dfrac{F^2 d}{\gamma\varphi} + \dfrac{\gamma}{2d})e^{-b\varphi^{3/2}/F} & for \quad g_D < 0.5 & (4.7a) \\ \alpha\dfrac{F}{\varphi^{1/2}} e^{-b\varphi^{3/2}/F} \coth(\dfrac{\gamma\varphi^{1/2}}{Fd}) & for \quad 0.5 < g_D < 2 & (4.7b) \\ \alpha\dfrac{F}{\varphi^{1/2}} e^{-b\varphi^{3/2}/F} & for \quad g_D > 2 & (4.7c) \end{cases}$$

for metallic tubes, where we used 0.5 and 2 as an approximation of the dimensionless parameter g_D to separate three regions. Similarly, for the semiconducting tubes, the emission current becomes

$$j_{CN,s} = \begin{cases} \alpha(\dfrac{F^2 d}{\gamma\varphi} + \dfrac{\gamma}{18d})e^{-b\varphi^{3/2}/F} & for \quad g_D < 0.5 \\ \alpha\dfrac{F}{\varphi^{1/2}} e^{-b\varphi^{3/2}/F} \dfrac{\cosh(\gamma\varphi^{1/2}/3Fd)}{\sinh(\gamma\varphi^{1/2}/Fd)} & for \quad 0.5 < g_D < 2 \\ \alpha\dfrac{F}{\varphi^{1/2}} e^{-b\varphi^{3/2}/F} e^{-2\gamma\varphi/3Fd} & for \quad g_D > 2 \end{cases}. \quad \begin{matrix} (4.8a) \\ (4.8b) \\ (4.8c) \end{matrix}$$

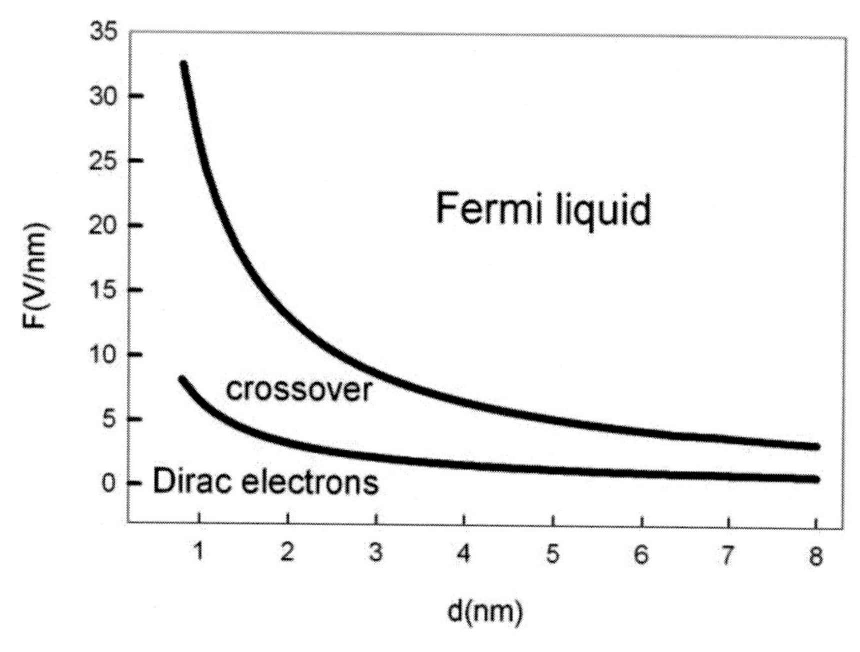

Figure 4.1. The phase diagram of the basic emission behavior of two typical electrons, Fermi liquid in the upper region, Dirac electrons in the lower region, and the crossover in the middle region.

For the region of $g_D < 0.5$, the emission current in the first terms in Eqs. (4.7a) and (4.8a) are still of the FN type, but the second term modifies the FN-type behavior. For $g_D > 2$, the emission current in (4.7c) and (4.8c) become non-FN type. In particular, for the semiconducting tubes, the emission current has an extra term, $e^{-2\gamma\varphi/3Fd}$, which reflects that the current decreases exponentially with the diameter of CNs. This is because the energy gap at the Fermi energy for the semiconducting tubes is inversely proportional to the diameter of the tubes.[Saito] In the intermediate region $0.5 < g_D < 2$, the I-V characteristic exhibits a more complicate correction to the FN-type behavior. The non-FN-type behavior of the I-V characteristic observed in experiments can be understood as arising from different physical mechanisms. In our opinion, the non-FN-type behavior should appear in the small-diameter tubes and low fields region.

From the basic many-electron system point of views, the FN theory developed based on the parabolic free-electron energy band. We may regard the FN-type behavior in Eq. (2.10) as the Fermi liquid (FL) behavior of field emission. Our generalized theory for CN is developed based on the low energy spectrum of SWCN, namely the linear energy band with the quantization transverse modes. This is a typical characteristic of Dirac electrons. The deviation from the FN-behavior is in the region of $g_D > 2$ corresponds to the Dirac-electron behavior. We give a phase diagram in Figure4.1, in which the Fermi liquid behavior is in the upper region, and the Dirac-electron behavior in the lower region and the crossover in the intermediate region.

4.4. Theory of Carbon Nanotube Thermionic Emission

For thermionic emission, the typical field F<5V/nm, temperature excites electrons overcoming the vacuum potential barrier and coming out. We also consider the linear energy band of CN to study the basic thermionic emission behavior of CN. The approach to obtain the thermionic emission equation basically follows the Murphy and Goods (MG) method. [Murphy, Liang 10]: The emitted electron energy is mainly near the top of the vacuum potential barrier for the thermionic emission, the transmission coefficient can be expressed approximately as [Murphy, Liang 08]:

$$D(\varepsilon_q(k),F) \approx \{1+\exp[-(F\hbar_0^4/m^2e^5)^{-1/4}\pi\varepsilon]\}^{-1}, \tag{4.9}$$

where $\varepsilon = 1-[\varphi-\varepsilon_q(k)]/(e^3F)^{1/2}$, and $h_0 = (F\hbar^4/m^2e^5)^{1/4}(e^3F)^{1/2}/(\pi k_B T)$. The Fermi-Dirac function can be expressed approximately to $f[\varepsilon_q(k)] \approx e^{-\varepsilon_q(k)/k_BT}$ for the thermionic emission. Notice that the emitted electron with the energy at the top of the vacuum potential barrier, we can assume that emitted electrons in different transverse channels have an equal contribution to emission current. Using the infinite integration technique [Liang 08], we can obtain the thermionic current

$$J_{CN} = \frac{eN}{\pi\hbar}k_BT\frac{\pi h_0}{\sin(\pi h_0)}\exp[-\frac{\varphi-(e^3F)^{1/2}}{k_BT}] \,. \tag{4.10}$$

It should be noted that the prefactor in the formula in Eq. (4.10) is linear to temperature T, which modifies the Richardson-Schottky (RS) thermionic emission equation [Murphy, Liang08]. For high temperatures and weak field $\pi h_0/\sin(\pi h_0) \approx 1$, and F≈0, the thermionic emission current in Eq. (4.9) reduces

$$J_{CN} = \frac{eN}{\pi\hbar}k_BT\exp[-\frac{\varphi}{k_BT}]. \tag{4.11}$$

This becomes the RS-like formula. Theoretically, we can plot $\ln(J_{CN}/T) \sim 1/T$ to compare the conventional RS formula with the experimental results. However, the difference between the RS-like formula in Eq. (4.11) and the conventional RS formula in Eq. (2.19) originates from different energy-band structures. If the energy-band structure of the emitter still plays a key role in thermionic electron emission, we should expect to observe experimentally new thermionic emission behavior in Eq. (4.11).

It should be remarked that as a typical quasi-one-dimensional material, CN contains some basic characteristic of low-dimensional materials. The theory of CN field and thermionic emissions should also describe the basic field and thermionic behaviors of low-dimensional systems, especially for Dirac electrons. However, the field and thermionic emission equations could not cover all detail properties of this class of materials due to ignoring some detail physical factors in our analytic derivation, such as the chiral effect of CN, the cap and defect effect. These detail effects can be taken into account by numerical calculation in the following section.

4.5. Theory of Luttinger Liquid Field Emission

On the other hand, CN has been also found theoretically and experimentally to exhibits Luttinger liquid behavior. [Bockrath, Komnik] Using the bosonization method, Gogolin and Komnik study the field emission behavior of one-dimensional interacting electrons based on the Luttinger liquid (LL) model. They give a FN-like formula of emission current, [Komnik]

$$j_{LL} = a_{LL}\left(\frac{F^2}{\varphi}\right)^{1/2g_{LL}} e^{-b_{LL}\varphi^{3/2}/F} , \qquad (4.12)$$

where $a_{LL} = \frac{a}{a_0}(4k_F D^2)^{-1/2g_{LL}}$ and $b_{LL} = \frac{4}{3}k_F^{1/2}$. The a is a factor responsible for the tip geometry; a_0 is the lattice constant; D is the band width of the LL; k_F is the wave vector at Fermi energy; g_{LL} is the LL parameter. The FN-like plot may be written as

$$\ln j_{LL} = \ln(a_{LL}) + \frac{1}{2g_{LL}}\ln\left(\frac{F^2}{\varphi}\right) - \frac{b_{LL}\varphi^{3/2}}{F} \qquad (4.13)$$

The I-V characteristic of the LL depends on the electron-electron interaction. For SWCN, the effective g_{LL} is estimated to around 0.5. [Bockrath] The emission current of the LL reduces to the form of the FN formula in Eq. (2.10). For the limit of weak interaction $g_{LL} \approx 1$, the emission current becomes, $j_{LL} = a_{LL}\frac{F}{\varphi^{1/2}}e^{-b_{LL}\varphi^{3/2}/F}$, which is qualitatively consistent with our result for small-diameter tubes and weak fields. In practice, we can plot the graph $\ln j_{LL} \sim x \equiv \ln\left(\frac{F^2}{\varphi}\right)$. The slope of the graph can be given by $S_{LL} = \frac{d\ln j_{LL}}{dx} = \frac{1}{2g_{LL}} + \frac{b_{LL}}{2}\frac{\varphi^{1/2}}{F}$. Thus the LL parameter can be obtained in the limit of strong field, $g_{LL} = \frac{1}{2\lim_{F\to\infty} S_{LL}}$. This result provides a way to experimentally identify the LL theory of field emission or to verify the LL property of the emitter.

4.6. General Empirical Theory of Field Emission

In general, the field emission behavior depends on the intrinsic properties of materials and the emission mechanism. From the empirical point of views, recently, R. Forbes proposed an empirical field emission formula to cover various field-emission behaviors, [Forbes09]

$$j_F = CV^{\kappa}e^{-B/V}, \tag{4.14}$$

where C and B are the material parameters, V is applied electric field and κ is a phenomenological parameter within $-1 \le \kappa \le 3$, which describes various emission mechanisms.[Forbes09] For example, the $\kappa = -1$ describes the electron emitted from a single atom and $\kappa = 2$ corresponds the FN-type field emission for metallic emitters [Forbes09]. The generalized field emission formula in Eq. (4.5) actually includes different κ values. This is consistent with the Forbes's empirical theory [Forbes09].

From the experimental point of views, we can also give the FN-like plot of this empirical formula $\ln j_F = \ln C + \kappa \ln V - B/V$. The slope of FN-like plot is given $S_F = \dfrac{d\ln j_F}{d\ln V} = \kappa + B/V$. We can obtain the parameter κ by $\kappa = \lim_{V\to\infty} S_F$. This provides a way to experimentally determine the phenomenological parameter.

General speaking, different κ describe different microscopic emission mechanisms. For a general emitter, it could involve different mechanisms. Thus, we propose a generalized empirical formalism of field emission [Liang 11]

$$j_{GE} = \sum_{\kappa} C_{\kappa} V^{\kappa} e^{-B_{\kappa}/V^{Q_{\kappa}}}, \tag{4.15}$$

where we introduce another parameter Q_{κ} to describe more complicated mechanism of field emission. The parameters C_{κ} and B_{κ} are related to the material properties of emitter. The κ ,and Q_{κ} describe different emission mechanisms. In general, the field electron emission includes different physical mechanisms, such as the bulk effect, the surface and tip effects, and the impurity and defect effects. The field emission currents originated from different mechanisms can be understood completely and empirically by the generalized empirical formula of field emission current in Eq. (4.15). For example $\kappa = -1$ describes the electron emitted from a single atom; $\kappa = 2$ corresponds to the FN-type field emission for metallic emitters, and $\kappa = 1$ means Dirac electron field emission. The other κ can cover the field electron emission from localized states of the impurities and defects at the tip. In other words, the generalized empirical formalism in Eq. (4.15), in principle covers all physical mechanisms of field emission. The further issues should try to find out the values of the phenomenological parameters C_{κ}, B_{κ}, κ ,and Q_{κ} from the specific cases.

4.7. General Empirical Theory of Thermionic Emission

Along the empirical way of field emission we can also construct an empirical thermionic emission theory. The empirical thermionic emission current may be written as [Liang 11]

$$J_G = \sum_{\nu} A_\nu T^\nu e^{-D_\nu / T^{q_\nu}} \tag{4.16}$$

where A_ν, D_ν, ν,and q_ν are phenomenological parameters, which describe different thermionic mechanisms. In principle, the empirical thermionic current in Eq. (4.16) can describe all thermionic emission mechanisms. For example, $\nu = 2$, and $q_\nu = 1$, the thermionic current formula in Eq. (4.16) reduces to RS formula. For $\nu = q_\nu = 1$, the formula becomes the formula of Dirac electronic thermionic emission in Eq. (4.11).

VII. New Development of Computer Simulation of Field Emission

5.1. Basic Idea on Computer Simulation

The computer simulation provides a powerful tool to investigate the field emission of NSM. Exact speaking, the computer simulation is based on the first-principle idea that consider a realistic system to the computational framework. However, a practical NSM, such as SWCN, contains at least 10^7 atoms. Actually, it is impossible to consider all detail of the system to the computational framework. The basic strategy of computer simulation is still to cover the main physical components of the system to build in the computational framework. There have been many works based on this idea.

5.2. The Tight-Binding Approach

5.2.1. Computational Framework

We develop a computational framework of CN field emission based on the tight-binding approach. [Liang03-06] In general, the emission current may be expressed in terms of two factors [Gadzuk, Flip, Liang 03]

$$j_{TB} = \sum_{q} \int_{BZ} N(\varepsilon,T) D(\varepsilon,F) dk, \tag{5.1}$$

where $\varepsilon = \varepsilon_q(k) - E_F$, where $\varepsilon_q(k)$ is the energy band of CN and E_F is the Fermi level. The $N(\varepsilon,T) = ev_g f(\varepsilon) > 0$ is the so-called supply function, where $f(\varepsilon)$ is the Fermi-

Dirac function and $v_g = \frac{1}{\hbar}\frac{\partial \varepsilon}{\partial k}$ is the group velocity of electrons. The $D(\varepsilon, F)$ is the transmission coefficient of emitted electrons through the vacuum potential barrier. We use the triangular-shape vacuum barrier with the image potential, $U(x,F) = \phi - \varepsilon - eFx - e^2/4x$, where φ is the work function and F is the local electric field. The transmission coefficient can be obtained approximately by WKB method [Liang03, 06],

$$D(\varepsilon,F) = \begin{cases} \exp[-b(\phi-\varepsilon)^{3/2} v(y)/F] & for \;\; y<1 \\ 1 & for \;\; y \geq 1 \end{cases}, \tag{5.2}$$

where the function $v(y)$ is related to certain elliptic function and $y = \sqrt{e^3 F}/|\phi - \varepsilon|$ [Modinos].

The energy band structure of NSM can be in principle taken into account for this formalism. For CN, the effective tight-binding Hamiltonian of MWCN can be written as [Liang 04]

$$H = \sum_{k,s} (\alpha_s(k) c^+_{k,s,A} c_{k,s,B} + h.c.) - \sum_{k,s} g_{s,s+1} (c^+_{k,s,A} c_{k,s+1,A} + c^+_{k,s,B} c_{k,s+1,B} + h.c.) \tag{5.3}$$

where

$$\alpha_s(k) = t(e^{ik_x^{(s)} a/\sqrt{3}} + 2e^{-ik_x^{(s)} a/2\sqrt{3}} \cos\frac{k_y^{(s)} a}{2}), \tag{5.4}$$

and t=2.7eV is the intralayer hopping constant. The interlayer hopping constant may be estimated by[7,9,10] $g_{s,s'} = g_0 e^{-d_{s,s+1}/a}$ where $d_{s,s+1}$ denotes the distance between the s and s+1 layers. As an approximate estimation of the interlayer coupling with the parameter, , $g_0 = t/8$.[Liang 05] The wave vector is

$$\begin{pmatrix} k_x^{(s)} \\ k_y^{(s)} \end{pmatrix} = \begin{pmatrix} \frac{a}{2L_s}(m_s - n_s)k + \frac{\sqrt{3}\pi a}{L_s^2}(n_s + m_s) q_s \\ \frac{\sqrt{3}a}{2L_s}(m_s + n_s)k + \frac{\pi a}{L_s^2}(n_s - m_s) q_s \end{pmatrix}. \tag{5.5}$$

The (k,q_s) in Eq.(8) are quantum numbers of MWCNs. The k values are within $-\frac{\pi}{T_s} < k < \frac{\pi}{T_s}$ and $q_s = 0,1,2..N_s - 1$ where $T_s = \frac{\sqrt{3}L_s}{d_{R,s}}$ is the length of the translational vector of the MWCN and $N_s = \frac{2L_s^2}{a^2 d_{R,s}}$ is the hexagon number per unit cell of the s tube[12]. The $d_{R,s}$ is the greatest common divisor of $2n_s+m_s$ and $2m_s+n_s$, and the $L_s = a\sqrt{n_s^2 + m_s^2 + n_s m_s}$ is the length of the chiral vector of the s tube, where a is the lattice vector of hexagons. Numerical solving the Hamiltonian (5.3) we can obtain the energy band of CN. Putting the energy band and the transmission coefficient in Eq. (5.2) into the Eq. (5.1) we can give the emission current.

The key point of this computational framework is that we take the energy band structure of CN into account by the tight-binding approximation. Based on this framework, we can investigate many field emission properties of CN, such as the current-voltage (I-V) characteristic, the chiral effect, size effect and some quantum effects.

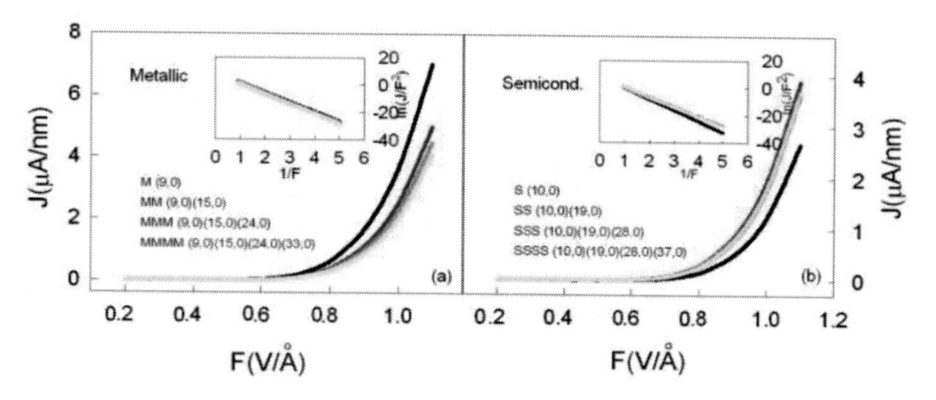

Figure 5.1. (On line color) : The current-voltage characteristic of field emission of single-wall and multi-wall carbon nanotubes in (a) for metallic tubes and in (b) for the semiconducting tubes. The FN plots are in the insets.

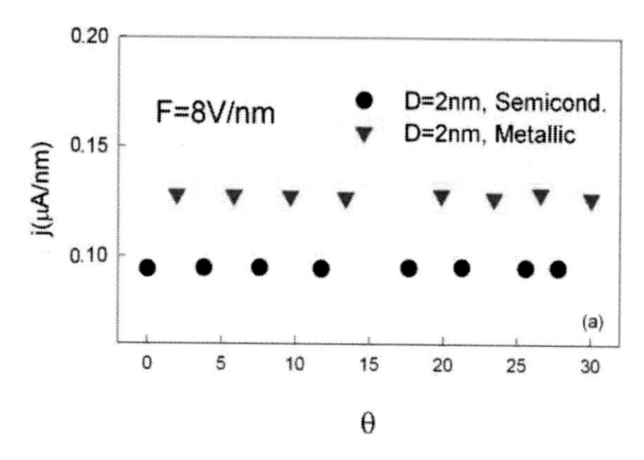

Figure 5.2. (On line color): The emission current density versus the chiral angle of the SWCN with similar diameters 2nm in a given electric field 8V/nm.

5.2.2. Current-Voltage Characteristic

The current-voltage (I-V) characteristic is an important feature of field emission. We investigate systematically the I-V characteristic of field emission of CN by means of the computational framework in the above section 5.2. In Figure 5.1 we present the I-V characteristics of two typical CNs.[Liang 06] One is composed of metallic CNs. The other is composed of semiconducting CNs. It can be seen that the I-V characteristic of all tubes shows a similar behavior. As the layer number increase the emission current densities decrease for metallic tubes, but have a different order for semiconducting tubes. To compare this I-V characteristic with FN type, we plot $\ln(j_L F^{-2}) \sim F^{-1}$ in the inset of the Figure 5.1. The FN plots still basically follow the FN type, but the slopes of different semiconducting tubes have slight difference.

5.2.3. Chiral Effect

The SWCN is identified by the chiral index The energy band structure of SWCN depends on the chiral index (n,m) and shows metallic or semiconducting. An interesting question is what role the chirality plays in the field emission of CN. Actually, the chiral vector determines two physical parameters, the chiral angle $\theta = \arctan\dfrac{\sqrt{3}m}{2n+m}$ and the diameter of CN. To examine the chiral effect of CNs on the emission current we investigate the emission current density versus the chiral angles at a given applied field for a group of SWNTs with very similar diameters, D≈2.0±0.03nm.[Liang03] In Figure 5.2, the current densities for both metallic and semiconducting CNTs are almost independent of the chiral angles. Physically, electrons are emitted from near Fermi energy in room temperature. The density of states (DOS) near Fermi surface is almost independent of the detail chiral angle of CN such that the current density is not sensitive to the chiral angle of CN at room temperature.

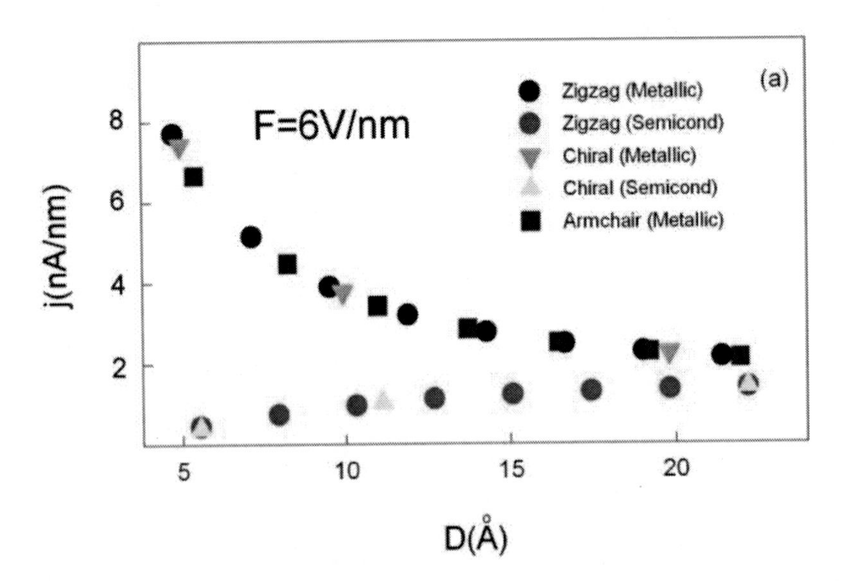

Figure 5.3. (On line color): The emission current densities of different kinds of SWCN versus the diameter of SWCN in a given electric field 6V/nm.

5.2.4. Size Effect

On the other hand, we examine the relationship between the current density and the diameter of CN at room temperature for a given applied field for several groups of SWNTs with different chiral angles in Figure 5.3.[Liang03] As the diameter increases, the current density decrease for metallic tubes, but increase for semiconductiong tubes. When the diameters of CNs are larger than 2nm the changes of the current densities become diminution and the current densities of the metallic and semiconducting tubes are very close. This size effect of CNs in field emission originates from the intrinsic properties of the energy band structure.

For the semiconducting tube the energy gap at Fermi surface is inverse proportional to the diameter of CN. The emission current is inverse proportional to the energy gap at the Fermi energy. Therefore, the larger diameter tube corresponds to the larger current density. This difference of the current densities for the metallic and semiconducting tubes results from their different energy band structures, which may be regarded as a quantum size effect in CN field emission.

For MWCN, most of MWCN become metallic due to the interlayer coupling (see the following section). Thus, the quantum size effect becomes small or disappears for MWCNs.

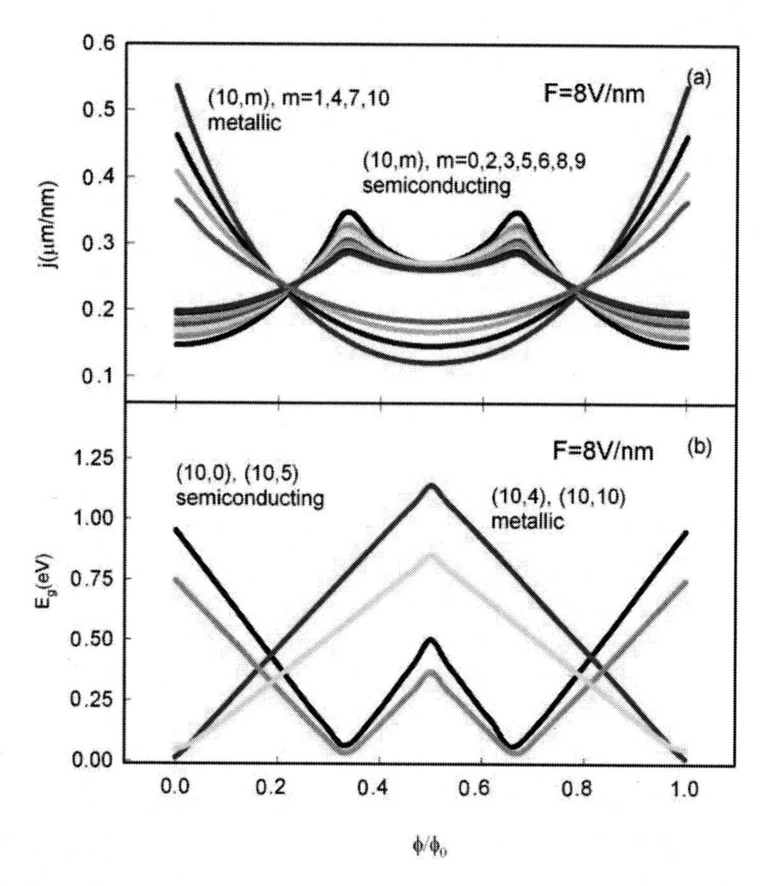

Figure 5.4. (On line color): (a) The emission current densities versus the ratio of the magnetic flux and flux quanta; (b) The energy gap at Fermi energy versus the magnetic flux and flux quanta.

5.2.5. *Aharovon-Bohm Effect and Quantum Phase Transition*

When a magnetic field is applied along the SWCN axes, the magnetic field can modify the electronic structure of CN, leading to the metal-semiconductor phase transition. We consider a magnetic field along the tube axes. Above formula Eqs.(5.1)-(5.4) are invariant (here we ignore the Zeeman effect (see 6.1)) except the wave vector in Eq. (5.5) [Jiang, Liang 05]

$$\begin{pmatrix} k_x \\ k_y \end{pmatrix} = \begin{pmatrix} \frac{a}{2L}k(m-n) + \frac{\sqrt{3}\pi a}{L^2}(q+\frac{\phi}{\phi_0})(n+m) \\ \frac{\sqrt{3}a}{2L}k(m+n) + \frac{\pi a}{L^2}(q+\frac{\phi}{\phi_0})(n-m) \end{pmatrix}, \tag{5.3}$$

where ϕ is the magnetic flux ϕ and ϕ_0 is the magnetic flux quanta. We investigate the emission current density versus the magnetic flux in a given local field F=8V/nm shown in Figure 5.4 (a). The current densities of these tubes exhibit obviously two branches, which correspond to the metallic and semiconducting tubes. Both of them are periodic to a period $\phi_0 = \mathrm{hc/e}$, and symmetric to a half of ϕ/ϕ_0, but they respond differently to the magnetic flux. Interestingly, the current densities of all tubes are approximately equal at $\phi/\phi_0 \approx 0.215$. [Note] The error results from the chiral and size effects of CN. Different local electric fields will not change qualitatively this behavior of the current density in the magnetic field. This may be regarded as a universal current density induced by the Abarovon-Bohm effect. In Figure 5.4 (b) we plot the energy band gaps of these tubes at the Fermi energy versus ϕ/ϕ_0. The energy gap can be tuned by the magnetic field and close to zero at $\phi/\phi_0 \approx 1/3$, which can be regarded as a semiconductor-metal quantum phase transition induced by the magnetic field. Actually the energy gap dominates the emission current density. When the magnetic flux is near $\phi/\phi_0 \approx 0.215$, the energy gap effect competes with the size effect such that the current densities of all tubes reach a very similar value. However, the same magnetic flux does not correspond to the same magnetic field for different diameter tubes. It is still not easy to observe experimentally the universal current density in magnetic field.

The emission energy spectrum is another important feature in field emission.[Liang05,06] In Figure 5.5 we show the energy spectra of the metallic and semiconducting tubes in the magnetic flux. The peaks of the energy spectra of metallic and semiconducting tubes exhibit different responses to the magnetic flux. At $\phi/\phi_0 = 1/4$ the peak of the semiconducting tube becomes high and shifts to the high-energy side (right) (see Figure 5.5 (a)), but the peak of metallic tube becomes low and shifts to the low-energy side (left) (see Figure 5.5(b)). This property may be used to distinguish the metallic and semiconducting tubes experimentally.

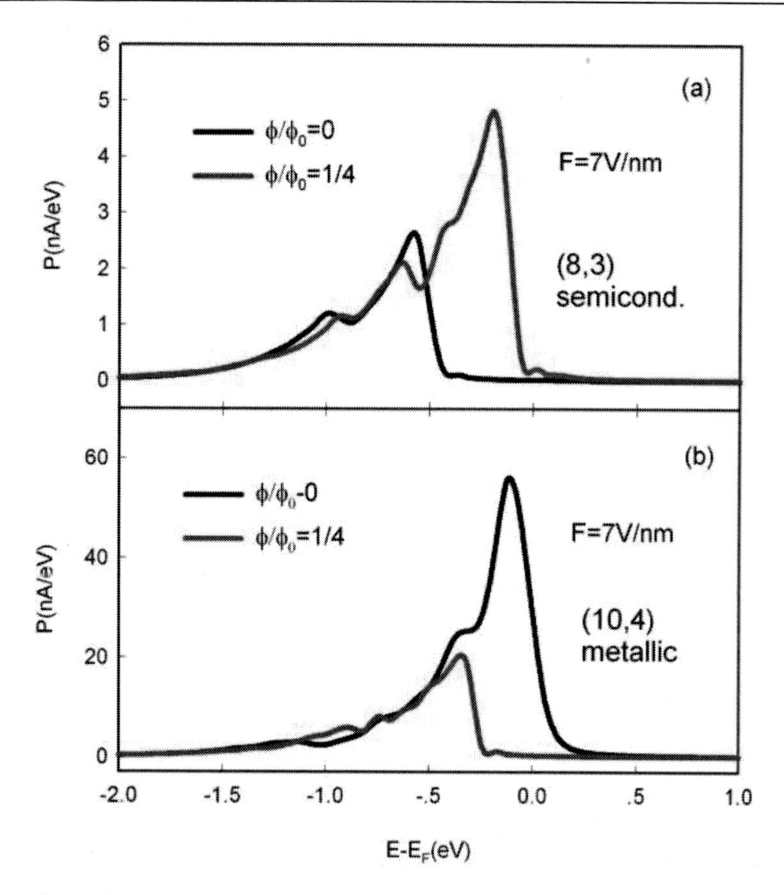

Figure 5.5. (On line color): The emission energy spectra of the semiconducting tubes (8,3) in (a) and metallic tube (10,4) in (b) for different magnetic flux.

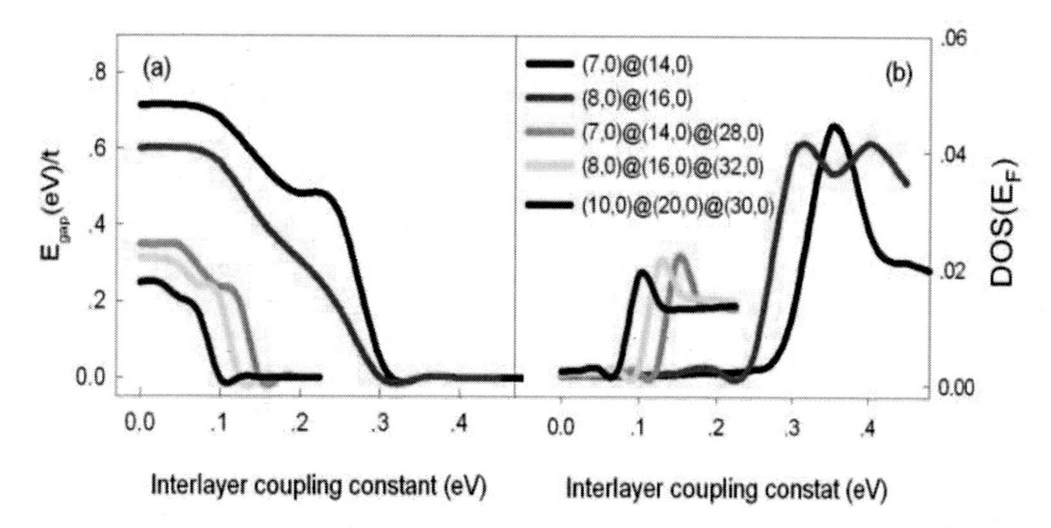

Figure 5.6. (On line color): (a) The energy gap at Fermi energy of semiconducting MWCN; (b) The density of states at Fermi energy of semiconducting MWCN.

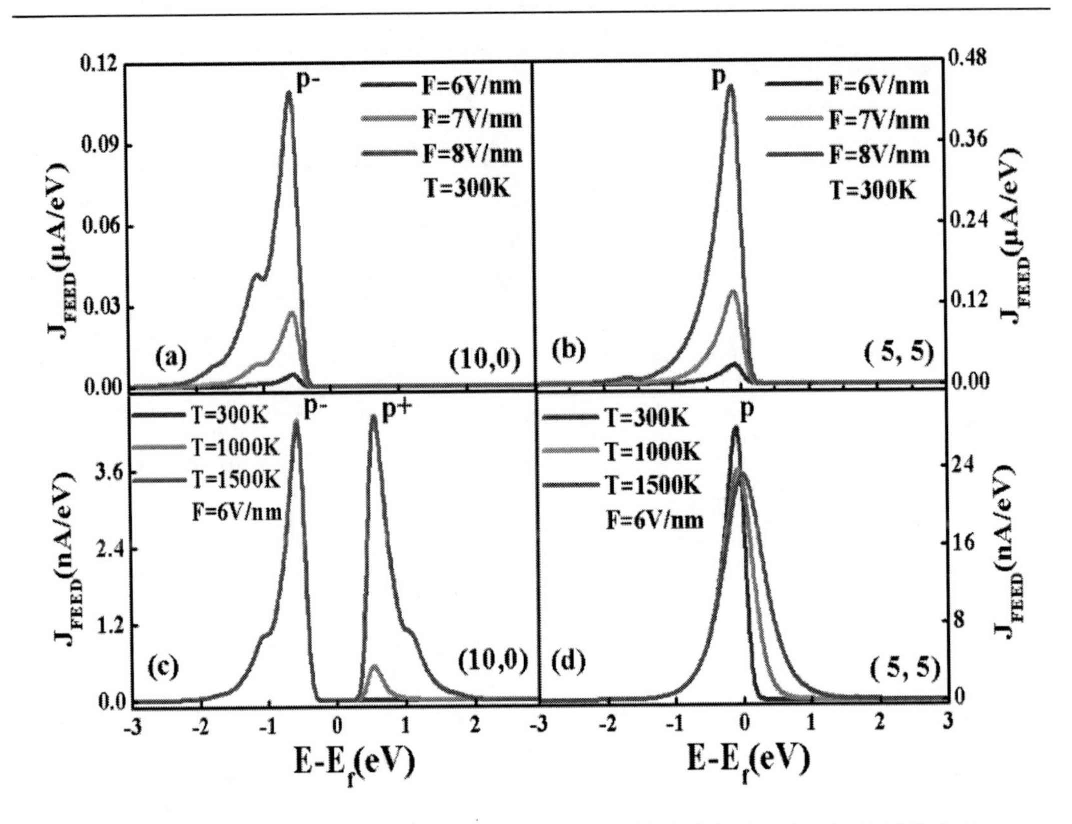

Figure 5.7. (On line color): The energy spectra of SWCN (10,0)/(5,5) in the electric field F=6, 7, and 8V/nm at room temperature in (a)/(b), and in temperature T=300, 1000, and 1500K at the electric field F=6 V/nm in (c)/(d).

5.2.6. Semiconductor-Metal Phase Transition Induced by Interlayer Coupling

For MWCN the interlayer coupling can modify the energy band structure. Using the tight-binding formalism, we investigate the energy gap and density of states of MWCN at Fermi energy versus the interlayer coupling of MWCN in Figure 5.6. [Liang 04] The energy gap decreases with the interlayer coupling increasing and the corresponding DOS at Fermi energy increases. This implies that the interlayer coupling can also drive the semiconductor-metal phase transition of MWCN. Therefore, most of MWCNs are actually metallic due to the interlayer coupling. This is also why the quantum size effect disappears in MWCN. Actually there have been the experimental evidences of transport indicating that MWCN is actually metallic, [Kociak] which is consistent with our theoretical prediction.

5.2.7. Field Emission Energy Distribution

Using the tight-binding formalism we investigate of emission energy spectra of CN. [Liang 06] Figure 5.7 shows the emission energy spectra of semiconducting tube (10,0) and a metallic tube (5,5) in different electric fields and temperatures. The stronger electric field makes the peak of the energy spectrum become higher and wide, and the energy distribution expand to the low energy direction. This is because the vacuum potential barrier in strong fields will become narrow enough such that more low energy electrons emitted. As temperature increases the peak of the spectrum also becomes high and wide, but the energy distribution expands to the high-energy direction. High temperature makes more electrons excite above Fermi energy and emitting. On the other hand, we can see that the energy gap at Fermi energy occurs for the semiconducting tube (10,0) when temperature reaches to 1500 K. These results are qualitatively consistent with the experimental results. [Dean, Fransen] This implies that the energy spectrum of CN exhibits some intrinsic properties of CN.

5.2.8. Maximum Emission Current Density of Carbon Nanotubes

Theoretically an interesting question is what component of MWCN has an optimal feature in field emission? For a given local field we compare the current densities of MWCN with different numbers of layers in Figure 5.8. [Liang 06] For SWCN the current density of the metallic tube is much larger than that of semiconducting tubes. As the layer number increase, their differences become small because the interlayer coupling induces the semiconductor-metal phase transition such that most of MWCN become metallic. Therefore, the emission current densities of MWCN are qualitatively invariant. [Liang 04, Kociak]

5.3. The First-Principle Methods

The detail properties of the tip could play important role in field emission for nano-scale material (NSM). The study of the detail effects of the tip of NSM and the exact potential barrier of NSM should start from the first-principle method. In general, the basic framework of the first-principle methods includes two basic steps. The first step is to calculate the electrostatic potential distribution and the energy states of NSM in an external electric field. The second step is to calculate the tunneling probability of emitted electrons [Adessi, Lu, Zheng, Peng, Mayer, Kim]. However, a practical emitter of NSM, for example SWCN, contains at least 10^7 atoms. It is still impossible to build in a computational framework of a realistic field emission system due to the limitation of computer ability. Usually one develops various techniques to implement the first-principle idea.

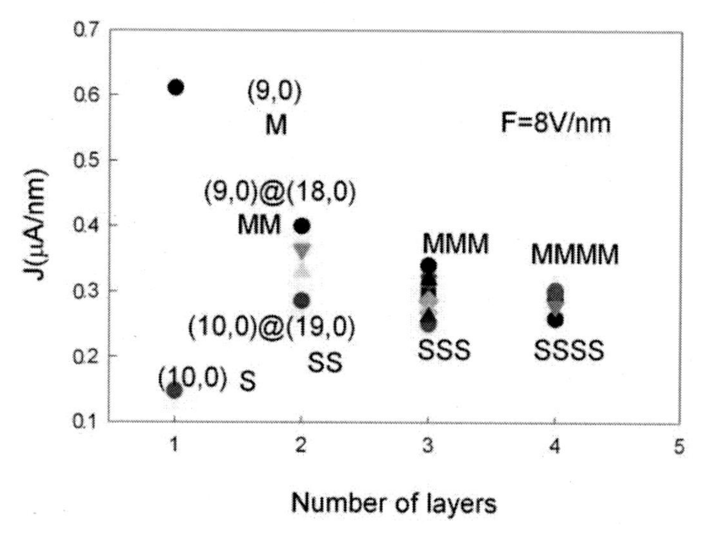

Figure 5.8. (On line color): The emission current densities versus the number of layers of MWCN in a given electric field 8V/nm.

5.3.1. The Multi-Scale Technique

The basic idea of the multi-scale strategy is to separate the emitter to different parts and to use different models and methods to construct the computational frameworks for different parts of the system. [Zheng, Peng] For example, the CN is divided to the tip and the long tube body. [Zheng, Peng] The tip usually contains several hundreds of carbon atoms, which can be simulated by the first-principle method of quantum mechanics. The other part can usually contain several thousands of carbon atoms, which is simulated by the molecule mechanics.

5.3.2. Field Penetration Effect and Field Enhancement Factor

From the microscopic point of views, the SWCN is not a metallic bulk material. Physically, the electric field can penetrate the SWCN and the field enhancement factor based on FN theory is not well-defined. There are several research groups to study this issue by different methods.

Using the first-principle method with the multi-scale technique, Peng et. al. investigate the field penetration effect and field enhancement factor of CN field emission. They found that the field penetration suppresses the charge potential near the tip of the tube and induces a few extra electrons. This effect suppresses equivalently the height of the vacuum potential barrier. In other words, this effect of the field penetration modifies the field enhancement factor. The effective field enhancement factor depends on the applied field and the length of SWCN, which can be expressed phenomenologically as [Peng]

$$\beta = \frac{F_0}{F} + 0.2\frac{\ell}{R}, \tag{5.7}$$

where F_0 is the field at the apex of SWCN in the absence of the applied field; F is the applied field; ℓ is the length of SWCN and R is the radius is SWCN. This result reveals some novel properties of the field emission beyond the FN theory.

5.4. Lippman-Schwinger Scattering Formalism

Adessi and Mayer develop a computational framework based on the Lippman-Schwinger scattering formalism.[Adessi, Mayer] They compare the emission currents of the open and capped (5,5) SWCN and find that the emission current of the open (5,5) CN is larger than that of the capped CN in a given applied field due to the effect of the field penetration and the dangling bond in the open CN. This suggests that the field enhancement factor should be reconsidered for the nanoscale emitters.

5.5. *ab-initio* Tight-Binding Method and Density Function Approximation

The other computation frameworks developed based on the first-principle method include Han's *ab-initio* tight-binding method [Han], Lu and Kim's density function approximation. Their attention is focused on the detail effects of the tip in the field emission behavior, including the electrostatic potential of CN, the charge distribution near the tip, the local density of states at the tip. [Han, Lu, Kim] However, most of them still use the WKB method to calculate the emission current. The I-V characteristic is still similar to FN type except a slightly saturation of the emission current in high current range.[Lu]

5.6. Cap and Doping Effects

Physically, the cap and defects in the tip of tube would lead to some localized states near the Fermi energy, which could play some roles in the field emission. For these nanoscale emitters, several research groups investigate the field emission of the capped and open SWCN by the first-principle idea with different techniques.[Peng,Mayer, Lu, Kim] They found that the electric potential distributions of the capped and opened SWCN are quite different. The charge dipole and excess charge occur near the cap of the tube.[Peng] Using the density function theory, Kim et. al. investigated the electronic structures of the open and capped (9,0) and (5,5) SWCN.[Kim] They found that the local density of states (LDOS) of the open zigzag SWCN has a large peak near the Fermi energy, which originated from the dangling bonds of the tube edge. This implies that the open zigzag SWCN is a more efficient emitter than the capped armchair SWCN. In particular, the SWCN chemically terminated by some atoms, such as hydrogen atoms (H), oxygen atoms (O), and hydroxyl groups (OH) modify sensitively the LDOS near the Fermi energy.[Kim] This could become a probe tip for selective chemical sensing.[Kim] Khazaei et. al. compared the local density of states and the emission currents of the pristine and Cs-doped capped (10,10) CN.[Khazaei] It is found that the emission current is enhanced strongly by the Cs doping from 5μA to 12~13μA.[Khazaei]

5.7. Electron Emission Mechanism

The electron emission follows mainly two mechanisms, quantum tunneling and thermoinic emission. The sharp emitter tip can enhance the local field of the tip, which helps electron emission. The experimental investigation indicates that CN has better field emission performance than the conventional emitter, such as low turn-on field and high current density, even though CN has higher work function.[Bonard] These novel phenomena have been puzzling experimentist. The comparative study of both capped and opened (5,5) SWCN by the first-principle calculation give some clues to understand the mechanism of the CN field emission. [Peng] The applied field modifies not only the vacuum potential barrier but also penetrate inside CN. This field penetration modifies the electrostatic potential of CN. The electric dipole and charge accumulation may occur at the apex in high applied field. These effects of the field penetration equivalently suppress the vacuum potential barrier, leading to the turn-on field of the CN field emission lowing, which can be regarded as a phenomenological field enhancement factor.

VIII. SPIN POLARIZED FIELD EMISSION

The spin polarized electron beam promises a potential application in nano-technology. The early studies mainly investigate the spin polarized field emission from some metallic ferromagnetic material, such as Ni and Fe. [Campagna] The main difficulty to implement the spin polarized field emission is that the spin polarization replies on very low temperature and strong magnetic field. Recently, the advance of the nanotechnology and computer simulation renews this issue. [Li, Gu, Chazalviel] The first-principle calculation exhibits that the Mn-doped GaN nanotube has the spin-split energy band structure. [Gu, Chazalviel] These results stimulate us to examine the possibility to improve the spin polarization of field emission from other nano-scale materials.

6.1. Spin Polarized Field Emission from Carbon Nanotubes

We consider the CN field emission model applied a magnetic field along the tube axis. The Zeeman effect of the magnetic field modifies the energy band structure of CN, leading to the spin-split energy band. For the SWCN, the energy band structure can be expressed as [Jiang, Liang10]

$$\varepsilon_q(k) = \pm t\sqrt{1+4\cos(\frac{\sqrt{3}k_x a}{2})\cos(\frac{k_y a}{2})+4\cos^2(\frac{k_y a}{2})} + \sigma_z \frac{g\hbar^2}{m_e^* r^2}\frac{\phi}{\phi_0} \qquad (6.4.2)$$

where the wave vector $k_x k_y$ satisfies the Eq.(5.3). The second term in Eq. (2) is the Zeeman effect of the magnetic field. The g factor may be 2 approximately; $\sigma_z = \pm 1/2$ for spin up and down respectively. The effective electron mass can be used to the effective electron mass in graphite as approximation $m_e^* \approx 2.222 m_e k_B T$.[Wallace] We set the work function φ = 4.7 eV for all SWCNs. Since our attention focus on the basic spin polarized properties of emission current, we use the local field to present our results. This framework of the SWCN field emission can be generalized to the multi-wall carbon nanotube field emission within an axial magnetic field.[Liang 06]

6.2. Spin Polarization of Emission Current

The spin polarization of emission currents is defined by $P = (j_\uparrow - j_\downarrow)/(j_\uparrow + j_\downarrow)$, where $j_{\uparrow(\downarrow)}$ is the emission current of the spin-up (down) electrons respectively. We investigate the emission current and their spin polarizations versus different electric fields and magnetic fields at room temperature in Figure 6.1.[Liang10] The spin polarization decreases with the electric field increasing in Figure6.1 (a) and (c). This is because the increase of total current is larger than the increase of difference between the spin-up and down components. The spin polarizations of the emission current reach 0.05~0.08 with 0.1~1μA emission current in low electric field. The spin polarizations of the emission currents increase linearly with the magnetic field increasing shown in Figure1 (b) and (d).

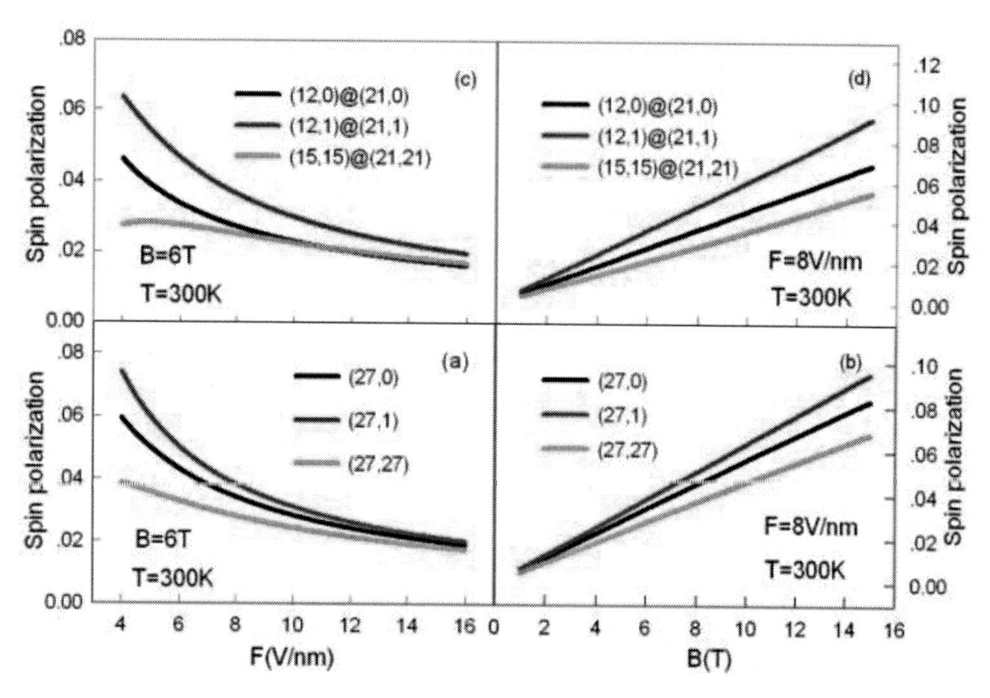

Figure 6.1. (On line color): The spin polarization of emission current versus the electric and magnetic fields in (a) and (b) for the typical SWCNs and in (c) and (d) for the typical MWCNs.

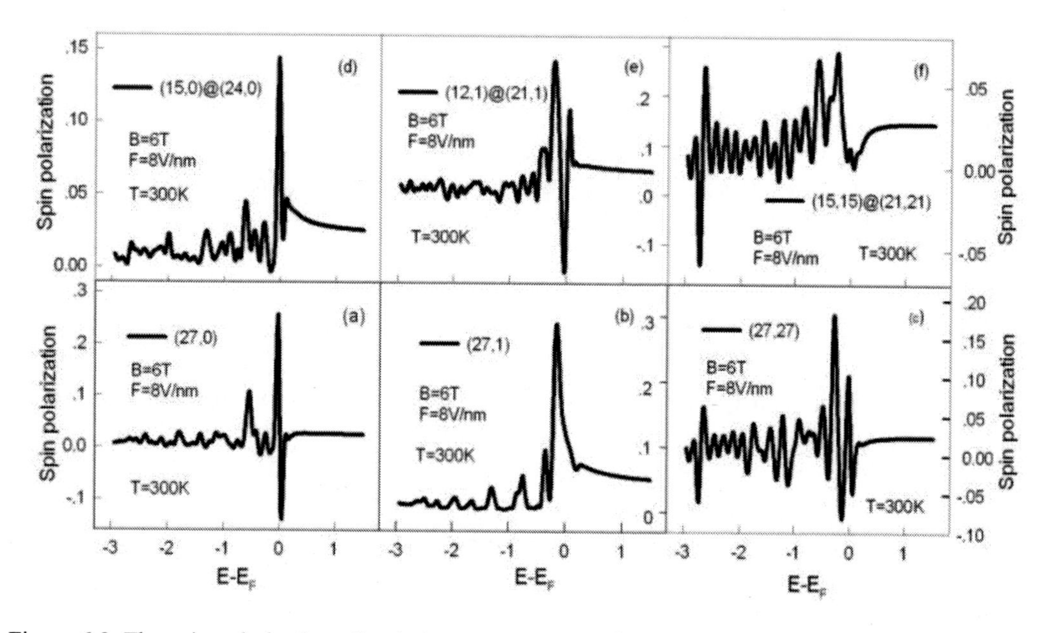

Figure 6.2. The spin polarization of emission energy spectra for three typical SWCMs in (a), (b) and (c) for three typical MWCNs in (d), (e) and (f) in given electric and magnetic field F=6V/nm and B=6T at room temperature.

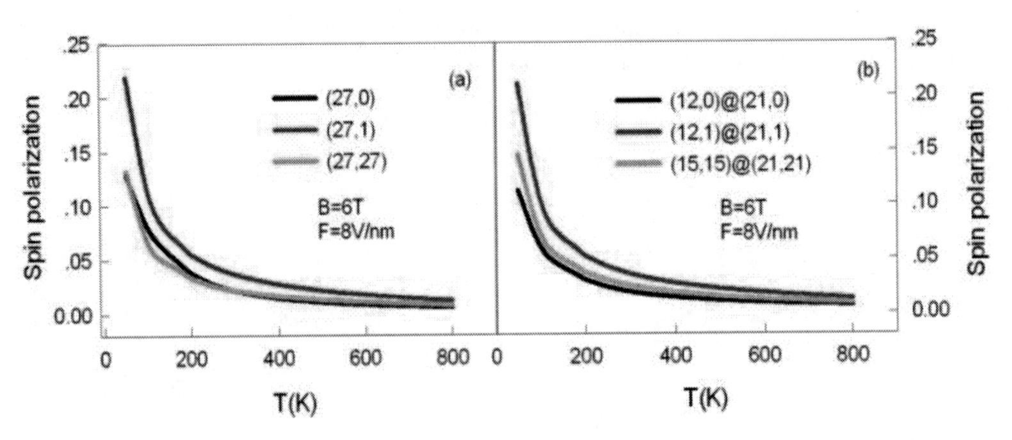

Figure 6.3. (On line color): The spin polarization of emission current versus temperature for three typical SWCMs and MWCNs in (a) and (b) in given electric and magnetic field, F=6V/nm and B=6T.

6.2. Spin Polarized Emission Energy Spectrum

In Figure 6.2, we plot the spin polarization of the energy distributions of three typical SWCNs and MWCNs at the magnetic field B=6T and the electric field F=8V/nm. [Liang10] The spin polarizations oscillate and the peaks reach 0.15~0.3eV near Fermi level.

6.3. Spin Polarized Emission Current Versus Temperature

Physically, temperature increasing will suppress the spin polarization of the field emission current. In Figure 6.3, we investigate the spin polarization of the emission currents versus temperature at the magnetic field B=6T and the electric field F=8V/nm for three typical SWCNs and MWCNs. [Liang10] The spin polarizations decrease with temperature increasing. The spin polarization reach to 0.1~0.2 with 0.01~0.02μA emission current at 100K and remain 0.05 at 400K. This implies that the spin polarization do not vanish around room temperature.

6.4. Basic Physical Factors of the Spin-Polarized Field Emission

In principle, the spin polarized field emission replies on the spin-split energy band of emitter. The spin-split energy band is usually realized by the ferromagnetic metals or applied magnetic field. [Campagna] The weak point of the ferromagnetic metallic emitters is that the ferromagnetic metals can provide only small emission current, while the applied magnetic field is difficult to be implemented in practice. Moreover, the spin-split energy band induced by ferromagnetic metals and applied magnetic field replies sensitively on temperature. This is the main physical encumbrance to realize the robust spin polarized field emission. The recent theoretical studies seem to open some new ways. The dilute magnetic semiconductor can generate spin current by the spin-orbit interaction, [Niu] which can inspire us to explore the possibility to yield spin-polarized field emission. Another possibility is that the quantum confinement effect can enhance the spin polarization of the emission current. [Niu] This is also challenging for us how to design the nanoscale emitter or how to seeking for new kinds of nanoscale material as emitters.

Actually, from the basic physical point of view, if we can generate a spin-dependent vacuum potential barrier we can yield the robust spin polarized field emission even though for conventional metallic emitters because the spin-dependent vacuum potential is more sensitive than the spin-split energy band for inducing the spin-polarized emission current. Thus, the challenging problem for theorist and experimentalist is how to generate a spin-dependent vacuum potential. [Liang10]

VII. Discussion And Perspective

The rapid progress of nanotechnology leads to new materials and new techniques emerging constantly, which upgrades theoretically and experimentally the field electron emission. In particularly, the novel phenomena observed in nano scale material (NSM) field emission stimulate many fundamental and challenging issues. The theoretical development expands along two directions. One is to build in new theory beyond the FN theory based on the new characteristics of NSM, such CN that has Dirac-electron and Luttinger liquid properties, and the quantum confinement effect in nanowires. This development not only provides a theoretical understanding of novel phenomena of NSM field emission, but also reveals the basic physics of these new kinds of material in field emission. The other direction is the computer simulation techniques in the field emission due to the rapid upgrade of the computer ability. The new algorithms and techniques allow one to face more and more realistic systems. This progress narrows the gap between the experimentalist design and theorist prediction.

Let us try to answer the question what is the new physics of NSM field emission. The first is that the energy band structure of NSM plays a crucial role in the basic field emission behavior. The new theories of CN and Luttinger liquid field emission tell us that the I-V characteristic beyond the FN-type originates from the new energy band structure of NSM. The second point we learn from the computer simulation is that the electric field can penetrate into the tip of emitter for NSM, which can lead to extra charges at the tip and equivalently suppressing the work function. This effect can be regarded as the effective field enhancement factor that depends on the applied field and the geometry of tip. The third is that the multi-peak energy distribution originates from the intrinsic energy band of NSM. In other words, so far we believe that the I-V characteristic of field emission is governed by the energy band structure of NSM, while the turn-on field mainly depends on the field penetration effect of NSM and the impurity and defect at the tip. However, actually we do not know from the computer simulation whether the field penetration effect can also modify the I-V characteristic of field emission. Hence, the true story should be expected to be further verified by the experimental investigation. Unfortunately, if the impurities and defects at the tip dominate the field emission behavior for NSM, we will never know the true story of field emission because the fingerprint effect of the samples disturbs the basic law we expect.

Actually there are still many unsolved problems when we face various NSMs. The first challenging problem is whether we can predict theoretically what physical factor dominates the basic field emission behavior for specific NSMs. This may be expected more interplay between theorist and experimentalist and more pure sample of emitter emerging. The other challenging issue is how to yield high coherent spectrum of electron emission energy. The third issue is how to yield robust spin polarized field emission. From the practical point of views, another meaningful issue is to find out the values of the phenomenological parameters of the empirical theory for different kinds of materials and different mechanisms. These parameters can become a standard data of field emission for the practical design of emitters.

Exact speaking, so far, most of the first-principle scheme can calculate only the electrostatic potential of CN. The electron tunneling through the vacuum potential is still calculated by the WKB approximation. Along this direction, it will be expected to develop a full first-principle computational framework to calculate numerically the transmission coefficient of electronic tunneling through the vacuum potential.

It should be emphasized that it is difficult and overfull expectation to fully understand of all novel phenomena in NSM field emission and to match completely the experimental data with the theoretical prediction based on the new theories because these theories are constructed based on some simplified models that is still far away from the experimental samples due to the samples being usually not pure and stable enough to make sure the data whether we are expecting. Actually, the values of these new theories do not reply on the matching between the experimental data and theoretical prediction because the new theories reveal new physics of field emission beyond the FN-type theory. Nevertheless, it is still reasonable to expect observing these new characteristics revealed by new theories as long as the basic factors in new theories dominate the physical behaviors.

Finally it should be remarked that this review could not cover all of new development of the field emission theory and contain ineluctably the author's taste.

CONCLUSION

The conventional field emission theory is facing new challenges due to rapid development of nanotechnology. The new phenomena in NSM field emission stimulate new fundamental interest and practical issues. The new theories and computer simulation open a new chapter of the field emission theory. The theory of Dirac electron and Luttinger liquid reveal some novel characteristics of field emission beyond the FN theory and provide the theoretical insights to understand the novel phenomena of NSM. The computer simulation and the empirical theory provide a powerful tool to narrow the gap between the experimental design and the theoretical prediction. These new developments bring us new thoughts, hints and methods to face challenges of new materials and new technologies.

ACKNOWLEDGMENTS

This work is supported financially by the National Natural Science Foundation of China (Grants No. 10774194), National Basic Research Program of China (973 Program: 2007CB935501).

REFERENCES

[Adessi] Adessi, Ch. and Devel, M., Phys. Rev. B62, R13314 (2000); Adessi, Ch. and Devel, M., *Ultramicroscopy* 85 215 (2000).

[Bonard] Bonard, J.M., Kind, H., Stockli, T., Nilsson, L. U., Solid-State Electronics, 45, 893 (2001); Bonard J. M., Salvetat, J. P., Stockli, T., Forre, L., and Chatelain, A., *Appl. Phys.* A 69, 245 (1999).

[Bockrath] Bockrath M., Cobden, D. H., Lu J., Rinzer A. G., Smalley R. E., Balents L. McEue P. L., Nature 397, 598 (1999); *Egger R. Phys. Rev. Lett.* 83, 5547 (1999).

[Campagna] Campagna M., Utsumi, T., and Buchanan, D. N. E. , J. Vac. Scie. Technol. 13, 193 (1976); Landolt, M., and Campagna, M., *Phys. Rev. Lett.* 38, 663 (1977); Gleich, W., et. al. Phys. Rev. Lett. 27, 1066 (1971); Landolt, M., et. al. *Phys. Rev. Lett.* 40, 1401 (1978); Chrobok, G., et. al. Phys. Rev. B 15, 429 (1977)

[Chazalviel] Chazalviel, J.N., and Yafet, Y., Phys. Rev. B 15, 1062 (1977); Ohwaki, T., Wortmann, D., Ishida, H., Blugel, S., and Terakura, K., Phys. Rev. B 73, 235424 (2006); Hao, S., Zhou, G., Wu, J., Duan, W., and Gu, B. L., *Phys. Rev.* B 69, 113403 (2004).

[Dean] Dean, K.A., Groening, O., Kuttel, O.M., and Schlapbach, L., Appl. Phys. Lett. 75, 2773 (1999); Fransen, M.J., Rooy, T. L., Kruit, P., *Appl. Surf. Scie.* 146, 312 (1999).

[Falvo] Falvo, M.R., Clary, G.. J., Taylor, II R.M., Chi, V., Brooks, Jr FP., Washutn, S., and Superfine, R., Nature 389 582 (1997); Frank, S., Poncharal, P., Wang, Z., L., Heer, W.A. de, *Science* 280, 1744 (1998).

[Filip] Filip, V., Nicolaescu, D., and Okuyama, O., J. Vac. Sci. Technol. B 19, 1016 (2001).

[Fowler] Fowler, R.H., and Nordheirm, L.W., Pro. R. Soc. London, Ser, A 119, 173(1928); Nordheirm, L.W., *Pro. R. Soc.* London, Ser, A 121, 626(1928).

[Forbes99] Forbes, Richard G., J. Vac. Sci. Technol., B17, (2), 526(1999); Forbes, Richards G., *Surf. And Interface Anal.*, 36, 395 (2004)..

[Forbes09]. Forbes, Richard G., *Technical Digest*: 22nd IVNC,13 (2009).

[Frank] Frank, S., Poncharal, P., Wang, Z.L., Heer, W.A. de, *Science* 280, 1744 (1998).

[Gadzuk] Gadzuk, J. W., and Plummer, E. W., *Rev. Mod. Phys*. 45, 487 (1973).

[Gu] Zhu, G., Duan, W., and Gu, B. L., Phys. Rev. Lett. 87 095504 (2001); Zhu, G., Duan, W., and Gu, B. L., *Phys. Rev. Lett.* 87, 095504(2001).

[Han] Han S., and Ihm, J., Phys. Rev. B 66, 241402 (2002); Park, N., Han, S., and Ihm, J., Phys. Rev. B 64, 125401(2001); Song, S. N., Wang, X. K., Chang, R. P. H., and Ketterson, J. B., *Phys. Rev. Lett.* 72, 697 (1994).

[Heer] de Heer, W. A. , Ch^atelain A., and Ugarte D. *Science* 270 1179 (1995); de Heer, W. Bacsa, Cha^telain, A., Gerfin, T., Humphrey-Baker, R., Forro, L., and Ugarte, D., *Science* 268, 845 (1995); Falvo, M.R., Clary, G.. J., Taylor, II R.M., Chi, V., Brooks, Jr F. P., Washutn, S., and Superfine, R., Nature 389, 582 (1997); Bachtold, A., Strunk, C., Salvetat, J. P., Bonard, J. M., Forró, L., Nussbaumer T., and Schönenberger, C., *Nature* 397, 673 (1999).

[Jiang] Jiang, J., Dong J. M., and Xing, D. Y., Phys. Rev. B 62, 13209 (2000); Lin M. K., and, Shung, K. W. K., *Phys. Rev.* B 52, 8423 (1995).

[Li] Li, B., Leung, T. C., Chan, C. T., Phys. Rev. Lett. 97, 087201 (2006); Alvarado, S.F., *Phys. Rev. Lett.* 75, 513 (1995).

[Liang 03] Liang Shi-Dong and Xu, N. S., Appl. Phys. Lett. 83, 1213 (2003); Liang Shi-Dong, Huang, N. Y., Deng, S. Z., and Xu, N. S., *Appl. Phys. Lett.* 85, 813 (2004).

[Liang 04] Liang Shi-Dong, *Physica* B352, 305 (2004); Liang Shi-Dong *Int. J. Mod. Phys.* B 21, 4377 (2007).

[Liang 05] Liang Shi-Dong, Huang, N. Y., Deng, S. Z., and Xu, N. S., J. Vac. Sci. Technol. B24,983 (2006); Huang, N. Y., Liang Shi-Dong, Deng, S. Z., and Xu, N. S., *Phys. Rev.* B 72 075412（2005）.

[Liang 06] Liang Shi-Dong, Deng, S. Z., and Xu, N. S.,, *Phys. Rev.* B 73, 245301 (2006);
Liang Shi-Dong, Deng, S. Z., and Xu, N. S., *Phys. Rev.* B 74, 155413 (2006); Tang, H., Liang Shi-Dong, Deng, S. Z., and Xu, N. S., *J. of Phys. Rev.* D39 , 5280(2006).

[Liang 08] Liang Shi-Dong, and Chen, L., *Phys. Rev. Lett.* 101, 027602 (2008); Liang Shi-Dong, and Chen, L., *J. Vac. Sci. Technol.* B28(2), C2A50 (2010).

[Liang 10] Liang Shi-Dong, *Technical Digest*: 23nd IVNC, (2010).

[Liang 11] Liang Shi-Dong, unpublished

[Lu] Buldum, A., and Lu, J. P., *Phys. Rev. Lett.* 91, 236801 (2003); Lu, J. P., *Phys. Rev. Lett.* 74, 1123 (1995).

[Mayer] Mayer, A., Miskovsky, N. M., and Cutler, P. H., *J. of Phys.* CM,15, r177(2003); Mayer A and Vigneron J. P., *Phys. Rev.* B 56 12599 (1997); Mayer A and Vigneron J. P., *Phys. Rev.* B 60 2875 (1999); Mayer A, Miskovsky, N. M., and Cutler P. H., *Appl. Phys. Lett.* 79 3338 (2001); Mayer A, Miskovsky, N. M., and Cutler, P. H., *Phys. Rev.* B 65 155420 (2002); Mayer A, Miskovsky, N. M., and Cutler, P. H., *Phys. Rev.* B 65 195416 (2002).

[Mintmire] Mintmire J. W., and White, C.T., *Phys. Rev. Lett.*, 81, 2506 (1998).

[Murphy] Murphy, E. L., and Good, R. H. Jr, *Phys. Rev.* 102, 1464 (1956).

[Modinos] Modinos, A., Field, thermionic, and secondary emission: Emission Spectroscopy, Plenum Press New York and London (2002).

[Niu] Niu, Y. R., et.al. Surf. Scien. 604, 1055 (2010); Ohwaki, T., et.al. *Phys. Rev.* B 73, 235424 (2006).

[Komnik] Komnik A., and Gogolin, A. O., Phys. Rev. B 66, 035407 (2002); Gogolin, A. O., and Komnik, A., *Phys. Rev. Lett.* 81, 256806 (2001).

[Kociak] Kociak, M., Suenaga, K., Hirahara, K., Saito, Y., Nakahira, T., and Iijima, S., *Phys. Rev. Lett.* 89, 155501 (2002); Ando, T., Zhao, X., Shimoyama, H., Sakai, G., Kaneto, K., *Int. J. Inorg. Mat.* 1, 77 (1999).

[Khazaei] Khazaei, M., Farajian, A. A., and Kawazoe, Y., Phys. Rev. Lett.95, 177602 (2005); Duana, X., Akdimb, B., and Pach, R., *Applied Surface Science* 243,11 (2005).

[Kim] Kim C., and Kim, B., Lee, S. M., Jo, C., and Lee, Y. H., *Phys. Rev.* B 65, 165418 (2002); Kim, C., Seo, K., and Kim, B., Park, N., Choi, S. Y., Park, K. A., and Lee, Y. H., *Phys. Rev.* B 68, 115403 (2003).

[Richardson] Richardson, O. W., *The Emission of Electricity from Hot Bodies,* Longmans, Green, New York, (1921); Dushman, S., *Rev. Mod. Phys.* 2, 381, (1930).

[Peng] Peng, J., Li, Z. B., He, C. S., Deng, S. Z., Xu, N. S., Zheng, X., and Chen, G. H., *Phys. Rev.* B 72, 235106 (2005).

[Rinzler] Rinzler, A.G., Hafner, J.H., Nikolaev, P., Lou, L., Kim, S.G., Tománek, D., Nordlander, P., Colbert, D.T., and Smalley, R.E., *Science* 269, 1550 (1995).

[Saito93] Saito, R., Dresselhaus, G., and Dresselhaus, M. S., *J. Appl. Phys.* 73, (2), 494 (1993); Kwon Y. K., and Tomanek, D., *Phys. Rev.* B58, R16001 (1998); Saito, R., Dresselhaus, G., and Dresselhaus, M. S., *Phys. Rev.* B61, 2981 (2000); Ando, Y., Zhao, X., Shimoyama, H., Sakai, G., Kaneto, K., *Int. J. Inorg. Mat.* 1, 77 (1999).

[Saito98] Saito, R., Dresselhaus, G., and Dresselhaus, M. S., Physical properties of carbon nanotubes (Imperial college press 1998).

[Wallace] Wallace, P. R. *Phys. Rev.*, 71, 622, (1947).

[Xu] Xu, N. S., Chen, J., Deng, S. Z., *Chin. Phys. Lett.* 18(9)(2001)1278;

[Zheng] Zheng, X., Chen, G. H., Li, Z. B., He, C. S., Deng, S. Z., and Xu, N. S, *Phys. Rev. Lett.* 92, 106803(2004).

In: Advances in Condensed Matter ... Volume 10
Editors: H. Geelvinck and S. Reynst

ISBN: 978-1-61209-533-2

Chapter 2

ORGANIC–INORGANIC HYBRID PROTON EXCHANGE MEMBRANE ELECTROLYTES FOR MEDIUM TEMPERATURE (100 - 200 ^{0}C) NON-HUMIDIFIED H_2/O_2 FUEL CELLS

G. Lakshminarayana,[1,*] V. S. Tripathi[1], Masayuki Nogami,[1,*] and I. V. Kityk[2†]
[1]Department of Materials Science and Engineering, Nagoya Institute of Technology, Showa, Nagoya, Japan
[2]Electrical Engineering Department, Czestochowa Technological University, Czestochowa, Poland

ABSTRACT

Anhydrous proton conducting inorganic-organic hybrid membranes were fabricated by sol-gel process with tetraethoxysilane/ tetramethoxysilane/ methyltriethoxysilane/ trimethylphosphate and 1-ethyl-3 methylimidazolium tetrafluoroborate [$EMIMBF_4$] ionic liquid as precursors. These hybrid membranes were studied with respect to their structural, thermal, conductivity, and hydrogen permeability properties. The Fourier transform infrared spectroscopy (FT-IR) and ^{31}P, ^{29}Si, ^{1}H, ^{13}C, and ^{19}F Nuclear magnetic resonance (NMR) measurements have shown good chemical stability, and complexation of $PO(OCH_3)_3$ with [$EMIMBF_4$] ionic liquid in the studied hybrid membranes. For all the prepared membranes, average pore sizes and specific surface areas were investigated by the Brunauer–Emmett–Teller (BET) method. The average pore size was increased proportionally with the ionic liquid weight percent ratio in the host phosphosilicate matrix within the range 1 - 9 nm, respectively. Thermogravimetric analysis (TGA) and differential thermal analysis (DTA) measurements confirmed that the hybrid membranes were thermally stable up to 290 ^{0}C. Thermal stability of the hybrid membranes was significantly enhanced due to the presence of inorganic SiO_2 framework and high

* Corresponding author: Tel: 81 52 735 5285; Fax: 81 52 735 5285. Email: glnphysics@rediffmail.com; nogami@nitech.ac.jp.
† Czestochowa Technological University, Al. Armii Krajowej 17, Czestochowa, Poland.

stability of [BF_4] anion. The effect of [$EMIMBF_4$] ionic liquid addition on the microstructure of the membranes was studied by Scanning electron microscopy (SEM) and Energy dispersive X-ray analysis (EDX) micrographs and no phase separation at the surfaces of the prepared membranes was observed and also homogeneous distribution of all the elements was confirmed. Conductivity of all the prepared membranes was measured within temperature range -25 to 155 ^{0}C, and high conductivity of 1 x 10^{-2} S/cm was obtained for 40 wt% [$EMIMBF_4$] doped 30TMOS-30TEOS-30MTEOS-10PO(OCH_3)$_3$ (mol%) hybrid membrane, at 155 ^{0}C under anhydrous conditions. The hydrogen permeability was found to decrease from 1.41 x 10^{-11} to 0.1 x 10^{-12} mol/cm.s.Pa for 40 wt% [$EMIMBF_4$] doped hybrid membrane as the temperature increases from 20 to 150 ^{0}C. For 40 wt% [$EMIMBF_4$] doped hybrid membrane, the membrane electrode assemblies were prepared and a maximum power density value of 11 mW/cm^2 at 26 mA/ cm^2 as well as a current density of 40.57 mA/ cm^2 were achieved at 150 ^{0}C under non-humidified conditions during utilization in a H_2/O_2 fuel cell.

Keywords: Inorganic-organic hybrids; electrolytes; conductivity; hydrogen permeability; H_2/O_2 fuel cells

1. INTRODUCTION

Recently, much attention has been focused upon the research and development of proton exchange membrane (PEM) fuel cells as power sources for stationary and portable applications because of their high energy conversion efficiency and near-zero pollutant emission [1-5]. Two key components determining the performance and cost of PEMFCs are the electrocatalysts and the proton exchange membrane (PEM). The latter is employed to provide proton conduction from the anode to the cathode, and to separate the anode (hydrogen) and cathode (oxygen) reactants. The desirable characteristics of the proton exchange membrane are: i) high proton conductivity; ii) absence of electronic conductivity; iii) low fuel or gas permeability; iv) appropriate chemical, thermal and mechanical stability under the operative conditions of the PEMFCs over long periods; and v) low cost. Also, improvement of duration of proton exchange membrane fuel cells (PEMFCs) is one of the tasks which should be attained before forthcoming practical wide applications. Factors affecting the duration of PEFC may be poisoning of catalysts, dissolution of Pt catalyst, and degradation of proton exchange membrane [6]. Currently, perfluorosulfonic acids (PFSA) such as Nafions produced by DuPont are widely used as the PEM in PEMFC because of their good mechanical features and high oxidative stability as well as their high proton conductivity compared to most hydrocarbon (HC) based membranes. However, high costs, loss of proton conductivity at high temperature (>80 ^{0}C) and low humidity, and relatively high gas permeability have limited PFSA membranes for further commercial applications [7, 8].

On the other hand, proton exchange membrane fuel cells (PEMFCs) operating at elevated temperature (>100 ^{0}C) have attracted recently considerable attention because they could benefit from enhanced tolerance to impurity of the fuel gas, simplified water and heat management, and increasing reaction rates at both cathode and anode compared with PEMFCs operating below 80 ^{0}C [9-12]. Therefore, there has been a great deal of research works concerning the development of anhydrous electrolyte membranes with high proton conductivities at temperatures above 100 ^{0}C [13-17]. Fuel cells operation above 100 ^{0}C under anhydrous conditions has been demonstrated using proton conducting electrolytes such as PBI/H_3PO_4 or $CsHSO_4$ [18-21]. However, such proton conductors still exhibit some drawbacks such as poor processability and mechanical properties and a limited operational temperature range.

Ormosil is known as an organic-inorganic hybrid material, in which organic functional groups can be chemically bonded to an inorganic silica matrix [22]. With a proper choice of organic functional groups and an optimization of preparing conditions, it is always possible to control the mechanical and thermal properties, and hydrophobicity of an ormosil [23]. Intermediate features of ormosils between those of inorganic silica and organic polymers have extended their potential applications [24].

Sol-gel chemistry enables to design hybrid inorganic organic materials, giving an access to the fabrication of glasses and films with novel promising properties [25]. The low reaction temperature on sol-gel processing facilitated incorporation of functional organic compounds into inorganic materials.

Tetraethoxysilane (TEOS), $Si(OC_2H_5)_4$ consisting of mono silicon, is a commonly used precursor for preparation of sol-gel-derived silica membranes. In the process of the hydrolysis and polymerization reaction of TEOS, the (≡Si-O-) unit can be the minimum for amorphous silica networks. Also, tetramethoxysilane (TMOS) is one of the most common alkoxides yielding a glassy silica network. In the present work, for the preparation of organically modified silicate structure, tetraethoxysilane (TEOS), tetramethoxysilane (TMOS) and methyltriethoxysilane (MTEOS) are chosen because of their relatively rapid hydrolysis under acidic conditions. In order to make the inorganic silica structure more hydrophobic, methyltriethoxysilane (MTEOS) was chosen. The hydrolysis/condensation rate at room temperature of MTEOS is ~7 times higher than that of TEOS [26]. This implies that the reaction time of MTEOS should be ~7 times shorter to obtain silica polymers with dimensions similar to those obtained with hydrolysis and condensation of TEOS. In addition, many factors such as temperature, condensation medium (acid/base), nature of the solvent and the type of alkoxide precursor, ageing, drying, stabilization and densification could influence the sol–gel structure since they are related to the rates of hydrolysis and condensation determining the structure of the gel [27].

One of the various approaches used to develop anhydrous proton exchange membranes above 100 ^{0}C was by replacing water as the proton solvent with a liquid that has a higher boiling point [28]. Room Temperature Ionic liquids (RTILs) nowadays are very fascinating for a growing number of scientists and engineers due to their unique physical and chemical properties with low melting points (<100 ^{0}C) that can exhibit intrinsically useful characteristics such as a wide liquid range, a negligible vapor pressure, thermal and chemical stability, and chemical tunabilities and recyclability [29-31]. In addition to offering many desirable features, such as chemical stability and low vapor pressure, their properties can be varied to meet exactly the needs of the process of interest. This tailoring of properties can be achieved by combining different cations and anions as well as by fine-tuning of the respective ions, e.g., by adjusting the length of an alkyl side chain on one of the ions. These unique properties of ILs make them appropriate candidates as advanced electrolyte materials in lithium ion batteries, fuel cells, dye-sensitized solar cells, and super capacitors [32-40]. Recently, many researchers focused on aromatic type ionic liquids, such as 1-ethyl-3-

methylimidazolium tetrafluoroborate ($EMIMBF_4$), which possess a relatively low viscosity and high ionic conductivity [41-43] for application in various electrochemical devices, including fuel cells, batteries and capacitors. In the present work, we assumed that the smaller size, lower viscosity and hydrophilic nature of the IL($EMIMBF_4$) can contribute well to the high conductivity of phosphosilicate /$EMIMBF_4$ hybrid membranes under anhydrous conditions at above 100 ^{0}C.

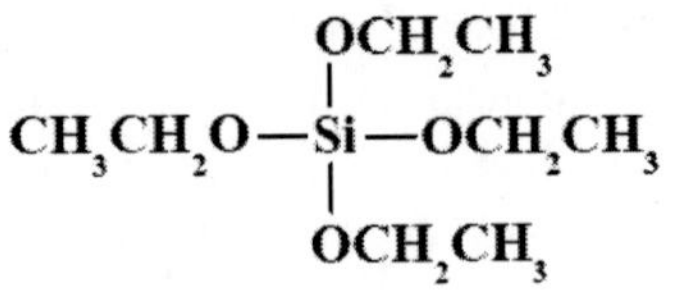

OCH_2CH_3 / $CH_3CH_2O-Si-OCH_2CH_3$ / OCH_2CH_3

TEOS

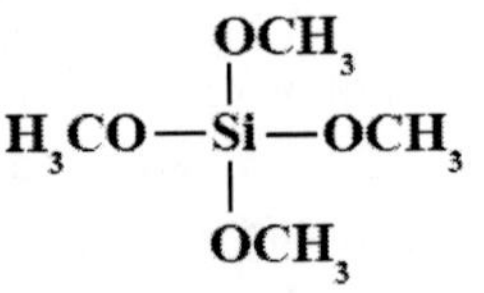

OCH_3 / $H_3CO-Si-OCH_3$ / OCH_3

TMOS

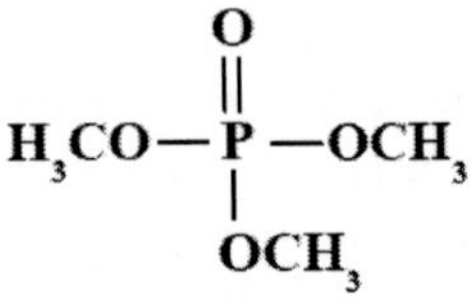

O / $H_3CO-P-OCH_3$ / OCH_3

$PO(OCH_3)_3$

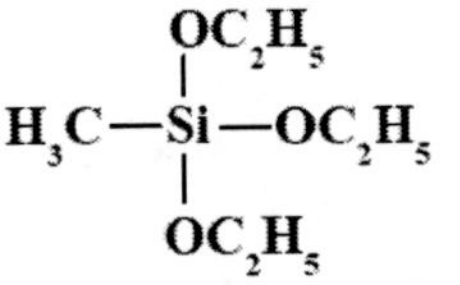

OC_2H_5 / $H_3C-Si-OC_2H_5$ / OC_2H_5

MTEOS

$[BF_4]^-$ (F, F—B—F, F); N, N⁺, H_3C, CH_2, CH_3

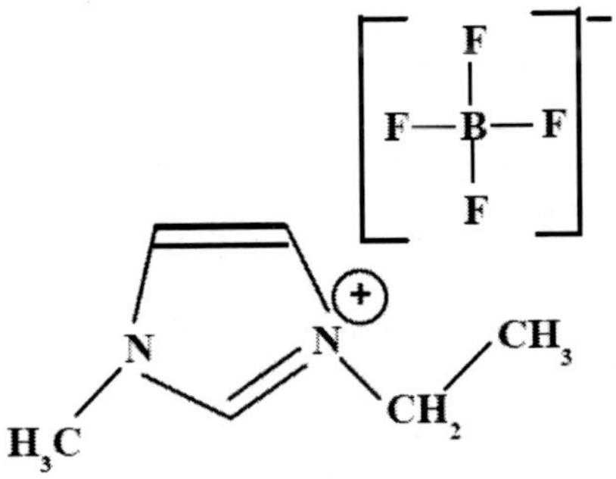

$EMIMBF_4$

TEOS

(Hydrolysis) $Si(OC_2H_5)_4 + xH_2O \longrightarrow Si(OH)_x(OC_2H_5)_{4-x} + xROH$

(condensation) $-O-Si-OH + H-O-Si-O- \longrightarrow O-Si-O-Si-O- + xH_2O$

$-O-Si-OH + H_5C_2-O-Si-O- \longrightarrow O-Si-O-Si-O- + xC_2H_5OH$

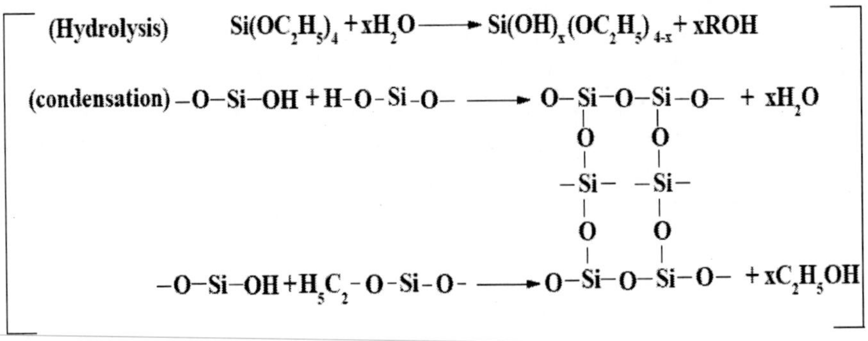

TMOS

(Hydrolysis) $\equiv Si-OCH_3 + H_2O \rightleftharpoons \equiv Si-OH + CH_3OH$

(water condensation) $\equiv Si-OH + \equiv Si-OH \rightleftharpoons \equiv Si-O-Si\equiv + H_2O$

(alcohol condensation) $\equiv Si-OCH_3 + \equiv Si-OH \rightleftharpoons \equiv Si-O-Si\equiv + CH_3OH$

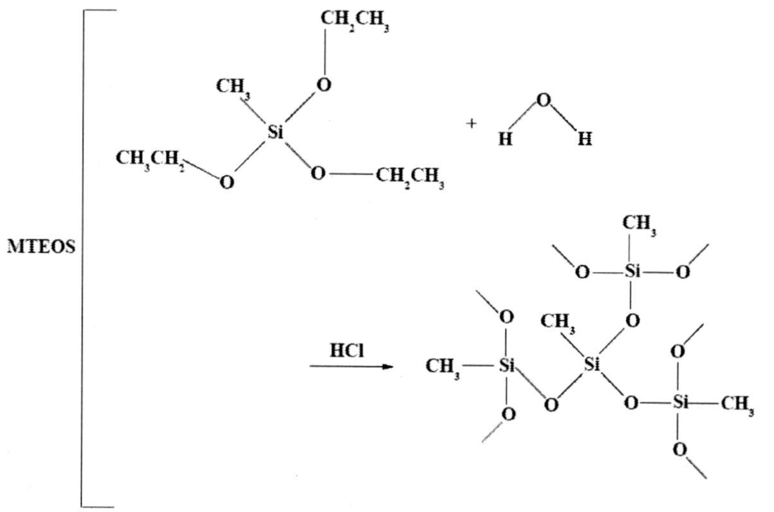

Figure 1. Chemical structures, hydrolysis and condensation reactions of all the sol-gel silica precursors used in the present work, including trimethyl phosphate and [$EMIMBF_4$] ionic liquid.

In the present work, we report on fabrication of hybrid membranes using tetraethoxysilane (TEOS)/ tetramethoxysilane (TMOS) / methyltriethoxysilane (MTEOS)]/ trimethyl phosphate ($PO(OCH_3)_3$) and 1-ethyl-3 methylimidazolium tetrafluoroborate ($EMIMBF_4$) ionic liquid as sol-gel precursors, and the conductivity of these composites under anhydrous conditions is measured. In our recent works [44, 45], trimethyl phosphate ($PO(OCH_3)_3$) was used as phosphorous dopant in phosphosilicate electrolyte membranes and proton conduction was observed in these materials. Figure 1 shows the chemical structures, hydrolysis and condensation reactions of all the sol-gel silica precursors used in this work, including trimethyl phosphate and [$EMIMBF_4$] ionic liquid. Under anhydrous conditions, the proton transport might be described by Grotthuss mechanism, in which only protons can move from site to site without the assistance of water molecules [46]. The prepared hybrid membranes are investigated with respect to their structural, thermal, and conductivity properties.

2. Experimental Studies

2.1. Preparation of Hybrid Membranes

The hybrid membranes were prepared by using tetraethyl orthosilicate ($Si(OC_2H_5)_4$, TEOS, 99.9%, Colcote, Japan), tetramethoxysilane ($Si(OCH_3)_4$, TMOS, 99.9%, Colcote, Japan), Methyltriethoxysilane (MTEOS, 98%, Aldrich Chemical Com.), trimethyl phosphate ($PO(OCH_3)_3$, 98%, Colcote, Japan), and [1-ethyl- 3-methylimidazolium tetrafluoroborate] (($EMIMBF_4$), 99%, Aldrich Chemical Com.) ionic liquid as precursors. All the initial solvents and materials were used as received. Water purified with a Milli-Q system from Millipore (AQUARIUS/GS-20R, Japan) was used for the experiments. All the hybrid membranes including host phosphosilicate matrix were prepared at ambient temperature under normal atmospheric pressure by sol-gel process. Several compositions of TMOS –

TEOS – MTEOS – PO $(OCH_3)_3$ – IL (i.e., 30TMOS – 30TEOS – 30MTEOS – 10 PO $(OCH_3)_3$ – x [$EMIMBF_4$] (x=10, 20, 30, and 40 wt %) were selected for the ionic liquid content optimization. It should be emphasized that we added [$EMIMBF_4$] ionic liquid in excess wt% to the host membrane 30TMOS – 30TEOS – 30MTEOS – 10PO $(OCH_3)_3$ (mol %).

Initially, the calculated amount of TMOS was hydrolyzed with water (as 0.15N-HCl aq.) and ethanol under magnetic stirring for 60 min. at room temperature. The molar ratio of TMOS/ EtOH/ H_2O/ HCl was equal to 1 /8 /4 /0.01. After one hour, calculated amount of TEOS was added to TMOS solution under vigorous stirring. The molar ratio of TEOS/ EtOH/ H_2O/ HCl was equal to1 /8 /4 /0.01. One hour later, MTEOS which was pre-hydrolyzed with water (as 0.15N-HCl aq) and ethanol were added to this TMOS/TEOS solution under constant stirring condition during 60 min.

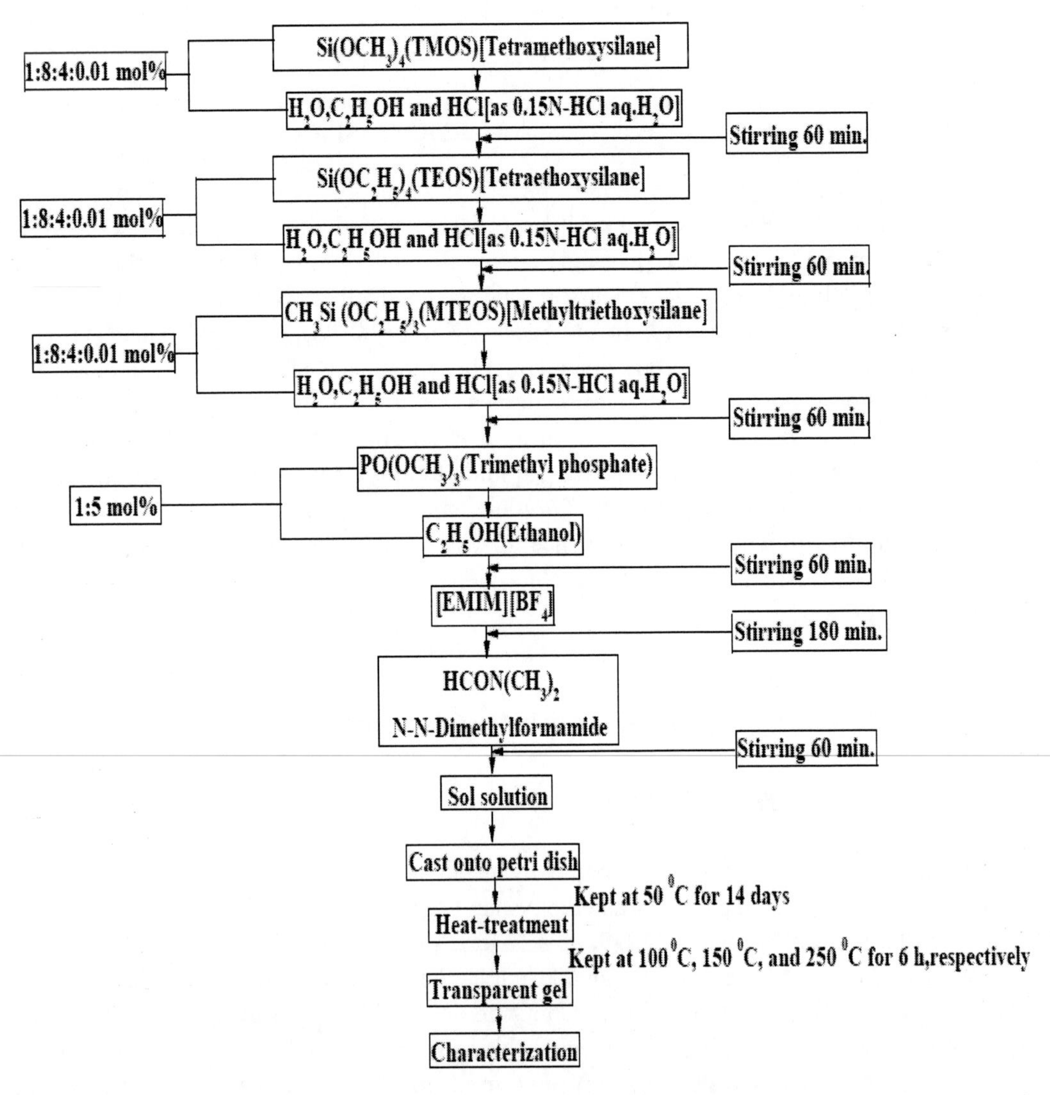

Figure 2. Synthetic procedure for obtaining the 30TMOS-30TEOS-30MTEOS-10 PO $(OCH_3)_3$-x [$EMIMBF_4$] (x=10, 20, 30, and 40 wt %) hybrid membranes.

The amount of MTEOS was taken in equal molar ratio to TMOS and TEOS, respectively. The value of R_w (molar ratio of H_2O/alkoxide) was fixed at 4 for the preparation of sols. Afterwards, when the above mixed solution becomes transparent, trimethyl phosphate dissolved in C_2H_5OH with five times moles of H_2O was added drop-wise under magnetic stirring for 1 h. Now, the 1- ethyl- 3- methylimidazolium tetrafluoroborate [$EMIMBF_4$] ionic liquid was added drop-wise to the hydrolyzed solution in argon atmosphere. In order to remove the water from $EMIMBF_4$, it was dried under vacuum at 100 ^{0}C for more than 24 h before use. After drying, it was handled in a glove box filled with dry argon. All of the experiments were performed at ambient temperature, and all handling operations of $EMIMBF_4$ were carried out under the argon gas atmosphere. With continuous stirring for 3 h, a homogeneous solution was obtained. Finally, (N-N-Dimethyl formamide) HCON $(CH_3)_2$ (1-3 ml) was added to standard solution followed by 60 min. stirring. Such obtained clear transparent solution was called as sol. The sol was cast onto Petri dishes and gelled at 50 ^{0}C for 14 days. Gel films obtained were then dried at 50 ^{0}C for 6 h, and consecutively at 100 ^{0}C for 6 h, 150 ^{0}C for 6 h, and at 250 ^{0}C for 6 h. Generally, pure ionic liquids show thermal stability upto 400 ^{0}C. So, we have selected the gel films heat-treatment temperature range within 50 to 250 ^{0}C, in steps of 50 ^{0}C increment, keeping the membranes for 6 h at each temperature. The sample thickness was varied from 0.1 mm to 0.5 mm. The host phosphosilicate membrane without [$EMIMBF_4$] ionic liquid addition was also prepared with the same procedure as described above for comparison. Generally, in TMOS/TEOS/MTEOS covalently bonded network, TEOS has controllable hydrolysis reaction rate and MTEOS possess good thermal stability. The whole synthetic process is presented in figure 2.

2.2. Characterization of Membranes

The optical transmission spectra of all the prepared membranes (with thickness about.2 mm) were recorded within the spectral range 200 - 800 nm by using JASCO Ubest 570 UV-vis-NIR spectrophotometer with 1 nm spectral resolution at ambient temperature. The Fourier transform infrared (FT-IR) spectra of the composite membranes were measured by JASCO FTIR-460 Plus spectrometer with spectral resolution about 4 cm^{-1}. The FT-IR spectra were measured within the spectral range 4000 - 400 cm^{-1} by using the KBr pellet for reference. The Brunauer- Emmett- Teller (BET) specific surface areas of all the prepared membranes were determined from N_2 adsorption-desorption isotherms at liquid nitrogen temperature. Composite membranes of approximately 0.05 g were degassed at 250 ^{0}C for a minimum of 6 h under vacuum before the measurement. The pore size distributions were analyzed with a Quantochrome- NOVA-1000 nitrogen gas sorption analyzer. The thermal degradation process and stability of the composites were investigated by thermo gravimetric analysis (TGA), differential thermal analysis (DTA) and differential scanning calorimeter (DSC) (Thermoplus 2, TG- 8120, Rigaku). The measurements were carried out under dry air with a heating rate of 5 ^{0}C/min. For all the prepared membranes, the surface morphology and homogeneous distribution of all the elements were monitored by scanning electron microscopy (JEOL-JSM-6301, NORAN instrument) and X-ray energy dispersive spectrometer (EDS, equipped on JEOL-JSM-6301 SEM), respectively. The ^{31}P MAS NMR spectra were measured with a Varian Unity Inova 300 spectrometer at 121.42 MHz with a sample spinning rate of about

5000 Hz, and chemical shifts were measured with reference to 85% aqueous H_3PO_4. The solid state high-resolution ^1H and ^{13}C magic-angle-spinning (MAS) NMR spectra were obtained using a Varion UNITY-400 plus spectrometer at the resonance frequency of 400 MHz, respectively. The ^1H and ^{13}C spectra were measured by a single-pulse sequence, and the ^{31}P spectra were traced by a single-pulse sequence with high power ^1H decoupling. »Si MAS NMR experiments were performed at frequency of 59.6 MHz. The spectrometer was equipped with a 7 mm CP/MAS probe and the samples were packed in zirconia rotors and spun at the magic angle at 4 kHz. The spectra were acquired using a spectral width of 100 kHz, a single pulse duration of 7.5 μs (π/2), a recycle delay of 30 s, and an acquisition time of 0.05 s, with 500 scans accumulated and referenced to tetramethylsilane (TMS). A line broadening of 10 Hz and a Gaussian function of 0.003 seconds was applied to process the spectra. The ^{19}F NMR spectra were measured by Varian Unity 400 plus with a 9.4-T vertical magnet of bore size 89 mm operated at 376.2 MHz, using trifluorotolune as reference material. The typical conditions were: spectral band width of 80 kHz, pulse width of 67.2 μs (50.4°), repetition time of 0.2 s, accumulation times of 128 scans. All the NMR measurements were conducted at 25 ^{0}C. The conductivity of the composite membranes was evaluated from the Cole–Cole diagrams by an AC method using Solartron SI-1260 impedance analyzer. Gold was evaporated on both sides of the composite membranes as electrodes, and conductivity of the composite membranes was determined from the impedance data obtained within the frequency range from 1 to 10^7 Hz with signal amplitude of 10 mV. The temperature was raised stepwise and the measurements were carried out after keeping the membrane at each temperature for 1 h under dry nitrogen. Conductivities were measured within the temperature range (-25) ^{0}C - 155 ^{0}C under a flow of dry nitrogen (non-humidified conditions). The conductivity (σ) was calculated from the electrolyte resistance (R) obtained from the intercept of the Cole–Cole plot with the real axis, the thickness (l) and the electrode area (A) according to the equation $\sigma = l/AR$. Hydrogen permeability of the composite membranes (0.5 mm thick) was measured by using a forced convection drying oven (DO-600FA), under dry conditions, consisting of two compartments with a capacity of approximately 50 cm^3, separated by a vertical membrane with an effective area of 20 cm^2. The contents of the compartments were under constant agitation. Gas concentrations were monitored by Varian CP-4900 Micro GC gas chromatography with relative standard deviation (RSD) of 0.13 %.

The membrane electrode assembly (MEA) fabrication and single fuel cell test conditions were described elsewhere [47].

3. RESULTS AND DISCUSSION

Figure 3(a) presents the FT-IR spectrum of pure [$EMIMBF_4$] ionic liquid. The observed spectral bands for pure ionic liquid can be assigned to the C-N, C-H stretching vibration modes of the imidazolium ring (3736, 3436, 3150, 3090, 2990 cm^{-1}), unresolved CH_3-N stretching vibration, CH_2-N stretching vibration, and ring in plane asymmetric stretching, and C-C stretching vibrations of the imidazolium ring (1636, 1574, 1508, 1458, 1388, 1339, 1300, 1170 cm^{-1}), asymmetric stretching vibrations of BF_4^-(1084, 1036 cm^{-1}), and out-plane C-H bending vibrations of imidazolium ring (842, 757, 701, 649, 622, 601, 534, 522 cm^{-1})[48,49]. Also, the small absorption bands observed around at 2359, and 2342 cm^{-1} can be assigned to atmospheric CO_2. Figure 3(b) presents the FT-IR spectra of 10-40 wt% $EMIMBF_4$ ionic liquid based hybrid membranes including host phosphosilicate matrix. For 30TMOS-30TEOS- 30MTEOS- 10PO $(OCH_3)_3$ (mol %) host matrix, spectral peaks at 3628, 3436, 2362, 1664, 1466, 1279, 1148, 1054, 955, 850, 799, 580, and 435 cm^{-1} are identified (figure 3 (b)). Interaction of water molecules with silica surfaces depends on the presence of functional surface groups. For silica surfaces hydroxyl groups are the most active sites for interaction with water molecules, so the concentration and nature of these groups largely determines the extent of water-membrane interaction. The broad absorption band centered at 1054 cm^{-1} is originated from ν_{as}Si-O-Si (TO mode) vibrations. The 1664 cm^{-1} band can be assigned to bending modes of the interstitial water molecules. Generally, in silica matrices the spectral bands within the spectral range 1960 - 1640 cm^{-1} correspond to combination of vibrations of the SiO_2 network (1664 cm^{-1} band should often hidden by molecular water band). Also, within the wavenumber range 4000 - 3000 cm^{-1}, the spectral bands will be caused mainly due to overtones or combinations of vibrations of Si-OH or H_2O. A broad OH stretching band with weak intensity is also observed within the spectral range 3500 - 3750 cm^{-1}. The two smaller absorption bands appeared around 3628, and 3436 cm^{-1} can be assigned to pair of surface Si-OH groups that are mutually linked by hydrogen bond or internal Si-OH bonds; and silanol groups linked to molecular water through hydrogen bonds, respectively [47]. The absorption bands within the wavenumber range 400 - 860 cm^{-1} can be attributed to the Si-O-Si bond bending vibrations and symmetric stretching vibrations of Si-O-Si bonds, respectively. The absorption bands observed at 1148, and 1279 cm^{-1} can be assigned to asymmetric stretching vibrations of Si-O-Si bridging sequences, and O=P bond stretching, respectively[50]. The 955 cm^{-1} band can be assigned to stretching vibration of free silanol groups on the surface of the amorphous silica. Absorption band centered at 799 cm^{-1} could be due to the symmetric stretching of the P-O-P and Si-O-Si (bridging oxygen atoms between the tetrahedral) bonds. Absorption bands centered at 3846, 3738, 3479, 3468, 3436, 3413, 3401, 2361, 2342, 1637, 1575, 1542, 1508, 1458, 1430, 1337, 1278, 1170, 1124, 1084, 834, 800, 756, 670, 650, 622, 601, 594, 533, 523, and 454 cm^{-1} are identified for 30TMOS-30TEOS- 30MTEOS- 10PO $(OCH_3)_3$ (mol %) membrane doped with 10-40 wt% [$EMIMBF_4$] ionic liquid (figure 3(b)). All these observed bands can be assigned similarly to the pure ionic liquid (figure 3(a)) spectral bands. Also, it should be emphasized that all the phosphosilicate matrices related absorption bands within the wavenumber range 400 - 1200 cm^{-1} are overlapped well with [$EMIMBF_4$] ionic liquid related spectral bands. The above results indicate that in the prepared hybrid membranes the [$EMIMBF_4$] ionic liquid interacted effectively in a molecular scale with phosphosilicate content through the selected sol-gel process. The alkyl C-H and ring C-H stretching modes may serve as a powerful probe tool for the study of hydrogen bond interactions in ionic liquids. Since slight changes in alkyl C-H and ring C-H stretching modes were observed (figure 3 (b)), the hydrogen bond interactions for the ionic liquid doped in the host phosphosilicate membrane is almost similar to that in the bulk phase. However, in the prepared hybrid membranes the intensity of asymmetric stretching vibrations of BF_4^- (1084 cm^{-1}) increases gradually with the ionic liquid weight percent ratio increment. These results suggest that the addition of ionic liquid to the phosphosilicate membrane does not significantly perturb the local structure of the imidazolium cations, while the local environment of the anions may be changed in the hybrid membranes.

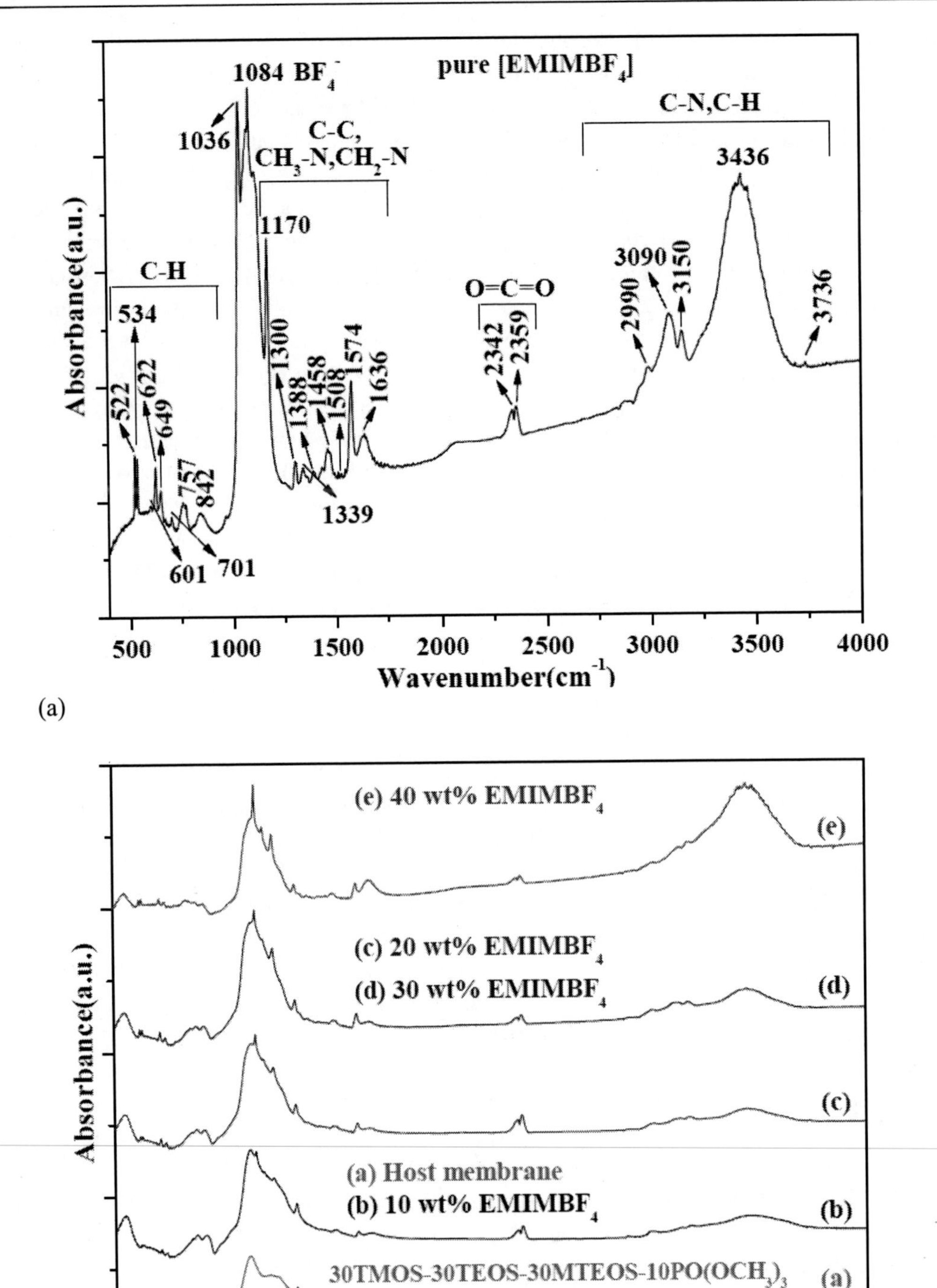

Figure 3. FT-IR spectra of (a) pure [$EMIMBF_4$] ionic liquid (b) 30TMOS- 30TEOS- 30MTEOS-10PO $(OCH_3)_3$-x [$EMIMBF_4$] (x=0, 10, 20, 30, and 40 wt %) hybrid membranes.

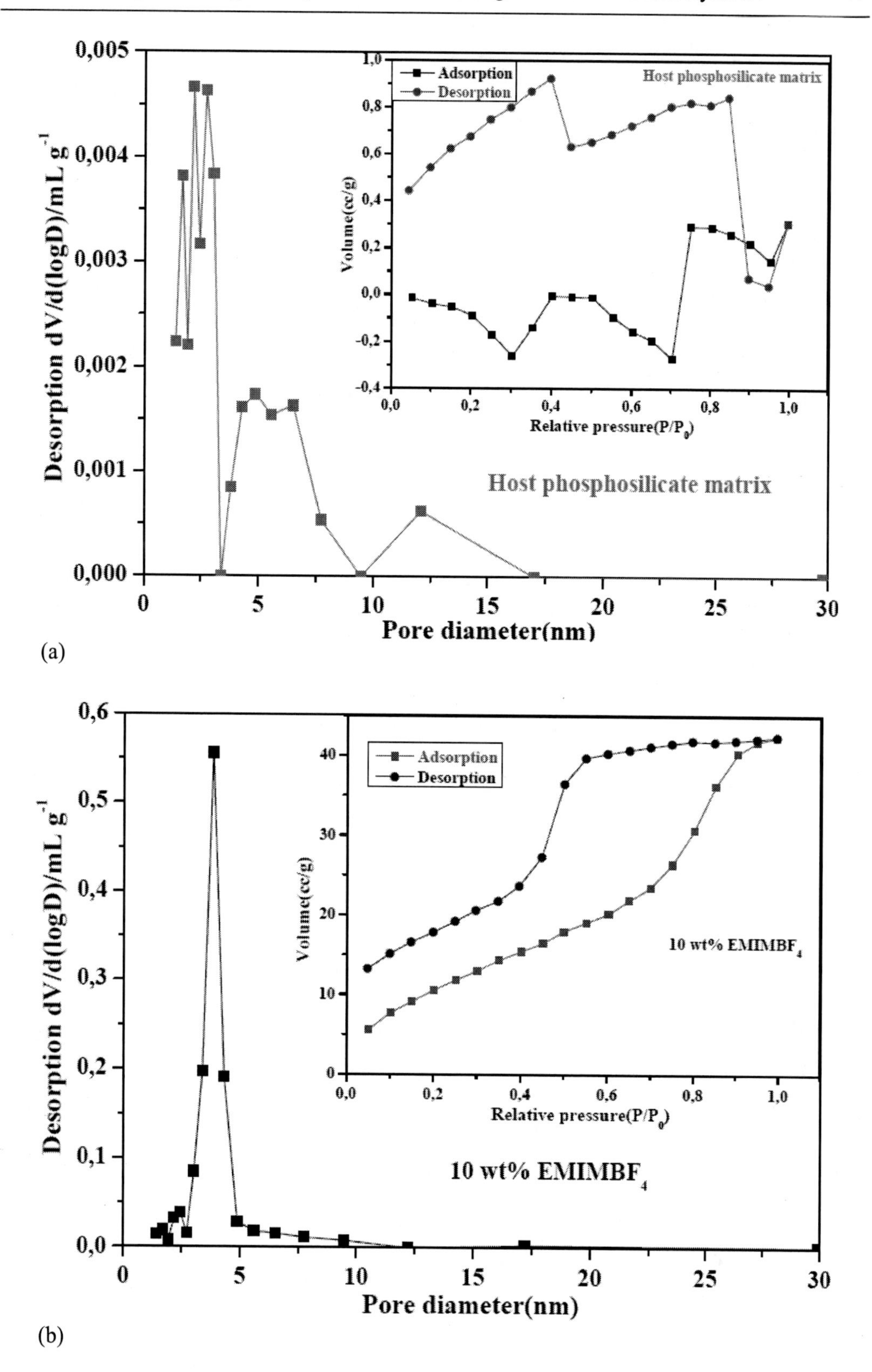

Figure 4. (Continued).

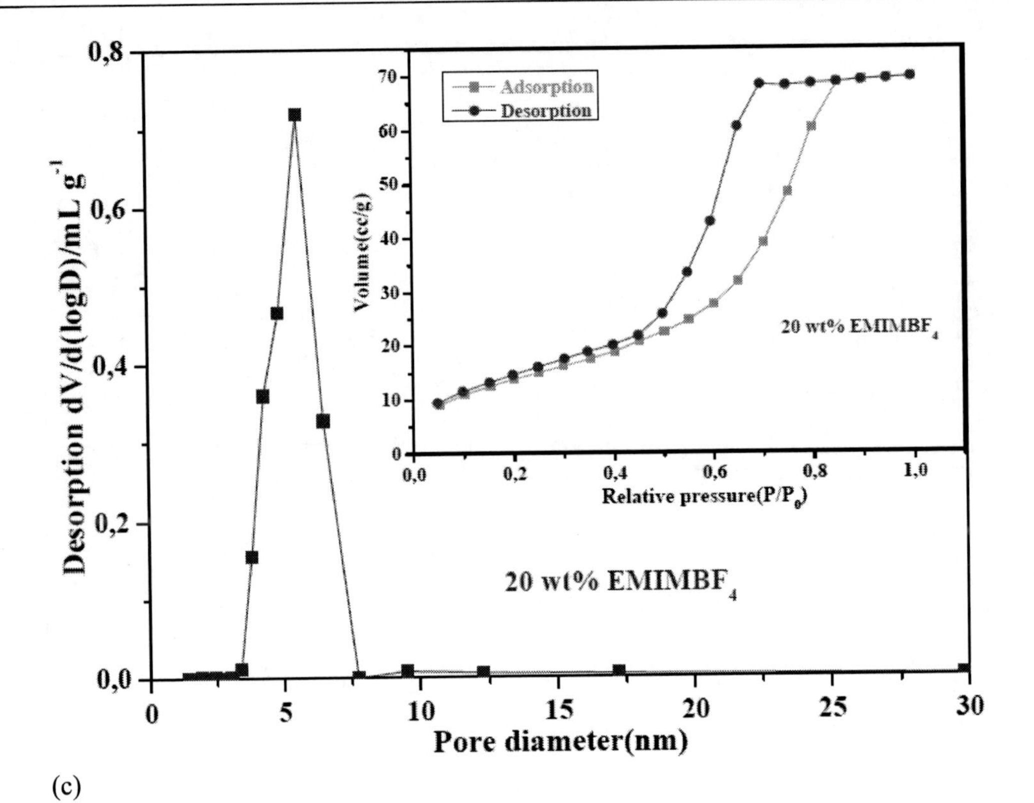
Desorption dV/d(logD)/mL g⁻¹
Pore diameter(nm)
20 wt% EMIMBF4
Adsorption
Desorption
Volume(cc/g)
Relative pressure(P/P0)
20 wt% EMIMBF4

(c)

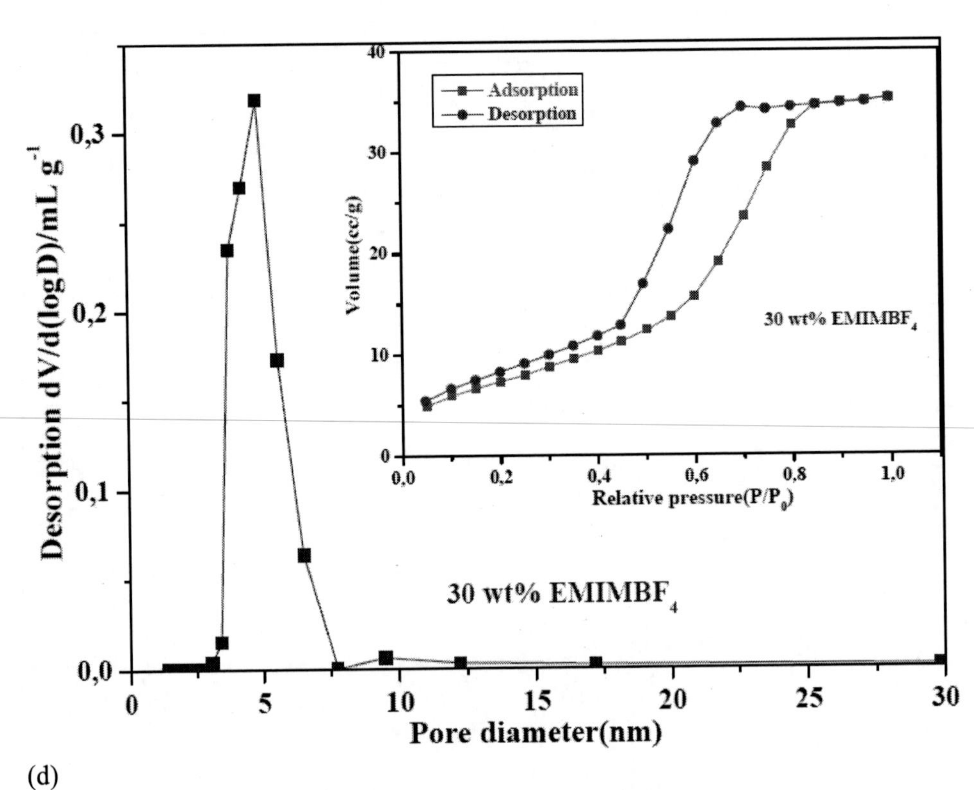
Desorption dV/d(logD)/mL g⁻¹
Pore diameter(nm)
30 wt% EMIMBF4
Adsorption
Desorption
Volume(cc/g)
Relative pressure(P/P0)
30 wt% EMIMBF4

(d)

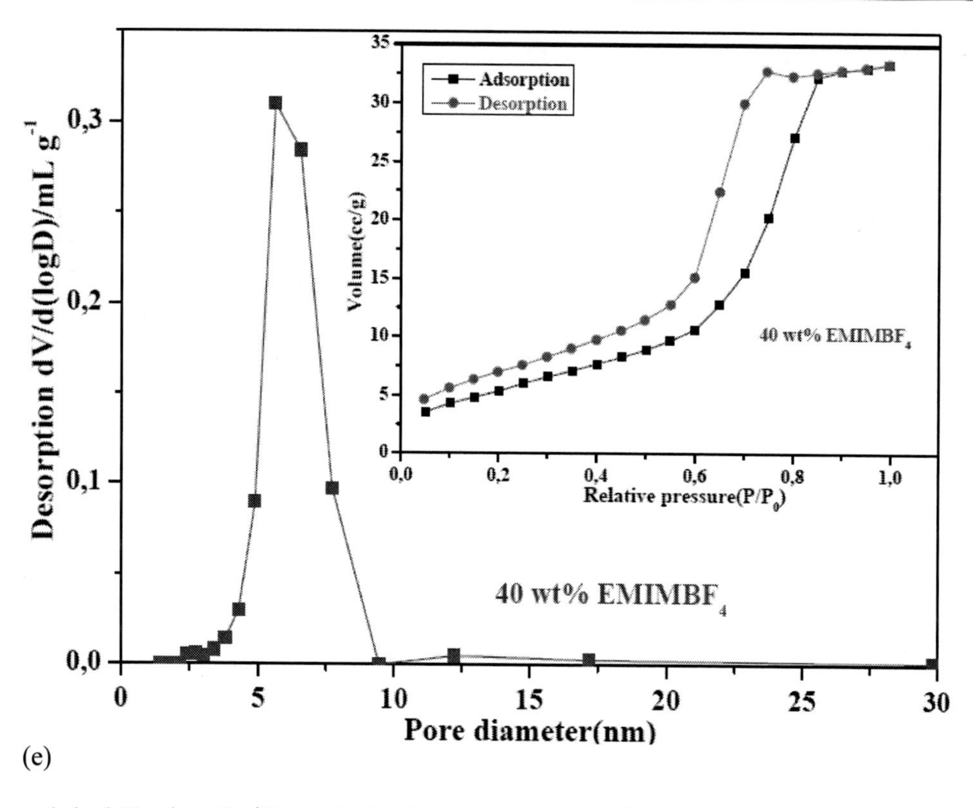

Figure 4. (a-e) N_2 adsorption/desorption isotherms, and corresponding pore size distribution curves (inset plots) of 30TMOS-30TEOS-30MTEOS- 10 PO $(OCH_3)_3$- x [$EMIMBF_4$](x=0,10, 20, 30, and 40 wt%) hybrid membranes.

Table 1. Textural properties of 30TMOS-30TEOS-30MTEOS-10(PO $(OCH_3)_3$)-x [$EMIMBF_4$] (x= 0, 10, 20, 30, and 40 wt %) hybrid membranes heat-treated at 250 ^{0}C

Sample	Composition of TMOS-TEOS-MTEOS-PO$(OCH_3)_3$)-x [$EMIMBF_4$]	Average pore size (nm)	Average pore volume (cm^3/g)	Specific surface area (m^2/g)
S1	(30-30-30-10) -0 wt %	1.039	0.00047	1.83
S2	(30-30-30-10)-10 wt%	5.927	0.065	44.24
S3	(30-30-30-10)-20 wt%	8.275	0.1077	52.06
S4	(30-30-30-10)-30 wt%	7.691	0.0544	28.29
S5	(30-30-30-10)-40 wt%	9.619	0.0517	21.49

N_2 adsorption measurements, which have been a powerful tool for nano- or mesoporous material characterization [51], were performed to attain more insight of the phosphosilicate based hybrid membranes prepared by using ionic liquid as dopant. For all the fabricated hybrid membranes, specific surface areas and pore-size distributions were determined from nitrogen adsorption/desorption isotherms as shown in figure 4 (a-e) inset plots. For all the ionic liquid based hybrid membranes, isotherms demonstrated a reversible part at low pressure and hysteresis loops at higher pressures. The shapes of the reversible parts as well as of the hysteresis loops were very similar for all samples. Table 1 presents the pore sizes and surface area properties for all the hybrid membranes. The BET surface areas of the composite membranes along with their average pore diameters calculated according to the Barret-Joyner-Halenda (BET) method are presented in the Table 1.

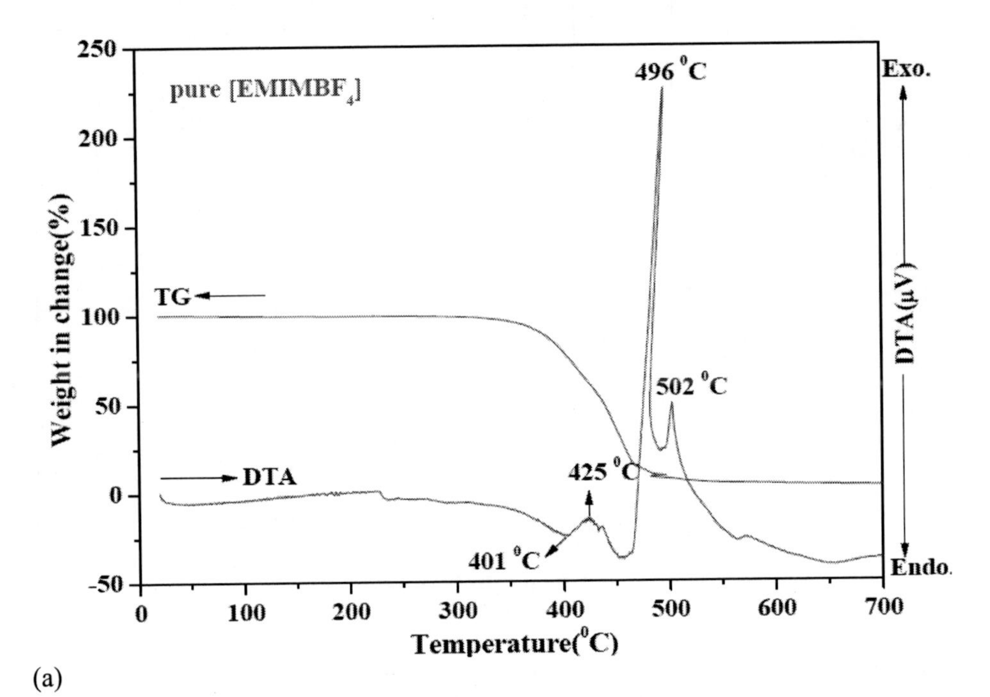

(a)

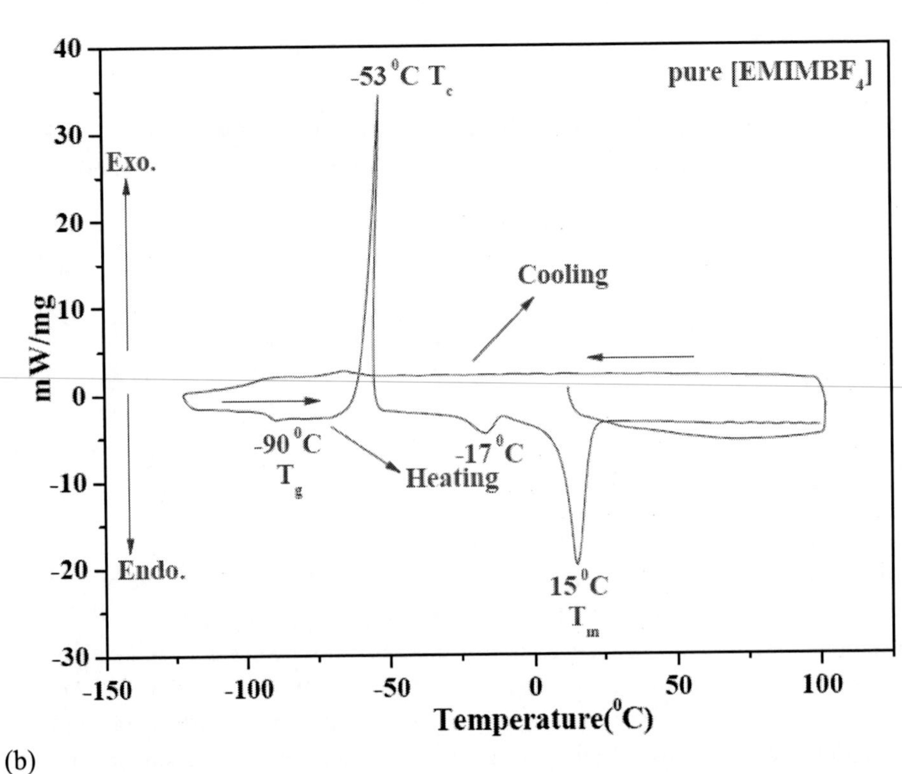

(b)

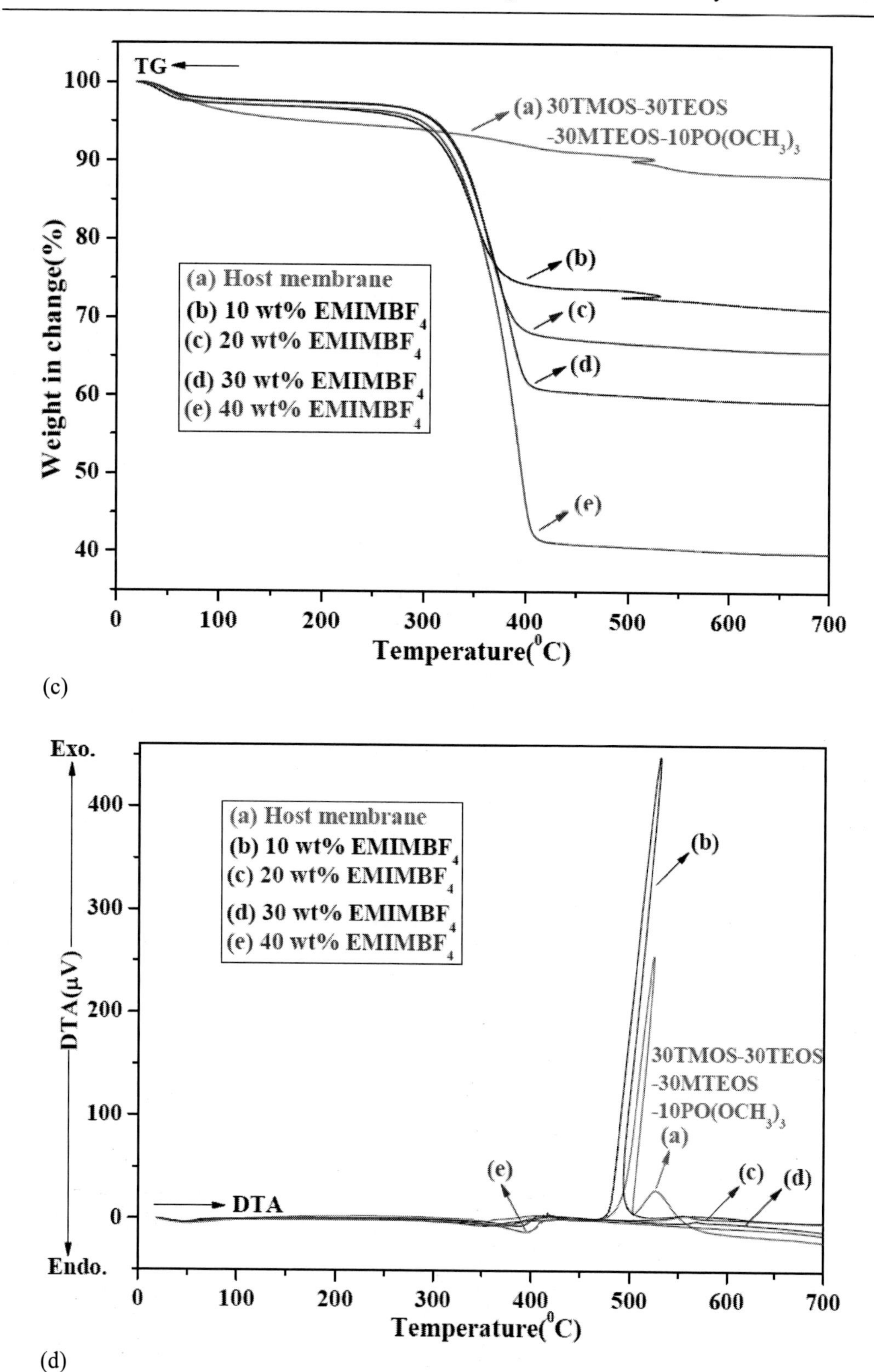

Figure 5. (Continued).

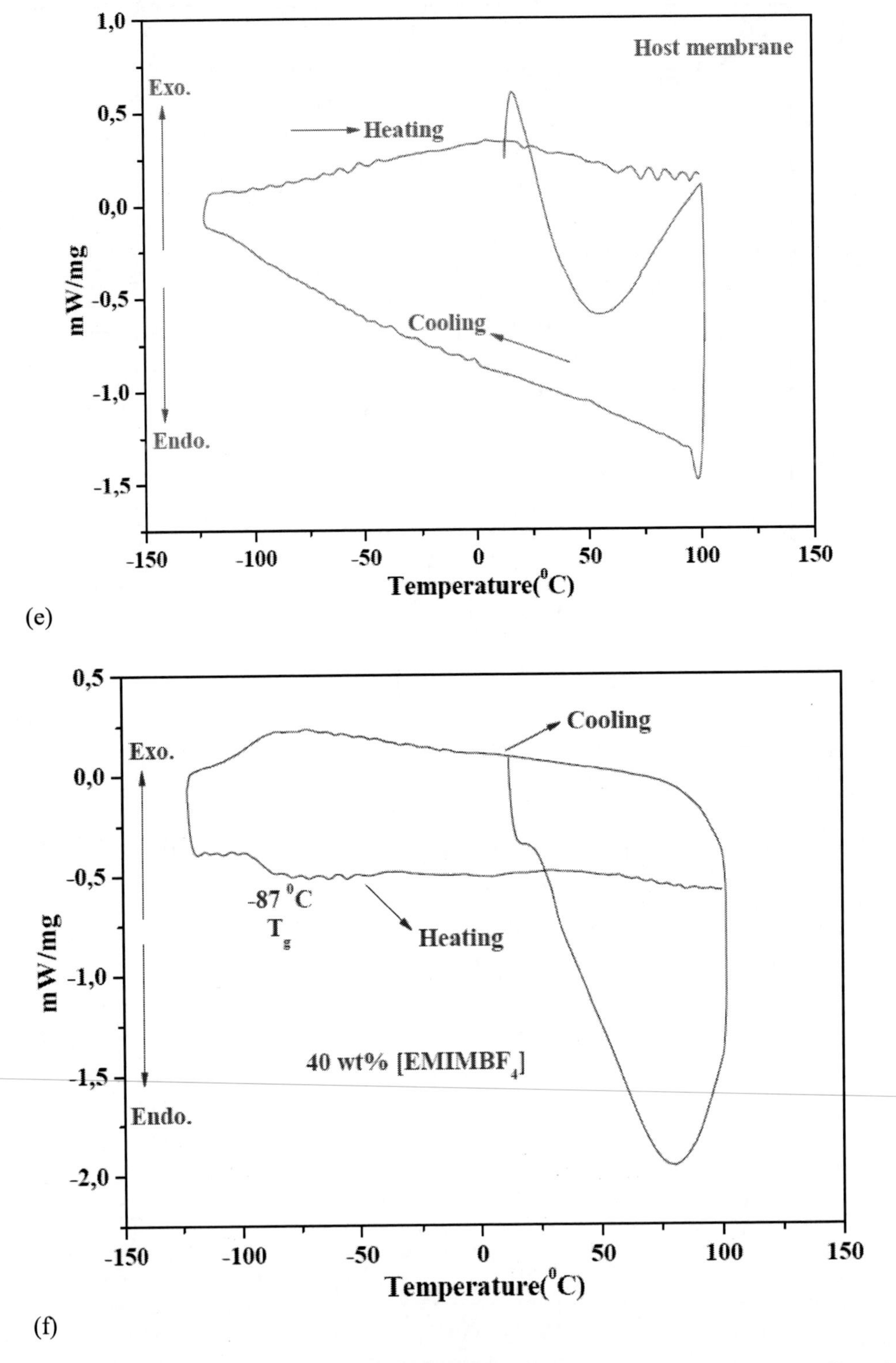

Figure 5. (a) TG-DTA (b) DSC profiles of pure [$EMIMBF_4$] ionic liquid; (c) TG (d) DTA profiles of 30TMOS-30TEOS-30MTEOS-10PO(OCH_3)$_3$ -x[EMIM BF_4] (x= 0, 10, 20, 30, and 40 wt%) membranes. (e) and (f) DSC profiles of host phosphosilicate and 40 wt% [$EMIMBF_4$] ionic liquid doped hybrid membranes.

The pores size distribution was calculated as a function of volume or specific surface area of the pores and presented similar curves for all samples as shown in figure 4 (a-e). Here, one can observe a decrease in specific surface area and pore volume and an increase in the pore size for all the ionic liquid based hybrid membranes. In all the hybrid membranes, the decrease in surface area may be due to an increase in the IL domains in the interconnected phosphosilicate inorganic structure. Under degassed conditions (250 ^{0}C), the ionic liquid based hybrid membranes resembled mesopores, and the desorptions were shifted to a relatively lower pressure, denoting large pores. This one represents a significant improvement for potential applications of such materials.

It is important to study the thermal stability of proton exchange membranes because a medium temperature operation at higher than 100 ^{0}C is eventually being aimed for the PEMFCs using these hybrid membranes. Thermal properties of all the prepared membranes are investigated by both thermogravimetry and differential thermal analysis. Figure 5 (a) and (b) presents the TG-DTA and DSC profiles of the pure [$EMIMBF_4$] ionic liquid. Normally a temperature corresponding to 10% mass loss is taken as the dissociation temperature (T_d) and in the present case, also temperatures corresponding to 5 and 10% mass loss have been determined for pure ionic liquid. The $EMIMBF_4$ ionic liquid used in the present study is thermally stable up to temperature above 300 ^{0}C, and the weight loss takes place by a one-step weight loss process. The temperatures at 5 and 10% mass loss are 352 and 377 ^{0}C (T_d) for $EMIMBF_4$, respectively, indicating its high thermal stability over a wide temperature range. From the corresponding DTA profile, one endothermic peak at 401 ^{0}C, and three exothermic peaks at 425, 496, and 502 ^{0}C are identified. The endothermic peaks indicate the decomposition of the side groups, side and main chains of the [$EMIMBF_4$] ionic liquid, respectively. From the DSC thermogram (figure 5(b)), an exothermic peak (-53 ^{0}C) based on the crystallization and a heat capacity change assigned to the glass transition was observed for pure [$EMIMBF_4$] during the heating scan. During the heating scan from -100 to 100 ^{0}C (reheating step) for [$EMIMBF_4$], a sharp endothermic peak corresponding to T_m (15 ^{0}C), together with a heat capacity change corresponding to T_g (-90 ^{0}C) are observed. The coexistence of the T_m and T_g indicates relatively slow crystallization kinetics of [$EMIMBF_4$], which has been widely observed in ionic liquids [52]. Also, a small endothermic peak at -17 ^{0}C was observed for the pure $EMIMBF_4$. The question of whether this endothermic peak is ascribed to essential characteristics of $EMIMBF_4$ or the existence of contaminants needs further exploration. A higher glass transition point is observed as the cation has a longer alkyl side chain. The thermal stability of the room temperature ionic liquids, therefore, depends on the anionic species. Figure 5 (c) and (d) presents the TG and DTA profiles of all the prepared hybrid membranes, including host phosphosilicate matrix. An initial loss, 3 wt. % of adsorbed water molecules was found around 82 ^{0}C for host matrix. Also 5 wt. % and 10 wt % losses were identified around at 215 ^{0}C and 522 ^{0}C due to the condensation of structural hydroxyl groups, respectively. On the other hand, the weight loss of the 10-40 wt. % [$EMIMBF_4$] ionic liquid doped hybrid membranes below 100 ^{0}C was 2.0 wt. % only, due to the high content of added ionic liquid. All the prepared IL based hybrid membranes are thermally stable upto 290 ^{0}C and can be used at temperatures in the 100-200 ^{0}C range for PEMFCs. Following the host phosphosilicate matrix DTA profile, several exothermic peaks at 408, 524, and 527 ^{0}C are identified. All these observed exothermic peaks suggest the decomposition or burning of the residual organic groups in the phosphosilicate matrix.

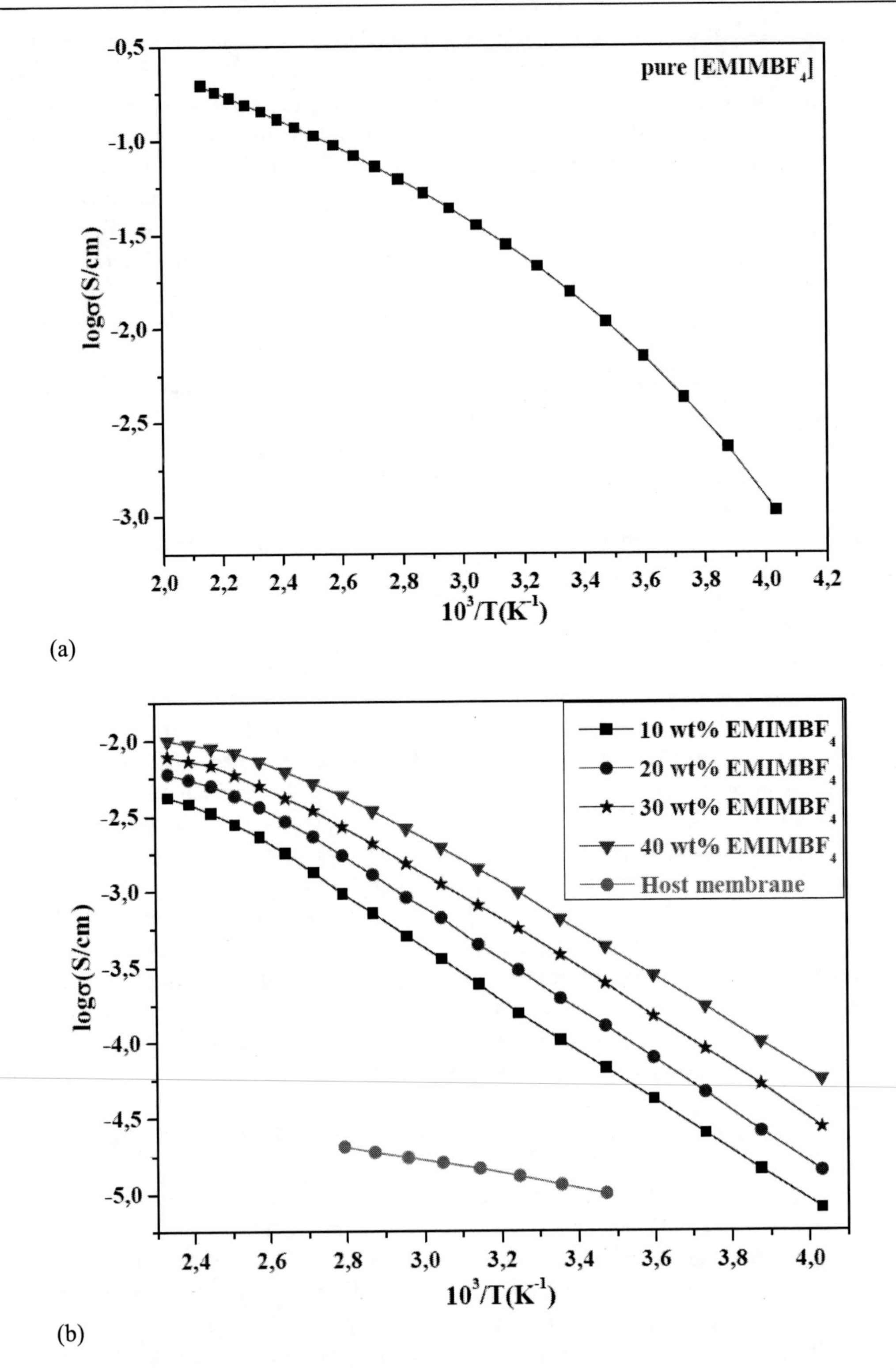

Figure 6. Conductivity of (a) pure [$EMIMBF_4$] ionic liquid (b) 30TMOS- 30TEOS- 30MTEOS-10PO $(OCH_3)_3$-x [$EMIMBF_4$] (x=0, 10, 20, 30, and 40 wt %) hybrid membranes.

Similarly, for 10-40 wt% [$EMIMBF_4$] ionic liquid based hybrid membranes, exothermic peaks at 530 ^{0}C; 414 and 555 ^{0}C; 417, 568 ^{0}C; and 420 ^{0}C are observed due to the decomposition of the ionic liquid. Moreover, small endothermic peaks at 352 ^{0}C; 367 ^{0}C; 387 ^{0}C; and 395 ^{0}C, respectively are also observed for 10-40 wt% ionic liquid based hybrid membranes. Above 600 ^{0}C, no additional exothermic peaks were revealed from all the prepared membranes DTA profiles indicating that the organic residues including any volatile species removal was completed below 600 ^{0}C. The observed decomposition temperature is relatively high compared with other C-H bonding-related polymers [53] despite of the high amount of added ionic liquid, and this high thermal stability can be attributed to both the inorganic-organic composite microstructure based on Si-O-Si backbones as well as to the stability of [BF_4] anion. Further, figure 5 (e) and (f) presents the DSC profiles of the host phosphosilicate and 40 wt % [$EMIMBF_4$] ionic liquid doped hybrid membranes within the temperature range -100 -100 ^{0}C, respectively. No any endothermic peak was observed within the range 80 - 90 ^{0}C for the host membrane. On the other hand, following the DSC profile of 40 wt% ionic liquid doped hybrid membrane the glass transition temperature (T_g) with a heat capacity change was found at around -87 ^{0}C.

In PEMFCs or DMFCs, proton conductivity of membranes is a crucial parameter because the cell performance is strongly dependent on this property. Figure 6 (a) presents the ionic conductivity of pure [$EMIMBF_4$] ionic liquid within the temperature range (-25) -195 ^{0}C. The conductivity increases from 1.061 x 10^{-3} S/cm to 1.97 x 10^{-1} S/cm with an increment in temperature from (-25) -195 ^{0}C in steps of 10 ^{0}C due to the high ionic conductivity and thermal stability including low viscosity of pure ionic liquid. At ambient temperature (25 ^{0}C), the measured conductivity was 1.548 x 10^{-2} S/cm. In general, the ionic conductivity of the membrane electrolytes containing hydrophilic ILs ($EMIMBF_4$) is higher than that of the membranes containing hydrophobic ILs ($EMIMPF_6$). Figure 6(b) presents the conductivity of all the prepared hybrid membranes, measured within the temperature range -25 ^{0}C to 155 ^{0}C under anhydrous conditions. For comparison, anhydrous conductivity of host phosphosilicate matrix from 15-85 ^{0}C is also presented in figure 6(b). The thickness of all the prepared hybrid membranes for the conductivity measurements was about 0.2 mm. The maximum conductivity of host membrane at 85 ^{0}C was 2.04 x 10^{-5} S/cm only due to the loss of water molecules that exist within the membrane matrix with an increase in the temperature under non-humidified conditions. In order to increase the conductivity of host matrix within the temperature range (-25) -155 ^{0}C, [$EMIMBF_4$] ionic liquid is added because ILs are able to act as protonic charge carriers at medium temperatures (100-200 ^{0}C), even under anhydrous conditions. For all the ionic liquid doped hybrid membranes conductivity was increased with an increase in the ionic liquid weight percent ratio from 10-40 wt. %. Maximum conductivity of 1 x 10^{-2} S/cm at 155 ^{0}C was obtained for 40 wt% [$EMIMBF_4$] ionic liquid doped membrane under anhydrous conditions. At 85 ^{0}C, the 40 wt% IL doped membrane has shown conductivity value of 4.29 x 10^{-3} S/cm which is two orders of magnitute higher than that of host matrix conductivity. In the [BF_4] anion, the highly electronegative fluorine atom contributes to the distribution of the anionic charge of borate. In addition to the anionic charge distribution, the effect of the surface covering of the anion backbone by fluorine atoms may be a significant factor for weak interaction with the [EMIM] cation. It should be emphasized that even though the ionic liquid contribute to the anhydrous conductivity, the all available protonic carriers would be solely responsible for the observed higher conductivity of the ionic liquid doped membranes. Using the prepared IL based hybrid membranes, the measured conductivity values at 155 ^{0}C under anhydrous state are encouraging for their application in H_2/O_2 fuel cells electrolytes. Also, from figure 6 (b), one can see that the plots of conductivity (σ) versus reciprocal temperature (T^{-1}) are not linear within the temperature range -25 to 155 ^{0}C for the prepared IL doped membranes. This result indicates that the proton conduction in the IL doped membranes does not exactly follow an Arrhenius-type relation, but also exhibits the Vogel-Tamman-Fulcher (VTF) type behavior. Previously, this Vogel-Tamman-Fulcher (VTF) type behavior in temperature dependence of electrical conductivity measurements was also observed for EMIM-based ionic liquids [54]. Gas permeability of a PEM and its impact on fuel cell operation is an important parameter that needs to be characterized under different, realistic conditions for use in the development and advancement of PEM fuel cells. When different concentrations of hydrogen or oxygen gas exist across a gas permeable membrane, the gases permeate through the membrane due to the partial pressure gradient. Figure 7 presents the measured hydrogen permeability values for 40 wt % [$EMIMBF_4$] ionic liquid doped hybrid membrane within the temperature range 20-150 ^{0}C (thickness=0.5 mm). The hydrogen permeability value was decreased from 1.41 x 10^{-11} to 0.1 x 10^{-12} mol/cm.s.Pa with increasing temperature from 20 ^{0}C up to 150 ^{0}C. We thought that the reason for the observed high H_2 permeability values of 40 wt% [$EMIMBF_4$] ionic liquid based hybrid membrane might be due to the Knudsen diffusion process that exists in phosphosilicate matrices [55] when the H_2 gas permeation decreases with temperature increment. Knudsen diffusion is crucial when the mean free path of the gas molecules is greater than the pore size. In such situations the collisions of the molecules with the pore wall are more frequent than the collisions among the molecules.

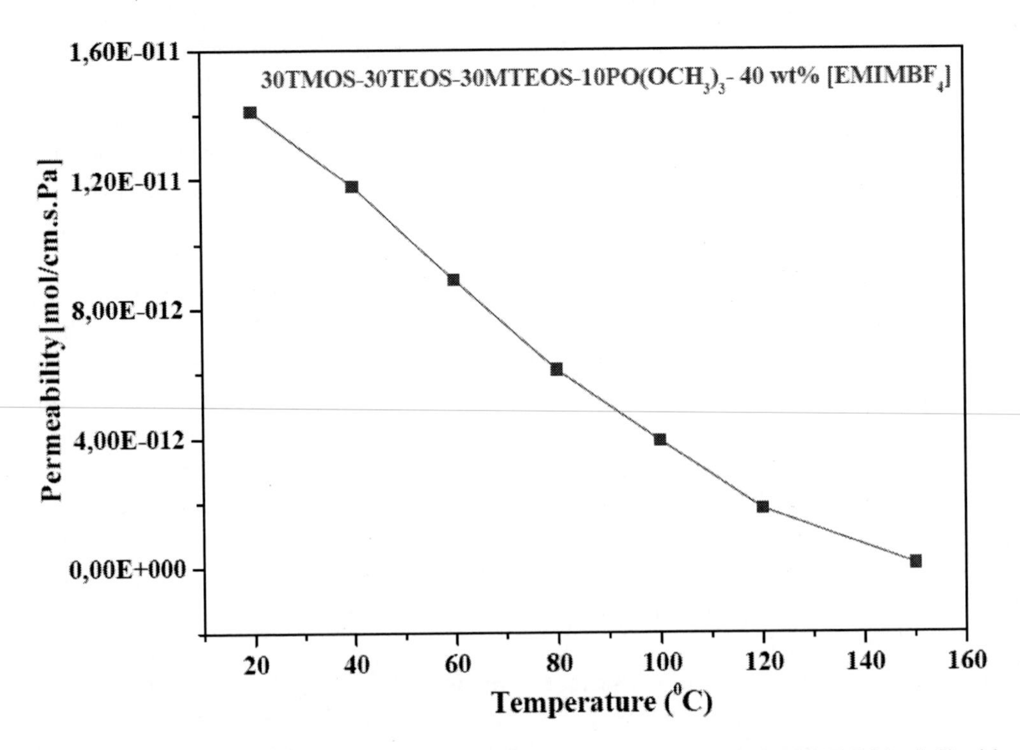

Figure 7. Hydrogen permeation rate as a function of temperature for 40 wt% [$EMIMBF_4$] ionic liquid doped hybrid membrane. The permeability measurements were performed under hydrogen feed in the temperature range 20-150 ^{0}C.

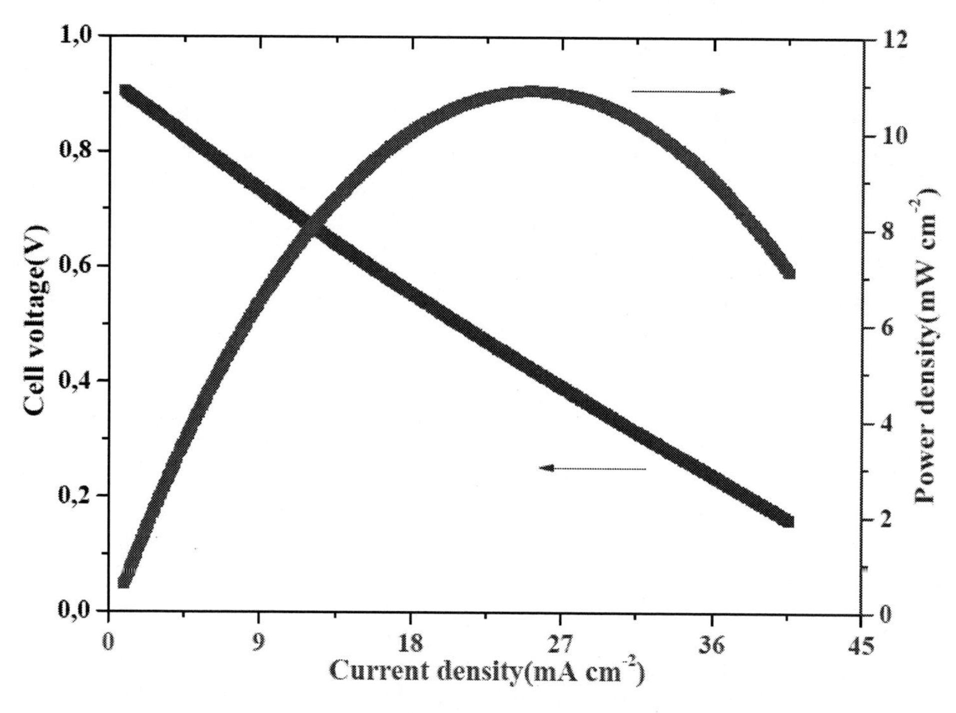

Figure 8. Polarization curve displaying the current and power densities for the 40 wt% [$EMIMBF_4$] ionic liquid doped hybrid membrane at 150 ^{0}C under anhydrous conditions.

Separation selectivities with this mechanism are proportional to the ratio of the inverse square root of the molecular weights. It is well known that the permeability coefficient of any pure gas through a membrane is directly related to the dimensions (size and thickness etc.) of the membrane including applied gas pressure. In the present work, it is possible by the selected sol-gel method to prepare thin membranes with desired thickness just by varying the volume of sol solutions of the studied membranes, in the Petri dishes. To increase the power density, decrease of the membrane resistance by means of reduction of the thickness of the electrolyte membrane is required without loss of mechanical strength.

Additionally, we performed the single fuel cell test at 150 ^{0}C for the 40 wt % [$EMIMBF_4$] ionic liquid doped 30TMOS-30TEOS-30MTEOS-10PO$(OCH_3)_3$ (mol %) hybrid membrane MEA to confirm the proton transfer in the membrane under nonhumidified H_2/O_2 gases (figure 8). The thickness of the selected hybrid membrane for the single fuel cell test was 0.2 mm. This figure presents the voltage versus current density dependence, for a gradual voltage decrease. Here, the voltage drop is the straight forward resistance to the flow of electrons through the electrode materials and the various interconnections, as well as the resistance to the flow of ions through the electrolyte. The open circuit voltage of the cell is about 0.9 V and is the conventional potential for the H_2/O_2 cell, and it indicates very small gas permeability through the hybrid membrane. The fuel cell voltage dropped with an increase in current density, and a maximum power density of 11 mW/cm^2 is obtained at current density of 26 mA/cm^2. This value may serve as a confirmation that the prepared hybrid membrane electrolytes may be an alternative for PEMFC application at intermediate temperatures, also considering that the test was run with a non-optimized, laboratory-type cell. However, the power outputs of the fabricated MEA cell under anhydrous conditions were lower than that expected from the conductivity value. The fuel cell performance prevailingly depends on the ohmic resistance of the electrolyte as well as on the electrode reaction resistance (polarization). The electrode resistance is, in turn dependent on the thickness of the electrolyte. For the present work, further, there obviously exists a need for continued improvements in the cell performance by different operating conditions (gas flow, temperature etc.) and determination of the long-term stability of fuel cell characteristics. It is then expected that by appropriate geometry optimization and membrane electrode assembly (MEA), the cell performance may be substantially improved. Such investigations will be a subject of our future work.

CONCLUSIONS

Novel 1-ethyl-3 methylimidazolium tetrafluoroborate [$EMIMBF_4$] ionic liquid based inorganic-organic hybrid phosphosilicate membranes have been successfully prepared by sol-gel process through hydrolysis and condensation of TMOS, TEOS, MTEOS, and $PO(OCH_3)_3$. These hybrid membranes possess good chemical stability and are thermally stable up to 290 ^{0}C in air due to inorganic SiO_2 framework and stable [BF_4] anion. The textural properties were investigated from nitrogen adsorption-desorption analysis and pore structure was found for all the prepared hybrids by Barrett-Joyner-Halenda (BJH) desorption method. For the fabricated composite membranes a distribution of average pore size within the range 1 - 9 nm was identified with the doped ionic liquid content weight ratio. The flexible, transparent and homogenous membranes have shown good conductivity under non-humidified conditions. Maximum conductivity of 1×10^{-2} S/cm was obtained for 40 wt% [$EMIMBF_4$] doped 30TMOS-30TEOS-30MTEOS-10PO $(OCH_3)_3$ (mol%) hybrid membrane at 155 ^{0}C under anhydrous conditions. The hydrogen permeability decreases from 1.41×10^{-11} to 0.1×10^{-12} mol/cm.s.Pa for 40 wt% ionic liquid doped hybrid membrane, as the temperature increases from 20 to 150 ^{0}C. For 40 wt% [$EMIMBF_4$] doped hybrid membrane, a maximum power density of 11 mW/cm^2 at 26 mA/ cm^2 as well as a current density of 40.57 mA/ cm^2 were achieved at 150 ^{0}C under anhydrous conditions. These hybrid membranes can be promising materials to be used for medium temperature fuel cells electrolytes and show good processibility for large area membrane. Also, we believe that this type of hybrid membranes should have an impact on further investigations in the field of proton conducting membranes for PEMPCs, and the methodology presented in this work could be extendable to other inorganic phosphosilicate networks and ILs with desirable properties for the preparation of anhydrous, nonfluorinated proton exchange membranes.

ACKNOWLEDGMENTS

The authors are grateful to the New Energy and Industrial Technology Development Organization (NEDO), Japan for the financial support.

Supporting Information available: Figure S1. UV-visible transmission spectra of 30TMOS-30TEOS-30MTEOS-10PO $(OCH_3)_3$-x [$EMIMBF_4$] (x=0, 10, 20, 30, and 40 wt %) membranes; Figure S2. (a) ^{31}P (b) ^{29}Si MAS NMR spectra of all the prepared hybrid membranes;(c) ^{1}H (d) ^{13}C (e) ^{19}F MAS NMR spectra of pure $EMIMBF_4$ ionic liquid; (f) ^{1}H,

(g) ^{13}C, and (h) ^{19}F MAS NMR spectra for 30TMOS-30TEOS- 30MTEOS-10 $PO(OCH_3)_3$ -x $[EMIMBF_4]$(x=0, 10, 20, 30, and 40 wt%) composite membranes; Figure S3. (a-e) SEM and (f-j) EDX micrographs of 30TMOS-30TEOS-30MTEOS- 10PO $(OCH_3)_3$- x $[EMIMBF_4]$ (x=0, 10, 20, 30, and 40 wt %) hybrid membranes; Figure S4. Working and reaction principles of H_2/O_2 fuel cell.

REFERENCES

[1] Steele, B. C. H.; Heinzel, A. Materials for fuel-cell technologies, *Nature* (2001), 414, 345-352.

[2] Hickner, M.A.; Ghassemi, H.; Kim, Y.S.; Einsla, B.R.; McGrath, J. E. Alternative Polymer Systems for Proton Exchange Membranes (PEMs), *Chemical Reviews* (2004), 104, 4587-4612.

[3] Peinemann, K.V.; Nunes, S.P. Eds., Membrane Technology: Membranes for Energy Conversion, Vol. 1, 2. Wiley-VCH, Weinheim, 2008.

[4] Rikukawa, K.; Sanui, K. Proton-conducting polymer electrolyte membranes based on hydrocarbon polymers, *Progress in Polymer Science* (2000), 25, 1463-1502.

[5] Mehta, V.; Cooper, J.S. Review and analysis of PEM fuel cell design and manufacturing, *Journal of Power Sources* (2003), 114, 32-53.

[6] Borup, R.; Meyers, J.; Pivovar, B.; Kim, Y.S.; Mukundan, R.; Garland, N.; Myers, D.; Wilson, M.; Garzon, F.; Wood, D.; Zelenay, P.; More, K.; Stroh, K.; Zawodzinski, T.; Boncella, J.; McGrath, J.E.; Inaba, M.; Miyatake, K.; Hori, M.; Ota, K.; Ogumi, Z.; Miyata, S.; Nishikata, A.; Siroma, Z.; Uchimoto, Y.; Yasuda, K.; Kimijima, K.; Iwashita, N. Scientific Aspects of Polymer Electrolyte Fuel Cell Durability and Degradation, *Chemical Reviews* (2007), 107, 3904-3951.

[7] Kerres, J.A. Development of ionomer membranes, *Journal of Membrane Science* (2001), 185, 3-27.

[8] Dimitrova, P.; Friedrich, K.A.; Stimming, U.; Vogt, B. Modified Nafion®-based membranes, *Solid State Ionics* (2002), 150, 115-122.

[9] Li, Q.; He, R.; Jensen, J. O.; Bjerrum, N. J. Approaches and recent development of polymer electrolyte membranes for fuel cells operating above 100 ^0C, *Chemistry of Materials* (2003), 15, 4896-4915.

[10] Hogarth, W. H. J.; Diniz da Costa, J. C.; Liu, G. Q. Solid acid membranes for high temperature (>140 ^0C) proton exchange membrane fuel cells, *Journal of Power Sources* (2005), 142, 223-237.

[11] Alberti, G.; Casciola, M. Composite membranes for medium-temperature PEM fuel cells, *Annual Review Materials Research* (2003), 33, 129-154.

[12] Garland, N.L.; Kopasz, J.P. The United , *Journal of Power Sources* (2007), 172, 94-99.

[13] Kim, E.K.; Son, S.W.; Won, J.; Kim, C.K.; Kang, Y.S. Effect of acidity, *Journal of Membrane Science* (2010), 348, 190-196.

[14] Hu, J.; Luo, J.S.; Wagner, P.; Conrad, O. Agert, C. Anhydrous proton conducting membranes, *Electrochemistry Communications* (2009), 11, 2324-2327.

[15] Alabi, C.A.; Chen, Z.; Yan, Y.S.; Davis, M.E. Insights into the Nature of Synergistic Effects in Proton-Conducting 4,4-1H,1H-Bitriazole-Poly(ethylene oxide) Composites, *Chemistry of Materials* (2009), 21, 4645-4652.

[16] Kim, S.-K.; Kim, T.-H.; Jung, J.-W.; Lee, J.-C. Polybenzimidazole containing benzimidazole side groups for high-temperature, *Polymer* (2009), 50, 3495-3502.

[17] Kim, J-D.; Mori, T.; Hayashi, S.; Honma, I. Anhydrous Proton-Conducting Properties of Nafion–1, 2, 4-Triazole and Nafion–Benzimidazole Membranes for Polymer Electrolyte Fuel Cells, *Journal of Electrochemical Society* (2007),154, A290-A294.

[18] Cheng, C. K.; Luo, J. L.; Chuang, K. T.; Sanger, A. R. Propane Fuel Cells Using Phosphoric-Acid-Doped Polybenzimidazole Membranes, *Journal of Physical Chemistry* B (2005), 109, 13036-13042.

[19] Jiang, F.J.; Pu, H.T.; Meyer, W. H.; Guan, Y.; Wan, D.C. A new anhydrous proton conductor, *Electrochimica Acta* (2008), 53, 4495-4499.

[20] Chisholm, C. R. I.; Haile, S. M. Entropy Evaluation of the Superprotonic Phase of $CsHSO_4$: Pauling's Ice Rules Adjusted for Systems Containing Disordered Hydrogen-Bonded Tetrahedra, *Chemistry of Materials* (2007), 19, 270-279.

[21] Matsuda, A.; Nguyen, V.H.; Daiko, Y.; Muto, H.; Sakai, M. Three-dimensional hydrogen, *Solid State Ionics* (2010), 181, 180-182.

[22] Mackenzie, J. D.; Bescher, E. P. "Structures, properties and potential applications of ormosils", *Journal of Sol-Gel Science and Technology* (1998), 13, 371-377.

[23] Eo, Y. J.; Lee, T. H.; Kim, S. Y.; Kang, J. K.; Han Y. S.; Bae, B. S. "Synthesis and molecular structure analysis of nano-Sized methacryl-grafted polysiloxane resin for fabrication of nano hybrid materials", *Journal of Polymer Science* Part B (2005), 43, 827-836.

[24] Yoo, H.S.; Han, J.Y.; Kim, S.W.; Jeon, D.Y.; Bae, B.S. Self-assembled SiO_2 photonic crystal infiltrated by Ormosil: Eu $(DBM)_3$ phen phosphor and its enhanced photoluminescence, *Optics Express* (2009), 17, 3732-3740.

[25] Lakshminarayana, G.; Nogami, M. Synthesis and Characterization of Proton Conducting Inorganic-Organic Hybrid Nanocomposite Membranes Based on mixed PWA- PMA- TEOS- GPTMS- H_3PO_4-APTES for H_2/O_2 Fuel Cells, *Journal of Physical Chemistry* C (2009), 113, 14540-14550.

[26] de Vos, R. M.; Maier, W.F.; Verweij, H. Hydrophobic silica membranes for gas separation, *Journal of Membrane Science* (1999), 158, 277-288.

[27] Tripathi, V.S.; Kandimalla, V.B.; Ju, H.X. Preparation, *Sensors and Actuators*: B Chemical (2006), 114, 1071-1082.

[28] Sekhon, S. S.; Park, J.-S.; Cho, E.K.; Yoon, Y.-G.; Kim, C.-S.; Lee, W-Y. Morphology Studies of High Temperature Proton Conducting Membranes Containing Hydrophilic/Hydrophobic Ionic Liquids, *Macromolecules* (2009), 42, 2054 -2062.

[29] Ohno, H. Electrochemical Aspects of Ionic Liquids, Wiley-Inter-Science: New York, 2005.

[30] Parvulescu, V. I.; Hardacre, C. Catalysis in ionic liquids, *Chemical Reviews* (2007), 107, 2615-2665.

[31] Nockemann, P.; Thijs, B.; Hecke, K.V.; V.Meervelt, L.; Binnemans, K. Polynuclear metal complexes obtained from the task-specific ionic liquid betainium bistriflimide, *Journal of Crystal Growth and Design*. (2008), 8, 1353-1363.

[32] Shin, J.-H.; Henderson, W. A.; Appetecchi, G. B.; Alessandrini, F.; Passerini, S. Recent developments in the ENEA lithium, *Electrochimica Acta* 50 (2005) 3859-3865.

[33] Tizzani, C.; Appetecchi, G. B.; Carewska, M.; Kim, G.-T.; Passerini, S. Investigation of the Electrochemical Properties of Polymer–LiX–Ionic Liquid Ternary Systems, *Australian Journal of Chemistry* (2007), 60, 47-50.

[34] Xu, W.; Angell, C. A. Solvent-Free Electrolytes with Aqueous Solution-Like Conductivities, *Science* (2003), 302, 422-425.

[35] Nakamoto, H.; Watanabe, M. Brønsted acid–base ionic liquids for fuel cell electrolytes, *Chemical Communications* (2007), 2539-2541.

[36] Nakamoto, H.; Noda, A.; Hayamizu, K.; Hayashi, S.; Hamaguchi, H.; Watanabe, M. Proton-Conducting Properties of a Brønsted Acid−Base Ionic Liquid and Ionic Melts Consisting of Bis(trifluoromethanesulfonyl)imide and Benzimidazole for Fuel Cell Electrolytes, *Journal of Physical Chemistry* C (2007), 111, 1541-1548.

[37] Kuang, D.; Uchida, S.; Baker, R.H.-; Zakeeruddin, S.M.; Grätzel, M. Organic Dye-Sensitized Ionic Liquid Based Solar Cells: Remarkable Enhancement in Performance through Molecular Design of Indoline Sensitizers, Angewandte Chemie International Edition (2008), 47, 1923 -1927.

[38] Cao, Y.; Zhang, J.; Bai, Y.; Li, R.; Zakeeruddin, S.M.; Gra¨tzel, M.; Wang, P. Dye-Sensitized Solar Cells with Solvent-Free Ionic Liquid Electrolytes, *Journal of Physical Chemistry* C (2008), 112, 13775-13781.

[39] Frackowiak, E.; Lota, G. J.Pernak, Room-temperature phosphonium ionic liquids for supercapacitor application, *Applied Physics Letters* (2005), 86, 164104/1-3.

[40] Wei, D.; Ng, T.W. Application of novel room temperature ionic liquids in flexible supercapacitors, *Electrochemistry Communications* (2009), 11, 1996-1999.

[41] Sekhon, S. S.; Park, J.-S.; Baek, J.-S.; Yim, S.-D.; Yang, T.-H.; Kim, C.-S. Small-Angle X-ray Scattering Study of Water Free Fuel Cell Membranes Containing Ionic Liquids, *Chemistry of Materials (*2010), *22,* 803-812.

[42] Tominaga, Y.; Asai, S.; Sumita, M.; Panero, S.; Scrosati, B. A novel composite polymer, *Journal of Power Sources* (2005), 146, 402-406.

[43] Yuyama, K.; Masuda, G.; Yoshida, H.; Sato, T. Ionic liquids containing the tetrafluoroborate anion have the best performance and stability for electric double layer capacitor applications, *Journal of Power Sources* (2006), 162, 1401-1408.

[44] Lakshminarayana, G.; Nogami, M. Synthesis and characterization of proton conducting inorganic–organic hybrid, *Electrochimica Acta* (2009), 54, 4731-4740.

[45] Lakshminarayana, G.; Nogami, M. Synthesis, Characterization and electrochemical properties of SiO_2-P_2O_5-TiO_2-ZrO_2 glass membranes as proton conducting electrolyte for low temperature H_2/O_2 fuel cells, *Journal of Physics* D: Applied Physics (2009), 42, 215501/1-11.

[46] Li, S.W.; Zhou, Z.; Zhang, Y.L.; Liu, M.L.; Li, W. 1*H*-1, 2, 4-Triazole: An Effective Solvent for Proton-Conducting Electrolytes, *Chemistry of Materials* (2005), 17, 5884-5886.

[47] Lakshminarayana, G.; Nogami, M. Inorganic−organic Hybrid Membranes with Anhydrous Proton Conduction Prepared from tetramethoxysilane/ methyl-trimethoxysilane /trimethylphosphate and 1-ethyl-3 methylimidazolium-bis (trifluoromethanesulfonyl) imide for H_2/O_2 Fuel Cells, *Electrochimica Acta* (2010), 55, 1160-1168.

[48] Wang, J.; Chu, H.; Li, Y. Why Single-Walled Carbon Nanotubes Can Be Dispersed in Imidazolium-Based Ionic Liquids, ACS Nano (2008), 2, 2540-2546.

[49] Shi, F.; Deng, Y. Abnormal FT-IR and FTRaman spectra of ionic liquids, *Spectrochimica Acta*-A (2005), 62, 239-244.

[50] Aronne, A.; Turco, M.; Bagnasco, G.; Pernice, P.; Serio, M.D.; Clayden, N.J.; Marenna, E.; Fanelli, E. Synthesis of High Surface Area Phosphosilicate Glasses by a Modified Sol-Gel Method, *Chemistry of Materials* (2005), 17, 2081-2090.

[51] Burkett, S.L.; Sims, S.D.; Mann, S. Synthesis of hybrid inorganic-organic mesoporous silica by co-condensation of siloxane and organosiloxane precursors, *Chemical Communications* (1996), 1367-1368.

[52] Noda, A.; Hayamizu, K.; Watanabe, M. Pulsed-Gradient Spin−Echo ^{1}H and ^{19}F NMR, *Journal of Physical Chemistry* B (2001), 105, 4603-4610.

[53] Xue, Y.H.; Fu, R.Q.; Wu, C.; Lee, J.Y.; Xu, T.W. Acid–base, *Journal of Membrane Science* 350 (2010) 148-153.

[54] Vila, J.; Gin´es, P.; Pico, J.M.; Franjo, C.; Jim´enez, E.; Varela, L.M.; Cabeza, O. Temperature dependence of the electrical conductivity in EMIM-based ionic liquids Evidence of Vogel–Tamman–Fulcher behavior, *Fluid Phase Equilibria* (2006), 242, 141–146.

[55] Uma, T.; Nogami, M. Structural and Transport Properties of Mixed Phosphotungstic Acid/Phosphomolybdic Acid/SiO_2 Glass Membranes for H_2/O_2 Fuel Cells, *Chemistry of Materials (2007), 19,* 3604-3610.

SUPPORTING INFORMATION

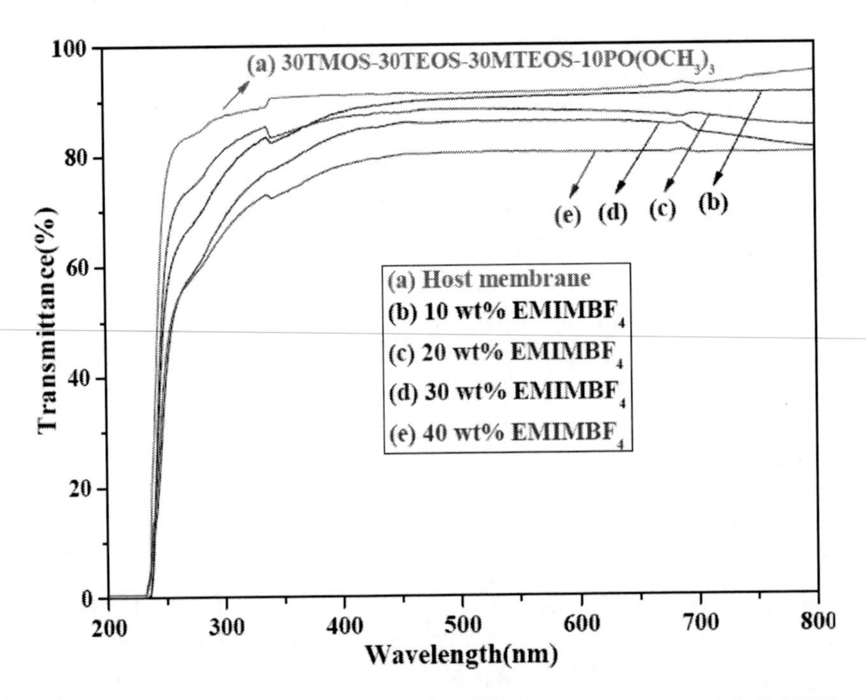

Figure S1. UV-visible transmission spectra of 30TMOS-30TEOS-30MTEOS-10PO $(OCH_3)_3$-x [$EMIMBF_4$] (x=0, 10, 20, 30, and 40 wt %) membranes.

Figure S1 presents the ultraviolet-visible optical transmission spectra of all the fabricated hybrid membranes. From these spectra, we have established that with [$EMIMBF_4$] ionic liquid weight percent ratio increment, the optical transmittance values decreased from 91% to 81% in the prepared composites, respectively. However, these hybrid membranes exhibit high optical transmittance value i.e.; above 80% at the wavelength of 800 nm. FTIR has proven to be a good tool for following the hydrolysis and condensation reactions in sol-gel processes.

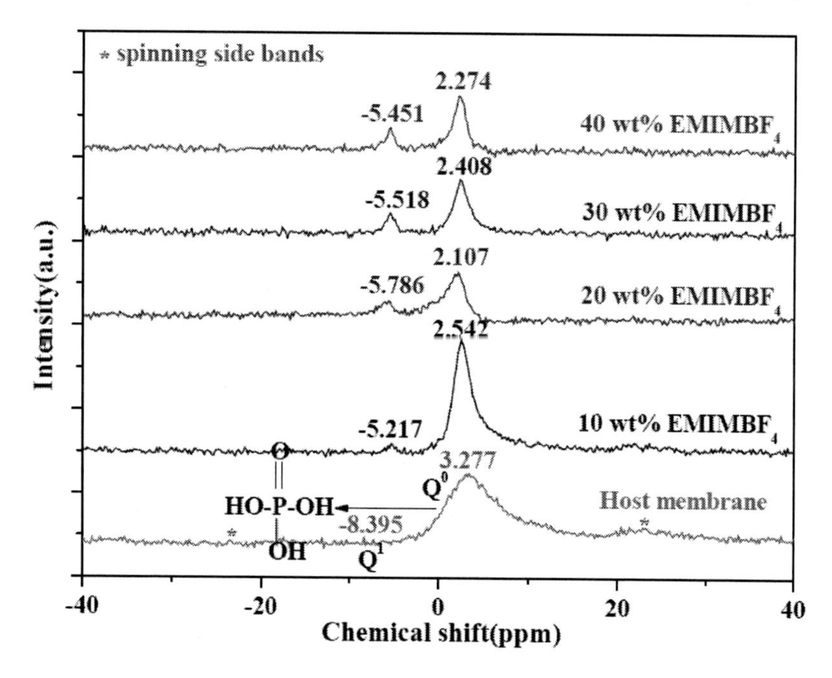

(a)

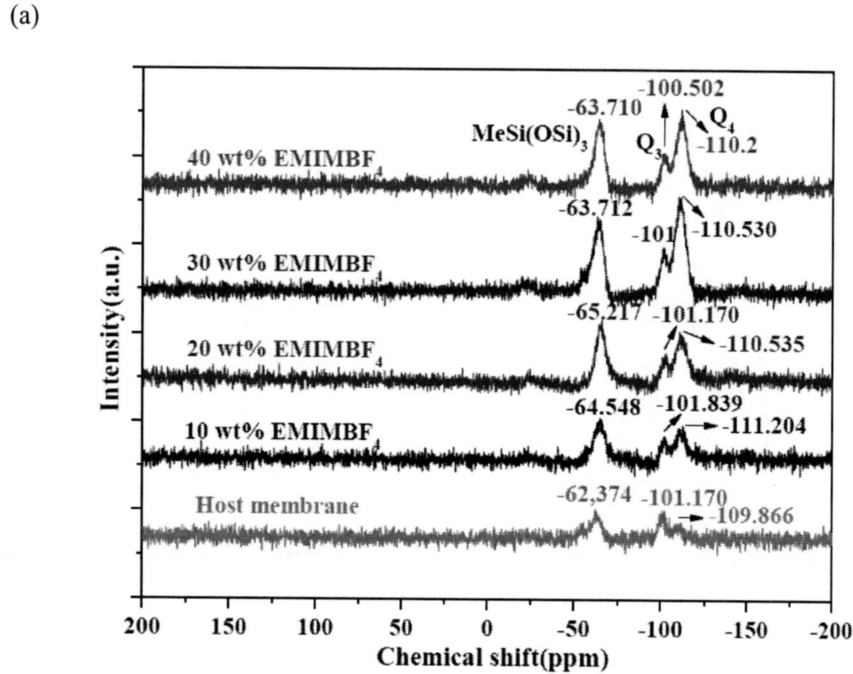

(b)

Figure S2. (Continued).

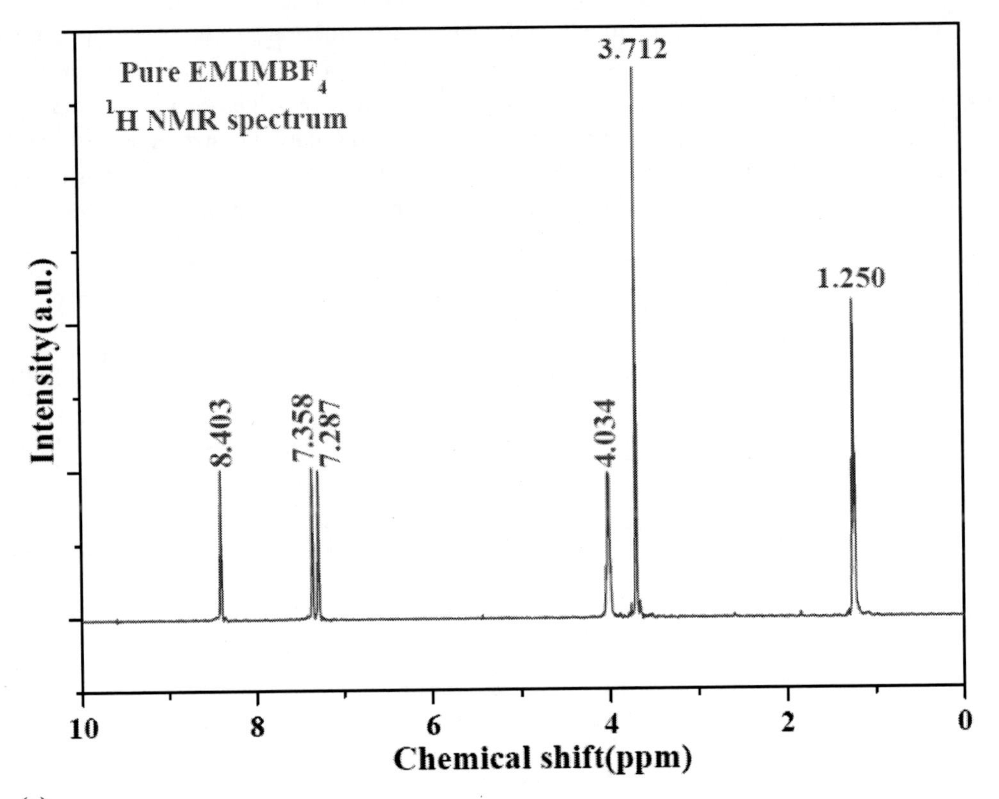
Pure EMIMBF4
1H NMR spectrum
3.712
1.250
8.403
7.358
7.287
4.034
Intensity(a.u.)
10
8
6
4
2
0
Chemical shift(ppm)

(c)

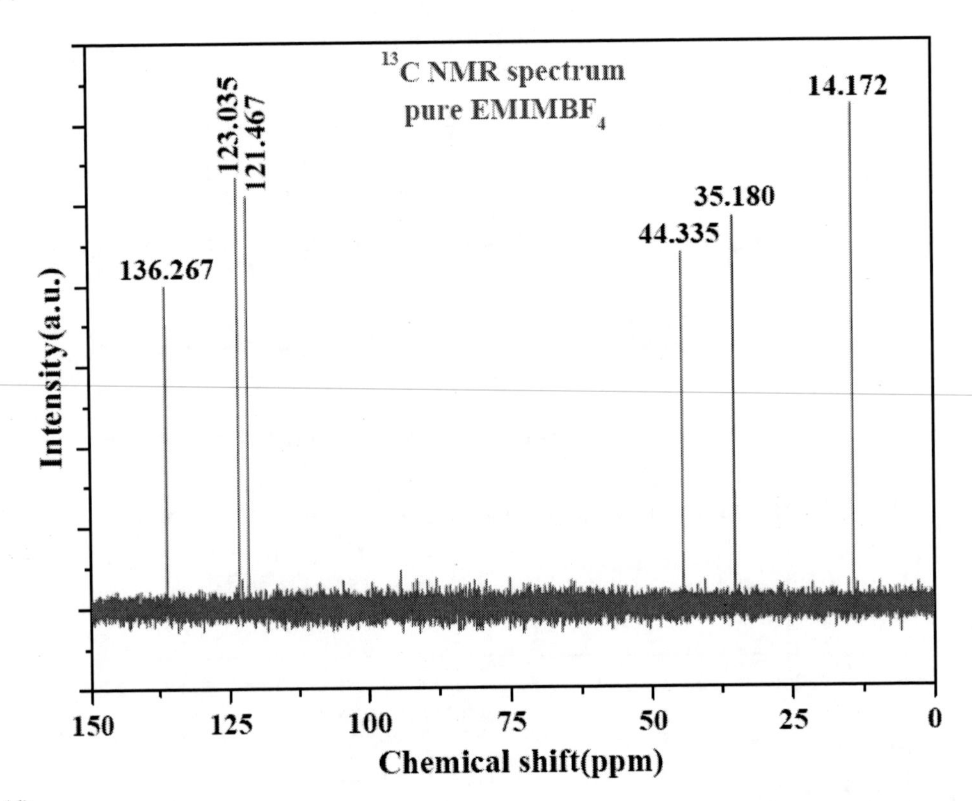
13C NMR spectrum
pure EMIMBF4
123.035
121.467
14.172
35.180
44.335
136.267
Intensity(a.u.)
150
125
100
75
50
25
0
Chemical shift(ppm)

(d)

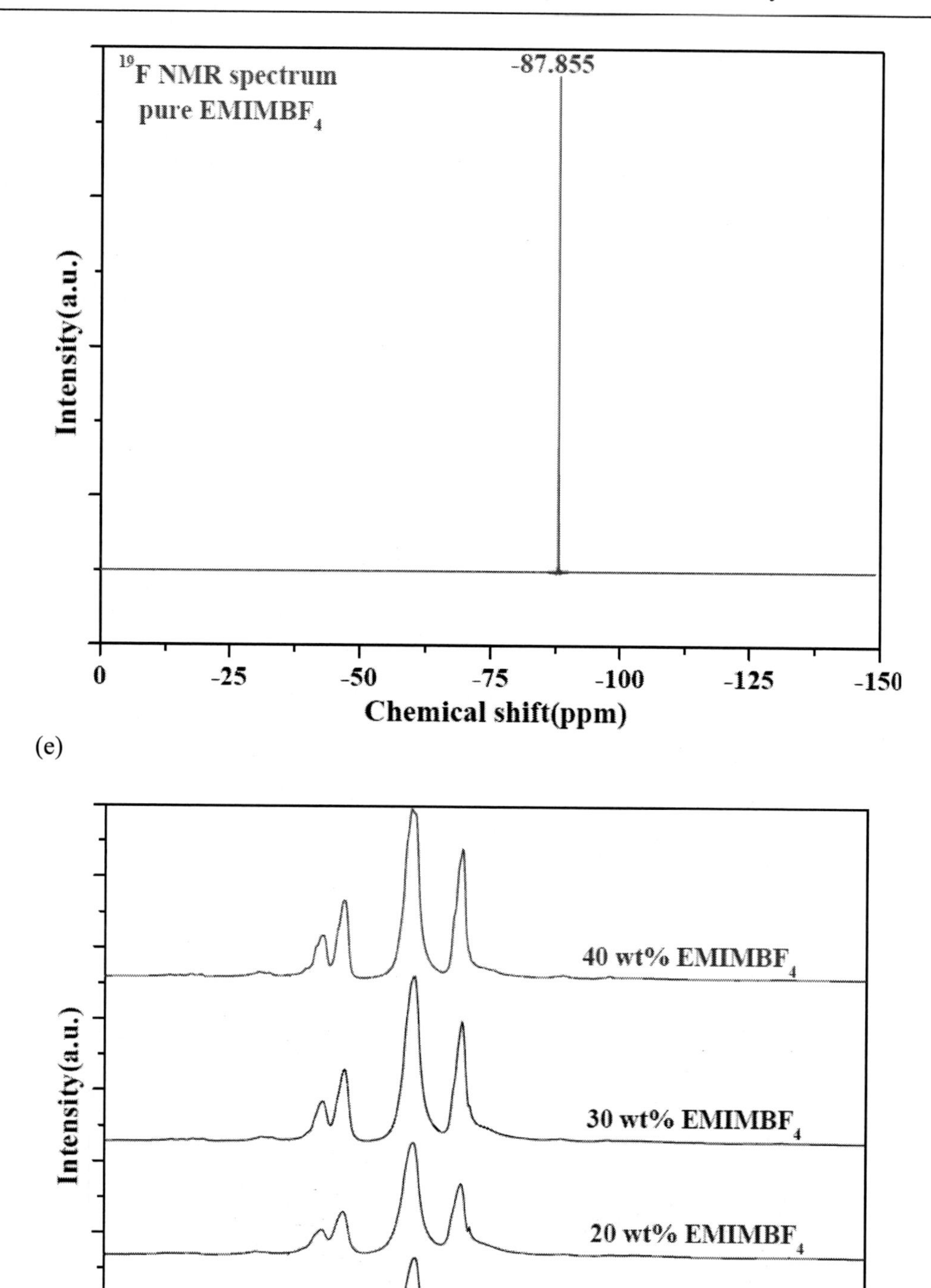

Figure S2. (Continued).

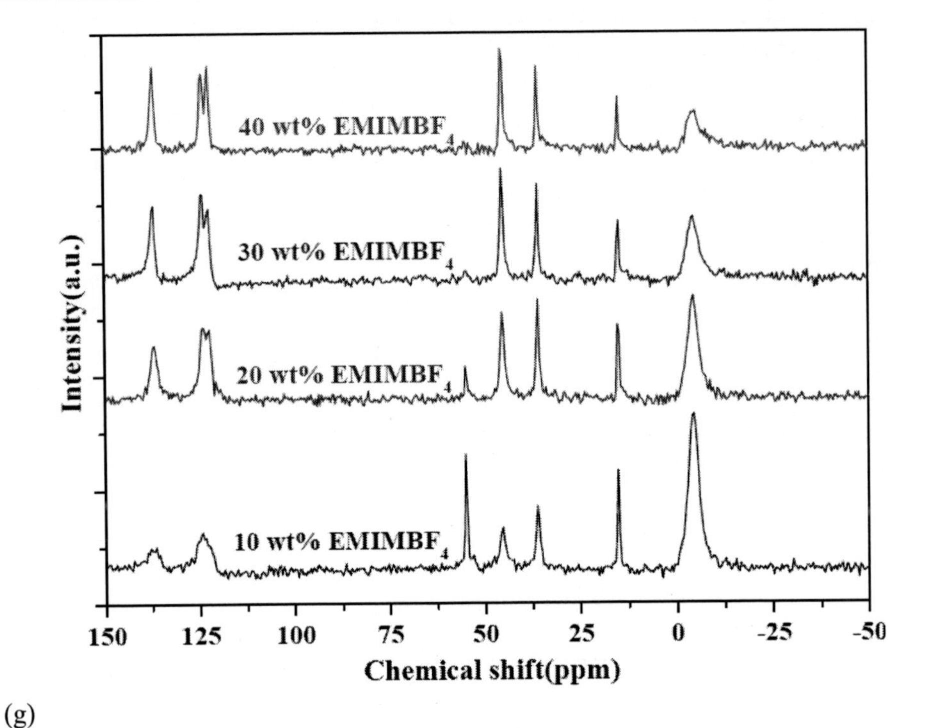

(g)

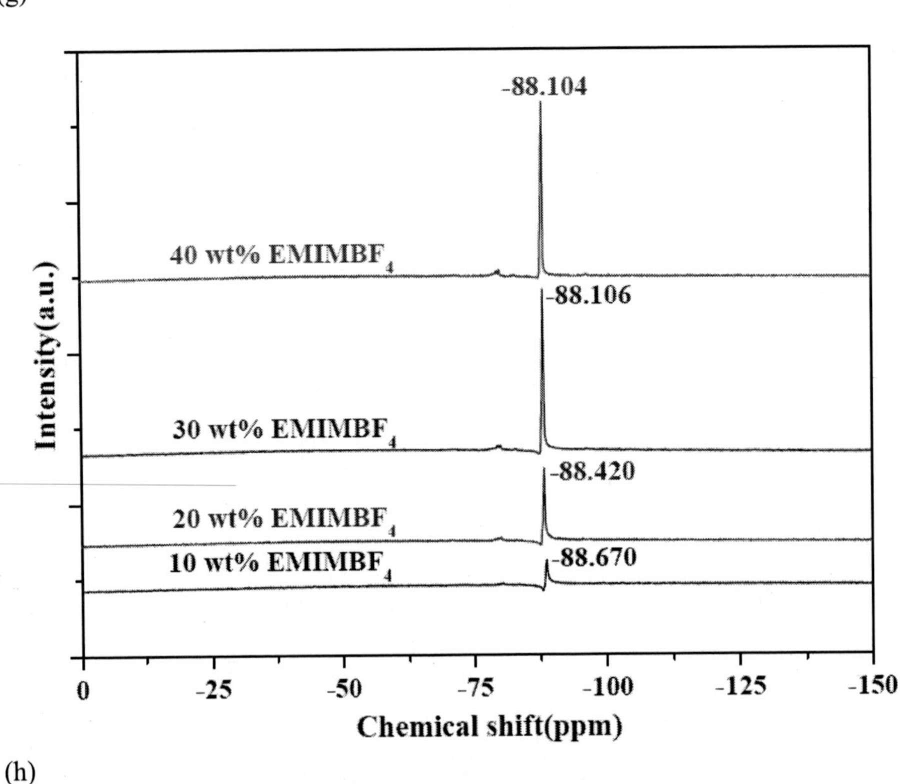

(h)

Figure S2. (a) ^{31}P (b) ^{29}Si MAS NMR spectra of all the prepared hybrid membranes;(c) ^{1}H (d) ^{13}C (e) ^{19}F MAS NMR spectra of pure $EMIMBF_4$ ionic liquid; (f) ^{1}H, (g) ^{13}C, and (h) ^{19}F MAS NMR spectra for 30TMOS-30TEOS- 30MTEOS-10 $PO(OCH_3)_3$ -x [$EMIMBF_4$](x=0, 10, 20, 30, and 40 wt%) composite membranes.

(a)

(b)

Figure S3. (Continued).

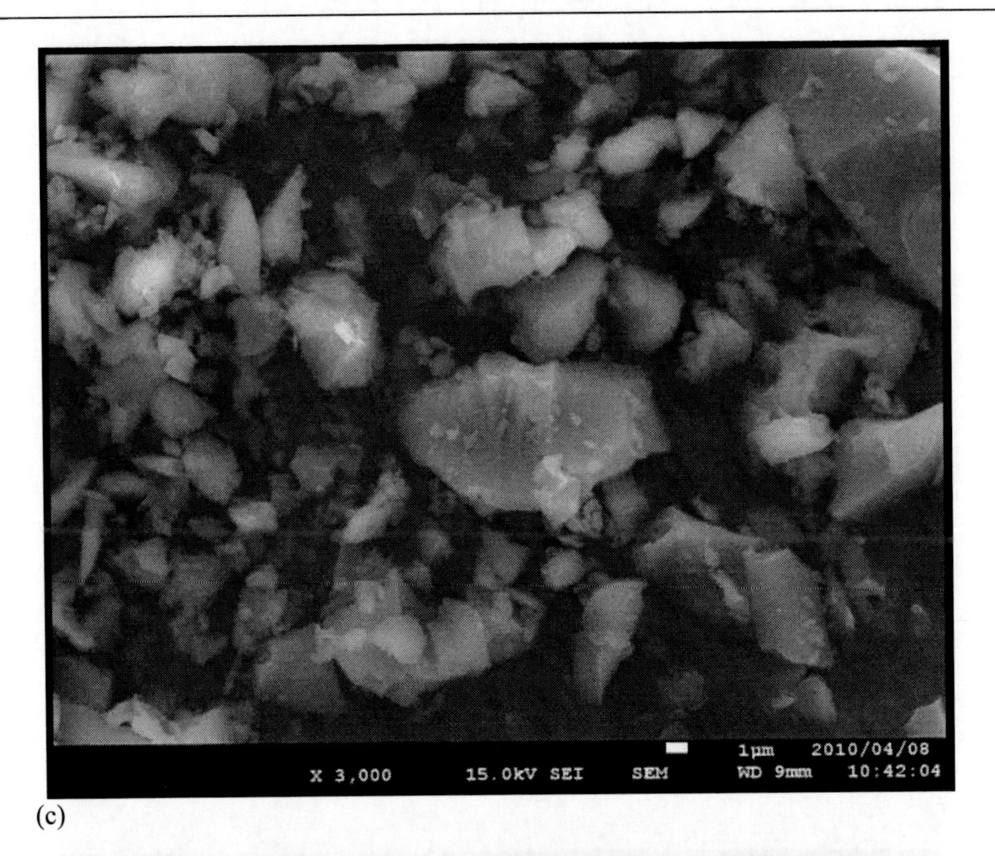
1μm 2010/04/08
X 3,000 15.0kV SEI SEM WD 9mm 10:42:04

(c)

1μm 2010/04/08
X 3,000 15.0kV SEI SEM WD 9mm 10:49:38

(d)

(e)

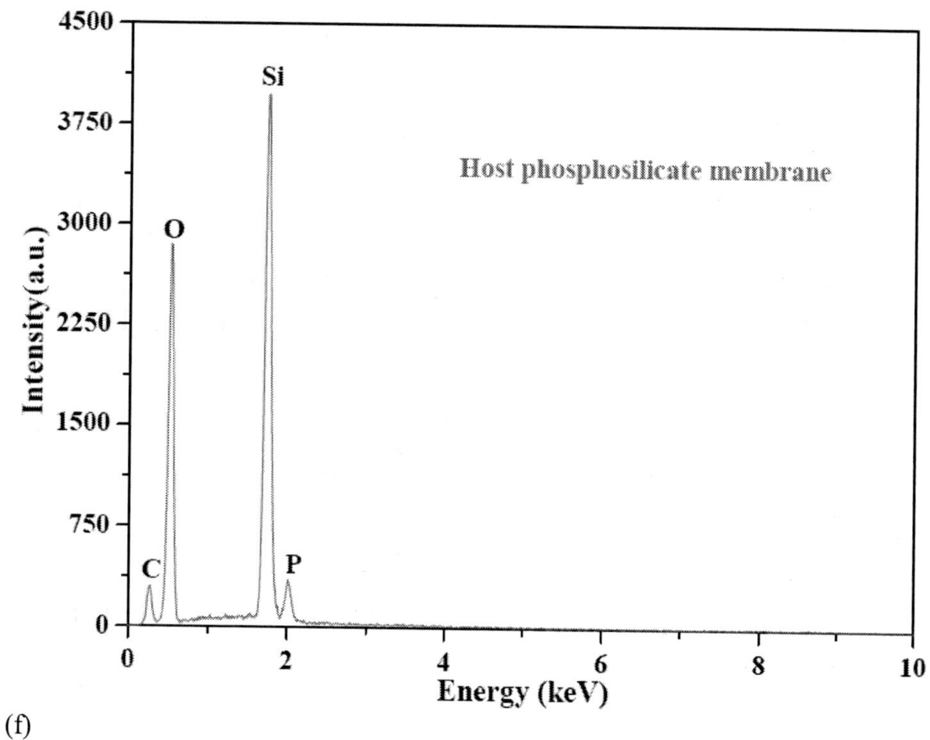

(f)

Figure S3. (Continued).

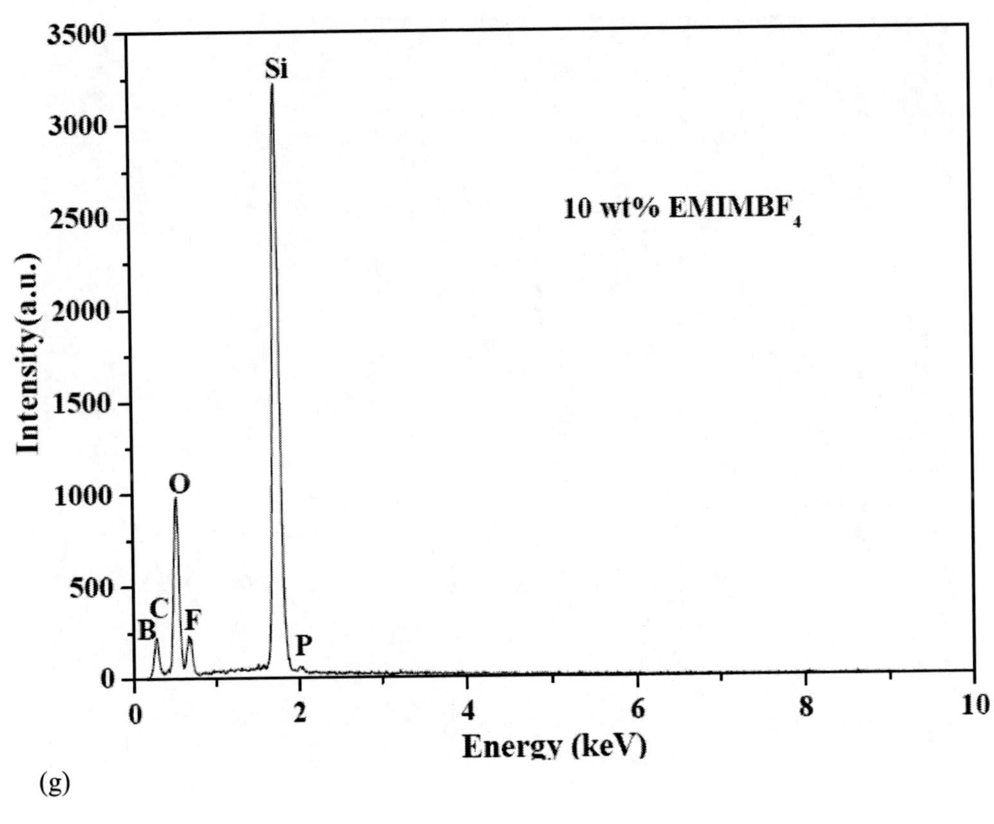
Si
10 wt% EMIMBF$_4$
O
B
C
F
P
Intensity(a.u.)
Energy (keV)
3500
3000
2500
2000
1500
1000
500
0
0
2
4
6
8
10

(g)

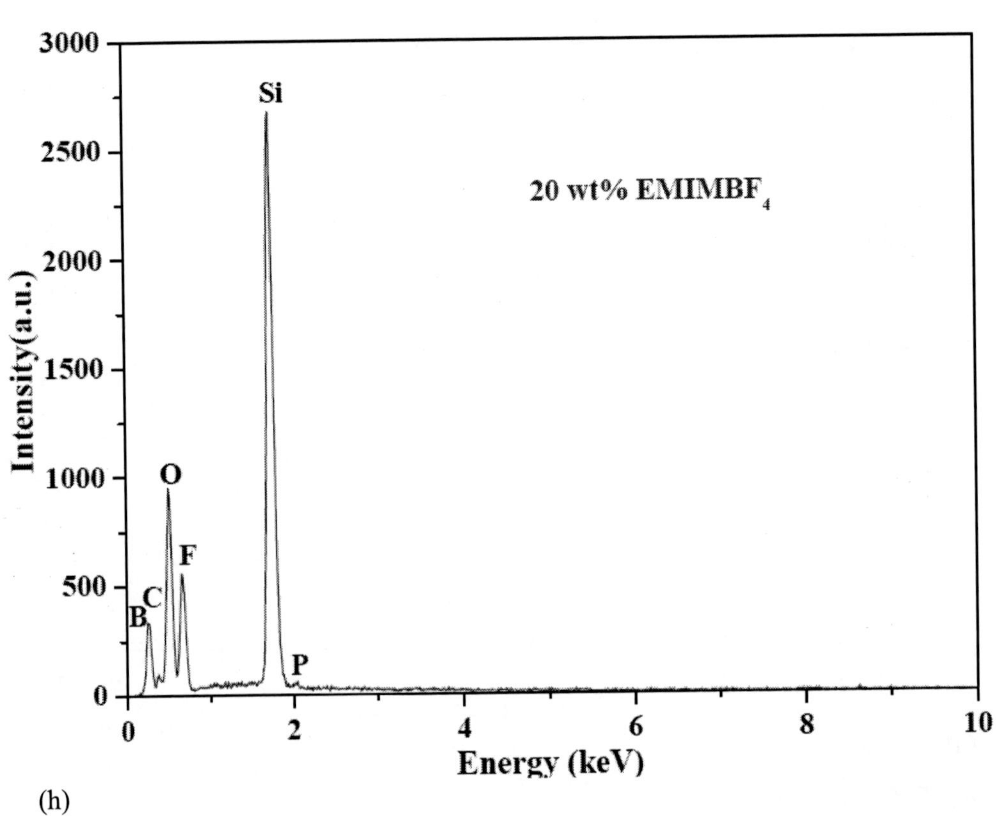
Si
20 wt% EMIMBF$_4$
O
F
B
C
P
Intensity(a.u.)
Energy (keV)
3000
2500
2000
1500
1000
500
0
0
2
4
6
8
10

(h)

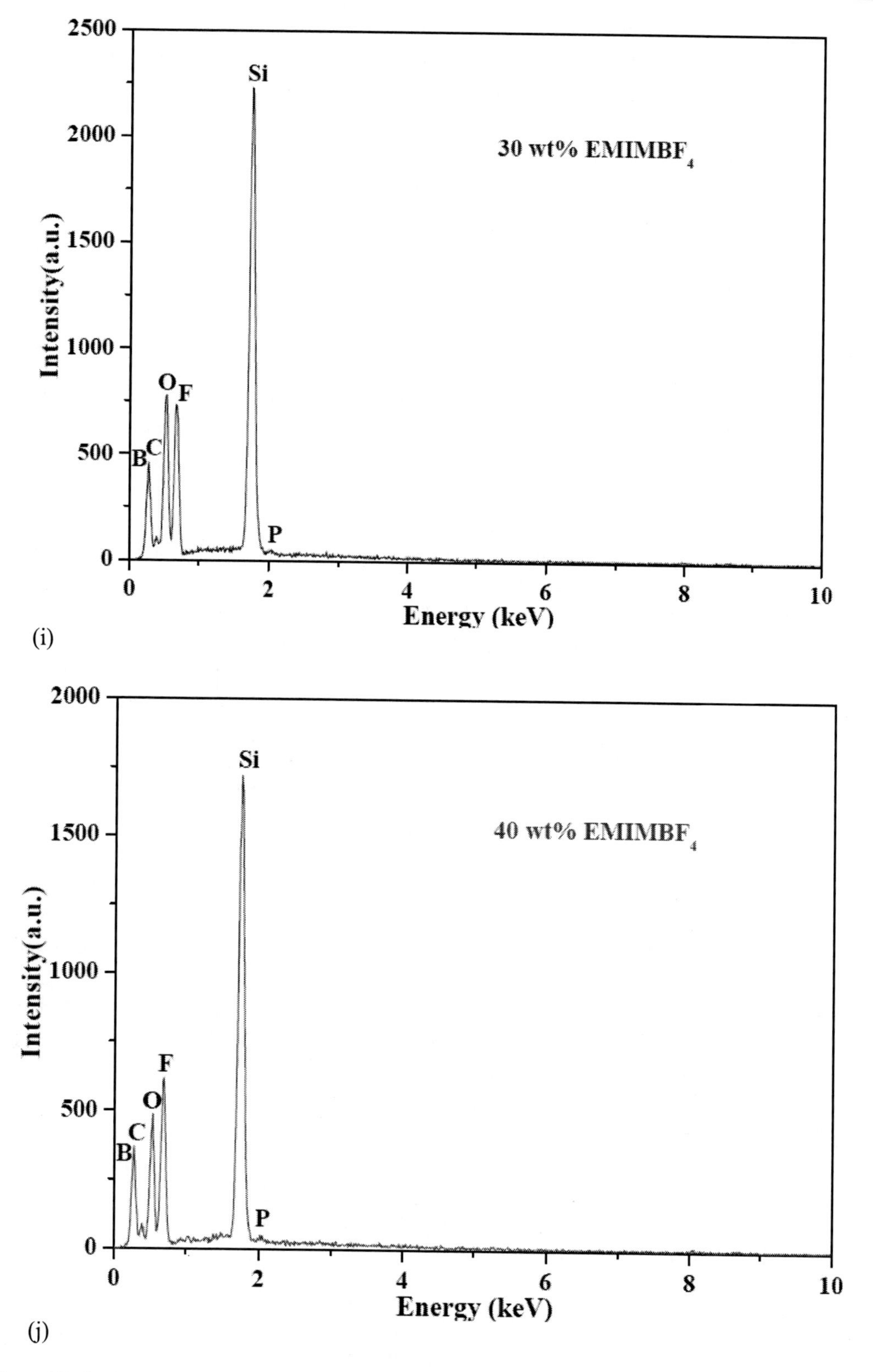

Figure S3. (a-e) SEM and (f-j) EDX micrographs of 30TMOS-30TEOS-30MTEOS- 10PO $(OCH_3)_3$- x $[EMIMBF_4]$ (x=0, 10, 20, 30, and 40 wt %) hybrid membranes.

Figure S2 (a) and (b) show the ^{31}P and ^{29}Si MAS NMR spectra of all the prepared hybrid membranes, respectively. From the ^{31}P NMR spectra (figure S2 (a)), a peak at 3.277 ppm is observed for host phosphosilicate membrane which is assigned to the isolated phosphorus content (Q^0 unit). A smaller peak due to condensed phosphorus content with a bridging oxygen (Q^1 unit) like Si-O-P and P-O-P is also observed at -8.395 ppm for the host membrane. The disappearance of the peaks which can be assigned to Q^3 and Q^4 units, respectively, corresponds to the hydrolysis of phosphosilicate network structure and $Si_5O(PO)_6$ during the heat-treatment. This result suggests that Q^3 and Q^4 units are readily hydrolyzed to form Q^0 units. For 10, 20, 30, and 40 wt% [$EMIMBF_4$] ionic liquid doped hybrid membranes, the peak (Q^0 unit) is identified at 2.542, 2.107, 2.408, and 2.274 ppm, respectively. Further, the Q^1 unit shifted to -5.217, -5.786, -5.518, and -5.451 ppm, respectively for 10-40 wt% ionic liquid doped hybrid membranes. For all the [$EMIMBF_4$] ionic liquid based hybrid membranes, the shifting of Q^0 and Q^1 units compared with host phosphosilicate membrane suggests the interaction of [$EMIMBF_4$] ionic liquid with trimethyl phosphate. The ^{29}Si NMR chemical shift is a powerful tool for identifying various surface sites. From figure S2 (b), for the host phosphosilicate membrane, a peak at -62.374 ppm which could be assigned to $MeSi(OSi)_3$ groups [1], and two more chemical shifts at -109.866 ppm, and -101.170 ppm are identified which can be assigned to the presence of quaternary SiO_4 units (Q_4) and $SiO_3(OH)$ units (Q_3), respectively. Similarly, for 10, 20, 30, and 40 wt% ionic liquid doped hybrid membranes, the bands assigned to $MeSi(OSi)_3$ groups, Q^4, and Q^3 units are identified at [-64.548, -111.204, and -101.839 ppm]; [-65.217, -110.535, and -101.170 ppm]; [-63.712, -110.530, and -101 ppm]; and [-63.710, -110.2, and -100.502 ppm], respectively. Figure S3 (c), (d), and (e) presents the ^{1}H, ^{13}C, and ^{19}F MAS NMR spectra of pure [$EMIMBF_4$] ionic liquid, respectively. The structure of $EMIMBF_4$ was identified by ^{1}H NMR spectrum (δ/ppm relative to $CDCl_3$): 8.403 (s, 1H, H(2)), 7.358 (s, 1H, H(4)), 7.287 (s, 1H, H(5)), 4.034 (q, 2H, and J = 21 Hz, H(6)), 3.712 (s, 3H, H(8)), and 1.250 (t, 3H, and J=14 Hz, H(7)). Similarly, for ^{13}C (δ/ppm relative to $CDCl_3$) chemical shifts at [136.267, 123.035, 121.467, 44.335, 35.180, and 14.172 ppm] are found. From the ^{19}F MAS NMR spectrum (using trifluorotolune as reference material), an intense and sharp signal at -87.855 ppm was observed for pure $EMIMBF_4$ ionic liquid. Further, figure S2 (f), (g), and (h) shows the ^{1}H, ^{13}C, and ^{19}F MAS NMR spectra of all the ionic liquid doped hybrid membranes, respectively. From figure S2 (f), ^{1}H chemical shifts at 8.583, 7.466, 3.675, 1.359, and 0.819 ppm are found for 10 wt. % [$EMIMBF_4$] ionic liquid doped hybrid membrane. Similarly, for 20, 30, and 40 wt% [$EMIMBF_4$] doped composites, signals at 8.511, 7.358, 3.783, 1.283, and 0.798 ppm; 8.491, 7.341, 3.746, 1.250, and 0.796 ppm; and 8.905, 8.528, 7.395, 3.800, 1.714, 1.229, and 0.870 ppm are found, respectively. In these prepared hybrid membranes, several ^{1}H chemical shifts shifting compared with pure [$EMIMBF_4$] ionic liquid could be due to the strong interaction between phosphosilicate content and ionic liquid. Also, from the ^{13}C NMR spectra, for 10-40 wt% [$EMIMBF_4$] ionic liquid doped hybrid membranes (figure S2 (g)), chemical shifts at [137.123, 124.164, 55.100, 45.067, 36.120, 15.217, and -4.515 ppm]; [136.706, 123.829, 121.990, 55.098, 45.068, 35.702, 15.215, and -4.513 ppm]; [136.705, 123.827, 121.989, 54.766, 45.067, 35.700, 14.883, and -4.512 ppm]; and [136.706, 123.828, 121.990, 45.066, 35.701, 14.882, and -4.933 ppm] are identified, respectively. Moreover, it should be emphasized that with the [$EMIMBF_4$] ionic liquid content increment the chemical shifts intensity within the range 120-140 ppm, and at 35 and 45 ppm is increased gradually while the chemical shifts intensity at -4, 15 and 55 ppm is decreased in the fabricated hybrid membranes. Compared with pure ionic liquid ^{13}C NMR signals the shifting of some ^{13}C NMR chemical shifts in the prepared composites strongly suggests the interaction of [$EMIMBF_4$] ionic liquid with inorganic phosphosilicate content. From figure S2 (h), ^{19}F chemical shifts at -88.670, -88.420, -88.106, and -88.104 ppm are identified for 10-40 wt% [$EMIMBF_4$] ionic liquid doped hybrids, respectively. The ^{19}F NMR spectrum generally shows the wide range distribution of chemical shifts due to the larger electron density than ^{1}H. Also, Fluorine-19 is a magnetically active nucleus providing excellent contrast capabilities.

Figure S3 (a-e) and (f-j) presents the SEM micrographs and the corresponding EDX spectra of 10-40 wt% [$EMIMBF_4$] ionic liquid based hybrid membranes, including host phosphosilicate membrane. Due to this strong interaction between phosphosilicate content and ionic liquid, all the ionic liquid based hybrid membranes mesoporous structure altered which are also confirmed by the Brunauer- Emmett- Teller (BET) method. From the EDX micrographs (figure S3 (f-j)), the presence of Si, P, F, O, and B elements are confirmed and they are all homogeneously distributed in the prepared hybrid membranes, respectively.

Generally, fuel cell construction consists of a fuel electrode (anode) and an oxidant electrode (cathode) separated by an ion-conducting membrane (electrolyte) as shown in figure S4. The PEMFC is a complex nonlinear, multiple input and output and big delay dynamic system; the working process is mixed flow transportation, heat conduction, and electrochemical dynamical reaction. Hydrogen fuel can be fed into the anode of the fuel cell. Oxygen (or air) enters the fuel cell through the cathode. At the anode, hydrogen will split into protons and electrons. The protons diffuse through the electrolyte to the cathode and the electrons create a current flow in the circuit, which can be utilized before they return to the cathode. They recombine with protons and oxygen to form water.

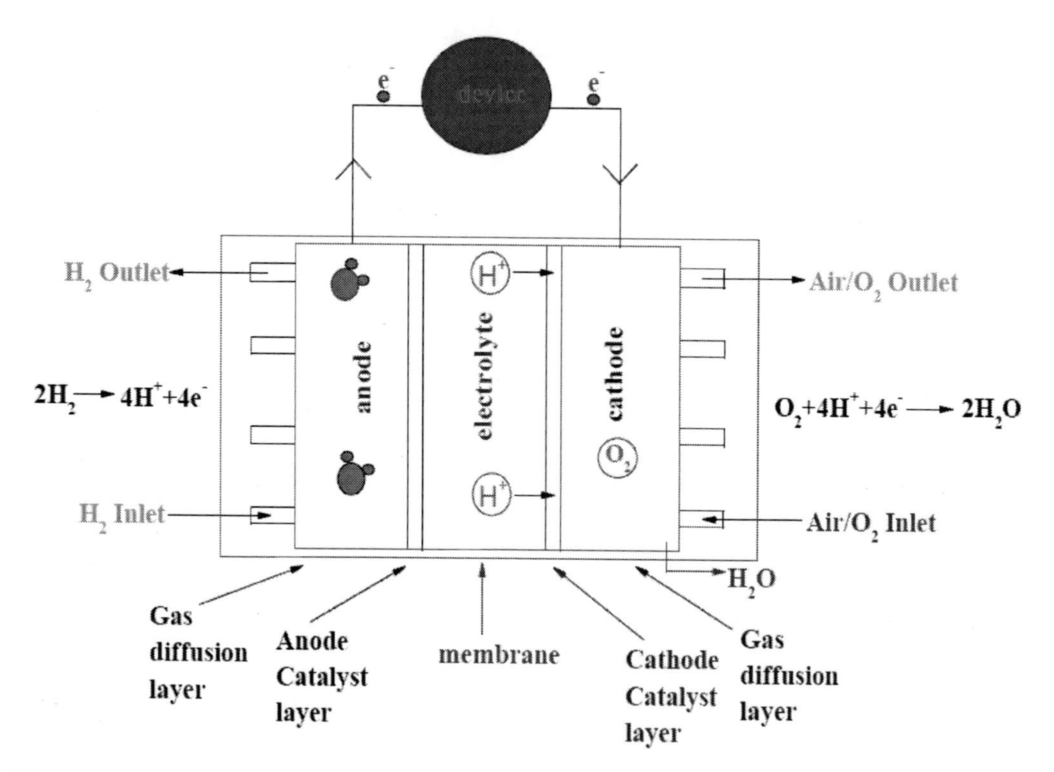

Figure S4. Working and reaction principles of H_2/O_2 fuel cell.

Anode: $2H_2 \rightarrow 4H^+ + 4e^-$ (1)

Cathode: $74\ O_2 + 4H^+ + 4e^- \rightarrow 2H_2O$ (2)

Overall: $74\ O_2 + 2H_2 \rightarrow 2H_2O$ (3)

The flow of ionic charge through the electrolyte must be balanced by the flow of electronic charge through an outside circuit, and it is this balance that produces electrical power. Even though the open circuit potential (OCP) of a fuel cell under ideal conditions is 1.23 V, in practice OCP (0.6-0.8 V) will be less than the ideal OCP and decreases with the increase in current density.

REFERENCES

[1]. T. Bein, R.F.Carver, R.D.Farlee and G.D. Stucky, Solid-state, J. Am. Chem. Soc. 110 (1988) 4546-4553.

In: Advances in Condensed Matter ... Volume 10
Editors: H. Geelvinck and S. Reynst

ISBN: 978-1-61209-533-2

Chapter 3

PRESSURE AND TEMPERATURE DEPENDENT FORM FACTORS OF THE EMPIRICAL PSEUDOPOTENTIAL METHOD FOR THE ELECTRONIC STRUCTURES OF NANOMATERIALS

A. M. Elabsy[*]
Department of Physics, Faculty of Science,
Mansoura University, Mansoura, Egypt

ABSTRACT

Electronic band structures for zinc blende (GaAs, AlAs) and diamond (Ge, Si) nanostructure semiconductors were performed. The calculations were based on the empirical pseudopotential method (EPM), which takes into account the effects of both hydrostatic pressure and temperature. The results showed an interesting behavior, that the pseudopotential form factors associated with the reciprocal lattice vectors of $|\vec{G'} - \vec{G}|^2 =$ 11 are more affected by both pressure and temperature than the other values of the form factors. Therefore, the effects of pressure and temperature must be considered when dealing with electronic band structure calculations based on empirical pseudopotential formalism.

INTRODUCTION

The empirical pseudopotential method (EPM) was developed in the last six decades [1-4] and the pioneer work of this method is that of Cohen and Bergstresser [4] who highlighted its use in performing band structure calculations for semiconductors of type diamond and zinc-blende structures. It is devoted to solve Schrödinger's equation for bulk crystals without

[*] E-mail: amelabsy@mans.edu.eg.

knowing exactly the potential experienced by an electron in the crystal lattice. In most work that based on EPM calculations authors considered only the room temperature and the normal pressure although both effects of temperature and pressure are important factors in fabricating and operating electronic devices. Elabsy et al.[5,6] have highlighted both effects and found that the large reciprocal Bravais lattice vectors are of important effect on the pseudopotential form factors and consequently on the behavior of the band structures for the concerned semiconductor. In the following sections we shall discuss these effects and the numerical analysis in details.

2. The Empirical Pseudopotential Method

In the EPM method, the core electrons are tightly bound to their nuclei, and the valence and conduction band electrons are influenced only by the remaining weak net effective potential called pseudopotential. The weak potential seen by the valence electrons is due to the cancellation of the large attractive (negative) core potential energy of the ion core and the large positive kinetic energy of the electron due to its rapid oscillations [1-4]. Since this potential can be expanded in plane waves Fourier series, an eigenvalue equation for determining an energy-momentum (E-k) relationship can be established. Although the Fourier coefficients for the potentials are not known, they can be empirically determined for a given crystal by fitting the calculated crystal energy gaps to the known measurements.

The aim of this section is to calculate the electronic band structure for the two types of crystals, the diamond and the zinc-blende structures.

To obtain the electronic structure of a semiconductor crystal, we solve the one electron Schrödinger's wave equation that based on the independent electron approximation which neglects the electron-electron interactions and takes the form

$$\left[\frac{-\hbar^2}{2m}\nabla^2 + V_c(r)\right]\psi_{n,\vec{k}} = E_{n,\vec{k}}(r)\psi_{n,\vec{k}} \tag{2.1}$$

In the above equation (2.1), $V_c(r)$ is the crystal potential which considers all interactions, n is the band index, $\vec{k}$ is the wave vector and $E_{n,\vec{k}}$ is the corresponding energy levels.

The solutions of the Schrödinger's wave equation $\psi_{n,\vec{k}}$ depend on the periodicity of the crystal potential that depends on the electron momentum $\vec{k}$. So, the electron wavefunction can labeled with both n and $\vec{k}$ and can be written in the Block form which separates the electron wave function into two parts one is the oscillatory plane wave and the other is periodic [7] as

$$\psi_{n,\vec{k}}(\vec{r}) = e^{i\vec{k}.\vec{r}}\xi_{n,\vec{k}}(\vec{r}) \tag{2.2}$$

where $e^{i\vec{k}.\vec{r}}$ is the plane wave solution (neglecting the interaction of the potential, i.e. free electron solution) and the function $\xi_{n,\vec{k}}(\vec{r})$ is a periodic function which stands for the periodicity of the crystal.

This periodic function can be expanded in terms of a complete set of normalized plane waves whose wave vectors are the reciprocal lattice vectors, $\vec{G}$ of the crystal [8] which takes the form

$$\xi_{n,\vec{k}}(\vec{r}) = \frac{1}{\sqrt{\Omega}} \sum_{\vec{G}} a_{n.\vec{k}}(\vec{G}) e^{i\vec{G}\cdot\vec{r}} \tag{2.3}$$

where $a_{n.\vec{k}}(\vec{G})$ is the Fourier coefficient, $\vec{G}$ is the reciprocal Bravais lattice vector which can be written as a linear combination of the primitive reciprocal Bravais lattice vectors $\vec{b}_j$ as

$$\vec{G} = g_1 \vec{b}_1 + g_2 \vec{b}_2 + g_3 \vec{b}_3 \tag{2.4}$$

where g_j (j = 1, 2, and 3) are integers and $\vec{b}_j$ are the primitive reciprocal Bravais lattice vectors and Ω is the total volume of the crystal.

Substituting equation (2.3) into equation (2.2) one gets:

$$\psi_{n,\vec{k}}(\vec{r}) = \frac{1}{\sqrt{\Omega}} \sum_{\vec{G}} a_{n.\vec{k}}(\vec{G}) e^{i(\vec{k}+\vec{G})\cdot\vec{r}} \tag{2.5}$$

Applying Dirac's brackets notation, the Dirac ket wave function has the form

$$|\vec{k}+\vec{G}\rangle = \frac{1}{\sqrt{\Omega}} e^{i(\vec{k}+\vec{G})\cdot\vec{r}}$$

Then the wave function (2.5) can be expressed as

$$\psi_{n,\vec{k}}(\vec{r}) = \sum_{\vec{G}} a_{n.\vec{k}}(\vec{G})|\vec{k}+\vec{G}\rangle \tag{2.6}$$

For bulk semiconductors of type zinc-blende and diamond structures there are two atoms (*cation* – *anion*) in each Bravais lattice point. Consider the lattice point is located at the translation lattice vector $\vec{R}$ which lies at the midway between the *cation* and the *anion* that constitute the lattice species which is given by

$$\vec{R} = n_1 \vec{a}_1 + n_2 \vec{a}_2 + n_3 \vec{a}_3 \tag{2.7}$$

where $\mathrm{n_i}$; $\mathrm{i} = 1,2,\text{and } 3$ integers are $\text{equal}0, \pm1, \pm2, \pm3, \ldots$, and $\vec{\mathrm{a}}_\mathrm{i}$ are the primitive translational lattice vectors in real space of the Bravais lattice.

Let the *cation* atom is located for example at - τ and the *anion* atom is located at τ, and $\vec{r}$ is the position vector of the electron the center lattice point (midway) of the cation – anion locations at $\vec{R}$ as displayed in figure 2.1.

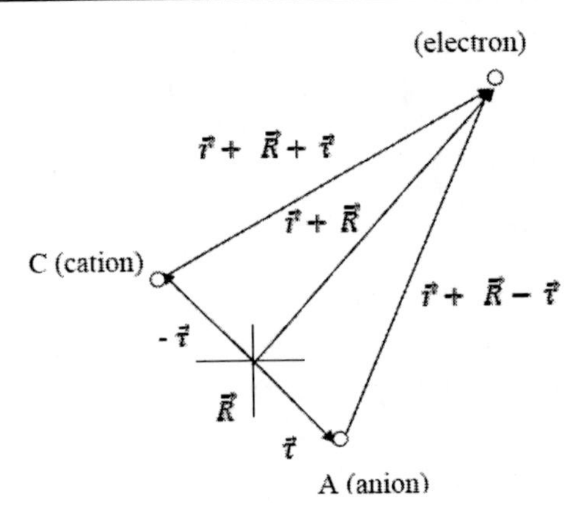

Figure 2.1. The locations ($\pm\tau$) of the atomic species (cation and anion) relative to the lattice point located at $\vec{R}$ and the electron position vector $\vec{r}$.

Therefore, the approximate smooth crystal potential (pseudopotential) is considered as a sum of atomic-like spherically symmetric local pseudopotentials centered on the atomic sites at ($\pm\tau$) which takes the form

$$V_\ell(r) = \sum_{\vec{R}} [V_C(\vec{r}+\vec{R}+\vec{\tau}) + V_A(\vec{r}+\vec{R}-\vec{\tau})] \tag{2.8}$$

where ℓ refers to the local, $\vec{R}$ stand for the coordinates of the lattice points, and $\pm\tau$ are the positions of the atoms that constitute the crystal relative to the lattice points.

$V_C(\vec{r}+\vec{R}+\vec{\tau})$ and $V_A(\vec{r}+\vec{R}-\vec{\tau})$ are the local pseudopotentials of the electron due to its interaction with the cation and anion, respectively.

From now on as an approximation, we shall replace the crystal potential $V_C(r)$ with the local potential $V_\ell(r)$ since we ignore the contributions arising from the non-local due to the atomic orbit and the spin-orbit interactions.

Substitute from equations (2.6) and (2.8) into equation (2.1) to solve the one electron Schrödinger equation using the pseudopotential and the pseudo wavefunction, as follows

$$\left[\frac{-\hbar^2}{2m}\nabla^2 + V_\ell(r)\right] \sum_{\vec{G}} a_{n,\vec{k}}(\vec{G})|\vec{k}+\vec{G}\rangle = E_{n,\vec{k}} \sum_{\vec{G}} a_{n,\vec{k}}(\vec{G})|\vec{k}+\vec{G}\rangle \tag{2.9}$$

Multiply equation (2.9) from left by the bra wavefunction $\langle\vec{k}+\vec{G'}|$ and employ the Kronecker delta function identity:

$$\langle\vec{k}+\vec{G'}|\vec{k}+\vec{G}\rangle = \delta_{\vec{G},\vec{G'}} = \begin{cases} 1 \; for \; \vec{G}=\vec{G'} \\ 0 \; for \; \vec{G}\neq\vec{G'} \end{cases}$$

we get

$$\frac{\hbar^2}{2m} a_{n,\vec{k}}(\vec{G'}) \left|\vec{k}+\vec{G'}\right|^2 + \sum_{\vec{G}} a_{n,\vec{k}}(\vec{G}) \langle\vec{k}+\vec{G'}| \sum_{\vec{R}} \left(V_C(\vec{r}+\vec{R}+\vec{\tau}) + V_A(\vec{r}+\vec{R}-\vec{\tau})\right) \left|\vec{k}+\right.$$
$$G=En,k\;an,kG' \tag{2.10}$$

To evaluate the LHS of the above equation (2.10), we first calculate the pseudopotential matrix element term as follows

$$\sum_{\vec{R}} \langle \vec{k}+\vec{G}'| \left(V_C(\vec{r}+\vec{R}+\vec{\tau}) + V_A(\vec{r}+\vec{R}-\vec{\tau})\right) |\vec{k}+\vec{G}\rangle$$

$$= \sum_{\vec{R}} \langle \vec{k}+\vec{G}'| \left(V_C(\vec{r}+\vec{R}+\vec{\tau})\right) |\vec{k}+\vec{G}\rangle + \sum_{\vec{R}} \langle \vec{k}+\vec{G}'| \left(V_A(\vec{r}+\vec{R}-\vec{\tau})\right) |\vec{k}+\vec{G}\rangle$$

$$\equiv I_C + I_A \tag{2.11}$$

where

$$I_C = \Sigma_{\vec{R}} \langle \vec{k}+\vec{G}' | V_C(\vec{r}+\vec{R}+\vec{\tau}) | \vec{k}+\vec{G}\rangle \tag{2.12}$$

and

$$I_A = \Sigma_{\vec{R}} \langle \vec{k}+\vec{G}' | V_A(\vec{r}+\vec{R}-\vec{\tau}) | \vec{k}+\vec{G}\rangle \tag{2.13}$$

Then

$$\mathrm{I_C} = \frac{1}{\Omega} \sum_{\vec{R}} \int e^{-i(\vec{k}+\vec{G}')\cdot\vec{r}}\, V_C(\vec{r}+\vec{R}+\vec{\tau})\, e^{i(\vec{k}+\vec{G})\cdot\vec{r}}\, d^3r$$

where Ω is the total volume of the crystal, then

$$I_C = \frac{1}{\Omega} \Sigma_{\vec{R}} \int e^{-i(\vec{G}'-\vec{G})\cdot\vec{r}}\, V_C(\vec{r}+\vec{R}+\vec{\tau})\, d^3r$$
$$= \frac{1}{\Omega} \Sigma_{\vec{R}} \int e^{-i\vec{q}\cdot\vec{r}}\, V_C(\vec{r}+\vec{R}+\vec{\tau})\, d^3r \tag{2.14}$$

where we replace the vectors difference $\vec{G}' - \vec{G}$ by $\vec{q}$, which is a new reciprocal lattice vector lies in the first Brillouin zone.

In the above integral (2.14) replace the vector $\vec{r}$ by $\vec{r} - \vec{R} - \vec{\tau}$, then

$$I_C = \frac{1}{\Omega} \sum_{\vec{R}} \int e^{-i\vec{q}\cdot(\vec{r}-\vec{R}-\vec{\tau})}\, V_C(r)\, d^3r$$
$$= \frac{1}{\Omega} \Sigma_{\vec{R}}\, e^{i\vec{q}\cdot(\vec{R}+\vec{\tau})} \int e^{-i\vec{q}\cdot\vec{r}}\, V_C(r)\, d^3r \tag{2.15}$$

Applying the Fourier transformation for the atomic pseudopotential

$$V(q) = \frac{1}{\Omega_0} \int_{\Omega_0} e^{-i\vec{q}\cdot\vec{r}}\, V(r)\, d^3r \tag{2.16}$$

where q is the magnitude (scalar value) of the reciprocal vector $\vec{q} = \vec{G}' - \vec{G}$ and $\Omega_0 = \vec{a}_1 \cdot (\vec{a}_2 \times \vec{a}_3)$, is the volume of the unit cell in real space.

Substitute from equation (2.16) into equation (2.15), we get

$$I_C = \frac{\Omega_0}{\Omega} \sum_{\vec{R}} e^{i\,\vec{q}\cdot(\vec{R}+\vec{\tau})}\, V_C(q)$$

$$= \frac{1}{N}\left[\sum_{\vec{R}} e^{i\,\vec{q}\cdot\vec{R}}\right]\left\{e^{i\,\vec{q}\cdot\vec{\tau}}\, V_C(q)\right\} \qquad (2.17)$$

where N is the total number of lattice point in the crystal of volume $\Omega = N\,\Omega_0$.

In the above equation (2.17) it is well known from the definition of the reciprocal Bravais lattice vector that

$$\vec{q}\cdot\vec{R} = 2\pi\, m;\ m = 0, 1, 2, 3, \ldots \text{ and consequently } e^{i\,\vec{q}\cdot\vec{R}} = 1.$$

The first term in the square brackets in equation (2.17) is called the *structure factor* and the sum over the translational lattice vectors $\vec{R}$ gives the total number N of the lattice points inside the crystal. The second term in the curved brackets in equation (2.17) that depends only on the atomic potential is called the *form factor*. Thus, equation (2.17) takes the form

$$I_C = e^{i\,\vec{q}\cdot\vec{\tau}}\, V_C(q) \qquad (2.18)$$

Therefore, expanding the exponential function in terms of the sinusoidal form, equation (2.18) can be rewritten as

$$I_C = e^{i\,\vec{q}\cdot\vec{\tau}}\, V_C(q) = V_C(q)\,[\cos(\vec{q}\cdot\vec{\tau}) + i\,sin\,(\vec{q}\cdot\vec{\tau})] \qquad (2.19)$$

Similarly, the integral concerns the anion potential, I_A in equation (2.13) can be written as

$$I_A = e^{-i\,\vec{q}\cdot\vec{\tau}}\, V_A(q) = V_A(q)\,[\cos(\vec{q}\cdot\vec{\tau}) - i\,sin\,(\vec{q}\cdot\vec{\tau})] \qquad (2.20)$$

From the above two equations (2.19) and (2.20), the local pseudopotential matrix element in equation (2.11) takes the form

$$\sum_{\vec{R}} \langle \vec{k}+\vec{G}' | \left(V_C(\vec{r}+\vec{R}+\vec{\tau}) + V_A(\vec{r}+\vec{R}-\vec{\tau})\right) | \vec{k}+\vec{G}\rangle$$

$$= \cos(\vec{q}\cdot\vec{\tau})\,[V_C(q) + V_A(q)] + i\sin(\vec{q}\cdot\vec{\tau})\,[V_C(q) - V_A(q)]$$

$$= V^s(q)\cos(\vec{q}\cdot\vec{\tau}) + i\,V^a(q)\sin(\vec{q}\cdot\vec{\tau}) \equiv V_\ell(q)$$

or

$$V_\ell(q) = V^s(q)\cos(\vec{q}\cdot\vec{\tau}) + i\,V^a(q)\sin(\vec{q}\cdot\vec{\tau}) \qquad (2.21)$$

where $V_\ell(q)$ denotes the local pseudopotential,

$V^s(q) = V_C(q) + V_A(q)$ is known as the *symmetric form factor,* and
$V^a(q) = V_C(q) - V_A(q)$ is known as the *antisymmetric form factor.*
Substituting from equation (2.21) into equation (2.10), we obtain

$$\frac{\hbar^2}{2m} a_{n,\vec{k}}(\vec{G}') \left|\vec{k} + \vec{G}'\right|^2 + \sum_{\vec{G} \neq \vec{G}'} a_{n,\vec{k}}(\vec{G})\, V_\ell(q) = E_{n,\vec{k}}\, a_{n,\vec{k}}(\vec{G}') \tag{2.22}$$

The above equation can be simplified by adopting the atomic units (a.u.) in which $\hbar = m = 1$, then equation (2.22) can be rewritten as

$$\frac{1}{2} a_{n,\vec{k}}(\vec{G}') \left|\vec{k} + \vec{G}'\right|^2 + \sum_{\vec{G} \neq \vec{G}'} a_{n,\vec{k}}(\vec{G})\, V_\ell(q) = E_{n,\vec{k}}\, a_{n,\vec{k}}(\vec{G}') \tag{2.23}$$

The above equation is a matrix equation can be solved numerically to obtain the required eigenvalues and eigenvectors. Before we solve equation (2.23) we have to know the $\vec{G}$ values and the pseudopotential form factors which depend on the structure of the semiconductor under investigation and also we have to determine the $\vec{k}$ values. In the next sections we present the procedures to evaluate these values.

2.2. Determination of the Allowed $\vec{G}$ Values

Since we are concerned with two types of structures, the zinc-blende (or what is called sphalerite) structure such as the III-V semiconductors (e.g. GaAs, AlAs, etc.) and the II-VI semiconductors (e.g. CdTe, ZnSe, etc.) and the diamond (form of carbon) structure such as Si, Ge, SiGe alloys, and α-Sn. These structures are cubic with a face-centered cubic Bravais (fcc) in real space with a basis of two atoms (cation-anion) at each Bravais lattice point and each atom is located from the center lattice point at $\frac{a}{8}(1,1,1)$, where a is the lattice constant [9].

To determine the allowed $\vec{G}$ values needed for calculations we calculate first the corresponding reciprocal lattice vector. It is known that the reciprocal lattice vectors of a face-centered cube (fcc) in real space are the body-centered cube (bcc) lattice. Therefore, the corresponding lattice vectors to the bcc lattice are of the form given by equation (2.4)

$$\vec{G} = g_1 \vec{b}_1 + g_2 \vec{b}_2 + g_3 \vec{b}_3 \tag{2.4}$$

where $\vec{b}_j$, j = 1, 2, and 3 are the primitive reciprocal Bravais lattice vectors which have the form

$$\vec{b}_1 = \frac{2\pi}{a}(-1,1,1), \vec{b}_2 = \frac{2\pi}{a}(1,-1,1), and\ \vec{b}_3 = \frac{2\pi}{a}(1,1,-1) \tag{2.24}$$

The above vectors can be derived from the relations

$$\vec{b}_1 = 2\pi \frac{\vec{a}_2\, x\, \vec{a}_3}{\vec{a}_1 \cdot (\vec{a}_2\, x\, \vec{a}_3)}, \vec{b}_2 = 2\pi \frac{\vec{a}_3\, x\, \vec{a}_1}{\vec{a}_1 \cdot (\vec{a}_2\, x\, \vec{a}_3)}, \text{ and } \vec{b}_3 = 2\pi \frac{\vec{a}_1\, x\, \vec{a}_2}{\vec{a}_1 \cdot (\vec{a}_2\, x\, \vec{a}_3)}. \tag{2.25}$$

where $\vec{a}_i$ (i = 1, 2, and 3) are the primitive translational vectors of the Bravais lattice.

It is found after inspection of equation (2.21) that, the magnitude of the pseudopotentials approach zero for absolute values of the reciprocal lattice vectors of difference $|\vec{q}| \leq \sqrt{11}$, so we restrict our choice of the $\vec{G}$ values for the set of $|\vec{q}| = 0, 3, 4, 8, and\ 11$ which give non-zero pseudopotential and we found that a number of 65 for the $\vec{G}$ values is enough to get a reasonable results.

These $\vec{G}$ values can be produced by using equations (2.4) and (2.24) as follows:

$|\vec{\boldsymbol{q}}| = \mathbf{0}$ Group:

For $g_1 = g_2 = g_3 = 0$, then $\vec{G} = \frac{2\pi}{a}(0,0,0)$.

$|\vec{\boldsymbol{q}}| = \sqrt{\mathbf{3}}$ Group:

This group can be produced from the commutations of $g_1 = g_2 = g_3 = \pm 1$ as

$$\vec{G} \equiv \frac{2\pi}{a}(1,1,1), \frac{2\pi}{a}(1,1,-1), \frac{2\pi}{a}(1,-1,1), \frac{2\pi}{a}(-1,1,1),$$
$$\frac{2\pi}{a}(1,-1,-1), \frac{2\pi}{a}(-1,1,-1), \frac{2\pi}{a}(-1,-1,1), \frac{2\pi}{a}(-1,-1,-1).$$

Similarly, we can proceed for the other groups and tabulate the 65 plane waves which include the 6th nearest neighbor reciprocal lattice vectors from the origin of the BZ as shown in table (2.1) [10,11].

Table 2.1. The 65 permutation values for the reciprocal lattice $\vec{G}$, vectors used in the calculations in units of $(2\pi/a)$ which include the 6th nearest neighbor reciprocal lattice vectors

$\|\vec{q}\|^2$	Permutations of $\vec{G}$ $(2\pi/a)$
0	(0, 0, 0)
3	(1,1,1), (1,1,-1), (1, -1,1), (-1,1,1), (-1, -1,1), (-1,1, -1), (1, -1, -1), (-1, -1, -1).
4	(2,0,0), (0, 2,0), (0,0, 2), (-2,0,0), (0, -2,0), (0,0,- 2).
8	(2,2,0), (2,0,2), (0,2,2), (-2,-2,0), (-2,0,-2), (0,-2,-2), (-2,2,0), (-2,0,2), (0,-2,2), (0, 2,-2), (2,-2,0), (2,0,-2).
11	(3,1,1), (1,3,1), (1,1,3), (-3,1,1), (1,-3,1), (1,1,-3), (3,-1,1), (3,1,-1), (1, 3,-1), (1,-1, 3), (-1, 1, 3), (-1,3,1), (3,-1,-1), (-1,3,-1), (-1,-1,3), (-3,-1,1), (-3,1,-1), (1,-3, -1), (1,-1,-3), (-1,-3,1), (-1,1,-3), (-3,-1,-1), (-1,-3,-1), (-1,-1,-3).
12	(2,2,2), (-2,2,2), (2,-2,2), (2,2,-2), (-2,-2,2), (-2,2,-2), (2,-2,-2), (-2,-2,-2).
16	(4,0,0), (0,4,0), (0,0,4), (-4,0,0), (0,-4,0), (0,0,-4,).

2.3. Determination of the Components of the Wave Vector $\vec{k}$

The components of the wave vector $\vec{k} \equiv (k_x, k_y, k_z)$ can be determined by considering a specimen of length $50a$ where a is the lattice constant, so the number of sampling points is twice the value of 50 by considering the whole interval of the BZ [-1, 1]. So, one obtains the incrimination $\Delta\vec{k}$ of $\vec{k}$ in units of $(2\pi/a)$. Thus, by knowing the $\vec{k}$ and $\vec{G}$ values, one gets the required energy eigenvalues for each $\vec{k}$ by solving the equation (2.23) numerically. For solving the equation numerically, we can use a routine based on MATLAB language [5,6], which is 65 x 65 matrix based on 65 bulk reciprocal lattice vector $\vec{G}$'s. Arrange the resulting eigenvalues for each $\vec{k}$ descendingly to obtain the valence and conduction bands. The energy gap can be calculated by setting the top of the valence bands to zero energy. The symmetric and antisymmetric pseudopotential form factors mentioned above in equation (2.21) are adjusted to yield the band structures of the investigated semiconductor by matching the calculated energy gap with the experiment data at some critical points of the first Brillion zone.

2.4. Choosing the Pseudopotential Form Factors

The choice of the pseudopotential form factors associated with each choice of the absolute value q of $\vec{q}$ depends on the type of structure of the semiconductor under investigation as follows

2.4.1. The Diamond Structure

Since the diamond structure is a special case of the type of zinc-blende structure with a basis of two atoms in the Bravais unit cell but the two atoms are the same such as Si and Ge, therefore the anti-symmetric form factor must vanish and the symmetric form factor remains, thus

$$\mathrm{V^a(q)} = 0 \; and \; V_\ell(q) = \mathrm{V^s(q)} \cos\,(\vec{q}\cdot\vec{\tau}). \tag{2.26}$$

Then equation (2.23) becomes

$$\tfrac{1}{2}\, a_{n,\vec{k}}(\vec{G}')\, \left|\vec{k} + \vec{G}'\right|^2 + \textstyle\sum_{\vec{G}\neq\vec{G}'} a_{n,\vec{k}}(\vec{G})\;\; \mathrm{V^s(q)} \cos\,(\vec{q}\cdot\vec{\tau}) = E_{n,\vec{k}}\; a_{n,\vec{k}}(\vec{G}') \tag{2.27}$$

The location of the cation-anion from the Bravais lattice site which is considered to be at the center of the line joining the two atoms and each atom is positioned at $\pm\tau \equiv \pm\frac{a}{8}\,(1,1,1)$. It is clear that symmetric form factor associated with $|\vec{q}|^2 = 4$ is vanished because the cosine term is zero as follows:

$$\vec{q}\cdot\vec{\tau} = \left[\left(\tfrac{2\pi}{a}\right)(\pm 2{,}0{,}0)\right]\cdot\left[\pm\tfrac{a}{8}\,(1,1,1)\right] = \left(\tfrac{\pi}{4}\right)(2) = \tfrac{\pi}{2}, \text{ thus}$$

$$\cos(\vec{q}\cdot\vec{\tau}) = \cos\left(\frac{\pi}{2}\right) = 0.$$

The other terms of q exist which satisfy the condition $|\vec{q}|^2 \leq 11$ for the first three non-zero pseudopotentials.

These terms are equivalent to $|\vec{q}|^2 = 3, 8, and\ 11.$ i.e. the pseudopotential form factors $V^s(\sqrt{3}), V^s(\sqrt{8}), V^s(\sqrt{11})$ must be chosen and adjusted to match the calculated energy gaps with the measured values.

2.4.2. The Zinc-Blende Structure

This type of structure is the same as we mentioned before for diamond but with two different atom bases such as GaAs and AlAs. So, for the calculation of the band structure all symmetry and anti-symmetry form factors remain and equation (2.23) will be solved to obtain the required pseudopotential form factors and their associated band structure for the investigated semiconductor.

The three terms of the anti-symmetric form factors associated with this type of structure are relevant to $|\vec{q}|^2 = 3, 4, and\ 11.$

The factor with $|\vec{q}|^2 = 8$ is excluded because the sine function term is zero since

$$\vec{q}\cdot\vec{\tau} = \left[\left(\frac{2\pi}{a}\right)(\pm 2, \pm 2, 0)\right]\cdot\left[\pm\frac{a}{8}(1,1,1)\right] = \left(\frac{\pi}{4}\right)(4) = \pi.$$

Thus $\sin(\vec{q}\cdot\vec{\tau}) = \sin(\pi) = 0$; or it may be $sin(0) = 0.$

Therefore, the symmetric pseudopotential form factors $V^s(\sqrt{3}), V^s(\sqrt{8})$, and $V^s(\sqrt{11})$ and the anti-symmetric the pseudopotential form factors $V^a(\sqrt{3}), V^a(\sqrt{4})$, and $V^a(\sqrt{11})$ must be chosen and adjusted to match the calculated energy gaps with the measured values.

2.4.3. Numerical Analysis

The energy band structure can be obtained by solving the N x N matrix equation (2.23) which can be rearranged as

$$\left[\frac{1}{2}\left|\vec{k} + \vec{G}'\right|^2 - E_{n,\vec{k}} + \sum_{\vec{G}\neq\vec{G}'}^{N} V_\ell(q)\right] a_{n,\vec{k}}(\vec{G}') = 0.$$

The above equation has only a nontrivial solution if the determinant of the coefficients equals zero. Thus,

$$\left\|\frac{1}{2}\left|\vec{k} + \vec{G}'\right|^2 - E_{n,\vec{k}} + \sum_{\vec{G}\neq\vec{G}'}^{N} V_\ell(q)\right\| = 0. \quad (2.28)$$

The above equation is called Secular equation which can be solved numerically obtain the required energy eigenvalues$E_{n,\vec{k}}$, $a_{n,\vec{k}}(\vec{G}')$ are the eigenvectors, and N is the number of plane waves used in the expansion of the eigenvectors which is taken to be 65.

The first two terms in equation (2.28) and equation (2.23) constitute the diagonal elements of the matrix and the second term which contains the local pseudopotential form factors gives the off-diagonal terms. These off-diagonals are simplified by considering only the elements for which $q \equiv \sqrt{3}, \sqrt{8}, and \sqrt{11}$ for the symmetric form factors that can be used in determining the band structure for the diamond structure type (Si and Ge) and $q \equiv \sqrt{3}, \sqrt{4}, and \sqrt{11}$ for the antisymmetric form factors that can be used with the symmetric ones in determining the band structure for zinc-blende structure type (GaAs and AlAs).

3. TEMPERATURE EFFECT ON ELECTRONIC BAND STRUCTURE OF BULK SEMICONDUCTORS

Although temperature is a basic physical parameter, its effect on the behavior and manufacturing of the electronic devices based on semiconductor materials still a crucial factor. Special interest concerns the advent of nanodevices based on compound semiconductors and it is well known that such devices are highly sensitive to the temperature fluctuations [12], where scattering of optical phonons play a central role in the determination of their properties. So understanding the behavior of these materials under an external effect such as temperature is becoming important.

This section is aiming to show the effect of temperature on the form factors and consequently the electronic band structures of bulk semiconductors for two types, diamond structure such as Si and Ge and zinc-blende structure such as GaAs and AlAs. We consider linear variations of the form factors as functions of temperature. Also, the variation of the electronic energy band, $E_{n,\vec{k}}(T)$ as a function of the wave vector $\vec{k}$ of the first Brillouin zone is taking into our consideration. Moreover, the temperature dependence of the lattice constant as a material parameter is taken into account. These calculations involved the local EPM that ignores the non-local and spin-orbit corrections.

Employing temperature effect, Secular equation (2.28) can be rewritten in the form

$$\left\| \frac{1}{2} \left| \vec{k} + \vec{G}' \right|^2 - E_{n,\vec{k}}(T) + \sum_{\vec{G} \neq \vec{G}'}^{N} V_\ell(q,T) \right\| = 0, \tag{3.1}$$

where

$$V_\ell(q,T) = \mathrm{V^s}(q,\mathrm{T}) \cos(\vec{q} \cdot \vec{\tau}) + i\, \mathrm{V^a}(q,\mathrm{T}) \sin(\vec{q} \cdot \vec{\tau}) \tag{3.2}$$

is the temperature dependent empirical local pseudopotential and $\mathrm{V^{s,a}}(q,\mathrm{T})$ are the symmetric, $\mathrm{V^s}(q,\mathrm{T})$ and anti-symmetric, $\mathrm{V^a}(q,\mathrm{T})$ temperature dependent form factors that are fitted empirically to obtain the required energy gap for the associated semiconductor and take the form

$$\mathrm{V^{s,a}}(q,\mathrm{T}) = \mathrm{V^{s,a}}(q,\mathrm{T_0}) - \delta^{s,a}\, T \tag{3.3}$$

Since $\mathrm{T_0} = 0$ K is the zero temperature, $\delta^{s,a}$ are the temperature coefficient for the symmetric and antisymmetric form factors. The superscripts $\boldsymbol{s}$ and $\boldsymbol{a}$ refer to symmetric and antisymmetric; respectively. $\vec{k}$ is the propagation wave vector, $\vec{q} = \vec{G}' - \vec{G}$, and $\vec{\tau} =$

$\pm \frac{a(T)}{8}(1,1,1)$ is the position vector of each atom from the center of the lattice point in the unit cell and $a(T)$ is the temperature dependent lattice constant, which is determined from the relation [13]

$$a(T) = a(T_R)\,[1 + \alpha_{th}(T - T_R)] \tag{3.4}$$

where T_R is the room temperature which is taken to be 300 K and α_{th} is the linear thermal expansion coefficient and its value corresponds to the associated semiconductor is listed in table (3.1).

The temperature dependent energy band structure, $E_{n,\vec{k}}(T)$ are obtained by solving numerically the Secular determinant given in equation (3.1). To solve equation (3.1) numerically, we used a routine which is based on MATLAB language. This routine is 65 x 65 matrix based on N = 65 bulk reciprocal lattice vectors, $\vec{G}$'s as shown in table (2.1).

3.1. Determination of the Temperature Dependent Energy Eigenvalues

Knowing the propagation wave $\vec{k}$ values, as we explained previously in section (2.3), one gets the required eigenvalue for each $\vec{k}$ by solving numerically the Secular determinant, equation (3.1). Arrange the obtained eigenvalues and label them. In the diatomic unit cell, there are two atoms each atom will share by four valence electrons and then there are eight electrons including spin degeneracy. The lowest four will be occupied first and are associated to the valence bands. Next states that are higher in energy are empty and associated to the conduction bands.

Set the top of the valence bands to zero energy by subtracting the value corresponding to the top valence band from all the eight sorted values and determine the energy gaps along the symmetry axis from the difference between the top of the valence band and the lowest bottom of the conduction bands. Matching the calculated energy gap with the corresponding experimental value by adjusting the values of the form factors until the best fitting with the experimental energy gaps is obtained.

Table 3.1. Values of the Varshni's parameters and the linear expansion coefficients for Si, Ge, GaAs, and AlAs

Varshni's Parameters[a]	Si	Ge	GaAs	AlAs
$\beta\ (10^{-4}\ K^{-1})$	4.730	4.800	5.405	7.000
γ (K)	636	235	204	530
$\alpha_{th}\ (10^{-6}\ K^{-1})$[b]	2.616	5.750	6.030	4.280

[a]Ref. [13], [b]Ref. [14].

The experimental temperature energy gaps are obtained from Varshni's [13] empirical formula

$$E_g(T) = E_g(T_0) - \beta\, T^2/(T + \gamma) \qquad (3.5)$$

The values of β and γ are listed in table (3.1) for the required semiconductor materials of diamond type (Si and Ge) and of zinc-blende type (GaAs and AlAs).

Since an assessment of the magnitude of an effect is addressed, we manipulate calculations of the energy band structures based on the local EPM for two semiconductors of two different structure types, diamond (Si and Ge) and zinc-blende (GaAs and AlAs).

The temperature coefficients for symmetric and antisymmetric form factors associated with Si, Ge, GaAs, and AlAs are listed in table (3.2).

Figure (3.1) displays the variation of the symmetric form factors for Si with temperature. It is seen from this figure that, all the form factors are linearly decreased with enhancing temperature and the magnitude of the temperature dependent symmetric form factor belonging to $q^2 = 11$ is positive and has the highest value while that belonging to $q^2 = 3$ is negative and has the lowest one. This is due to the fact that, raising temperature increases the dimension of the crystal, as observed from the variation of the lattice constant $a(T)$ in table (3.3) which yields decreasing in the potential energy seen by the electron. It is found that, the fundamental energy gap for Si is more sensitive to the temperature dependent symmetric form factor associated with the reciprocal lattice vectors of $q^2 = 11$.

Good agreement is seen from table (3.3) when comparing the calculated fundamental energy gap for Si, which is an indirect energy gap (at X-point) with the experimental values at different temperatures.

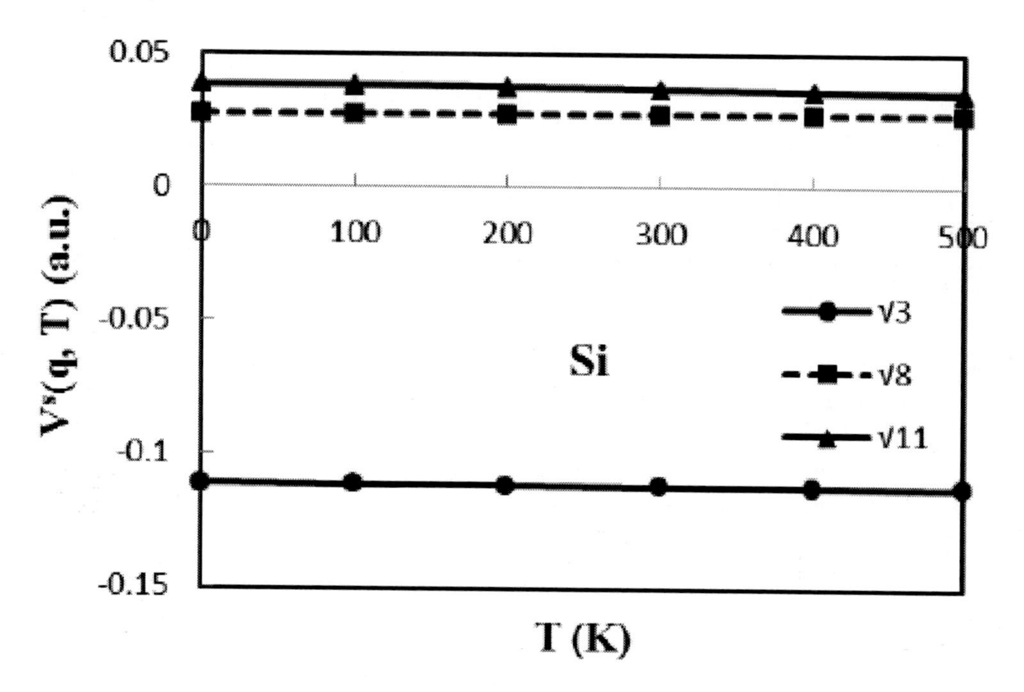

Figure 3.1. Variation of the symmetric form factors for Si with temperature.

Table 3.2. Temperature coefficients symmetric (3, 8, and 11) and antisymmetric (3, 4, and 11) form factors, $\delta^{s,a}$ associated with Si, Ge, GaAs, and AlAs as functions of the square of magnitude of the reciprocal lattice vector, $\vec{q}$

$\delta^{s,a}$ 10^{-6} (a.u. K^{-1})				
q^2	Si	Ge	GaAs	AlAs
3	3.0	3.0	0.3	0.3
8	0.3	3.0	0.3	0.3
11	6.0	3.0	140.7889	8.0
3	-	-	0.3	0.3
4	-	-	0.3	0.3
11	-	-	0.3	0.3

Table 3.3. Variations of the lattice constant and the fundamental indirect energy gap for Si with temperature

Si						
Parameters	T (K)					
	0	100	200	300	400	500
$a(T)$ (a.u.)	10.254512	10.257197	10.259881	10.262566	10.265251	10.267935
$E_g^X(T)$ (eV)	1.170000 1.170000[a] 1.170000[b]	1.156000 1.163573[a]	1.141900 1.147368[a]	1.127800 1.124519[a] 1.120000[b]	1.113500[a] 1.096949[a]	1.099100 1.065907[a]

[a]Ref. [13], [b]Ref. [15].

Figure (3.2) exhibits the temperature dependent electronic energy band structure, $E_{n,\vec{k}}(T)$ given in eV, for Si as a function of the wave vector $\vec{k}$ of the Brillouin zone at two different temperatures 0K and 500K. The solid lines represent the energy bands at T = 0 K, while the dashed lines represent the energy bands at T = 500 K. Figure (3.2) shows also that, the temperature dependent conduction energy bands are more affected by temperature than the valence energy bands. This behavior is due to the fact that increasing temperature leads to a decrease in the conduction bands, i.e. they move downward toward the top of the valence band. It is seen also from this figure, that the discrepancies between the two curves for the two different temperatures are more pronounced at the L-point which is about 165 meV and both the two symmetry Γ-(about 87meV) and X-(about 71 meV) points.

The direct and indirect energy band gaps for Si as functions of temperature are plotted in figure (3.3). We can see that the direct energy gap at the Γ-point, E_g^Γ the indirect energy band gaps at the L-point, E_g^L and at the X-point, E_g^X are decreased by increasing the temperature. So, Si stills an indirect band gap semiconductor with increasing temperature, but its fundamental energy gap decreases with increasing temperature.

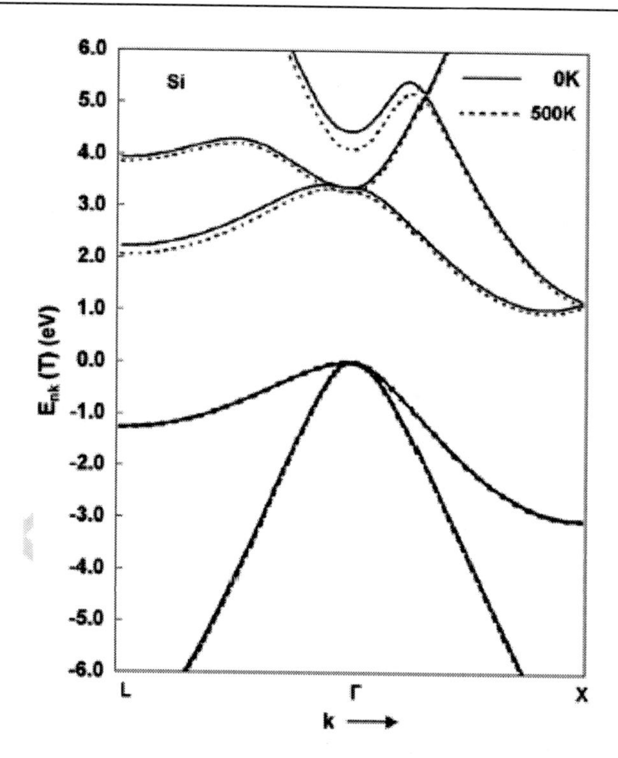

Figure 3.2. The energy band structure for Si at two different temperatures, 0 K (solid lines) and 500 K (dashed lines).

Figure (3.4) displays the variation of the symmetric form factors for Ge with temperature. It is seen from this figure that, all the form factors are linearly decreased with enhancing temperature, which is the same behavior as of Si. Also, the value of the temperature dependent symmetric form factor belonging to $q^2 = 11$ is positive and has the highest value while that belonging to $q^2 = 3$ is negative and has the lowest one. This is also due to the fact that, raising temperature increases the dimensions of the crystal, as observed from the variation of the lattice constant $a(T)$ presented in table (3.4) which yields decreasing in the potential energy seen by the electron. It is found that, the fundamental energy gap for Ge is more sensitive to the temperature dependent symmetric form factor associated with the reciprocal lattice vectors of $q^2 = 11$.

Table (3.4) illustrates the dependence of the lattice constant and fundamental energy gap for Ge on temperature. It is seen from table (3.4) the increasing of the lattice constant and the decreasing of the fundamental energy gap with increasing temperature.

It is also seen from table (3.4) a good agreement between the indirect fundamental energy gap for Ge (at L–point) for different temperatures and the experimental values which are determined from Varshni's formula [13]. Also, an excellent agreement is obtained at T = 0 K between the calculated energy gap and the reported one [13, 14]. Furthermore, one can see from table (3.4) that there are little discrepancies in the temperature dependent fundamental energy gap for the calculated values and those determined from Varshni's formula, since they are getting lower for temperatures less than room temperature and turn over for temperatures above room temperature. These deviations of the calculated energy gap values from the experimental ones are attributed to ignoring of the non-local and spin-orbit effects in the present local model.

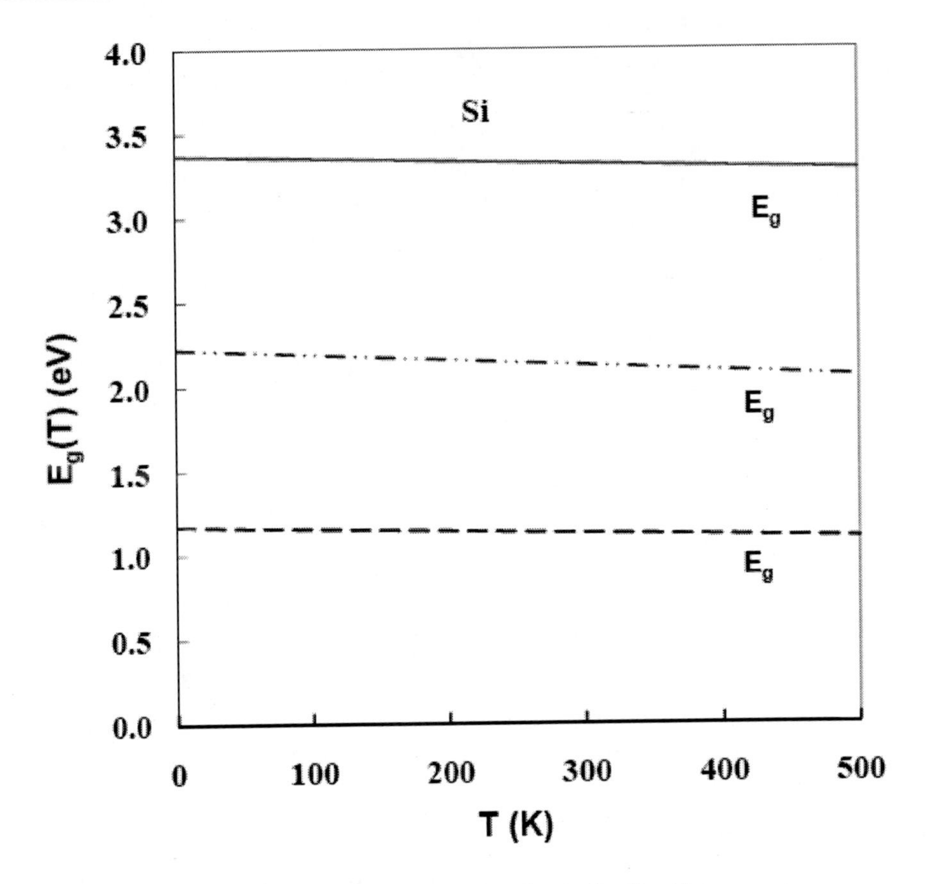

Figure 3.3. Variation of the direct and indirect energy band gaps for Si with temperature.

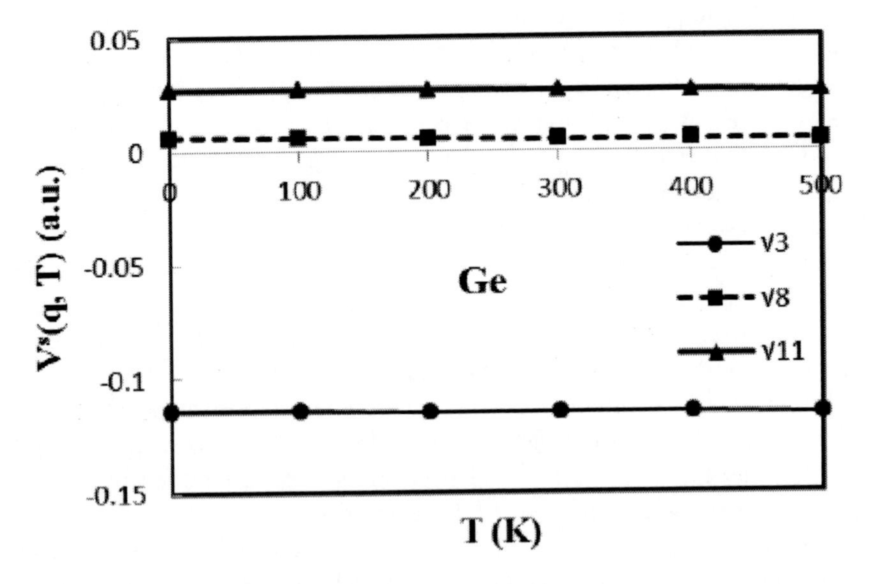

Figure 3.4. Variation of the symmetric form factors for Ge with temperature.

Table 3.4. Dependence of the lattice constant and the fundamental energy gap for Ge on temperature

Ge						
Parameters	T (K)					
	0	100	200	300	400	500
$a\ (T)$ (a.u.)	10.650775	10.656910	10.663045	10.669179	10.675314	10.681449
$E_g^L(T)$ (eV)	0.742000 0.742000[a] 0.741000[b]	0.714200 0.727672[a]	0.686300 0.697862[a]	0.658400 0.661252[a] 0.660000[b] 0.760000[c]	0.630400 0.621055[a]	0.602300 0.578735[a]

[a]Ref. [13], [b]Ref. [15], [c]Ref. [16].

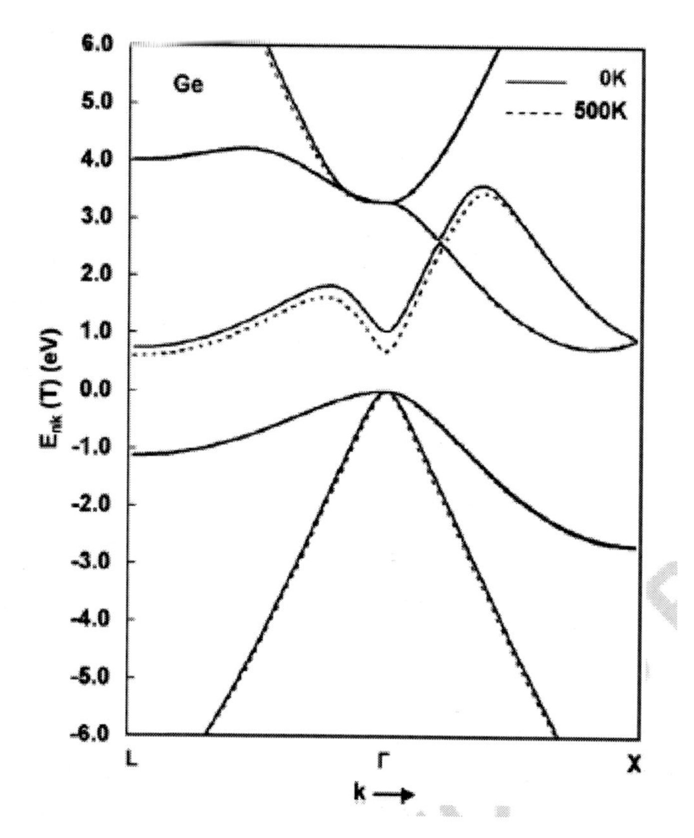

Figure 3.5. The energy band structure for Ge at two different temperatures, 0 K (solid lines) and 500 K (slid line).

Figure (3.5) presents the variation of the electronic energy band structure, $E_{n,\vec{k}}(T)$ given in eV, for bulk Ge at two different temperatures (T = 0 K and 500 K). One observes from figure (3.5) that, the temperature dependent conduction energy bands are more affected by temperature than the valence bands. It is also seen from the figure that the energy differences between the calculated electronic energies at 0 K and 500 K are about 356 meV at the X-point, 140 meV at the L-point and 6 meV at Γ–point. It is more pronounced at the symmetry X-point than any other $\vec{k}$ –point in the first Brillouin zone.

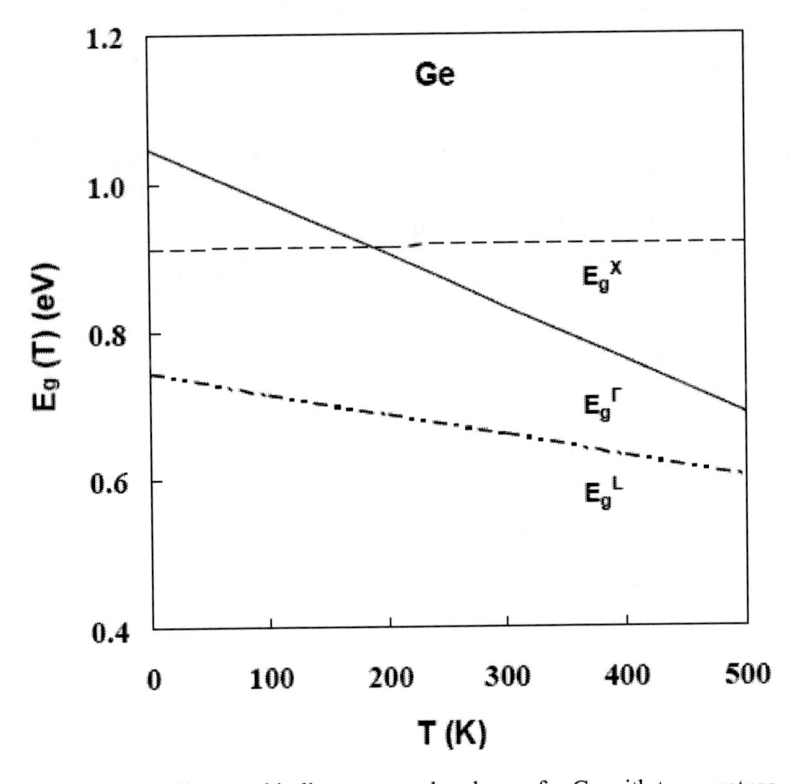

Figure 3.6. Variations of the direct and indirect energy band gaps for Ge with temperature.

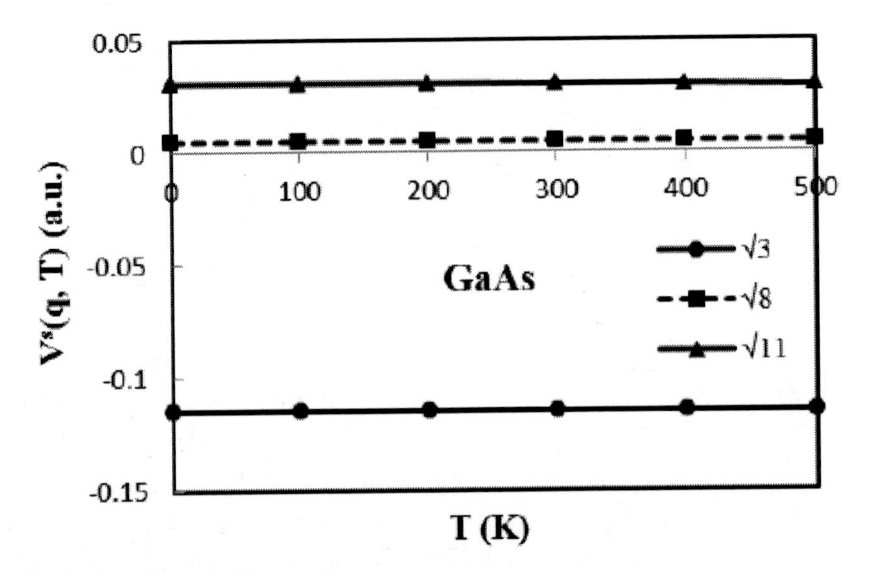

Figure 3.7. Variation of the symmetric form factors for GaAs with temperature.

The direct and indirect energy band gaps for Ge as functions of temperature are plotted in figure (3.6). We see that the direct energy gap at the Γ-point, E_g^Γ and the indirect energy gap at the L-point, E_g^L are decreased fast by increasing the temperature while the indirect energy gap at the X-point, E_g^X is slowly decreased. These changes in the energy band gaps for Ge lead to a crossover energy band gaps of 0.91 eV between E_g^Γ and E_g^X at a temperature of 186 K. But Ge remains an indirect band gap semiconductor at the L-point in spite of the crossing between the Γ- and the X- bands, this is because the L- band is lower in energy than both of them. It is seen also from both figure (3.6) and table (3.4) that, the fundamental energy gap for Ge decreases with increasing temperature.

Both figures (3.7) and (3.8) display the linearly decreasing of the local symmetric for q^2 = 3, 8, and 11 and the antisymmetric for q^2 = 3, 4, and 11 pseudopotential form factors by increasing the temperature of GaAs. From figure (3.7) it is seen that, the symmetric form factor belongs to q^2 = 11 is positive and has the highest value while that corresponding to q^2 = 3 is negative and has the lowest value in analogue to the behaviors of both Si and Ge. From figure (3.8) we can see an interesting result that the behavior of both q^2 = 11 and 3 are flipped over. It means that the antisymmetric form factor associated to q^2 = 3 is positive and has the highest value while that corresponding to q^2 = 11 stills positive but has the lowest value. This behavior is in fact due to the nature of the antisymmetric form factors in which the contributions to the local pseudopotential of both atoms constitute the unit cell oppose each other.

Figure (3.9) displays the variation of the temperature dependent electronic energy band structure for GaAs which is a zinc-blend type semiconductor as a function of the propagation wave vector, $\vec{k}$ and temperature (T = 0 K and 500 K). It is seen that the temperature dependent first conduction energy band is more affected by temperature than the other conduction and valence bands and exhibits more enhancement at the Γ-point of symmetry than any other k-point in the Brillouin zone. Conduction bands are also more affected by temperature than the valence bands. The temperature effect becomes less pronounced for higher conduction bands and nearly most valence bands.

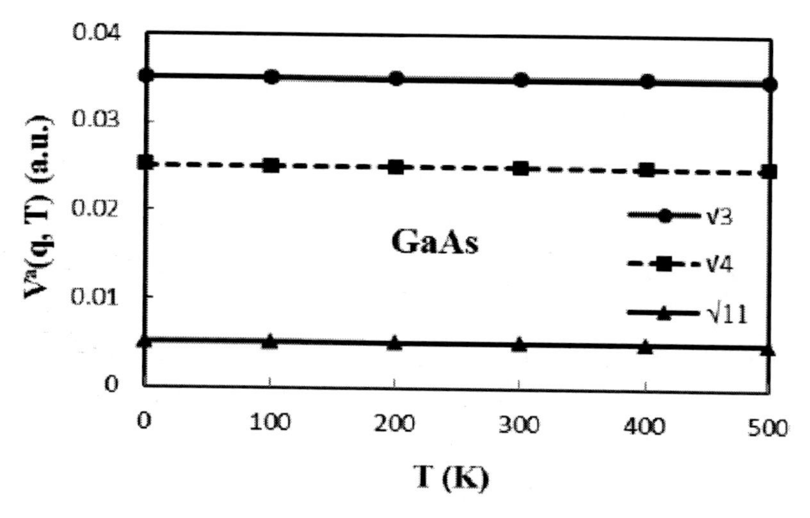

Figure 3.8. Variation of the antisymmetric form factors for GaAs with temperature.

Table 3.5. Dependence of the lattice constant and the fundamental energy gap for GaAs on temperature

GaAs						
Parameters	T (K)					
	0	100	200	300	400	500
$a\ (T)$ (a.u.)	10.663404	10.669845	10.67628 7	10.682729	10.689171	10.695612
$E_g^{\Gamma}(T)$ (eV)	1.519000 1.519000[a] 1.519000[b] 1.550000[c]	1.490000 1.501220[a]	1.460900 1.465485 a	1.431800 1.422482[a] 1.425000[d] 1.420000 [b]	1.402700 1.375821[a]	1.373500 1.327061[a]

[a]Ref. [13], [b]Ref. [15], [c]Ref. [16], [d] Ref. [17].

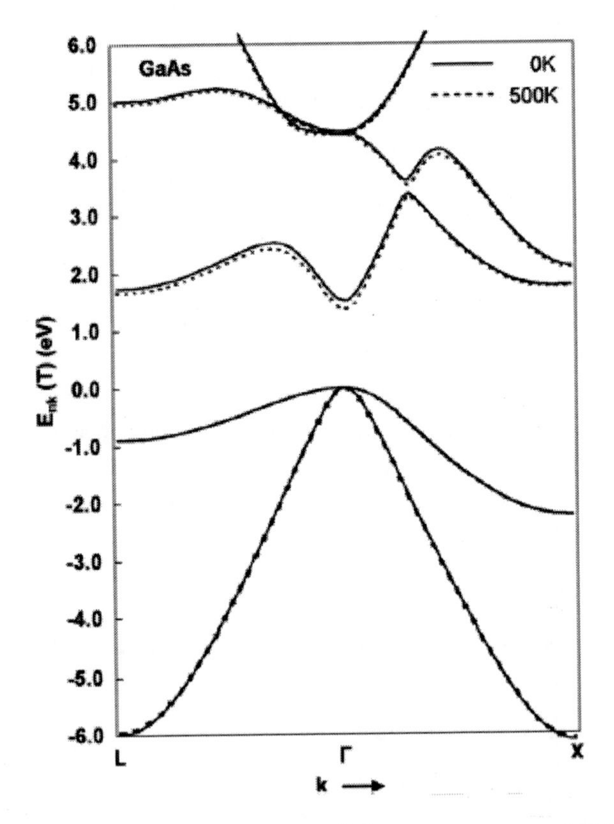

Figure 3.9. The energy band structure for GaAs at two different temperatures, 0 K (solid lines) and 500 K (slid line).

The energy differences between the calculated electronic energies at 0 K and 500K are about 146 meV at Γ-point, 79 meV at the L-point, and 26 meV at the X-point. Good agreement is obtained between the computed fundamental energy gaps for the GaAs which is a direct energy gap (at Γ-point) and those corresponding values that determined from Varshni's formula [13] at different temperatures, as seen from table (3.5). An excellent agreement is obtained when comparing the energy gap at 0 K with those of Ref. [13] and Ref. [15].

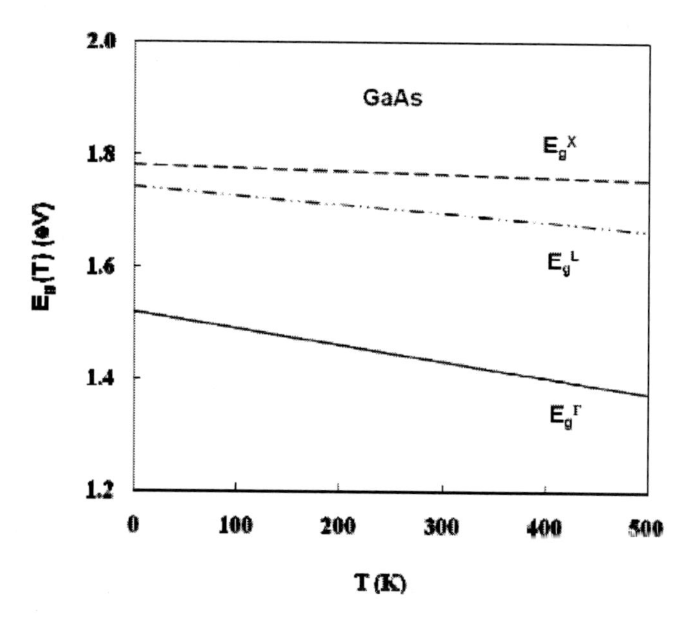

Figure 3.10. Dependence of the direct and indirect energy gaps for GaAs on the temperature.

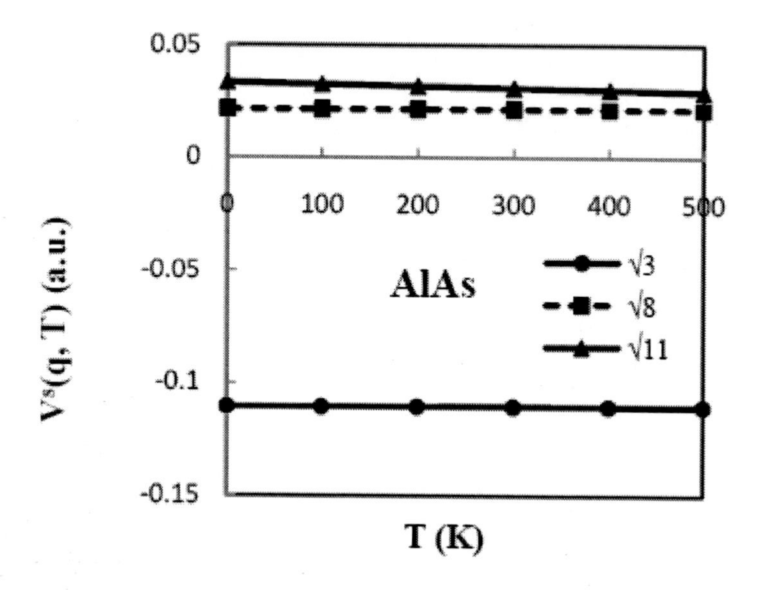

Figure 3.11. Variation of the symmetric form factors for AlAs with temperature.

The direct and indirect energy band gaps as functions of temperature for GaAs are plotted in figure (3.10). We see that the direct energy gap, E_g^Γ and both indirect energy band gaps E_g^L and E_g^X are decreased by increasing the temperature. Also, it is seen from both figure (3.10) and table (3.5) that the fundamental energy gap for GaAs decreases with increasing the temperature and GaAs remains a direct energy band semiconductor under the enhancement of temperature.

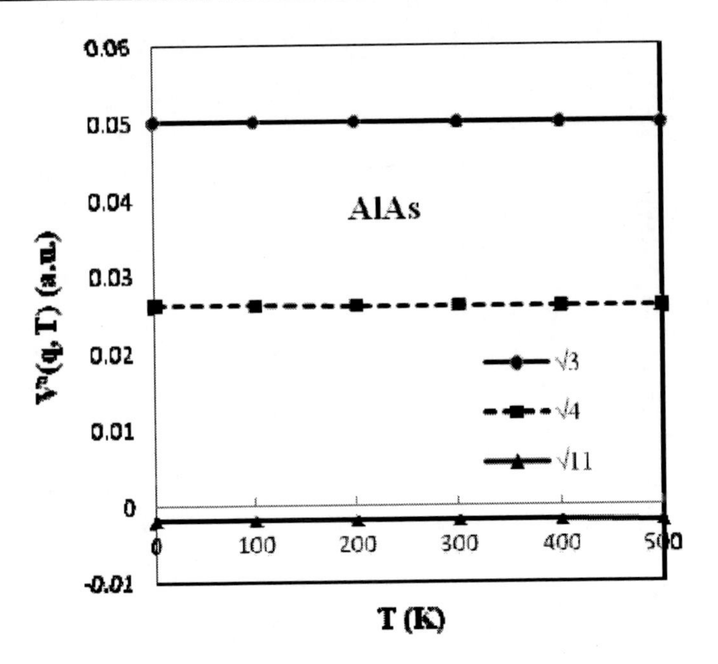

Figure 3.12. Variation of the antisymmetric form factors for AlAs with temperature.

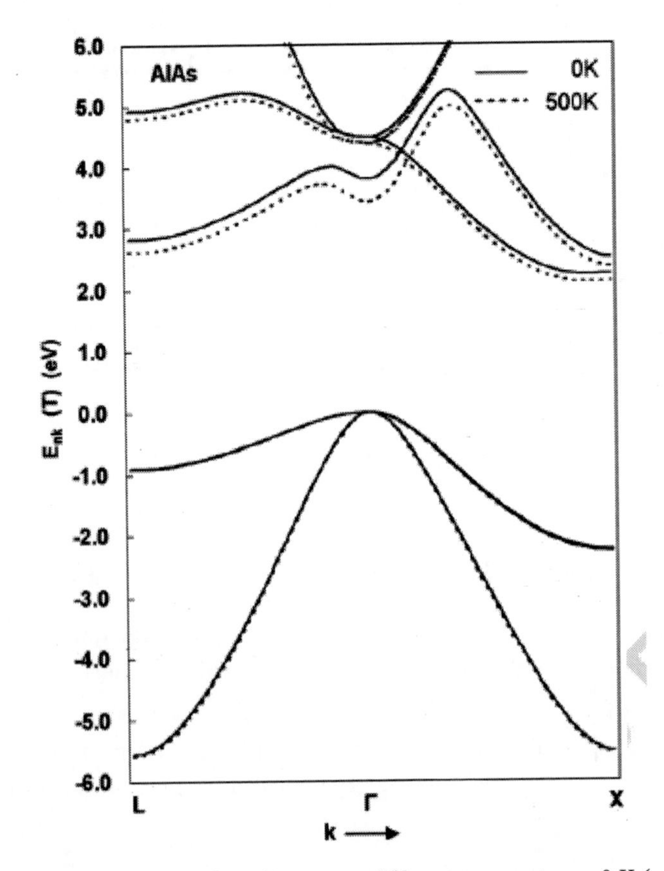

Figure 3.13. The energy band structure for AlAs at two different temperatures, 0 K (solid lines) and 500 K (slid line).

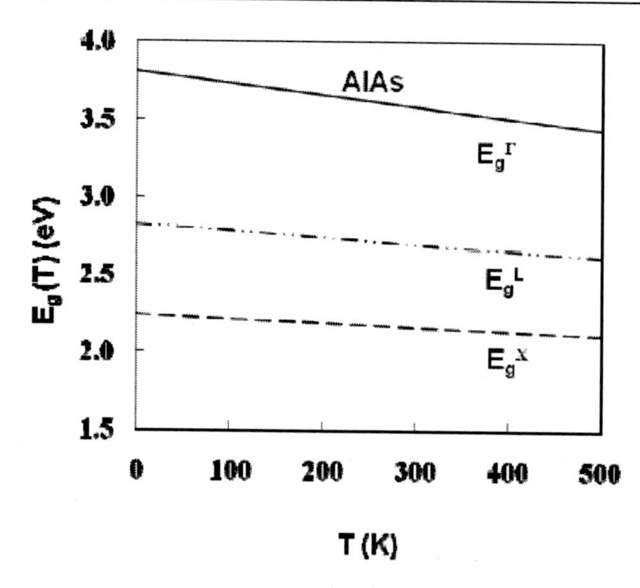

Figure 3.14. Dependence of the direct and indirect energy gaps for AlAs on temperature.

Both figures (3.11) and (3.12) display the linearly decreasing of the local symmetric for $q^2 = 3$, 8, and 11 and the antisymmetric for $q^2 = 3$, 4, and 11 pseudopotential form factors for AlAs which is a zinc-blende type by increasing the temperature. It is seen from both figures that all the symmetric and antisymmetric form factors decrease with raising temperature in a similar behavior like that of GaAs. From figure (3.11) it is seen that, the symmetric form factor belongs to $q^2 = 11$ is positive and has the highest value while that associated to $q^2 = 3$ is negative and has the lowest value in analogue to their behavior for GaAs. From figure (3.12) we can see also a very interesting result that the behaviors of both form factors belong to $q^2 =$ 11 and 3 are flipped over in both magnitude and sign. It means that the antisymmetric form factor associated to $q^2 = 3$ is positive and has the highest value while that corresponding to $q^2 = 11$ is negative and has the lowest value. This behavior is in fact due to the antisymmetric form factors in which the contributions to the local pseudopotential of both atoms constitute the unit cell oppose each other and the effect of $q^2 = 11$ is more pronounced due to their high number of permutations.

Figure (3.13) exhibits the variation of the temperature dependent electronic energy band structure for AlAs which is a zinc-blende type semiconductor as a function of both propagation vector $\vec{k}$ and temperature. It is seen that, the effect of temperature on the direct energy gap at the Γ-point of symmetry is more pronounced than any other k-point in the first Brillouin zone. It is also seen that, most conduction energy bands are more affected by temperature than the valence bands. It is also seen from figure (3.13) that, the energy differences between the calculated electronic energies at 0 K and 500 K are about 368 meV at the Γ-point, 206 meV at the X-point, and 123 meV at the L-point.

The direct and indirect energy band gaps for AlAs as functions of temperature are plotted in figure (3.14). It is seen from figure (3.14) that, the direct energy gap, E_g^{Γ} and both indirect energy band gaps, E_g^{L} and E_g^{X} are decreased by increasing the temperature. Also, one can see that AlAs remains an indirect energy band gap semiconductor although its fundamental energy gap decreases with raising temperature as seen from both figure (3.14) and table (3.6).

Table 3.6. Dependence of the lattice constant and the fundamental energy gap for AlAs on temperature

AlAs						
Parameters	T (K)					
	0	100	200	300	400	500
a (T) (a.u.)	10.682599	10.687178	10.691756	10.696334	10.700912	10.705490
E_g^X (T) (eV)	2.240000 2.240000[a] 2.240000[b] 2.260000[c]	2.215500 2.228889[a]	2.191000 2.201644[a]	2.166500 2.164096[a] 2.161000[d] 2.150000[b]	2.142000 2.119569[a]	2.117500 2.070097[a]

[a]Ref. [13], [b]Ref. [15], [c]Ref.[16], [d]Ref. [17].

In table (3.6) the computed indirect fundamental energy gap for AlAs at the X–point is compared with those associated with the experimental values that determined from Varshni's formula [13] and those of Lee et al. [17] at 300 K. An excellent agreement is obtained at T = 0 K. The little deviations between the computed and the experiment at higher temperatures are attributed to ignoring both effects arising from the non-local and the spin-orbit corrections.

Furthermore, it is found that AlAs semiconductor is the most affected by temperature than any other presented semiconductors.

4. Pressure Effect on Electronic Structure of Bulk Semiconductors

Pressure is an important factor in studying electronic devices based on semiconductors in both fabrication and operation. It affects the performance of these semiconductors which changes the electronic structure and leads to new manmade structures for specific purposes. The recent modern technology and the development in the diamond anvil cell [18, 19] made it possible to study the electronic and optical properties of semiconductors under the required hydrostatic pressure. Several authors have reported the effect of pressure on the energy gaps for some semiconductors, theoretically and experimentally [20, 21, 14, 15, and 17].

The aim of this section is to illustrate the effect of pressure on the form factors, the lattice constants, and mainly its effect on the electronic energy band structures for two types of structures, the diamond such as Si and Ge and the zinc-blende such as GaAs and AlAs semiconductors. In these calculations we considered a linear variation of the pseudopotential form factors as function of pressure to calculate the variation of the electronic energy bands, $E_{n,\vec{k}}(P)$ as function of the propagation wave vector $\vec{k}$ of the first Brillouin zone (BZ) from normal (P = 0 kbar) up to 120 kbar pressure. Comparisons are made between the obtained results for the fundamental energy band gaps and the experimental data which show a good agreement.

4.1. Determination of the Pressure Dependent Energy Eigenvalues

Employing the empirical local pseudopotential method (EPM), the pressure dependent eigenvalues and the eigenvectors are found by solving the matrix equation (2.23) which can be rewritten in the form

$$\frac{1}{2}\, a_{n,\vec{k}}(\vec{G}',P)\,|\vec{k}+\vec{G}'|^2 + \sum_{\vec{G}\neq\vec{G}'} a_{n,\vec{k}}(\vec{G},P)\; V_\ell(q,P) = E_{n,\vec{k}}\, a_{n,\vec{k}}(\vec{G}',P) \qquad (4.1)$$

where

$$V_\ell(q,P) = V^s(q,P)\cos(\vec{q}\cdot\vec{\tau}) + V^a(q,P)\,\sin(\vec{q}\cdot\vec{\tau}) \qquad (4.2)$$

is the pressure dependent empirical local pseudopotential and $V^{s,a}(q,P)$ are the symmetric, $V^s(q,P)$ and antisymmetric, $V^a(q,P)$ pressure dependent form factors that are fitted empirically to obtain the required energy gap for the associated semiconductor and take the form

$$V^{s,a}(q,P) = V^{s,a}(q,\,P_0) + \mu^{s,a}\,P \qquad (4.3)$$

where $P_0 = 0\ kbar$ is the normal pressure and $\mu^{s,a}$ are the pressure coefficient form factors as listed in table (4.2) for the Si, Ge, GaAs, and AlAs. In equation (4.1), $\vec{k}$ is the propagation wave vector, $\vec{G}$ and $\vec{G}'$ are the reciprocal lattice vectors, with $\vec{q} = \vec{G}' - \vec{G}$ and $\tau = \frac{a(P)}{8}\,(1,1,1)$ is the position vector of each atom in the unit cell and $a(P)$ is the pressure dependent lattice constant, which is determined from the relation given by Adachi [14],

$$a(P) = a(P_0)\,[1 + P\ b'/b]^{-1/3b'} \qquad (4.4)$$

where b and b' are the bulk modulus and its pressure derivative, respectively. Their values are listed in table (4.1), $a(P_0)$ and $a(P)$ are the lattice parameters at normal pressure and a pressure P, respectively.

The pressure dependent energy band structure are obtained by solving the Secular determinant, equation (2.26) which can be rewritten as

$$\left\|\frac{1}{2}|\vec{k}+\vec{G}'|^2 - E_{n,\vec{k}}(P) + \sum_{\vec{G}\neq\vec{G}'}^{N} V_\ell(q,P)\right\| = 0 \qquad (4.5)$$

As mentioned previously, the first two terms in the above equation (4.5) constitute the diagonal elements of the matrix, while the third term constitutes the off-diagonal elements.

For the diamond type structure, such as Si and Ge, the antisymmetric pseudopotential form factors vanish for the reason mentioned before since the two atoms forming the unit cell are identical. But for semiconductors that have the zinc-blende type structure, such as GaAs and AlAs, the symmetric and anti-symmetric pseudopotential form factors must be considered, since the two atoms that constitute the Bravais unit cell are different.

Equation (4.5) can be solved numerically to give the energy band structure as a function of pressure (P) at different $\vec{k}$ values.

4.2. Determination of the Pressure Dependent Energy Eigenvalues

to solve equation (4.5) numerically, we use a routine based on MatlaB language with N = 65 matrix elements of the bulk reciprocal lattice vectors, $\vec{G}$'s. These values are corresponding to $q^2 = 3$, 8 and 11 for symmetric form factors (e.g. diamond type structure) and equal to 3, 4, and 11 for antisymmetric form factors (e.g. zinc-blende type structure) which satisfy the condition, $q^2 \leq 11$ that gives non-zero pseudopotential and they are taken to the 6th nearest neighbors to the center of the first BZ. The corresponding $\vec{G}'s$ values are taken in units of $2\pi/_a$ and are listed in table (2.1).

We proceed in choosing the $\vec{k}$ (k_x, k_y, k_z) values as we did in the previous section (3.1) by considering the same length of **50*a*** and taking a number of sampling points of 100 for the whole interval of the BZ [-1, 1]. Then the incrimination $\Delta\vec{k}$ of $\vec{k}$ in units of $2\pi/a(P)$ is calculate and consequently the required eigenvalues for each $\vec{k}$ are computed and then arranged descendingly to obtain the valence and conduction bands.

The top of the valence bands is shifted to zero energy by subtracting its value from all obtained energy values and then one can determine the required energy gap. The best fitting energy gap is obtained by adjusting the pressure dependent symmetric and antisymmetric form factors, $V^{s,a}(q, P)$ until the determined energy gap match the corresponding experimental value.

The experimental pressure dependent energy gaps are obtained from the empirical relation [14],

$$E_g^{d,id}(P) = E_g^{d,id}(P_0) + \rho\, P + \sigma\, P^2 \qquad (4.6)$$

where ρ and σ are the hydrostatic pressure coefficients listed in table (4.1), d and id stand for direct and indirect energy bands, respectively.

The calculations of the energy band structures based on the local EPM for Si, Ge, GaAs, and AlAs semiconductors as functions of hydrostatic pressure are performed by a routine based on the MATLAB language.

The pressure coefficients symmetric and antisymmetric form factors used in the computations are listed in table (4.2).

Figure (4.1) displays the variation of the local symmetric pseudopotential form factors for Si as a diamond structure type with pressure ranging from zero up to 120 kbar. It is seen from figure (4.1) that, the symmetrical form factor at $q^2 = 11$ is more sensitive to the pressure than the other values. It has also the highest value than the other q-values. Table (4.3) shows the pressure dependence of the lattice constant and the fundamental energy band gap for Si on pressure. It is seen from table (4.3) that the lattice constant for Si decreases linearly with increasing pressure, which is a well known physical phenomenon.

Table 4.1. Values of the bulk modulus, *b* its pressure derivative, *b'* and the hydrostatic pressure parameters ρ and σ for Si, Ge, GaAs, and AlAs semiconductors [14]

Pressure Parameters	Si	Ge	AlAs	GaAs
ρ (10^{-2} eV.Gpa^{-1})	-1.43	4.80	11.50	-1.53
σ (10^{-4} eV.Gpa^{-2})	0.00	0.00	-24.50	0.00
b (Gpa)	97.84	74.70	75.50	77.90
b'	4.24	4.55	4.49	4.40

Table 4.2. The hydrostatic pressure coefficient symmetric (3, 8, and 11) and antisymmetric (3, 4, and 11) form factors, μ^s for Si, Ge, GaAs, and AlAs as functions of the square of magnitude of the reciprocal lattice vector, $\vec{q}$

μ^s 10^{-6} (a.u. / kbar)				
q^2	Si	Ge	GaAs	AlAs
3	3.3333	6.6667	0.3333	0.3333
8	0.3333	8.8889	6.6667	0.3333
11	47.4000	97.3880	140.7889	66.4111
3	-	-	0.0333	0.3333
4	-	-	34.4444	0.3333
11	-	-	83.5556	0.3333

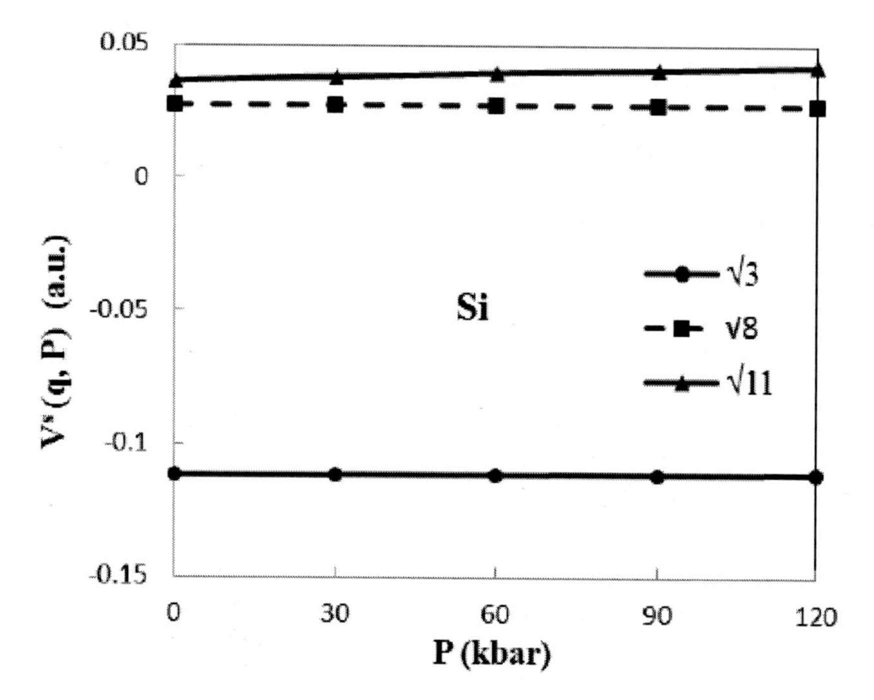

Figure 4.1. Variation of the symmetric form factors for Si with hydrostatic pressure.

Table 4.3. Dependence of the lattice constant and the fundamental energy gap for Si on pressure

Si					
Parameters	P (kbar)				
	0	30	60	90	120
$a(P)$ (a.u.)	10.262566	10.164430	10.077780	10.000280	9.930231
$E_g^X(P)$ (eV)	1.1100 1.1100[a]	1.0671 1.0671[a]	1.0242 1.0242[a]	0.9813 0.9813[a]	0.9384 0.9384[a]

[a]Ref. [14].

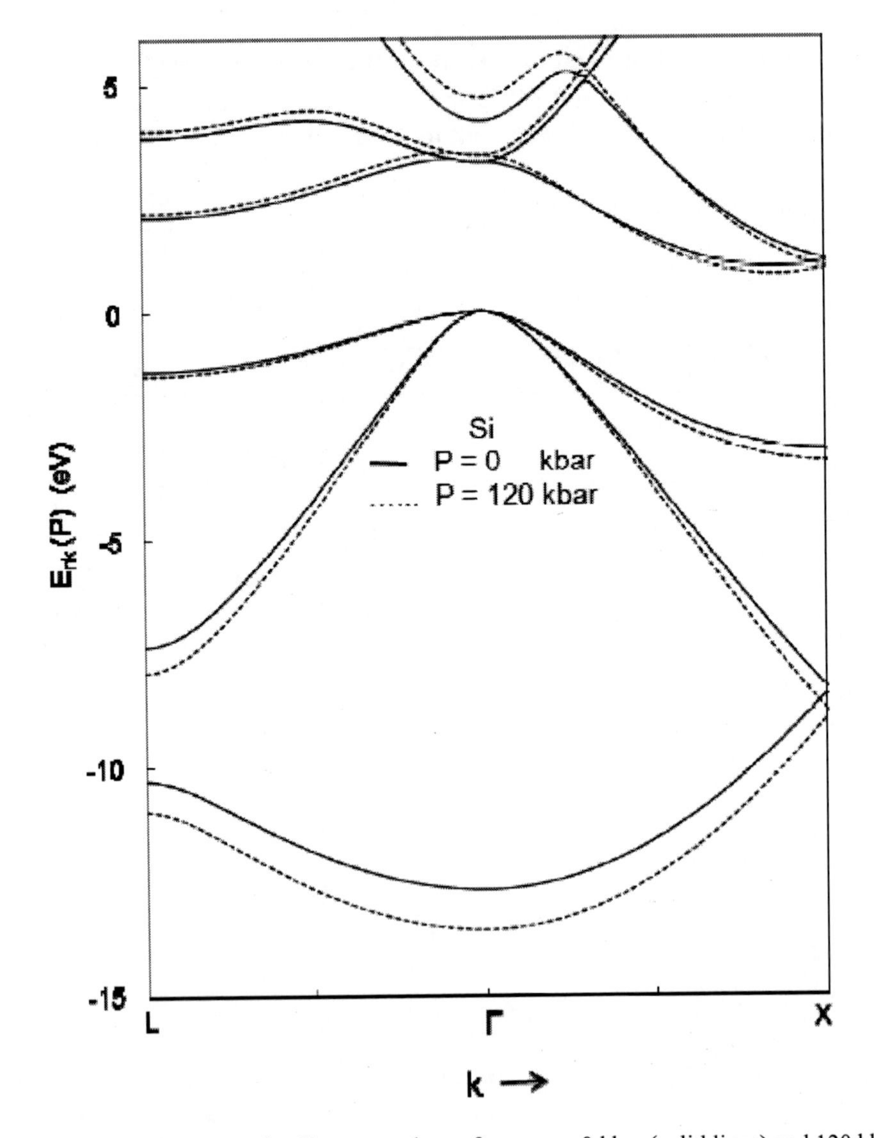

Figure 4.2. Energy band structure for Si at two values of pressure, 0 kbar (solid lines) and 120 kbar (dashed lines).

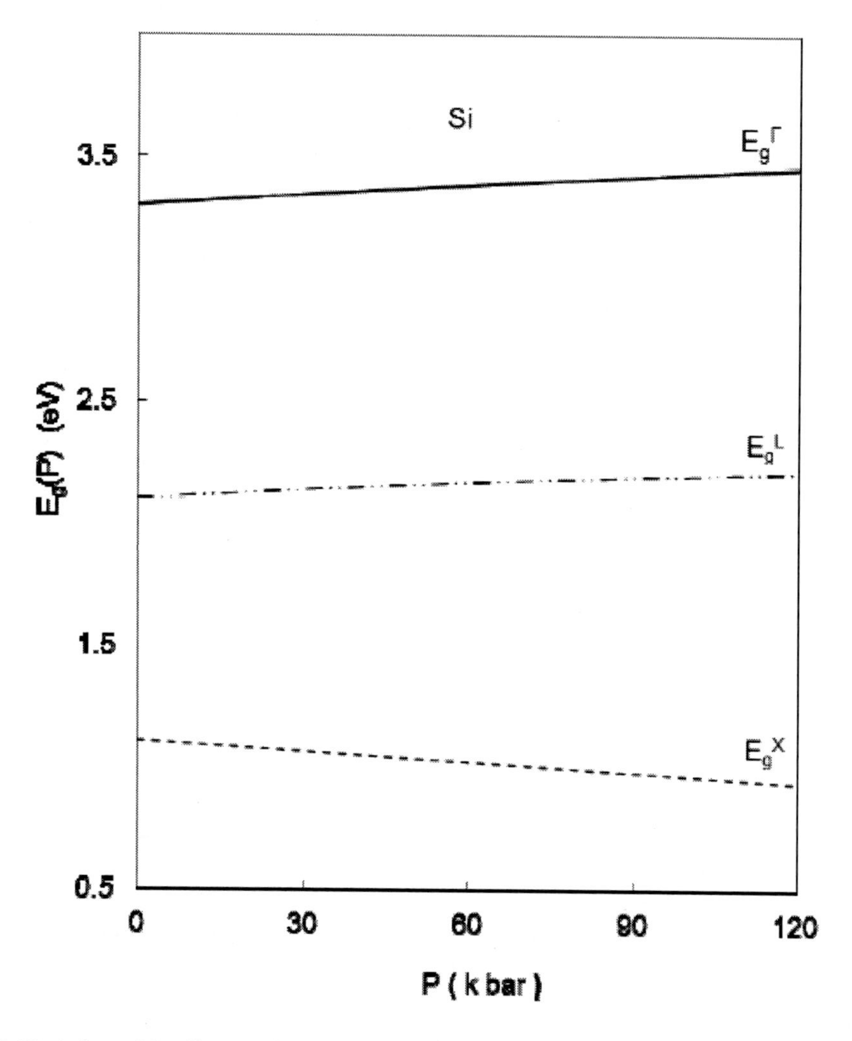

Figure 4.3. Variation of the direct and indirect energy band gaps for Si with pressure.

Figure (4.2) displays the pressure dependent electronic energy band structure, $E_{n,\vec{k}}(P)$ given in eV, for Si as a function of the propagation wave vector, $\vec{k}$ of the BZ at two different pressures, normal pressure (solid lines) and 120 kbar pressure (dashed lines). This figure shows that, both the conduction and valence energy bands are affected by pressure, except at the top of the valence band. Also one can see that, the low lying valence bands are highly sensitive to the pressure. The valence bands are shifted downward while most of the conduction bands are shifted upward except for the first conduction band near and at the X-symmetry point. It is seen from figure (4.2) that, the energy differences between the calculated electronic energies at 0 kbar and 120 kbar are about 172 meV at the X-point, 108 meV at the L-point, and 149 meV at the Γ–point.

The direct and indirect energy band gaps as functions of pressure for Si are plotted in figure (4.3). We see that the direct energy gap at the Γ-point, E_g^{Γ} and the indirect band gap at the L-point, E_g^{L} are increased by increasing the pressure while the indirect band gap at the X-point, E_g^{X} is decreased. Thus, Si remains as an indirect energy band semiconductor and its fundamental energy gap decreases with enhancing pressure as seen from both figure (4.3) and table (4.3).

Figure (4.4) displays the variation of the local symmetric pseudopotential form factors for Ge with pressure. It is seen that, the symmetric form factor of q^2 equals 11 is more sensitive and has the highest value than the other q-values. This result is due to the fact that the number of possible permutation is large and all of the reciprocal waves contribute and add up to make their associated form factor large. It is also seen that all the symmetric form factors for Ge are increased linearly by increasing the pressure.

Table (4.4) presents the variations of the lattice constant and the fundamental energy band gap for Ge with pressure. It is noticed that the lattice constant of Ge decreases linearly by increasing the pressure, which is a well known physical phenomenon.

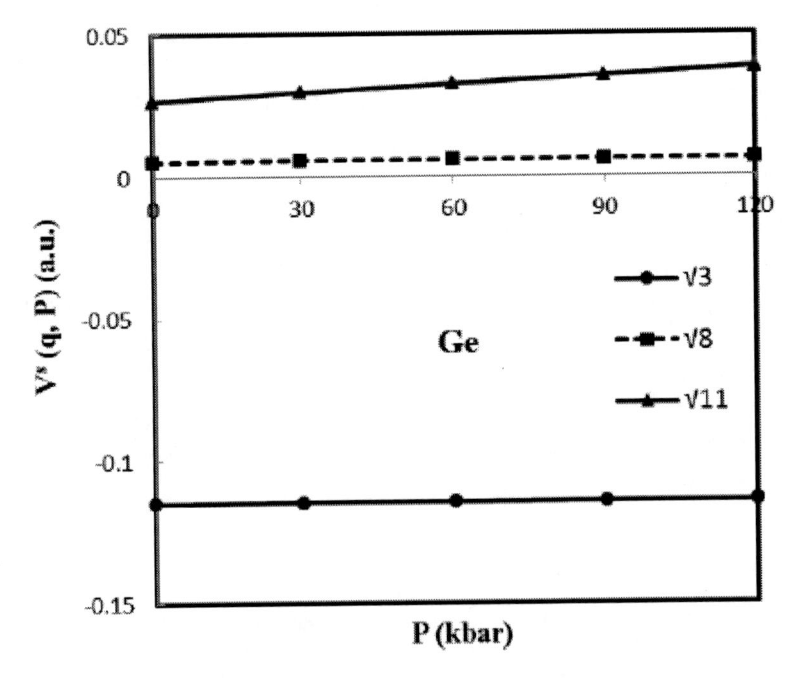

Figure 4.4. Variation of the symmetric form factors for Ge with hydrostatic pressure.

Table 4.4. Dependence of the lattice constant and the fundamental energy gap for Ge on pressure

Ge					
Parameter	P (kbar)				
	0	30	60	90	120
$a(P)$ (a.u.)	10.669179	10.538810	10.428470	10.332950	10.248840
E_g (P) (eV)	0.665700 0.665700[a]	0.809800 0.809700[a]	0.953700 0.953700[a]	1.097800 1.097700[a]	1.241600 1.241700[a]

[a]Ref. [14].

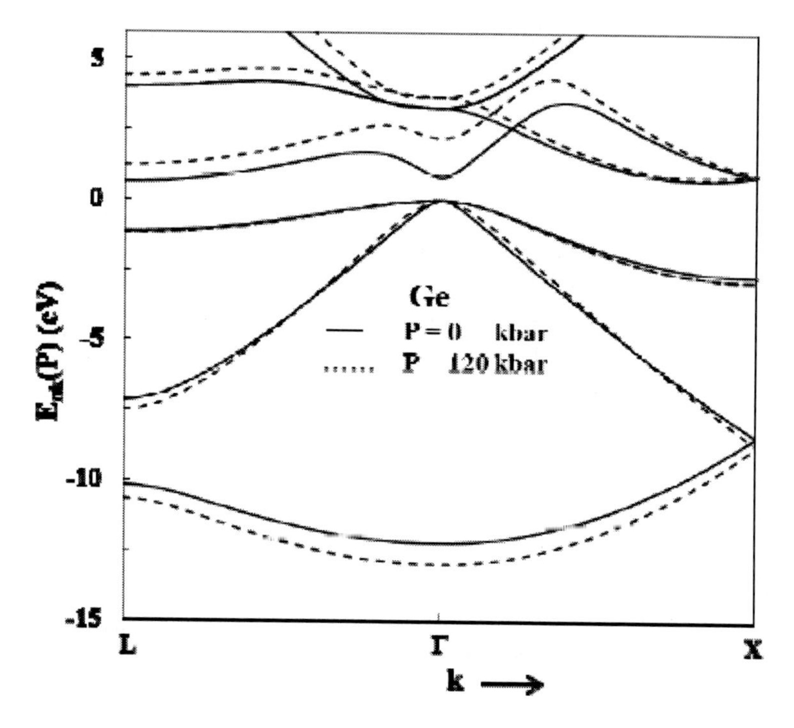

Figure 4.5. Energy band structure for Ge at two values of pressure, normal 0 kbar (solid lines) and 120 kbar (dashed lines).

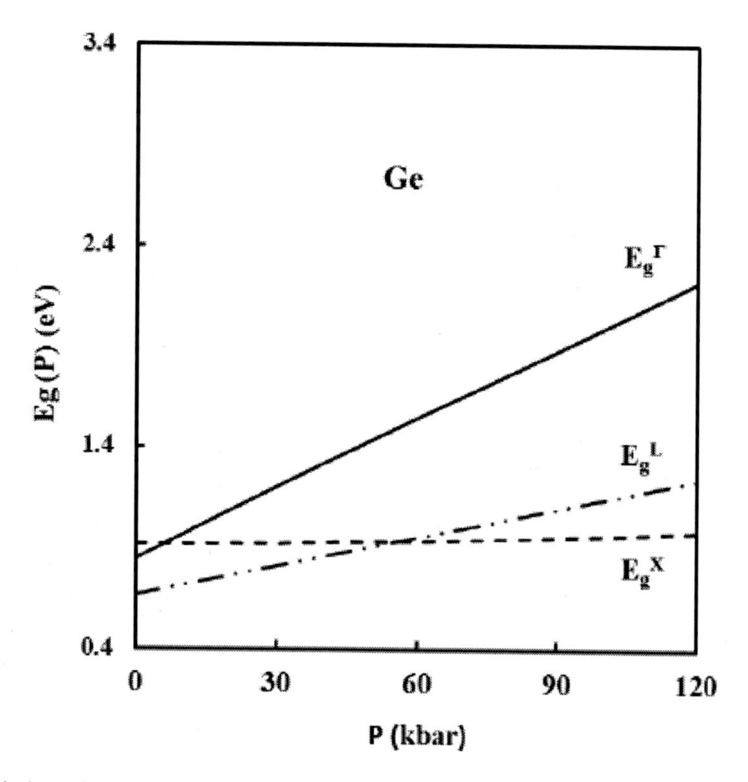

Figure 4.6. Variation of the direct and indirect energy band gaps for Ge with pressure.

Table 4.5. Dependence of the lattice constant and the fundamental energy gap for GaAs on pressure

GaAs					
Parameters	P (kbar)				
	0	30	60	90	120
$a(P)$ (a.u.)	10.6822	10.5528	10.44293	10.34759	10.26348
E_g (P) (eV)	1.4270 1.4270[a]	1.7499 1.7499[a]	2.0287 2.0288[a]	2.2636 2.2636[a]	2.4542 2.4542[a]

[a]Ref. [14].

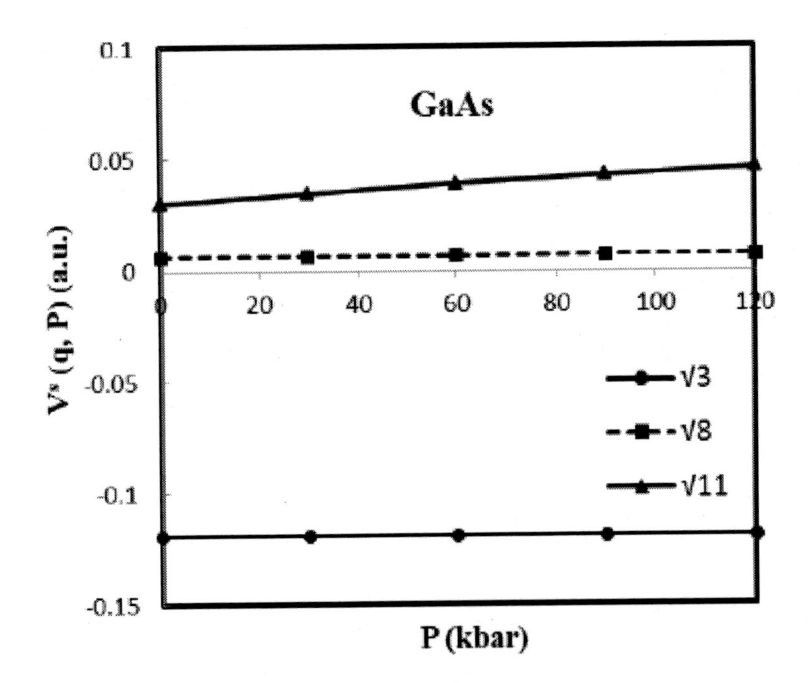

Figure 4.7. Variation of the symmetric form factors for GaAs with hydrostatic pressure.

Figure (4.5) exhibits the pressure dependent electronic energy band structure, $E_{n,\vec{k}}(P)$ given in eV, for Ge as a function of the propagation wave vector, $\vec{k}$ of the BZ at two different pressures, normal pressure (solid lines) and 120 kbar pressure (dashed lines). This figure shows that, both the conduction and valence energy bands are affected by pressure. Some of valence bands are shifted downward while most of the conduction bands are shifted upward except for the first and second conduction bands near and at the X symmetry point. It is seen from figure (4.5) that the energy differences between the calculated electronic energies at 0 kbar and 120 kbar are about 60 meV at the X-point, 574 meV at the L-point and 1.362 eV at Γ–point.

The direct and indirect energy band gaps for Ge as functions of temperature are plotted in figure (4.6). We see that the direct energy gap E_g^{Γ} and the indirect band gap E_g^{L} are increased by increasing the pressure while E_g^{X} is decreased. It seen an interesting result from this figure, that the fundamental energy gap for Ge changes its symmetry form L-band at a pressure nearly less than 60 kbar to the X-band due to the energy band crossing L-X.

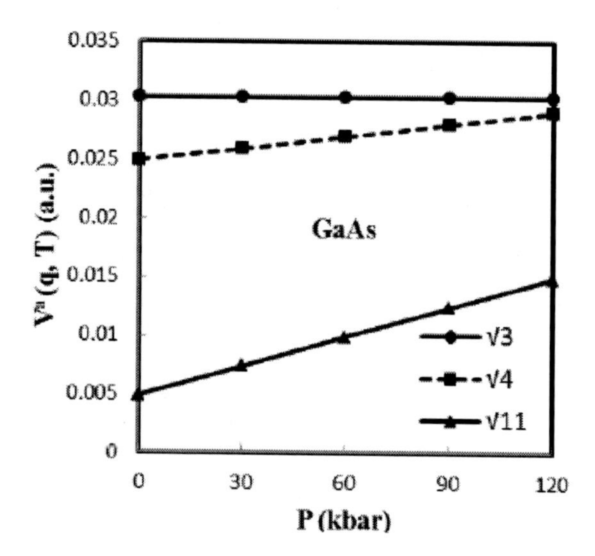

Figure 4.8. Variation of the antisymmetric form factors for GaAs with hydrostatic pressure.

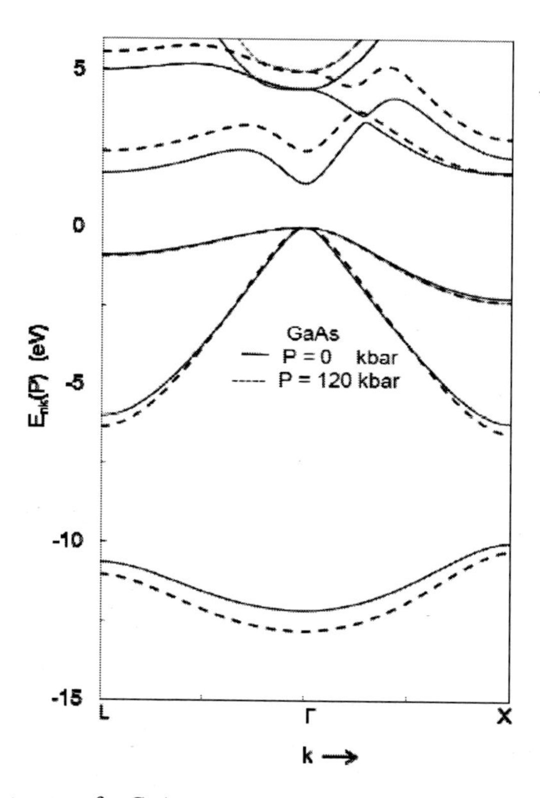

Figure 4.9. Energy band structure for GaAs at two pressures values, 0 kbar (solid lines) and 120 kbar (dashed lines).

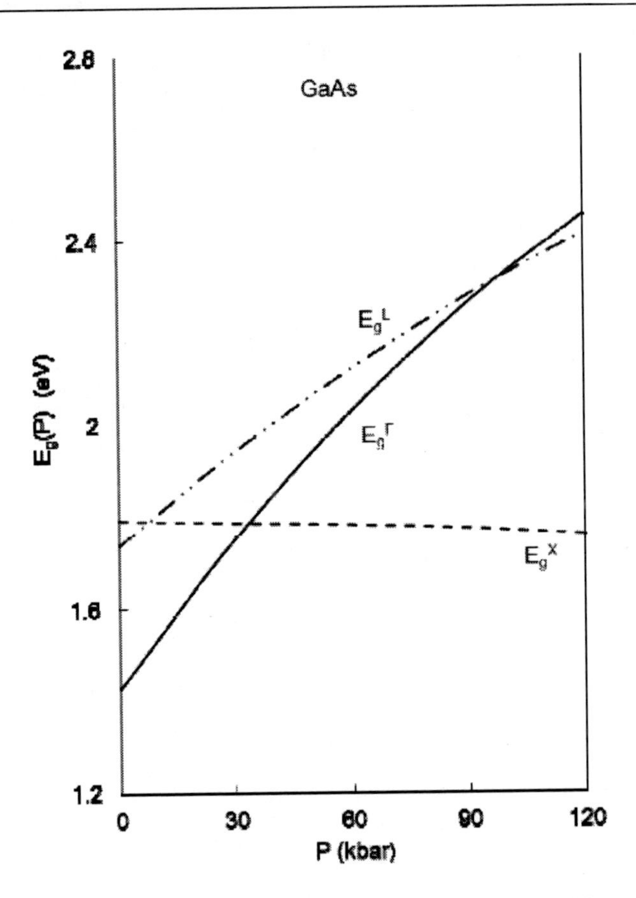

Figure 4.10. Variation of the direct and indirect energy band gaps for GaAs with pressure.

Figures (4.7) and (4.8) display the variations of the local symmetric and antisymmetric pseudopotential form factors for GaAs with pressure, respectively. It is seen that from figure (4.7) that, the symmetric form factor for $q^2 = 11$ is more sensitive to the pressure effect than the others and it is positive in sign and has the highest value while that of $q^2 = 3$ is negative and has the lowest value. It is seen from figure (4.8) that the two antisymmetric form factors associated to $q^2 = 11$ and 3 change their positions in a way that the form factor of 3 is the largest and positive and that of 11 is the lowest but its sign is positive. The reason for that is due to the antisymmetric nature of their contribution in the local pseudopotential since the potentials of the two atoms cancel each other and the contribution of the largest q value is high.

Table (4.5) illustrates the variations of the lattice constant and the fundamental energy gap for GaAs at various pressures from zero up to 120 kbar. It is seen from this table the decrease in value of the lattice constant and the increase in value of the fundamental energy gap with increasing pressure.

Figure (4.9) displays the pressure dependent electronic energy band structure, $E_{n,\vec{k}}(P)$ given in eV, for GaAs as a function of the propagation wave vector, $\vec{k}$ of the BZ at two different pressures, normal pressure (solid lines) and 120 kbar pressure (dashed lines). It is seen from figure (4.9) that the minimum of the conduction band at normal pressure (0 kbar) is located at the high-symmetry Γ-point. Hence, GaAs is considered as a (Γ–Γ) direct-gap semiconductor. When the pressure is increased to 120 kbar the first conduction band at Γ and L points are shift upward, while that near the X-point is moved slightly down relative to the valence-band maximum. This makes the lowest minimum of the first conduction band for GaAs semiconductor lies at the X-point. Therefore, the GaAs becomes an indirect (Γ–X) energy gap semiconductor under a pressure of 120 kbar. It is also seen from figure (4.9) that, the energy differences between the calculated electronic energies at 0 kbar and at 120 kbar are about 34 meV at the X-point, 1239 meV at the L-point and 1027 meV at Γ–point.

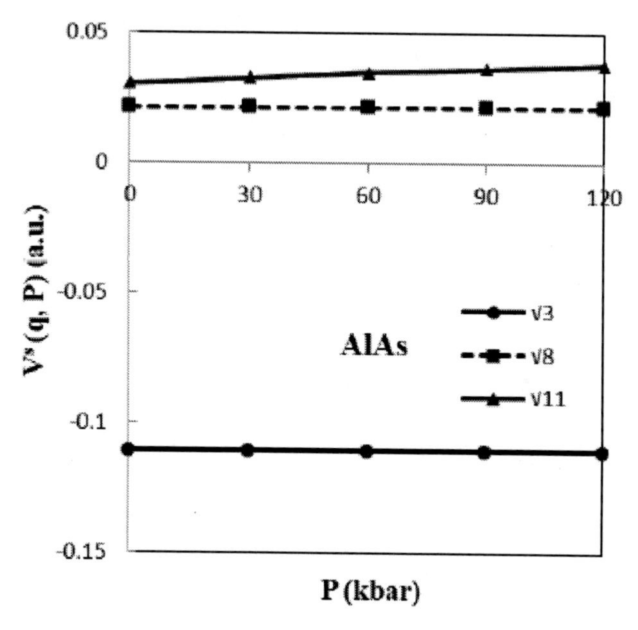

Figure 4.11. Variation of the symmetric form factors for AlAs with hydrostatic pressure.

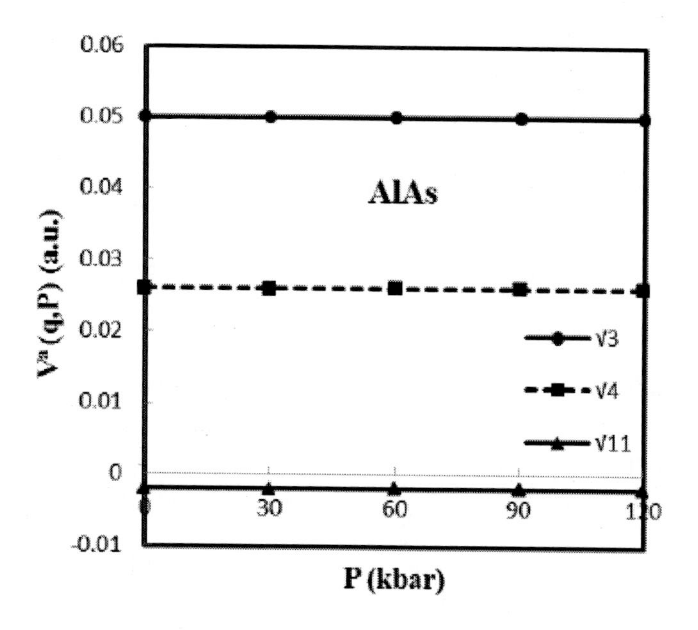

Figure 4.12. Variation of the antisymmetric form factors for AlAs with hydrostatic pressure.

Table 4.6. Dependence of the lattice constant and the fundamental energy band gap for AlAs on pressure

AlAs					
Parameters	P (kbar)				
	0	30	60	90	120
a (P) (a. u)	10.696334	10.570240	10.462440	10.368410	10.285120
E_g (P) in eV	2.1500 2.1500[a]	2.1040 2.1041[a]	2.0583 2.0582[a]	2.0123 2.0123a	1.9663 1.9664[a]

[a]Ref. [14].

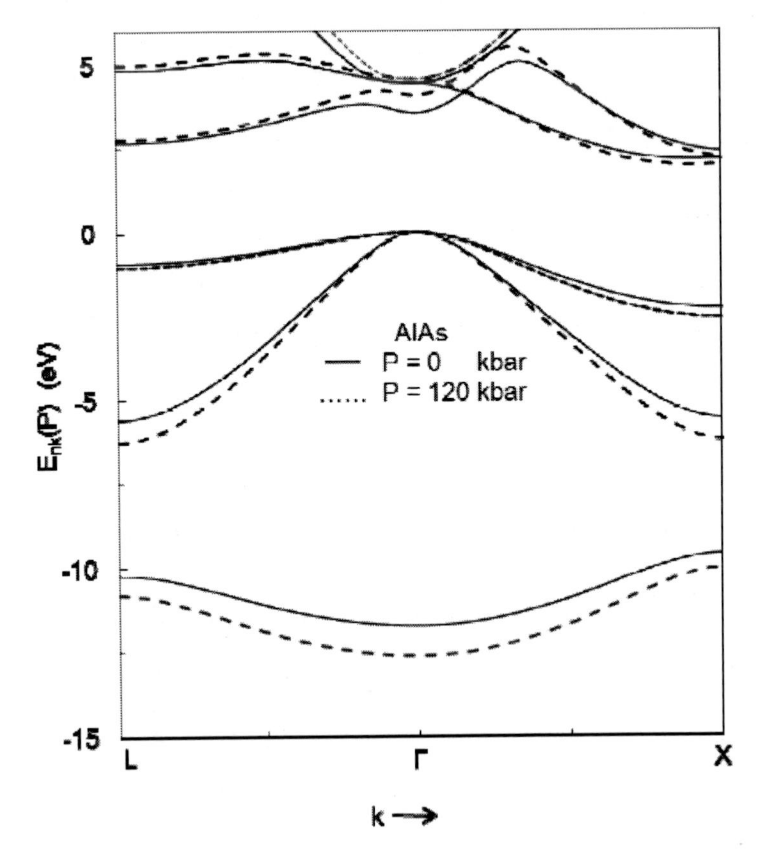

Figure 4.13. Energy band structure for AlAs at two different values of pressure, 0 kbar (solid lines) and 120 kbar (dashed lines).

Figure (4.10) displays the variations of the direct and indirect energy band gaps for GaAs with pressure. It is seen from figure (4.10) that the both direct at the Γ-point, E_g^{Γ} and the indirect at the L-point, E_g^{L} energy band gaps are increased by increasing the pressure while the indirect energy band gap at the X-point, E_g^{X} is decreased and bands crossing occur among them.

As a consequence, GaAs goes over from a direct band gap (Γ_v–Γ_c) semiconductor to an indirect band gap (Γ_v–X_c) semiconductor at a pressure of 30 kbar which corresponds to an energy value of 1.75 eV, which is in a good agreement with that reported in Ref. [21].

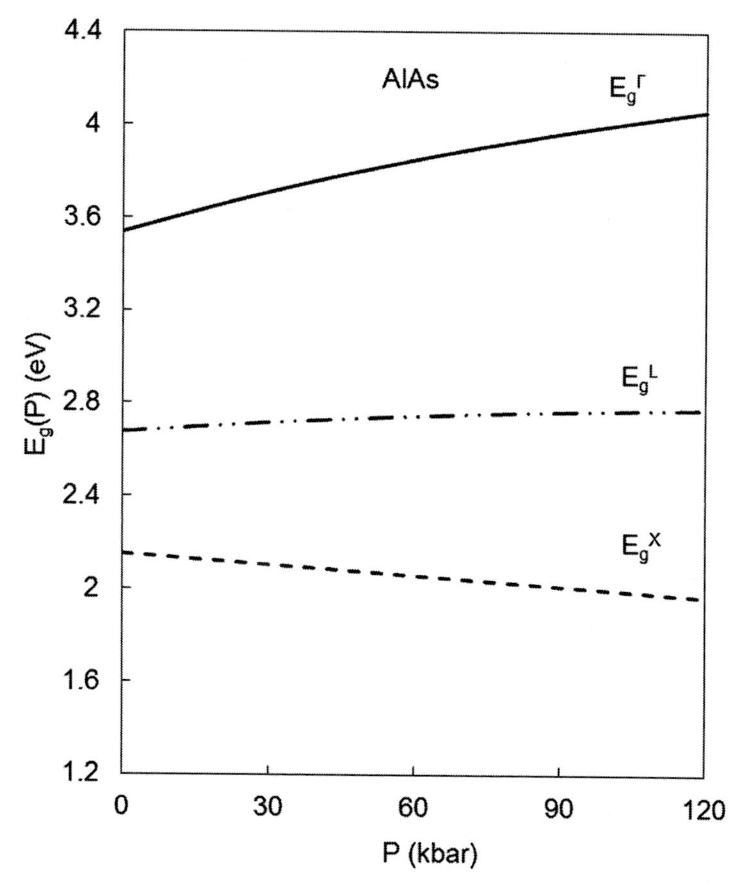

Figure 4.14. Variation of the direct and indirect energy band gaps for AlAs with pressure.

Both figures (4.11) and (4.12) illustrate the variations of the local symmetric and the antisymmetric pseudopotential form factors for AlAs with pressure.

It is seen from figure (4.11) that the symmetric factor of q^2 equals 11 is the most sensitive to the pressure than the others. It is positive in sign and has the highest value while that of q^2 equals 3 is negative and has the lowest value. This result is due to the nature of the symmetric form factor of q2 = 11 which contributes to the local pseudopotential more than the other values since it has the highest permutation. In figure (4.12) it is seen that, the two form factors of q^2 = 11 and 3 are flipped over in a way that the form factor associated to 3 is positive in sign and has the highest value while that of 11 has the lowest value but with negative sign. The reason for this result arises from the antisymmetric nature of them and the high cancellation in the magnitude of the local pseudopotential for 11 than 3.

Figure (4.13) presents the pressure dependent electronic energy band structure, $E_{n,\vec{k}}(P)$ given in eV, for AlAs as a function of the propagation wave vector, $\vec{k}$ of the BZ at two different pressures, normal pressure (solid lines) and 120 kbar pressure (dashed lines). We see from this figure that most conduction and valence energy bands are affected by pressure. At normal pressure, AlAs is an indirect energy band-gap semiconductor with its lowest minimum conduction band lies at the X-valley. The shift of the first conduction band is upward at the Γ- and L-points, but it shifts downward at the X-point. This indicates that AlAs remains as an indirect energy band gap ($\Gamma \rightarrow X$) semiconductor even at a pressure of 120 kbar, since there is no crossing occurs for the energy bands even at high pressure of 120 kbar. In general, as the pressure increases, the lattice constant decreases while the fundamental energy gap for AlAs semiconductor is decreased as shown in table (4.6).

It is also seen from figure (4.13) that, the energy differences between the calculated electronic energies at the two values of pressure, 0 kbar and 120 kbar are about 184 meV at the X-point, 100 meV at the L-point, and 520 meV at the Γ–point.

The energies of the direct and indirect energy band gaps as functions of pressure for AlAs are plotted in figure (4.14). One can see that both the direct energy gap at the Γ-point, E_g^{Γ} and the indirect energy band gap at the L-point, E_g^{L} are increased by increasing the pressure while the indirect energy band gap at the X-point, E_g^{X} is decreased.

CONCLUSION

It is found that the effects of temperature and pressure on the symmetric form factors belonging to the absolute value of the reciprocal lattice vectors q of $\sqrt{11}$ are more pronounced than other form factors with absolute value $q = \sqrt{3}, 2, \sqrt{8}$. This is due to the fact that the number of parameters included in the permutations for $\vec{G}$ group associated with $q = \sqrt{11}$ is greater in magnitude than those related to the other factors of q differ than 11. This leads to the fact that, the contribution of the reciprocal lattice vector for $q = \sqrt{11}$ in the pseudopotential energy term is great which leads to more pronounced in both symmetric and antisymmetric. As a matter of fact, higher values of permutations for $\vec{G}$ group may lead to more enhancements of their effects on the form factors, which in turn need more investigation.

REFERENCES

[1] Z.G. Zhu, T. Low, M.F. Li, W.J. Fan, P. Bai, D.L. Kwong, and G. Samudra, Semicond. *Sci. Technol.* 23 (2008) 025009.

[2] M.H. Cohen, V. Heine, *Phys. Rev.* 122 (1961) 1821.

[3] B.J. Austin, V. Heine, L.J. Sham, *Phys. Rev.* 127 (1962) 267.

[4] M.L. Cohen, T.K. Bergstresser, *Phys. Rev.* 141 (1966) 789.

[5] A.M. Elabsy, E.B. Elkenany, *Physica* B 405 (2010) 266.

[6] A.M. Elabsy, A.R. Degheidy, H.G. Abdelwahed, E.B. Elkenany, *Physica* B 405 (2010) 3709.

[7] P. Harrison, Quantum Wells, Wires, and Quantum Dots, second edition, Wiley, New York, (2005).

[8] B. K. Ridley, Quantum Processes in Semiconductors, fourth ed., Clarendon Press, Oxford, (1999).

[9] Charles Kittel, introduction to solid state physics, Wiley, Sons, Inc. (2005), 8th Ed.

[10] E.B. Elkenany, MSc. thesis, University of Mansoura, Egypt (2010).

[11] P. Harrison, Quantum Wells, Wires and Dots, Second Edit., J. Wiley and Sons, Ltd. (2005), Ch. 11.
[12] Y. F. Tsay, B. Gong, S. S. Mitra, and J. F. Vetelino, *Phys. Rev.* B 6 (1972).
[13] Y. P. Varshni, Physica 34 (1967) 149.
[14] S. Adachi, Properties of Group-IV, III-V and II-VI Semiconductors, Wiley and Sons, Inc. (2005).
[15] J. Piprek, Semiconductor Optoelectronic Devices, Academic Press, (2003).
[16] G. J. Ackland, *Rep. Prog. Phys.* 64 (2001) 483.
[17] H. J. Lee , L. V. Juravel, J. C. Woolley, *Phys. Rev.* B 21 (1980) 659.
[18] G.J. Ackland, Rep. *Prog. Phys.* 64 (2001) 483.
[19] Jayaraman, *Rev. Mod. Phys.* 55 (1983) 65.
[20] U. D. Sharma and M. Kumar, *Physica* B 405 (2010) 2820.
[21] M. Boucenna and N. Bouarissa, *Materials Chemistry and Physics* 84 (2004) 375.

In: Advances in Condensed Matter ... Volume 10
Editors: H. Geelvinck and S. Reynst

ISBN: 978-1-61209-533-2

Chapter 4

A PARALLEL MODEL BASED COMPENSATOR FOR RATE-DEPENDENT HYSTERESIS

Yonghong Tan[1*], Xinliang Zhang[2•] and Ruili Dong[1]
[1]College of Information, Mechanical and Electrical Engineering,
Shanghai Normal University, Shanghai, China
[2]School of Electrical Engineering and Automation,
Henan Polytechnic University, Henan, China

ABSTRACT

In this chapter, the phenomenon of rate-dependent hysteresis in piezoelectric actuators is described based on an experimental result. Then, a brief analysis on this phenomenon is presented. After that, an expanded input space based modeling approach previously proposed by the authors is presented. On this constructed expanded input space, the multi-valued mapping of rate-independent hysteresis can be transformed to a one-to-one mapping. Thus, any traditional modeling methods can be applied to modeling of the behavior of rate-independent hysteresis on this expanded input space. Then, a model with parallel structure is proposed to approximate both hysteretic behavior and rate-dependent feature of the piezoelectric actuators. In the parallel model, the rate-independent hysteretic performance is approximated by a neural network based on the expanded input space method, while the rate-dependent behavior of the hysteresis is approximated by a network with the sum of weighted first-order difference operators. For the compensation of the effect of hysteresis, the corresponding inverse model for rate-dependent hysteresis is obtained. Furthermore, a compensator is constructed to compensate for the effect of the rate-dependent hysteresis in piezoelectric actuators. Moreover, the experimental results of applying the proposed method to a piezoelectric actuator are presented.

* Shanghai Normal University, No. 100, Guilin Road, Shanghai 200234, China.
• Henan Polytechnic University, Henan 454003, China.

I. INTRODUCTION

Nowadays, the piezoelectric actuators (PEAs) have been extensively applied to micro-positioning systems such as micro-machining, disk drive, and micro-robot in electronics assembly etc. due to the excellent characteristics of fast response, nanometer displacement resolution and high stiffness inherent in the PEAs. Nevertheless, the existence of the hysteresis in the PEAs would lead to the degradation of the control performance and even instability in the case of closed-loop control. Figure 1 shows a typical hysteresis which has the non-differentiable feature and multi-valued mapping. From Figure 1, we can see that the turning points, e.g. *A, B, C, E, F,* and *G* are non-differentiable points whilst the input may be corresponding to more than one single value, e.g. the input v_1 is corresponding to three output values, i.e. $H_B(v_1)$, $H_C(v_1)$, and $H_D(v_1)$.

On the other hand, the input signal fed to the PEA usually covers a wide range of frequencies, thus, it may yield the rate-dependent hysteretic phenomenon especially when the frequency of the input signal exceeds a certain threshold. In this situation, it will result in a larger phase-lag if the frequency is higher. Thus, the corresponding response of the actuator will be rather sluggish. Usually, in high-speed positioning systems, the input signal will be changed very quickly that implies the frequency of the input will be very high. Hence, the performance of the actuator may be deteriorated due to the effect of the rate-dependent behavior of the hysteresis in the case of the input with higher frequency. Therefore, the compensation on the effect of rate-dependent hysteresis in the piezoelectric actuators is necessary in order to derive high speed and high precision positioning performance.

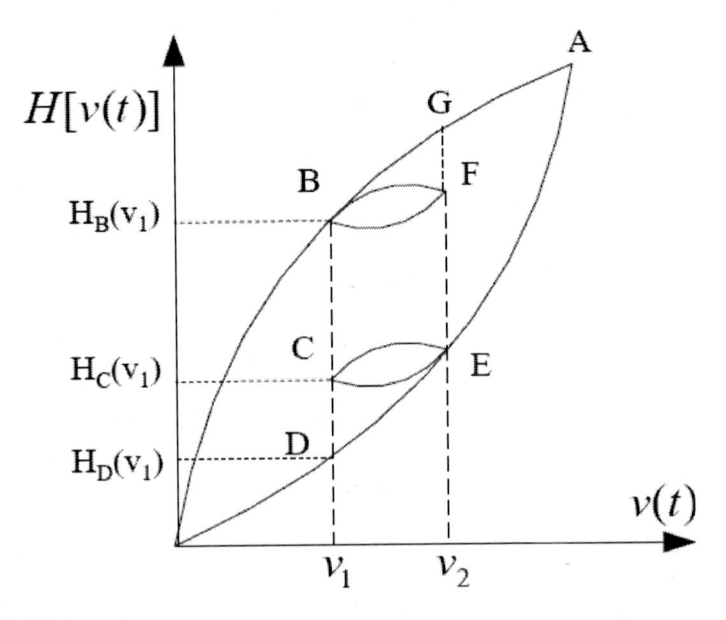

Figure 1. A typical hysteresis.

In order to compensate for the effect of rate-dependent hysteresis, the model-based compensation strategies are usually used [5, 7-10]. Hence, the construction of an accurate model to describe the rate-dependent hysteresis is very important. Up till now, there have been different models proposed to describe the characteristic of rate-dependent hysteresis. Ang et. al. [1] proposed Prandtl-Ishlinskii (PI) plus creep model to describe the rate-dependent behavior of hysteresis. Dong et. al. [2] proposed a neural network based model with generalized gradients to approximate the rate-dependent hysteresis in piezoelectric actuators. Moreover, a recurrent diagonal neural network based model as well as a nonlinear auto-regressive moving average with exogenous input (NARMAX) model for rate-dependent hysteresis were proposed by Refs. [3] and [19], respectively.

Due to the complex characteristic of rate-dependent hysteresis, the modeling of its behavior usually meets the following problems, i.e.

1) multi-valued mapping between the input and output of the hysteresis;
2) non-smoothness at the turning points where the gradients of the hysteresis output with respect to its input don't exist;
3) dynamic effect relying on the change of the input frequency.

In order to construct a model based compensator to suppress the effect of rate-dependent hysteresis in PEAs, we should consider how to overcome the above mentioned problems in the modeling procedure.

Therefore, this Chapter is organized as follows: In Section 2, the phenomenon of rate-dependent hysteresis in piezoelectric actuators will be described. Then, in Section 3, the expanded input space modeling approach will be presented. On the constructed expanded input space, the multi-valued mapping of rate-independent hysteresis can be transformed to a one-to-one mapping. Thus, any traditional modeling methods can be applied to modeling of the behavior of rate-independent hysteresis on this expanded input space. Then, in Section 4, a novel model with parallel structure is proposed to approximate both hysteretic behavior and rate-dependent feature of the piezoelectric actuators. In the parallel model, the rate-independent hysteretic performance is approximated by a neural network based on the expanded input space method, while the rate-dependent behavior of the hysteresis is approximated by a network with the sum of weighted first-order difference operators. In Section 5, the corresponding inverse model for rate-dependent hysteresis is obtained. After that, in Section 6, a compensator is constructed to compensate for the effect of the rate-dependent hysteresis in piezoelectric actuators. In Section 7, the experimental results of applying the proposed method to a piezoelectric actuator are presented. Finally, the conclusion will be given in Section 8.

II. THE PHENOMENON OF RATE-DEPENDENT HYSTEREIS IN PIEZOELECTRIC ACTUATORS

In this section, an experimental example will be presented to show the phenomenon of the hysteresis relying on the frequency-varying input. In the experiment, the piezoelectric actuator (PZT-753.21C) is excited by a frequency-varying sinusoidal signal. The frequency is varying from 10Hz to 2000Hz. Figure 2 shows the displacement against voltage curve of the hysteresis when the input frequency changes from 10Hz to 2000Hz

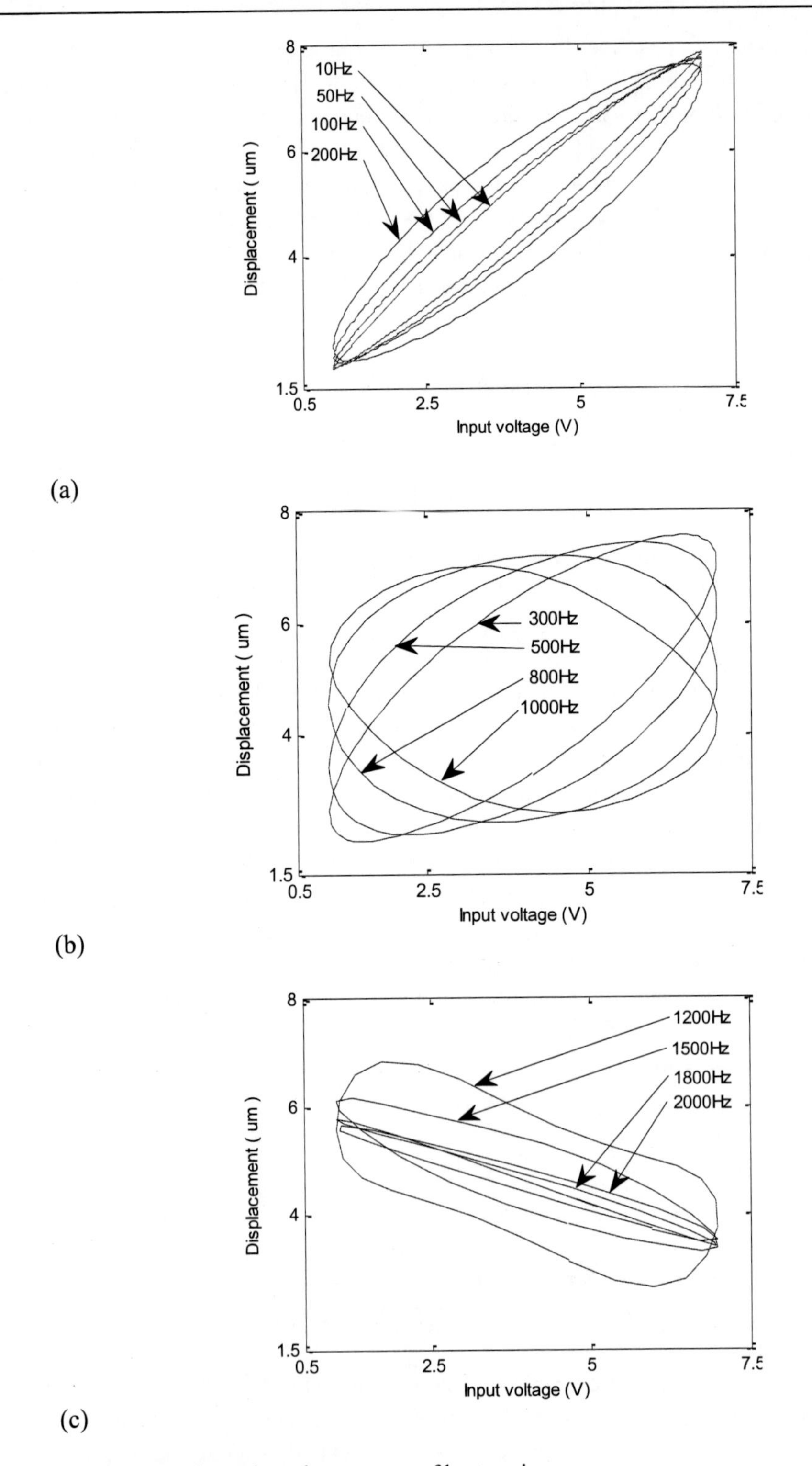

Figure 2. The rate-dependent phenomenon of hysteresis.

It has been found that, with the increase of the frequency in the input signal, the width of the hysteretic curve increases but the height decreases. Moreover, the hysteretic curve rotates in the clockwise direction. Obviously, the PEA would demonstrate a rather complicated performance in this case.

As the piezoelectric actuator is actually a low-pass filter at the low-frequency segment, the input with low frequency can pass through it without any distortion. Figure 2 (a) illustrates this phenomenon. However, Figure 2 (b) shows that if the input with higher frequency exceeds the cut-off frequency of the piezoceramics, the magnitude of the displacement of the piezoceramics will be reduced and the response rate of the actuator will also be decreased. On the other hand, Figure 2 (c) shows that, as the frequency of the input continuously increases, which is much higher than the cut-off frequency of the piezoceramics, the corresponding gain of the piezoceramics will greatly degrade and the phase lag of the actuator will also be increased. From those experimental results, it can be concluded that the performance of the piezoelectric actuator will be significantly affected by the input frequency.

III. The Expanded Input Space Method for Hysteresis

Usually, the models often used to describe the behavior of hysteresis are the so called Preisach model[13, 14], Krasnosel'skii-Pokrovskii (KP) model[15], and Prandtl-Ishlinskii (PI) model[16] etc. Those models are actually the integral of the weighted combination of backlash-like operators. The problem when we use the above-mentioned modeling methods is that the derived model has a very complex structure, in which hundred of operators have to be utilized. It may cause much computational burden in applications. Recently, neural networks have been applied to the approximation of hysteresis [17, 18]. However, it is known that the conventional neural networks cannot be used to approximate any systems with multi-valued mapping [12]. Therefore, we should consider an option to transform the multi-valued mapping of hysteresis into a one-to-one mapping. Then, the neural networks can be applied to hysteretic modeling. In this section, an operator which can roughly extract the characteristic of hysteretic movement of the system is constructed. The constructed operator is used as one of the coordinates of the expanded input space. The input of the hysteresis is another coordinate in the space as well. On this constructed expanded input space, the output of the hysteresis can be uniquely determined.

3.1. The Hysteretic Operator

The hysteretic operator is described as

$$f_H[u](t) = (u - u_p)(1 - \exp(-|u - u_p|)) + f_H[u_p] \tag{1}$$

where $u(t)$ is the current input; $f_H[u](t)$ is the current output of the operator; $f_H[u_p]$ is the output of the operator at the current dominant extremum u_p.

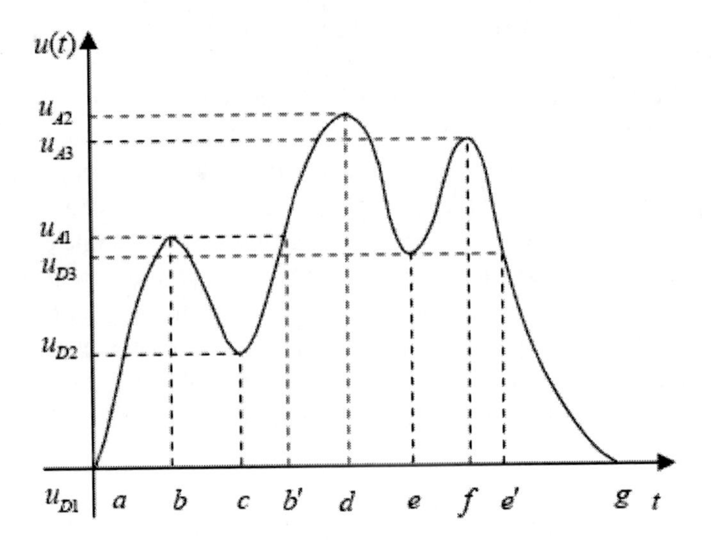

(a)

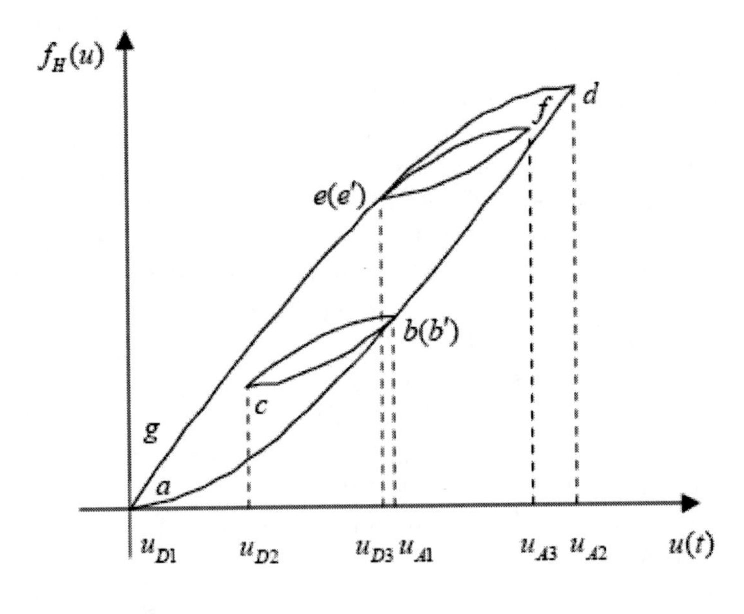

(b)

Figure 3. Input-output curve of the hysteretic operator.

Note that the proposed hysteretic operator keeps the wiping-out property [11]. Figure 3(a) and (b) demonstrate the function of the proposed hysteretic operator. The detailed explanation is given by remarks 1 and 2.

Remark 1: To realize the wiping-out property of the operator, two memory stacks, i.e. M_A and M_D are introduced to store the local maxima and the local minima respectively[20]. The input set of the operator is normalized within [0, 1]. The initial local minimum, being equal to zero, is pre-stored in M_D, while the stack M_A is initialized to be empty.

Two stacks are respectively described as

$$M_D = [u_{D1}, u_{D2}, \ldots, u_{Dj}, \ldots, u_{Dm}] \tag{2}$$

and

$$M_A = [u_{A1}, u_{A2}, \ldots, u_{Ai}, \ldots, u_{An}] \quad (3)$$

where m and n denote the length of M_D and M_A respectively. Moreover,

$u_{D1} = 0$, also $f_H(u_{D1}) = 0$;

$u_{D1} < u_{Dj-1} \leq u_{Dj} \leq u_{Dm}$, for $3 \leq j \leq m$; and

$u_{Ai-1} \geq u_{Ai} \geq u_{An}$, for $2 \leq i \leq n$;

The stacks grow simultaneously during the construction of the closed-loop and shrink simultaneously in the case where the input exceeds the local extremum of the past input.

As shown in Figure 3(a) and (b), at the initial time instance a, it has

$M_D = [u_{D1}]$, and

$M_A = [\,]$.

At the time instance b when the sign of the input rate changes, it leads to

$M_D = [u_{D1}]$, and

$M_A = [u_{A1}]$.

Similarly, at the time instance c, it has

$M_D = [u_{D1}\ u_{D2}]$, and

$M_A = [u_{A1}]$.

At b' when the input exceeds the local maximal extremum of the input, it results in

$M_D = [u_{D1}]$, and

$M_A = [\,]$.

Likewise, at the time instances d, e, f, e', and g, it yields

$M_D = [u_{D1}]$, and $M_A = [u_{A2}]$ at d,

$M_D = [u_{D1}\ u_{D3}]$, and $M_A = [u_{A2}]$ at e,

$M_D = [u_{D1}\ u_{D3}]$, and $M_A = [u_{A2}\ u_{A3}]$ at f,

$M_D = [u_{D1}]$, and $M_A = [u_{A2}]$ at e', and

$M_D = [u_{D1}]$, and $M_A = [\ ]$ at g.

Remark 2: For a pair of local extrema, i.e. maximum u_A and minimum u_D respectively, in terms of (1),

if $\dot{u}(t) > 0$, then

$$f_{H+}[u](t) = (u - u_D)(1 - \exp(u_D - u)) + f_H[u_D]; \text{ and} \tag{4}$$

if $\dot{u}(t) < 0$, then

$$f_{H-}[u](t) = (u - u_A)(1 - \exp(u - u_A)) + f_H[u_A] \tag{5}$$

where $\dot{u}(t)$ denotes the derivative of the input $u(t)$.

It can be concluded that both $f_{H+}[u](t)$ and $f_{H-}[u](t)$ are strictly point-symmetrical to each other. Moreover, the area of the closed-loop is solely determined by the distance between u_A and u_D, irrelevant to time t. It means that all the minor-loops which are constructed in different transition time-segments are identical in shape. The hysteretic operator also satisfies the property of minor-loop congruency.

From what mentioned above, the proposed hysteretic operator can represent the characteristics of the classical Preisach-type hysteresis. It keeps the properties including minor-loop congruency and wiping-out properties as well as the change-tendency at ascending, descending sub-loops and the turning points. In the following, it will be proved that, the multi-valued mapping between the output of the classical Preisach-type hysteresis and its input can be transformed into a one-to-one mapping on the expanded input space with the introduced hysteretic operator. Moreover, the obtained mapping is a continuous one-to-one mapping on a compact set.

Lemma 1: The hysteretic operator $f_H[u](t)$ is a continuous function.

Proof: The proof can be divided into two cases.

Case 1: When the present input doesn't exceed the local extremum of the past input, according to the definition of the operator described in (1), it is obviously that $f_H[u](t)$ is a continuous function.

Case 2: When the current input exceeds the local extremum of the past inputs.

Considering the case that the value of the current input exceeds the local maximum of the past inputs at the time instances shown in Figure3 (a) and (b), i.e. segment $b \to c \to b' \to d$, where the input value exceeds the local maximum u_{A1} which occurs at time instance b'.

Define $f_H[u_{A1}]$ as the output of the operator at the time instance b. Then, we can obtain $f_H[u_{D2}]$ at the time instance c when the input decreases monotonically to u_{D2}. From (5), it has

$$f_H[u_{D2}] = (u_{D2} - u_{A1})(1 - \exp(u_{D2} - u_{A1})) + f_H[u_{A1}] \tag{6}$$

Define $f_{H*}[u_{A1}]$ as the output of the operator at b' when the input $u(t)$ increases monotonically to u_{A1} again. Thus, it yields

$$f_{H^*}\left[u_{A1}\right]=\left(u_{A1}-u_{D2}\right)\left(1-\exp(u_{D2}-u_{A1})\right)+f_H[u_{D2}] \tag{7}$$

Also, there exists

$$\lim_{u(t)\to u_{A1}^-} f_H\left[u\right](t)=f_{H^*}\left[u_{A1}\right]. \tag{8}$$

Then, substituting (6) into (7) leads to

$$f_{H^*}\left[u_{A1}\right]=f_H\left[u_{A1}\right]. \tag{9}$$

On the other hand, in terms of (4), it has

$$\lim_{u(t)\to u_{A1}^+} f_H\left[u\right](t)=f_H\left[u_{A1}\right]. \tag{10}$$

Therefore, it leads to

$$\lim_{u(t)\to u_{A1}^-} f_H\left[u\right](t)=\lim_{u(t)\to u_{A1}^+} f_H\left[u\right](t) \tag{11}$$

i.e. $f_H[u](t)$ is continuous at the local extremum u_{A1}. Considering case 1, it can be concluded that $f_H[u](t)$ is continuous in the case that the input exceeds the local maximum.

The same conclusion can be drawn for the case where the input passes the local minimum, from Figure 2 (a) and (b), we see that the corresponding input moves along segment $e\to f\to e'$.

So the proposed hysteretic operator $f_H[u](t)$ is a continuous function.

Lemma 2: If there exist two different time instances t_1 and t_2, it results in $u(t_1) = u(t_2)$ but the corresponding output of the hysteresis will be $H[u](t_1) \neq H[u](t_2)$, where $u(t_1)$ and $u(t_2)$ are not the extrema, then

$$(u(t_1), f_H[u](t_1)) \neq (u(t_2), f_H[u](t_2)).$$

Proof: Based on the point-symmetric feature described in remark 2, it can be known that, for a pairs of extrema, the constructed ascending sub-loop and descending sub-loop of the hysteretic operator are strict point-symmetric to each other.

On the other hand, for any input $u(t)$ which is not the extremum, from (1), it can be derived that,

$$\frac{df_H[u]}{du}>0. \tag{12}$$

So the intersection of the ascending sub-loop and descending sub-loop at any closed-loop only occurs at the local minimum or maximum.

Also, according to the wipe-out property of the proposed hysteretic operator stated in remark 1, it can be concluded that all the closed-loops never intersect to each other except for the extrema.

Therefore, we can conclude that if $u(t_1) = u(t_2)$, also $u(t_1)$ and $u(t_2)$ are not the extrema but $H[u](t_1) \neq H[u](t_2)$, then $(u(t_1), f_H[u](t_1)) \neq (u(t_2), f_H[u](t_2))$.

3.2. The Construction of the Expanded Input Space

In this subsection, an expanded input space is constructed with the above-proposed hysteretic operator. The purpose of the construction of such expanded input space is to transform the multi-valued mapping between the hysteresis output and its input into a one-to-one mapping. Then, the neural network can be applied to the modeling of hsyeresis on this expaned input space. Based on the discussion presented in above section, we have

Theorem 1: Suppose the input space of the hysteresis is defined as $R^{2:}$ $U \times F$ .where $u(t) \in U$ and $f_H[u](t) \in F$. Then, there exists a continuous one-to-one mapping Γ: $R^2 \to R$, such that the classic Preisach-type hysteresis can be represented by

$$H[u](t) = \Gamma(u(t), f_H[u](t)).$$

Proof: In terms of lemma 2, it is known that the coordinate $(u(t), f_H[u](t))$ is corresponding uniquely to $H[u](t)$ except for the extrema.

On the other hand, for the case that $u(t_1)$ and $u(t_2)$ are the extrema and $u(t_1) = u(t_2)$, according to the property of the hysteresis, there exists

$$H[u](t_1) = H[u](t_2) \tag{13}$$

At the same time, for the operator, it has

$$f_H[u](t_1) = f_H[u](t_2) \tag{14}$$

So in this case, the coordinate $(u(t), f_H[u](t))$ is also corresponding uniquely to $H[u](t)$. Moreover, for the continuous property of the hysteresis, it leads to

$$\left.\begin{array}{r} t_1 - t_2 \to 0 \\ u(t_1) - u(t_2) \to 0 \end{array}\right\} \Rightarrow H[u](t_1) - H[u](t_2) \to 0. \tag{15}$$

Considering lemma 1 yields

$$u(t_1) - u(t_2) \to 0 \Leftrightarrow f_H[u](t_1)) - f_H[u](t_2) \to 0. \tag{16}$$

Therefore, there exists

$$\left.\begin{array}{r} u(t_1) - u(t_2) \to 0 \\ f_H[u](t_1) - f_H[u](t_2) \to 0 \end{array}\right\} \Rightarrow H[u](t_1) - H[u](t_2) \to 0 \tag{17}$$

i.e. there exists a continuous one-to-one mapping Γ: $R^2 \rightarrow R$, so that $H[u](t) = \Gamma(u(t), f_H[u](t))$.

Remarks 3: According to the properties of the proposed hysteretic operator illustrated respectively in remark 1 and lemma 1, it can be concluded that, for the continuous and bounded input, the proposed operator is also continuous and bounded. The expanded input space consisting of $u(t)$ and $f_H[u](t)$ is a continuous compact set.

Figure 4 demonstrates how a one-to-one mapping can be derived on the expanded input space. From Figure 4, it is seen that the expanded input space consists of two coordinates, one is the input u and another is the hysteretic operator $f_H(u)$. In the following, we use an example to explain how the multi-valued mapping of the hysteresis is transformed into a one-to-one mapping on this constructed expanded input space. Suppose $u=u_x$. Then, it yields the corresponding outputs of the hysteretic operator, i.e. $f_{Ha}(u_x)$ and $f_{Hb}(u_x)$, respectively. Then, we have two points on the constructed expanded input space, i.e. $a(u_{x,} . f_{Ha}(u_x))$ and $b(u_{x,} . f_{Hb}(u_x))$, respectively. In this case, the corresponding output values of the hysteresis, i.e. $A=H(u_{x,} . f_{Ha}(u_x))$ and $B=H(u_{x,} . f_{Hb}(u_x))$ can be uniquely specified.

In terms of Theorem 1, the feedforward neural networks can be used to approximate the one-to-one mapping Γ on the expanded input space and it gives

$$\left\| \Gamma\left(u(t), f_H[u](t)\right) - NN\left(u(t), f_H[u](t)\right) \right\| < \varepsilon \tag{18}$$

where $NN(\cdot)$ represents the neural network which is used for the approximation of the hysteresis. ε is the approximation error and for any $\varepsilon_N > 0$, there exists $\varepsilon < \varepsilon_N$.

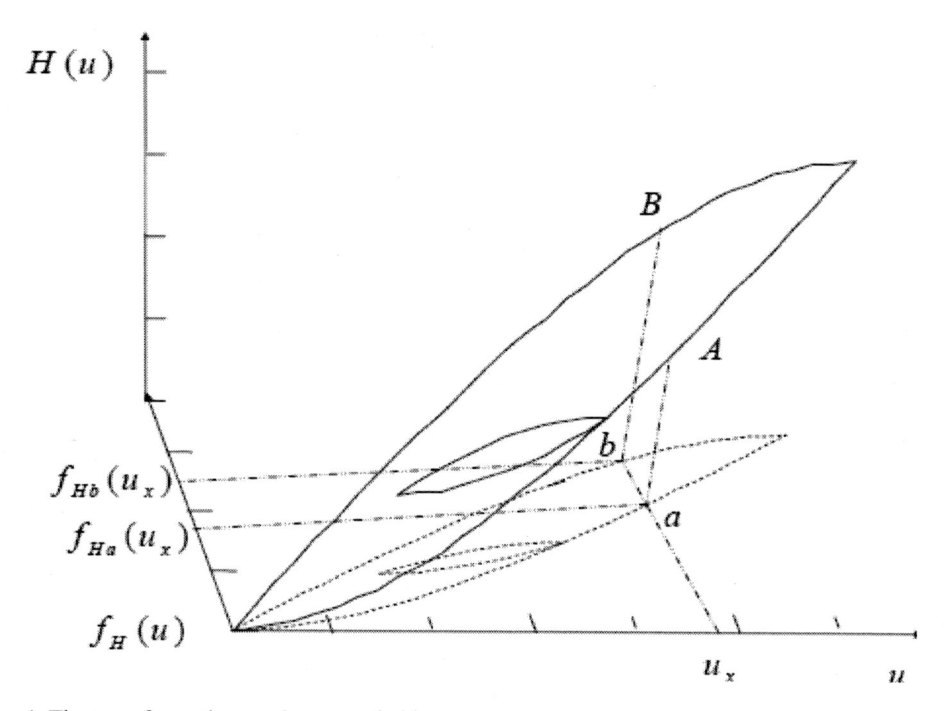

Figure 4. The transformation on the expanded input space.

IV. THE PARALLEL MODEL FOR RATE-DEPENDENT HYSTERESIS

Consider the universal approximation capability of the neural networks. The obtained neural model from (18) can be used for the modeling of the rate-independent characteristics of hysteresis. At the same time, considering the following facts:

1) the proposed hysteresis operator is a rate-independent function, i.e. the output of the operator is independent of the frequency of the input; and
2) the used feedforward neural network is static, i.e. it is irrelevant to the input frequency,

it can be concluded that the derived neural model from (18) is also a static function, i.e. it cannot be used to describe the rate-dependent characteristics of a dynamic hysteresis.

In this section, the first-order differential operator, i.e.

$$G_i(s) = e^{-\tau s}\frac{k_i}{\lambda_i s+1} \tag{19}$$

where $\lambda_i > 0$ is the time-constant and τ is the time delay of the operator, is introduced into the model. The introduced first-order differential operators are used to construct a submodel parallel to the neural submodel mentioned above. In this scheme, the first-order differential operators based submodel is used to describe the dynamics of the hysteresis. Based on those two submodels, a hybrid model to describe the rate-dependent hysteresis is constructed. The architecture of the hybrid model is shown in Figure 5 [21].

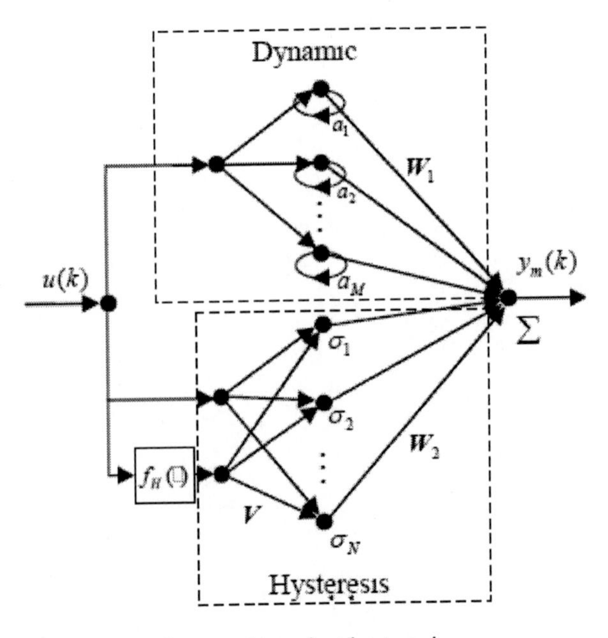

Figure 5. The hybrid model structure for rate-dependent hysteresis.

The corresponding first-order difference operator can be described as

$$G_i(z^{-1}) = \frac{w_{1i}}{1-a_i z^{-1}} z^{-d} \tag{20}$$

where $a_i = \exp(-\lambda_i/T_s)$. $d = \tau/T_s$, T_s is the sampling period, $w_{1i} = k_i(1-a_i)$ is the weighting factor, $i = 1, 2 \ldots M$ and M is the number of the operators. Meanwhile, due to the fact that $a_i \in (0, 1)$, the logarithmic sigmoid function (Logsig) is introduced to realize it, i.e.

$$a_i = \mathrm{logsig}(c_i) = \frac{1}{1+\exp(-c_i)}. \tag{21}$$

Thus, the corresponding output of the hybrid model can be described as

$$y_m(k) = \boldsymbol{W}^{\mathrm{T}} \boldsymbol{O}(k) + \theta_2 \tag{22}$$

where

$$\boldsymbol{W} = \begin{bmatrix} \boldsymbol{W}_1 \\ \boldsymbol{W}_2 \end{bmatrix},$$

$$\boldsymbol{O}(k) = \begin{bmatrix} \boldsymbol{O}_1(k) \\ \boldsymbol{\sigma}(k-d) \end{bmatrix} \text{ and}$$

$$\boldsymbol{O}_1(k) = \mathrm{diag}(\boldsymbol{A}) \boldsymbol{O}_1(k-1) + u(k-d) \tag{23}$$

$$\boldsymbol{\sigma}(k) = \boldsymbol{\sigma}\left(\boldsymbol{V}^{\mathrm{T}} \boldsymbol{X}(k) + \boldsymbol{\theta}_1\right) \tag{24}$$

and

$$\boldsymbol{A} = \mathrm{logsig}(\boldsymbol{C}) \tag{25}$$

where, k means the kth sampling time interval. $O_1(k) = [O_{1i}(k)]^{\mathrm{T}}$, is the output vector of the first-order difference operator, $i = 1, 2 \ldots M$; $W_1 = [w_{1i}]^{\mathrm{T}}$, is the corresponding weights vector; $A = [a_i]^{\mathrm{T}}$ is the self-recurrent weight vector; $\sigma(k) = [\sigma_j(k)]^{\mathrm{T}}$ is the output of the hidden layer of NN, $j = 1,2 \ldots N$; $W_2 = [W_{2j}]^{\mathrm{T}}$, is the weight vector between the hidden layer and output layer; V is the weight matrix between the first hidden layer and the input layer; $\theta_1 = [\theta_j]^{\mathrm{T}}$, is the thresholds for the hidden layer; θ_2 is the threshold of the output layer of the hybrid model; $X(k)$ is the input vector of NN, i.e. $X(k) = [u(k), f_H(k)]^{\mathrm{T}}$.

The parameters W, V, θ_2, θ_1 and $C(A)$ can be identified by use of the nonlinear optimization method such as Levenberg-Marquardt algorithm [4] for batch training, i.e.

Define the cost function as

$$E = \frac{1}{2}\sum_{k=1}^{L}(y(k) - y_m(k))^2 = \frac{1}{2}\boldsymbol{e}^{\mathrm{T}}\boldsymbol{e} \tag{26}$$

where L is the length of the input/output data set; $y(k)$ is the actual output sequence of the rate-dependent hysteresis. The weight vector is defined by

$$\boldsymbol{\theta}^{\mathrm{T}} = \begin{bmatrix} \boldsymbol{W}_1^{\mathrm{T}} & \boldsymbol{C}^{\mathrm{T}} & \boldsymbol{W}_2^{\mathrm{T}} & \theta_2 & \boldsymbol{V}_1 & \boldsymbol{V}_2 & \boldsymbol{\theta}_1^{\mathrm{T}} \end{bmatrix}. \tag{27}$$

Letting $d = 0$ for simplicity gives

$$\frac{\partial e(k)}{\partial \boldsymbol{W}_1^{\mathrm{T}}} = -\boldsymbol{O}^{\mathrm{T}}(k); \tag{28}$$

$$\frac{\partial e(k)}{\partial \boldsymbol{C}^{\mathrm{T}}} = -\sum_{i=1}^{N}\left[\left(\mathrm{diag}(\boldsymbol{A})\right)^{i}\mathrm{diag}(1-\boldsymbol{A})\,\mathrm{diag}\left(\boldsymbol{W}_1\right)\boldsymbol{O}(k-i)\right]^{\mathrm{T}} \tag{29}$$

$$\frac{\partial e(k)}{\partial \boldsymbol{W}_2^{\mathrm{T}}} = -\boldsymbol{\sigma}^{\mathrm{T}}(k); \tag{30}$$

$$\frac{\partial e(k)}{\partial \theta_2} = -1; \tag{31}$$

$$\frac{\partial e(k)}{\partial \boldsymbol{V}_1} = -\left[\mathrm{diag}\left(\boldsymbol{W}_2\right)\boldsymbol{\sigma}'(k)u(k)\right]^{\mathrm{T}}; \tag{32}$$

$$\frac{\partial e(k)}{\partial \boldsymbol{V}_2} = -\left[\mathrm{diag}\left(\boldsymbol{W}_2\right)\boldsymbol{\sigma}'(k)p(k)\right]^{\mathrm{T}}; \tag{33}$$

and

$$\frac{\partial e(k)}{\partial \boldsymbol{\theta}_1^{\mathrm{T}}} = -\left[\mathrm{diag}\left(\boldsymbol{W}_2\right)\boldsymbol{\sigma}'(k)\right]^{\mathrm{T}}; \tag{34}$$

where $\boldsymbol{\sigma}'(\cdot)$ denotes the derivative of $\boldsymbol{\sigma}(\cdot)$. The Jacobian matrix is defined by

$$\begin{aligned}\boldsymbol{J} = [&\left[\frac{\partial e(k)}{\partial \boldsymbol{W}_1^{\mathrm{T}}}\right]^{\mathrm{T}}\left[\frac{\partial e(k)}{\partial \boldsymbol{C}^{\mathrm{T}}}\right]^{\mathrm{T}}\left[\frac{\partial e(k)}{\partial \boldsymbol{W}_2^{\mathrm{T}}}\right]^{\mathrm{T}} \\ &\left[\frac{\partial e(k)}{\partial \theta_2}\right]^{\mathrm{T}}\left[\frac{\partial e(k)}{\partial \boldsymbol{V}_1}\right]^{\mathrm{T}}\left[\frac{\partial e(k)}{\partial \boldsymbol{V}_2}\right]^{\mathrm{T}}\left[\frac{\partial e(k)}{\partial \boldsymbol{\theta}_1^{\mathrm{T}}}\right]^{\mathrm{T}}]\end{aligned}. \tag{35}$$

Then, the corresponding weight-update algorithm is

$$\boldsymbol{\theta}^{(l+1)} = \boldsymbol{\theta}^{(l)} + \Delta\boldsymbol{\theta}^{(l)} \tag{36}$$

where

$$\Delta\boldsymbol{\theta}^{(l)} = -[\boldsymbol{J}^{\mathrm{T}(l)}\boldsymbol{J}^{(l)} + \mu\boldsymbol{I}]^{-1}\boldsymbol{J}^{\mathrm{T}(l)}\boldsymbol{e}^{(l)} \ . \tag{37}$$

The damping factor μ is updated by

$$\mu = \begin{cases} 0.5\mu & if\ E^{(l+1)} \le E^{(l)} \\ 2\mu & if\ E^{(l+1)} > E^{(l)} \end{cases} \tag{38}$$

where l is the iteration index. Note that the initial value of μ is chosen to be a large positive number, e.g. $\mu = 1.0\text{x}10^6$ whilst the internal variable $\boldsymbol{O}(k)$ can be estimated by (23) in each training epoch with the previously estimated weight matrix $\boldsymbol{A}$.

V. Inverse Neural Hysteresis Submodel

It is noted that the relation between the hysteresis and its inverse can be shown in Figure 6. From Figure 6, we note that the inverse hysteresis is asymmetric to its original image on line $H=u$. Similar to what we discussed above, we can also develop an inverse hysteretic operator to extract the main feature of the inverse hysteresis. Then, we can follow the above-mentioned method to construct an expanded input space with the introduced inverse hysteretic operator. On this expanded input space, the inverse hysteresis can also be transformed to a one-to-one mapping. The inverse hysteretic operator can be obtained based on the hysteretic operator, i.e.

$$f_v[H(u)] = \ln(\sqrt{[H(u)-H(u_p)]^2+1}+[H(u)-H(u_p)]+f_v[H(u_p)] \tag{39}$$

where $H(u)$ is the input of the inverse hysteretic operator, which is actually the output of the hysteresis.

Similarly, the same result can be concluded that, with the introduction of the inverse hysteretic operator, the multi-valued mapping of the inverse hysteresis can also be transformed into a one-to-one mapping on the corresponding expanded input space. Moreover, the obtained mapping is a continuous one-to-one mapping on a compact set. Hence, the neural sub-model for the inverse rate-independent hysteresis can be derived, i.e.

$$\left\| \Gamma_v[H(t), f_v(H(t))] - NN_v[H(t), f_v(H(t))] \right\| < \varepsilon_v \tag{40}$$

where $NN_v\,(\cdot)$ is the neural network to approximate $\Gamma_v\,(\cdot)$, i.e. the mapping of the inverse hysteresis on the corresponding expanded inverse input space; ε_v is the approximation error; for any $\varepsilon_{Nv} > 0$, there exists $\varepsilon_v < \varepsilon_{Nv}$.

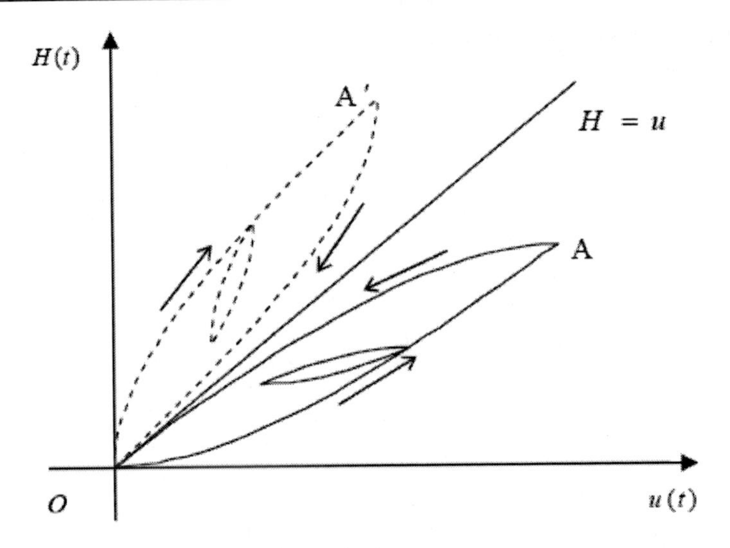

Figure 6. The relation between the hysteresis and its inverse.

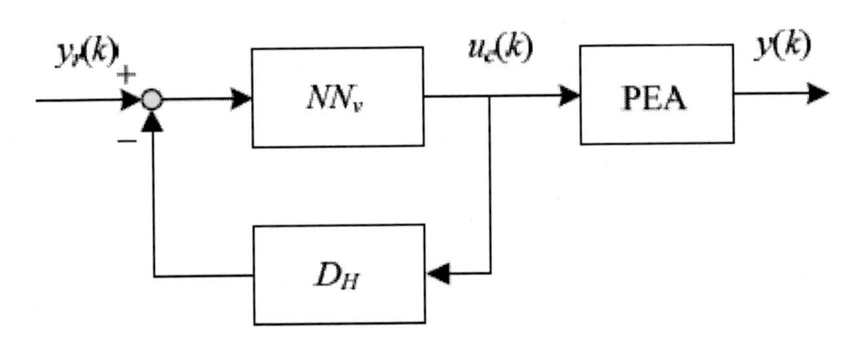

Figure 7. The structure of the compensator based on the hybrid model.

VI. The Model Based Compensator for Rate-Dependent Hysteresis

It is known that the compensation for the effect of the hysteresis in PEA is one of the important issues in precision positioning systems. Ref. [7] proposed an adaptive variable structure control method for hysteresis while Ref. [8] proposed an adaptive output feedback control strategy for hysteresis systems. However, all those methods are for rate-independent hysteresis. For the issue of compensation for the effect of rate-dependent hysteresis, the rate-dependent hysteretic compensator should be constructed [10]. In terms of the proposed model, the corresponding feedforward compensator to suppress the effect of rate-dependent hysteresis in PEA can be derived. The architecture of the compensator is shown in Figure 7. It can be seen that the compensator also has a parallel structure. In this architecture, NN_v is the inverse neural sub-model to suppress the effect of the rate-independent hysteretic performance in the PEA while D_H represents the linear dynamic sub-model used to compensate for the effect of rate-dependent behavior of the hysteresis inherent in the PEA.

Therefore, the corresponding control law to compensate for the rate-dependent hysteresis in the PEA can be described by

$$u_c(k) = NN_v[y_r(k) - D_H[u_c(k)]] \tag{41}$$

where u_c is the output of the compensator and y_r is the reference trajectory. Suppose the rate-dependent hysteresis can be decomposed into two parts, i.e. the rate-independent hysteresis and rate-dependent part. It can be described by

$$y(k) = H_{ri}[u_c(k)] + H_{rd}[u_c(k)] \tag{42}$$

where y is the output of the PEA, H_{ri} denotes the rate-independent part of the hysteresis and H_{rd} represents the rate-dependent part of the hysteresis. It is assumed that $H_{ri}NN_v = 1$ and $H_{rd}[u_c(k)] - D_H[u_c(k)] = 0$. Then, it leads to

$$\begin{aligned} y(k) &= H_{ri}[u_c(t)] + H_{rd}[u_c(t)] \\ &= H_{ri}NN_v[y_r(k) - D_H[u_c(k)]] + H_{rd}[u_c(k)]. \\ &= y_r(k) - D_H[u_c(k)] + H_{rd}[u_c(k)] \\ &= y_r(k) \end{aligned} \tag{43}$$

Obviously, the effect of the rate-dependent hysteresis in the PEA has been compensated. The output of the PEA can track the reference trajectory of the system. Of course, the accuracy of the model has a significant effect on the performance of the compensation.

VII. Experimental Results

In this section, the proposed approach is applied to the modeling of the dynamic hysteresis in the piezoelectric actuator. The experimental setup is composed of the following two parts, i.e.

1) the piezoelectric actuator (PZT-753.21C with the nominal displacement expansion of 0–20μm under the input voltage 0–120Volt.), voltage amplifier module (model E505.00) and a non-contact capacitive-type sensor/controller module (model E-509. CxA from PI Corp., Germany);
2) the personal computer, PC-7484 card (from Hotec Corp., Beijing, China) with 12 bits A/D converter, 8 bits D/A converter.

A dynamic voltage excitation with the frequency changing evenly from 1.2Hz to 300Hz is fed to the PZT to obtain the input (I) data and the output (O) data. The sampling frequency is 6000Hz.The first 1000 pairs of the measured data are used for model training while the next 600 I/O pairs of data are used for model validation. The time-delay in (24) is set to be $d = 1$.

The number of first-order difference operators is chosen as $M = 3$ while the hidden neurons of the neural network is $N = 5$. The Logsig function is adopted as activation function for the hidden layer of the neural network. The I/O data are normalized within [0, 1] used for model training. The Levenberg-Marquardt algorithm [4] is applied to the training of the model. After 400 epochs of training, the obtained mean squared error (MSE) and the relative maximal modeling error are 1.8274e-5 and 0.0103 respectively. On the other hand, the corresponding prediction results of the hybrid model for displacement expansion are demonstrated in Figs. 8(a), (b) and (c). From Figs. 8(a) and 8(b), we can see the comparison between the output of the hybrid model and the real data while Figure 8(c) illustrates the corresponding model validation residual. Note that the relative maximal model error is 0.0113 which is corresponding to 0.1062μm. The output of the hybrid model presents a well agreement with the actual measured data. In the obtained hybrid neural model, the corresponding parameter matrices are

$$\boldsymbol{W}_1 = [0.2375 \quad 0.7456 \quad -0.6413]^{\mathrm{T}};$$

$$\boldsymbol{W}_2 = [-0.0778 \quad 0.0221 \quad -1.0663 \quad 0.5635 \quad 0.8026]^{\mathrm{T}};$$

$$\theta_2 = 0.0363;$$

$$\boldsymbol{\theta}_1 = [0.4759 \quad 0.9059 \quad 0.3181 \quad 0.7600 \quad 0.5528]^{\mathrm{T}};$$

$$\boldsymbol{V} = \begin{bmatrix} 0.2040 & 0.4822 & 1.0632 & 0.8309 & -0.3238 \\ 0.5499 & 0.5372 & -0.2452 & 1.3230 & 0.8844 \end{bmatrix};$$

and

$$\boldsymbol{A} = [0.7098 \quad 0.6079 \quad 0.6730]^{\mathrm{T}}, \text{ respectively.}$$

For comparison, the PI-Creep model [10] with 6 PI operators and 5 creep operators is also implemented to approximate the data measured from the piezoelectric actuator. The thresholds of the PI operator are evenly distributed in (0.01, 0.5). The eigenvalues of creep operators as in (19) are chosen as, $\lambda_i = 10^{i-1} T_s$, T_s is the sampling period. The Levenberg-Marquardt algorithm was also used for the optimization of the parameters. After 400 epochs of training, the obtained modeling MSE is 5.1896e-005, and the relative maximal modeling error is 0.0181. On the other hand, the corresponding prediction results of the PI-Creep model are shown in Figs. 9(a) and (b) while the corresponding relative prediction error is shown in Figure 9(c). Note that the relative maximal prediction error is 0.0212, corresponding to 0.1993μm. Moreover, the identified parameters are shown as follows:

The weight of $u(k)$, i.e. $a = 2.5543$; the weight vector of the PI operators, i.e.

$\boldsymbol{b} = [0.4014 \quad 0.0524 \quad 0.0763 \quad 0.1345 \quad 0.3455 \quad 0.0890]^{T}$; and the weight vector of first-order differential operators, i.e. $\boldsymbol{c} = [-5.5090 \quad 2.7194 \quad -2.6573 \quad 4.6879 \quad -13.2193]^{T}$.

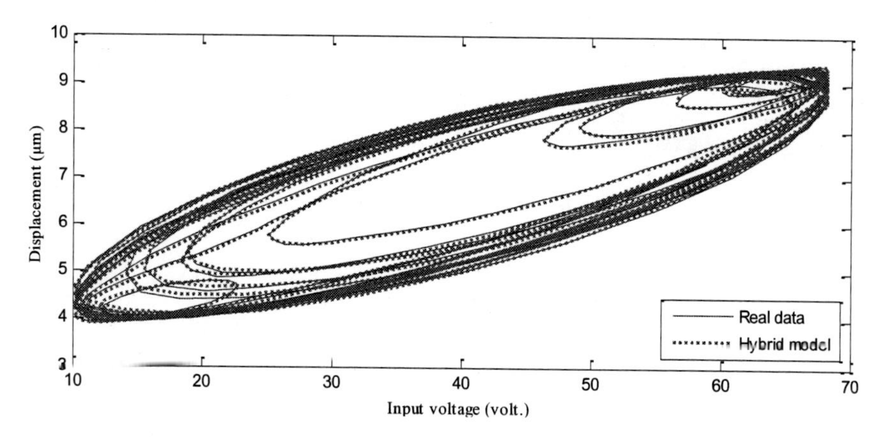

Figure 8(a). Model validation result of the hybrid model.

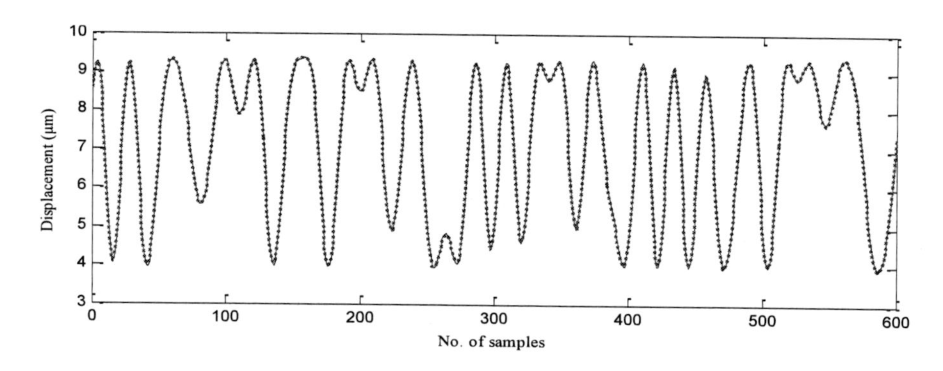

Figure 8(b). The prediction result of the hybrid model: real data(solid line), model output (dotted line).

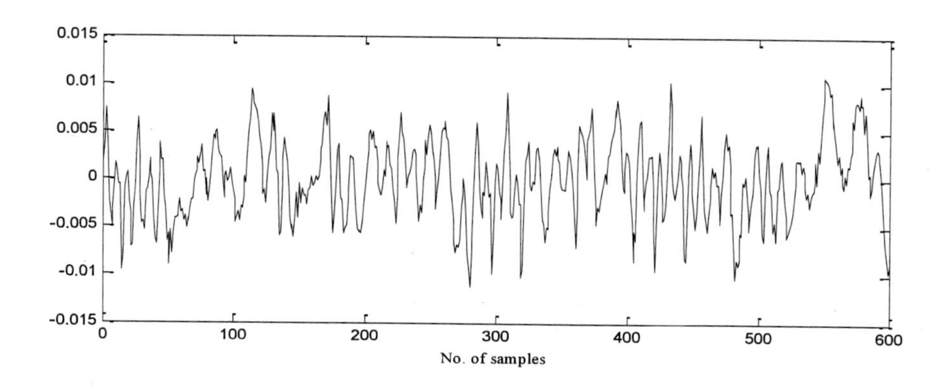

Figure 8(c). The relative prediction error of the hybrid model for displacement.

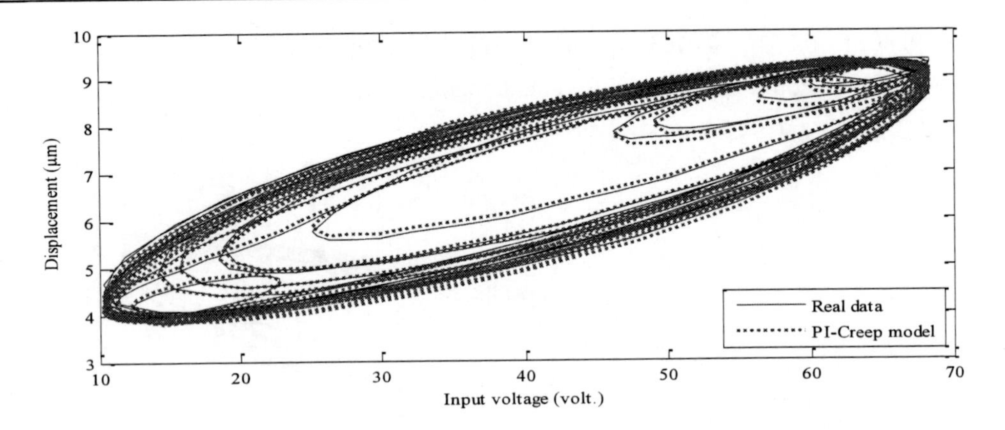

Figure 9(a). Model validation result of the PI-Creep model.

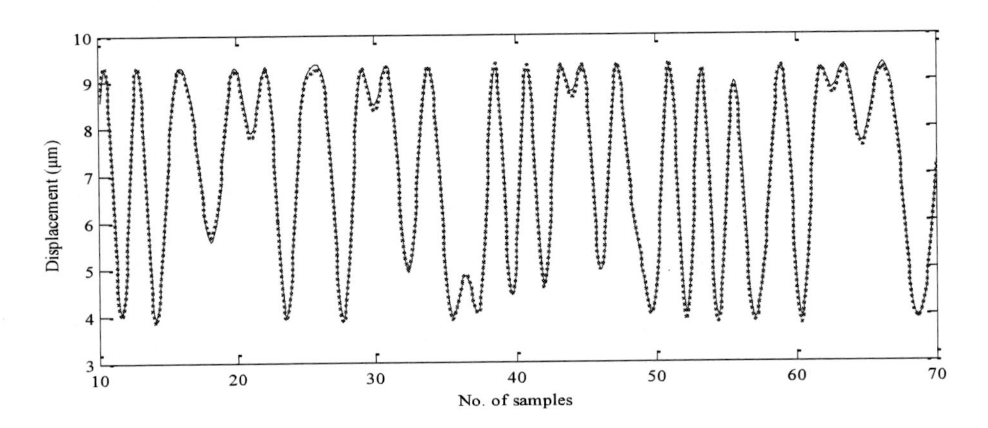

Figure 9(b). The prediction result of the PI-Creep model: real data (solid line), model output (dotted line).

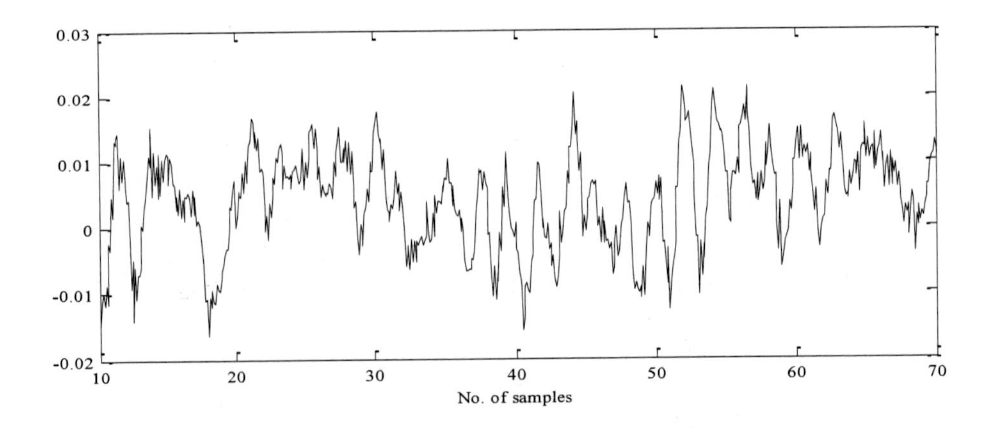

Figure 9(c). The relative prediction error of the PI-Creep model for displacement.

Based on the obtained neural inverse hysteresis sub-model, the corresponding compensator based on (41) is constructed to compensate for the effect of the rate-dependent hysteresis. The reference trajectory of the system is defined as

$$y_r(kT) = 4 + 2\exp(-2kT)\left[\sin\left(800\exp(-8kT)kT\right) + 1\right].$$

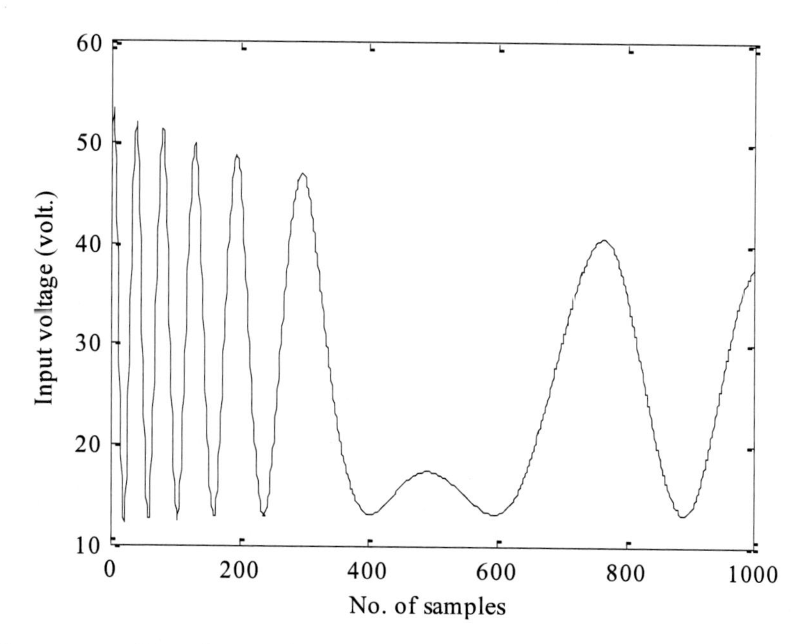

Figure10. The compensation voltage uc(k).

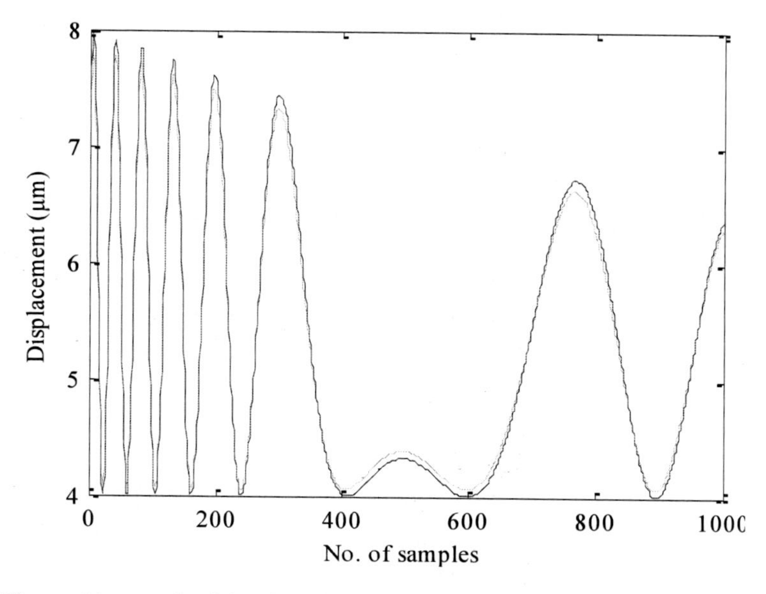

Figure 11(a). The tracking result of the piezoelectric actuator.

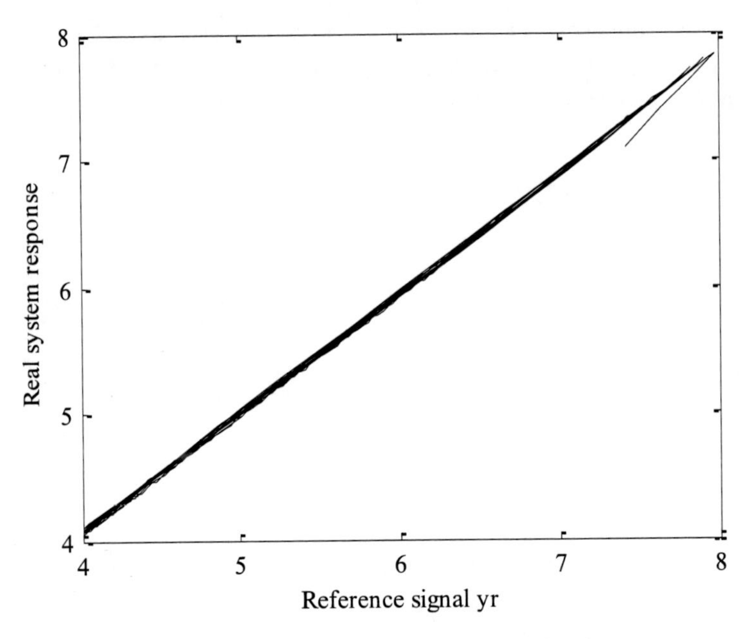

Figure11(b). The plot of the output of the PEA against the reference signal.

The generated compensation input voltage $u_c(t)$ is shown in Figure 10. Then, Figure 11(a) illustrates the corresponding tracking result. In Figure 11(a), the output of the actuator (dotted line) and the reference signal (solid line) are presented. Furthermore, Figure 11 (b) shows the plot of the output of the PEA against the reference signal. It is seen that with the compensation of the proposed method, the output of the PEA approaches the reference trajectory satisfactorily. It demonstrates that the response of the PEA is nearly free of the effect of the hysteresis with the compensation of the proposed method.

CONCLUSION

A model based compensation approach for rate-dependent hysteresis inherent in the piezoelectric actuators is proposed in this Chapter. The proposed model has a parallel structure which consists of a neural network based hysteresis sub-model and a linear dynamic sub-model. The neural sub-model constructed on an expanded input space is used to describe the rate-independent hysteretic behavior whilst the linear dynamic sub-model which is the sum of weighted difference functions is employed to approximate the rate-dependent feature of the hysteresis.

By use of the introduced hysteretic operator, the multi-valued mapping of the hysteresis can be transformed to a one-to-one mapping. Subsequently, the corresponding neural inverse model for the hysteresis is derived for the purpose of the inverse feedforward compensator design. The presented experimental results on a piezoelectric actuator have verified the effectiveness of the proposed compensation method for the rate-dependent hysteresis.

However, the performance of the proposed method depends upon the accuracy of the model to describe the rate-dependent hysteresis. The compensation errors may occur due to the modeling residuals. Therefore, an adaptive modeling strategy should be considered to improve the accuracy of the model or even more advanced control scheme should be developed based on the proposed compensation approach.

ACKNOWLEDGMENT

This research is partially supported by the Leading Academic Discipline Project of Shanghai Normal University (Grant Nos.: DZL811 and DRL904), the Innovation Program of Shanghai Municipal Education Commission (Grant Nos.:09ZZ141 and 11YZ92), the Advanced Research Grant of Shanghai Normal University (Grant Nos.: DYL201005 and DYL201006), the National Science Foundation of China (NSFC Grant No.: 60971004), the Natural Science Foundation of Shanghai (Grant No.: 09ZR14234000) and the Science and Technology Commission of Shanghai Municipality (Grant Nos.: 09220503000 and 10JC1412200).

REFERENCES

[1] Ang, W. T., Khosla P. K. and Riviere C. N., Feedforward controller with inverse rate-dependent model for piezoelectric actuators in trajectory-tracking applications, *IEEE/ASME Trans. Mechatron.*, Vol. 12, No. 2, 2007, 134-142.

[2] Dong R., Tan Y., Chen H. and Xie Y., "A neural networks based model for rate-dependent hysteresis for piezoceramic actuators", *Sensors and Actuators A*, 143, 2008,.370-376.

[3] Deng L. and Tan Y., Diagonal recurrent neural network with modified backlash operators for modeling of rate-dependent hysteresis in piezoelectric actuators, *Sensors and Actuators A: Physical*,Vol.148, No. 1, 4, 2008, 259-270

[4] Zhao X., Tan Y., Modeling hysteresis , *IEEE Transactions on Control Systems Technology,* Vol. 16, No.3, 2008, 484-490.

[5] Ge P., *Modelling and control of hysteresis in piezoceramic actuato*r, Ph. D. Dissertation, University of Rhode Island, 1996

[6] Hagan M., Menhaj M., Training feedforward networks with the Marquardt algorithm, *IEEE Trans. Neural Networks*, Vol. 5, No. 6, 1994, 889-993.

[7] Su C., Wang Q., Chen X., and Rakheja S., Adaptive variable structure control of a class of nonlinear systems with unknown Prandtl- Ishlinskii hysteresis, *IEEE Trans. Automatic Control,* Vol. 50, No.12, 2005, 2069-2074.

[8] Li C., Tan Y., Adaptive output feedback control of systems preceded by the Preisach-type hysteresis, *IEEE Trans. Sys. Man Cybern. B*, Vol. 35 No. 1, 2005,130-135.

[9] Janaideh M., Su C. and Rakheja S., Development of the rate-dependent Prandtl-shlinskii model for smart actuators, *Smart Materials and Structures,* Vol.17, No. 3, 2008, 1-11

[10] Oh J., Bernstein D., Semilinear Duhem Model for Rate-Independent and Rate-Dependent Hysteresis, *IEEE Trans. Auto. Control,* Vol. 50 No. 5, 2005, 631-645

[11] Armstrong-Helouvry B., *Control of machines with friction*, New York: Kluwer, 1991

[12] Wei J D, Sun C T. Constructing hysteretic memory in neural networks. *IEEE Trans. Systems, Man and Cybernetics Part B: Cybernetics*, Vol.30, No. 4, 2000, 601-609

[13] Mayergoyz I.D., *Mathematical models of hysteresis*, New York: Springer-Verlag, 1991.

[14] Mayergoyz I.D., Friedman G., Generalized Preisach model of hystereisis, *IEEE Trans. Magnetics*, Vol. 24, No. 1, 1998, 212-217

[15] M. A. Krasnosel'skii and A. V. Pokrovskii, *Systems with Hysteresis*. New York: Springer-Verlag, 1989

[16] Su C Y, Wang Q, Chen X, et al. Adaptive variable structure control of a class of nonlinear systems with unknown Prandtl-Ishlinskii hysteresis, *IEEE Trans. Automatic Control*, Vol. 50, No.12, 2005, 2069-2074

[17] Adly A.A., Abd-El-Hafiz S.K., Using neural networks in the identification of Preisach-type hysteresis models, *IEEE Trans. Magnetics*, Vol. 34, No.3, 1998, 629-635

[18] Claudio Serpico, Ciro Visone, Magnetic Hyteresis Modeling via feed-forward neural networks, *IEEE Trans.Magnetics*, Vol. 34, No.3, 1998, 623-628

[19] Deng L., Tan Y., Modeling hysteresis in piezoelectric actuators using NARMAX models, *Sensors and Actuators A-Physical,* Vol.149, No. 1, 2009, 106-112

[20] Zhang X., Tan Y. and Su M., Modeling of hysteresis in piezoelectric actuators using neural networks, *Mechanical Systems and Signal Processing*, Vol. 23, No. 8, 2009, 2699–2711

[21] Zhang X., Tan Y., A hybrid model for rate-dependent hysteresis in piezoelectric actuators, *Sensors and Actuators A: Physical*, Vol.157, No.1, 54–60

In: Advances in Condensed Matter ... Volume 10
Editors: H. Geelvinck and S. Reynst
ISBN: 978-1-61209-533-2

Chapter 5

Dynamic Space Charge Gratings in Wide-Gap Semiconductors and Nanostructured Materials

I. A. Sokolov * **and** ***M. A. Bryushinin*** †
A. F. Ioffe Physical Technical Institute, 194021,
Politekhnicheskaya 26, St. Petersburg, Russia

Abstract

The process of light interaction with condensed matter is an important problem of the modern solid state physics and quantum electronics. Holographic recording in wide-gap photorefractive materials is of particular interest from the scientific and practical point of view. Such a recording includes the stage of space charge formation, which is associated with the spatial redistribution of photoinduced carriers. A number of new methods studying the dynamics of space charge and photoconductivity in wide-gap semiconductors were proposed during recent years.

The approach based on non-steady-state photocurrent investigation is considered to be very attractive and powerful technique for characterization of wide-gap semiconductors, molecular crystals, amorphous materials, photorefractive polymers. The development of the new experimental and theoretical approaches for investigation of wide-gap semiconductors and nanostructured materials including photorefractive sillenites grown in the air and argon atmosphere, molecular crystal SnS_2, boron nitride crystal and nanosized GaN and Se within porous matrices is the main goal of this work.

1. Introduction

The process of light interaction with condensed matter is an important problem of the modern solid state physics and quantum electronics [1]- [5]. Holographic recording in wide-gap semiconductors and nano-structured materials is of particular interest from the scientific

*E-mail address: i.a.sokolov@mail.ioffe.ru
†E-mail address: mb@mail.ioffe.ru

and practical point of view. Practical applications are associated with the detection of low-amplitude acoustic vibrations of real objects, such as ultrasonic transducers, composite materials, biological objects, metal production, MEMS [6]- [13]. Conventional techniques using homodyne laser vibrometers [7] are suitable for practical applications in this regard for their high sensitivity which is limited in principle only by shot-noise of the laser used. Their utilization is restrained by several problems such as slow phase drifts in the interferometer arms due to environmental reasons, necessity of fine optical adjustment and suppression of laser amplitude noise. Photorefractive wide-gap semiconductor materials offer a novel elegant way to solve the problem of keeping operation interferometer point constant [2,14–17]. The crystal replaces the conventional beam-splitter and can be controlled not only electrically but also optically, i.e., based on the principles of nonlinear optics. In addition, it can be multilayered. Such a multilayered adaptive beam-splitter is nothing else but a volume dynamic hologram recorded by the reference and signal waves.

Holographic recording includes the stage of space charge formation, which is associated with the spatial redistribution of photoinduced carriers [18]. The nonstationary holographic photocurrent [19] (or non-steady-state photo-EMF [20]) is a holography related effect and it appears in a semiconductor material illuminated by an oscillating light pattern. Such illumination is usually created by two coherent light beams one of which is phase modulated with frequency ω. The alternating current is resulted from the periodic relative shifts of the photoconductivity and space charge gratings which arise in the crystal's volume under illumination. Like the holographic recording in photorefractive crystals this effect demonstrates adaptive properties that promote its application in such areas as vibration monitoring, velocimetry, etc. [21]. The sensitivity of adaptive photodetector is comparable with the sensitivity of classical schemes using photodiodes. The amplitude of the non-steady-state photo-EMF can be increased by an external ac field so one can expect the improvement of the sensitivity in certain ranges of temporal and spatial frequencies. In contrast to the holographic methods this technique allows the direct transformation of phase modulated optical signals into the electrical current and can be applied for characterization of centrosymmetrical and even amorphous materials [22]- [31]. Since the photocurrent is originated from the interaction of both the photoconductivity and space charge gratings a lot of photoelectric parameters can be measured [18,20].

In this Chapter we report space-and-time current spectroscopy for characterization of wide-gap semiconductors (photorefractive sillenites, BN, SnS_2, CdTe, CdZnTe). The approach is based on illumination of semiconductor material with an oscillating interference pattern formed of two light waves one of which is phase modulated with frequency ω. The non-steady-state photocurrent flowing through the short-circuited semiconductor is the measurable quantity in this technique.

Photorefractive sillenites $Bi_{12}Si(Ge,Ti)O_{20}$ are promising materials for optical signal processing due to their high sensitivity [2]. Large electrooptic coefficient allows an easy observation of the internal electric field. Shallow traps play significant role in the process of hologram formation. Recording and erasing processes have several time constants [32, 33]. Many attempts have been started in order to determine the charge carrier mobility, which is among the most important parameters of the light-induced charge transport. For photorefractive sillenites electron mobility values measured by various techniques from 5.2×10^{-9} to 3.4×10^{-4} $m^2V^{-1}s^{-1}$ have been reported [34–38]. This scattering of the

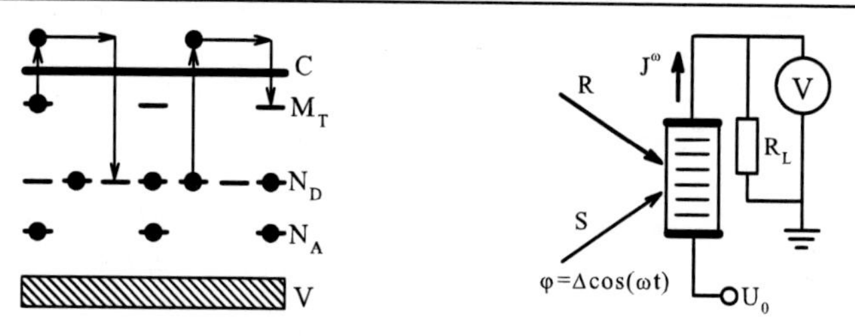

Figure 1. Model of a semiconductor with shallow levels (left figure) and the conventional scheme of holographic photocurrent measurements (right figure).

data is usually attributed to the influence of shallow traps which reduce the average velocity of excited carriers. Let us analyze the effect of non-steady-state photo-EMF for the model of photoconductor with a shallow level and consider its application for determination of material parameters.

2. Non-steady-State Photo-EMF in Semiconductors with Shallow Traps: Diffusion Regime

2.1. Electric Field and Photocurrent Amplitudes

This section presents calculations of the non-steady-state photo-EMF amplitude for the simplest two-level model of semiconductor with electron photoconductivity [39]. Photocarriers are excited from the deep donor levels (with the total density N_D) and from the shallow levels (with the total density M_T) [Fig. 1(a)]. The deep donors are ionized by light radiation only, for the shallow traps both the light and thermal mechanisms of activation are possible. The shallow level is assumed to be empty in the dark and the deep donor level is partially filled (with the density of vacant states N_A). The balance equations for this model can be written as follows [40]:

$$\frac{\partial N}{\partial t} = S_D I(N_D - N_A - N) - \gamma_D n(N_A + N), \tag{1}$$

$$\frac{\partial M}{\partial t} = -(\beta + S_T I)M + \gamma_T n(M_T - M). \tag{2}$$

Here N is the density of deep centers emptied by light ($N_D^+ = N_A + N$ is the total concentration of ionized donors), M is the density of shallow centers filled with electrons, n is the density of free electrons, S_D and S_T are the light excitation cross sections for deep and shallow levels, β is the thermal excitation rate from shallow level, γ_D and γ_T are the recombination constants for deep and shallow traps respectively, I is the light intensity.

For calculation of the non-steady state photocurrent amplitude kinetic equations (1), (2) should be added by the continuity and Poisson's equations and by the expression for the current density:

$$\frac{\partial}{\partial t}(N - M - n) + \frac{1}{e}\,\mathrm{div}\,\mathbf{j} = 0, \tag{3}$$

$$\operatorname{div}\mathbf{E} = \frac{e}{\epsilon\epsilon_0}(N - M - n), \tag{4}$$

$$\mathbf{j} = e\mu n\mathbf{E} + eD\,\mathrm{grad}\,n. \tag{5}$$

Here $\mathbf{j}$ is the current density, $\mathbf{E}$ is the electric field, μ is the electron mobility, $D = (k_BT/e)\mu$ is the diffusion coefficient, ϵ is the dielectric constant of the medium, ϵ_0 is the permittivity of free space, e is the electron charge.

Let us consider the excitation of the non-steady-state photo-EMF with an oscillating interference pattern formed by two coherent plane waves one of which is phase modulated with frequency ω and amplitude Δ [Fig. 1(b)]:

$$I(x,t) = I_0\{1 + m\cos[Kx + \Delta\cos(\omega t)]\}. \tag{6}$$

Here I_0 is the average light intensity, m is the contrast and K is the spatial frequency of the interference pattern [20]. For small contrast of the interference pattern m and small amplitude of phase modulation Δ ($m \ll 1$, $\Delta \ll 1$) the intensity distribution (6) can be written as [41]:

$$\begin{aligned} I(x,t) &= I_0 + I^{+0}\exp(iKx) + I^{-0}\exp(-iKx) \\ &\quad + I^{++}\exp[i(Kx+\omega t)] + I^{+-}\exp[i(Kx-\omega t)] \\ &\quad + I^{-+}\exp[i(-Kx+\omega t)] + I^{--}\exp[i(-Kx-\omega t)], \end{aligned} \tag{7}$$

where

$$\begin{aligned} I^{+0} &= I^{-0} = \frac{m}{2}I_0, \\ I^{++} &= I^{+-} = -I^{-+} = -I^{--} = \frac{im\Delta}{4}I_0. \end{aligned} \tag{8}$$

We shall look for the solution of Equations (1) - (5) in the form of identical decompositions for $N(x,t)$, $M(x,t)$, $n(x,t)$, $E(x,t)$ as it was done for the intensity distribution. All the necessary complex amplitudes for N, M, n and E have been found and the most important of them can be written as follows:

$$n^{\pm 0} = \frac{m}{2}\frac{S_DI_0(N_D - N_A - N_0)T_D + S_TI_0M_0T_T}{1 + T_D/\tau_D + T_T/\tau_T + \tau_M DK^2}, \tag{9}$$

$$n^{\pm +} = \frac{\pm im\Delta}{4}\frac{S_DI_0(N_D - N_A - N_0)T_D/(1+i\omega T_D) + S_TI_0M_0T_T/(1+i\omega T_T)}{1 + T_D/[\tau_D(1+i\omega T_D)] + T_T/[\tau_T(1+i\omega T_T)] + \tau_M DK^2/(1+i\omega\tau_M)}, \tag{10}$$

$$E^{\pm 0} = \frac{\mp imE_d}{2n_0}\frac{S_DI_0(N_D - N_A - N_0)T_D + S_TI_0M_0T_T}{1 + T_D/\tau_D + T_T/\tau_T + \tau_M DK^2}, \tag{11}$$

$$E^{\pm +} = \frac{m\Delta E_d}{4n_0}\frac{S_DI_0(N_D - N_A - N_0)T_D/(1+i\omega T_D) + S_TI_0M_0T_T/(1+i\omega T_T)}{\{1 + T_D/[\tau_D(1+i\omega T_D)] + T_T/[\tau_T(1+i\omega T_T)]\}(1+i\omega\tau_M) + \tau_M DK^2}. \tag{12}$$

Here N_0, M_0, n_0 are the stationary and spatial uniform concentrations, $E_d = (k_BT/e)K$ is the diffusion field, $\tau_M = \epsilon\epsilon_0/\sigma_0$ is the Maxwell relaxation time, $\sigma_0 = e\mu n_0$ is the average photoconductivity,

$$\begin{aligned} T_D &= (S_DI_0 + \gamma_D n_0)^{-1}, \\ T_T &= (\beta + S_TI_0 + \gamma_T n_0)^{-1} \end{aligned} \tag{13}$$

are the inverse values of the sum of the generation and ion recombination rates for deep and shallow levels [42],

$$\begin{aligned} \tau_D &= [\gamma_D(N_A + N_0)]^{-1}, \\ \tau_T &= [\gamma_T(M_T - M_0)]^{-1} \end{aligned} \tag{14}$$

are the recombination times of the electrons to the deep and shallow traps. Note, that the obtained relation for the electric field (12) can also be useful for the analysis of two-wave mixing in photorefractive materials.

The total photocurrent through the short-circuited sample (for cyclic boundary conditions) is defined by the drift component of current density averaged over the interelectrode spacing L [20]:

$$j(t) = \frac{1}{L}\int_0^L e\mu n(x,t)E(x,t)dx. \tag{15}$$

As follows from Eq. (15) the contribution to the first harmonic (with frequency ω) of the non-steady-state photocurrent density amplitude is produced only by the following combination of coefficients:

$$j^\omega = 2e\mu(n^{++}E^{-0} + n^{-+}E^{+0} + n^{+0}E^{-+} + n^{-0}E^{++}). \tag{16}$$

Finally, for the complex amplitude of the current density we obtain the following expression:

$$\begin{aligned} j^\omega &= -\frac{m^2\Delta\sigma_0 E_d i\omega\tau_M}{2n_0^2}\frac{S_D I_0(N_D - N_A - N_0)T_D + S_T I_0 M_0 T_T}{1 + T_D/\tau_D + T_T/\tau_T + \tau_M DK^2} \\ &\times \frac{S_D I_0(N_D - N_A - N_0)T_D/(1+i\omega T_D) + S_T I_0 M_0 T_T/(1+i\omega T_T)}{\{1 + T_D/[\tau_D(1+i\omega T_D)] + T_T/[\tau_T(1+i\omega T_T)]\}(1+i\omega\tau_M) + \tau_M DK^2}. \end{aligned} \tag{17}$$

It is important to point out that all expressions for the electric field and photocurrent amplitude were obtained with the only approximation of small contrast m and amplitude of phase modulation Δ (no assumptions were made about the occupancy of the deep and shallow levels and the prevalence of any activation mechanism for the shallow level).

2.2. Low Intensity Approximation

The expression for the complex amplitude of the photo-EMF [Eq. (17)] contains stationary and spatially uniform concentrations N_0, M_0 and n_0 that appear in the sample under uniform illumination with the average intensity I_0. The values N_0, M_0 and n_0 are calculated from (1), (2) and the neutrality condition:

$$N_0 = M_0 + n_0. \tag{18}$$

In this section we consider the situation of low illumination levels, i.e. slight filling of the shallow level:

$$M_0 \ll M_T. \tag{19}$$

The strong filling of the shallow level makes the effective mobility and lifetime tend to their real values [43], this situation is considered separately in section 2.3..

The simplest solution of equations (1), (2), (18) can be obtained in the case of linear generation from the deep level

$$N_0 \ll N_D - N_A \tag{20}$$

and predominance of the thermal activation of shallow traps under light excitation

$$S_T I_0 \ll \beta. \tag{21}$$

The assumption (20) means that we neglect light-induced changes in absorption [40]. If conditions (19) - (21) are satisfied the concentrations N_0, n_0 equal

$$N_0 = \sqrt{\frac{N_A^2}{4} + \frac{S_D I_0 (N_D - N_A)}{\gamma_D}\left(1 + \frac{M_T}{N_{CM}}\right)} - \frac{N_A}{2}, \tag{22}$$

$$n_0 = \frac{N_0}{(1 + M_T/N_{CM})}, \tag{23}$$

where

$$N_{CM} = \beta/\gamma_T = N_C \exp[-E_T/(k_B T)] \tag{24}$$

is the effective density of states in the conduction band calculated with respect to the energy position of shallow traps [43], N_C is the effective density of states in the conduction band, E_T is the energy depth of the shallow level, k_B is the Boltzmann constant, T is the absolute temperature.

The concepts of real and effective photoelectric parameters were considered elsewhere [32,43]. For the considered model, the relation between effective and real parameters is the following:

$$\mu/\mu' = \gamma_D/\gamma'_D = (1 + M_T/N_{CM}). \tag{25}$$

This expression denotes that all the values containing N_0 (e.g. Debye screening length L_S), μn_0-product (e.g. average photoconductivity σ_0) and $\mu\gamma_D^{-1}$-product (e.g. diffusion length L_D) can be written using either real or effective parameters.

There are two significant types of intensity dependencies of n_0 and N_D^+ (and as a consequence, photoconductivity and effective lifetime of carriers). If the intensity of the incident radiation is small enough, so that $N_0 \ll N_A$, the effective lifetime $\tau' = (\gamma'_D N_D^+)^{-1}$ is independent on light intensity and photoconductivity is the linear function of light intensity ($\sigma_0 \propto I_0$). For the opposite situation, i.e. for $N_0 \gg N_A$ (the case of quadratic recombination), the quantities σ_0 and τ' depend on the intensity as: $\sigma_0, 1/\tau' \propto I_0^{0.5}$.

As seen the general expression (17) is rather complicated for further analysis, so we shall consider the frequency transfer function of the photo-EMF under certain simplifying assumptions. Three of them, i.e. Eqs. (19) - (21), were made when we calculated the spatially uniform concentrations. The fourth condition can be written as follows:

$$S_T M_0 \ll S_D (N_D - N_A). \tag{26}$$

This is the case when the light excitation of free carriers from the deep level is more effective than the light excitation from the shallow one. Another condition can be written as

$$\beta T_D \gg 1. \tag{27}$$

This condition states that the number of free carriers excited from the shallow level during characteristic time T_D is larger than the number of carriers captured at this level (for $N_D \gg N_A + N_0$ the value T_D is nearly equal to the lifetime of ionized deep level, i.e. $T_D \approx (\gamma_D n_0)^{-1}$). Under these simplifying assumptions the expression for the photocurrent amplitude is the following:

$$j^\omega = \frac{-0.5m^2\Delta\sigma_0 E_d(1+\Theta+K^2L_S^2)^{-1}i\omega\tau_M}{1+\Theta+K^2L_S^2-\omega^2\tau_M\tau'+i\omega[\tau'+\tau_M(1+\Theta+K^2L_D^2)]}. \tag{28}$$

Here

$$L_D = \sqrt{\mu\tau_D k_B T/e} \tag{29}$$

is the average diffusion length of photocarriers (corresponding to deep traps),

$$\tau' = \tau'(\omega) = \tau_D\left(1+\frac{M_T/N_{CM}}{1+i\omega\beta^{-1}}\right) \tag{30}$$

is complex and frequency-dependent effective electron lifetime,

$$L_S = \sqrt{\frac{\epsilon\epsilon_0 k_B T}{e^2(N_A+N_0)(1-N_A/N_D)}} \tag{31}$$

is the Debye screening length [40,41],

$$\Theta = [(1+N_A/N_0)(1-N_A/N_D)]^{-1} \tag{32}$$

is the dimensionless parameter ($0 < \Theta < 1$). This parameter characterizes the type of captures to the deep level: $\Theta \ll 1$ for linear recombination ($N_0 \ll N_A$, $\sigma_0 \propto I_0$) and Θ tends to 1 for quadratic recombination ($N_0 \gg N_A$, $\sigma_0 \propto I_0^{0.5}$).

For the numerical calculations of the photo-EMF we use the following parameters [2, 32,38,44]:

$$\begin{aligned}
&N_D = 1\times10^{25}\ \mathrm{m^{-3}}, \qquad N_A = 0.95\times10^{22}\ \mathrm{m^{-3}},\\
&M_T = 10^{20}-10^{21}\ \mathrm{m^{-3}}, \quad \mu = 3.4\times10^{-4}\ \mathrm{m^2V^{-1}s^{-1}},\\
&S_D = 1.06\times10^{-5}\ \mathrm{m^2J^{-1}}, \quad \beta^{-1} = 2\times10^{-7}\ \mathrm{s},\\
&\gamma'_D = 1.65\times10^{-17}\ \mathrm{m^3s^{-1}}, \quad \tau_T = 4\times10^{-9}\ \mathrm{s}.
\end{aligned} \tag{33}$$

Conditions (19) - (21), (26), (27) are satisfied for moderate light intensities $I_0 = 0-3\times10^4$ W/m^2. It is useful to write the values contained in (28) and calculated for $I_0 = 10^4$ W/m^2:

$$\begin{aligned}
\tau_D &= 1.3\times10^{-7}\ \mathrm{s}, & \tau'(0) &= 6.4\times10^{-6}\ \mathrm{s},\\
\tau_M &= 6.9\times10^{-5}\ \mathrm{s}, & T_D &= 9.0\times10^{-3}\ \mathrm{s},\\
T_T &= 2.0\times10^{-7}\ \mathrm{s}, & L_D &= 1.0\times10^{-6}\ \mathrm{m},\\
L_S &= 9.1\times10^{-8}\ \mathrm{m}, & \Theta &= 7.1\times10^{-4}.
\end{aligned} \tag{34}$$

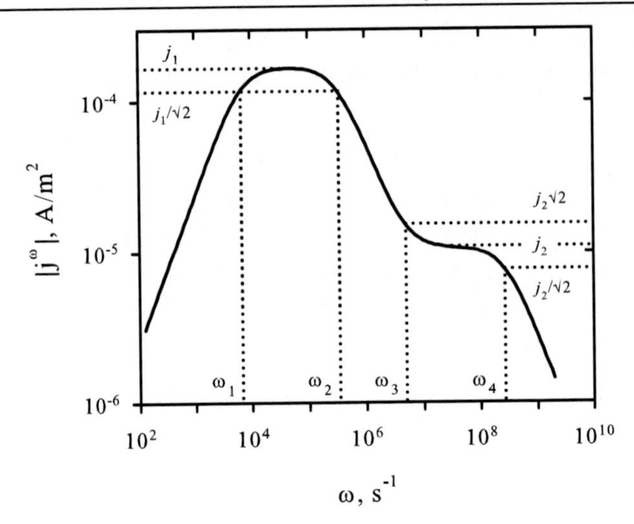

Figure 2. Theoretical dependence of the photocurrent amplitude j^ω versus frequency of phase modulation ω ($I_0 = 10^4$ W/m^2, $K = 10^6$ m^{-1}, $m = 0.2$, $\Delta = 0.1$).

Note that $T_T = \beta^{-1}$ if assumptions (19), (21) take place. As seen conditions $\tau_T \ll \beta^{-1} \sim \tau_D \ll \tau'(0) \ll \tau_M \ll T_D$, $L_S \ll L_D$, $\Theta \ll 1$ are satisfied.

The frequency dependencies of the photocurrent amplitude and its phase are shown in Figs. 2, 3. The curve $|j^\omega(\omega)|$ has one growing, two frequency independent and two decreasing regions.

The linear growth at low frequencies following by the frequency independent region has the standard explanation [20, 41]. For low oscillation frequencies ω both the photoconductivity and electric field gratings follow the movements of the interference pattern. This results in a small value of periodical spatial shifts between these distributions, and as a consequence, in a small amplitude of the output electrical signal.

In order to clarify this situation let us consider the frequency dependencies of complex amplitudes n^{-0}, n^{++}, E^{-0}, E^{++} (Fig. 4) and estimate their relative contribution to the total current (16). As seen from Fig. 4(a,b) for $\omega \ll \omega_1$ the following relation between corresponding complex amplitudes $n^{-0}, n^{++}, E^{-0}, E^{++}$ is valid: $|n^{++}E^{-0}| \approx |n^{-0}E^{++}|$ (or $|n^{-0}/n^{++}| \approx |E^{-0}/E^{++}|$). However the phases of products $n^{++}E^{-0}$ and $n^{-0}E^{++}$ are opposite [Fig. 4(c)] and these components compensate each other resulting in a small photocurrent amplitude. If the frequency of phase modulation is increased the oscillation amplitudes n^{++} and E^{++} become smaller with the faster decrease of the latter component [Fig. 4(a,b)]. Besides an additional phase shift appears between them [Fig. 4(c)]. Finally we have $n^{++}E^{-0} + n^{-0}E^{++} \propto -i\omega\tau_M$ for $\omega < \omega_1$. The first cut-off frequency ω_1 is defined as

$$\omega_1 \approx \frac{1 + \Theta + K^2 L_S^2}{\tau'(0) + \tau_M(1 + \Theta + K^2 L_D^2)}. \tag{35}$$

For the "relaxation type" photoconductor ($\tau'(0) \ll \tau_M$) and for oscillation frequencies $\omega > \tau_M^{-1}$ the electric field grating can be considered as "frozen in" (i.e. $|n^{++}E^{-0}| > |n^{-0}E^{++}|$), the amplitude of the photocurrent reaches its maximum $|j^\omega| \propto |n^{++}E^{-0}|$ and

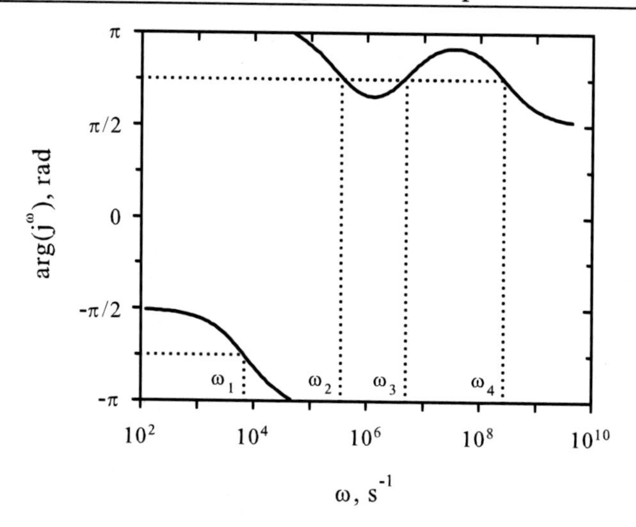

Figure 3. Frequency dependence of the photocurrent phase ($I_0 = 10^4$ W/m^2, $K = 10^6$ m^{-1}, $m = 0.2, \Delta = 0.1$).

becomes frequency independent up to the second characteristic frequency ω_2 [Figs. 2, 4(a)]:

$$\omega_2 \approx \frac{1}{\tau_M} + \frac{1+\Theta+K^2L_D^2}{\tau'(0)} + \frac{1+\Theta+K^2L_S^2}{\tau'(0)+\tau_M(1+\Theta+K^2L_D^2)}. \tag{36}$$

In the "lifetime" regime ($\tau'(0) \gg \tau_M$) relation $|n^{++}E^{-0}| \approx |n^{-0}E^{++}|$ maintain for $\omega < \omega_2 \approx \tau_M^{-1}$ and dependence $n^{++}E^{-0} + n^{-0}E^{++} \propto -i\omega\tau_M$ is expected but both complex amplitudes n^{++} and E^{++} decrease as $1/\omega$ for $\omega > \omega_1 \approx \tau'(0)^{-1}$ giving a plateau between ω_1 and ω_2.

The photocurrent amplitude for the first plateau for both "relaxation" and "lifetime" regimes can be written as follows:

$$j_1 \approx \frac{-0.5m^2\Delta\sigma_0 E_d(1+\Theta+K^2L_S^2)^{-1}}{1+\Theta+\tau'(0)/\tau_M+K^2L_D^2}. \tag{37}$$

The photo-EMF signal peaks at modulation frequency equal to

$$\omega_{m1} \approx \sqrt{\frac{1+\Theta+K^2L_S^2}{\tau_M\tau'(0)}}. \tag{38}$$

For high excitation frequencies $\omega > \omega_2$ the photocurrent amplitude falls down because of corresponding decrease of the photoconductivity grating amplitude n^{++} [Fig. 4(a)]. There are two regions on the frequency dependence of photocurrent where the signal amplitude decreases inversely proportional to the frequency of phase modulation ($\propto 1/\omega$). This behavior is associated with the frequency dependence of the effective electron lifetime

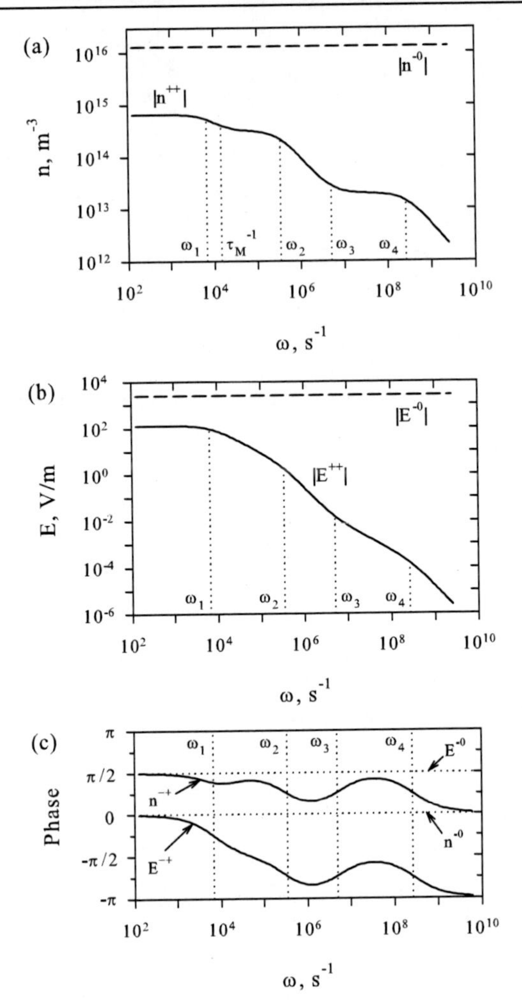

Figure 4. Amplitudes of electron concentration (a), electric field (b) and their phases (c) versus frequency of phase modulation ω ($I_0 = 10^4$ W/m^2, $K = 10^6$ m^{-1}, $m = 0.2$, $\Delta = 0.1$).

τ'. As seen from Eq. (30) there are two frequency regions where the effective lifetime τ' is real and frequency-independent:

$$\tau' \approx \tau'(0) = \tau_D(1 + M_T/N_{CM}) \tag{39}$$

for excitation frequencies much smaller than the thermal excitation rate from the shallow level ($\omega \ll \beta$) and

$$\tau' \approx \tau_D \tag{40}$$

for high excitation frequencies $\omega \gg \tau_T^{-1}$. The parameter $\tau'(0)$ was introduced earlier in Ref. [43] as a time constant describing the process of photoconductivity relaxation in semiconductor with shallow traps. When the crystal is illuminated with square modulated light the photocurrent pulse response has two characteristic regions: fast and slow [43].

The fast part starts immediately after illumination is switched on and associated with the process of establishment of thermal equilibrium between the shallow level and conduction band. The second (slow) region is observed under quasi steady-state condition and has characteristic time $\tau'(0)$.

The frequency dependence of the photo-EMF signal for $\omega > \beta$, as seen from Fig. 2, has another frequency independent region lying between two characteristic frequencies. The first one is equal to

$$\omega_3 \approx \beta \tag{41}$$

and the second one (for $\beta \gg \tau_M^{-1}$) can be written as:

$$\omega_4 \approx \frac{1+\Theta+K^2L_D^2}{\tau_D} + \frac{1}{\tau_T}. \tag{42}$$

One can see that the expressions for ω_2 and ω_4 have similar forms for the case of "relaxation type" photoconductor ($\tau'(0) \ll \tau_M$), linear photoconductivity ($\Theta \ll 1$) and low spatial frequencies ($KL_D < 1$). The second cut-off frequency ω_2 equals the inverse relaxation time of photoconductivity, i.e. $\omega_2 \approx \tau'(0)^{-1}$, whereas the fourth one ω_4 equals the inverse actual lifetime of electrons in the conduction band, i.e. $\omega_4 \approx \tau_D^{-1} + \tau_T^{-1}$.

The photocurrent amplitude in the frequency range $\omega_3 < \omega < \omega_4$ can be written as:

$$j_2 \approx \frac{-0.5m^2\Delta\sigma_0 E_d(1+\Theta+K^2L_S^2)^{-1}}{1+\Theta+\tau_D/\tau_M+\tau_D/\tau_T+K^2L_D^2}. \tag{43}$$

The existence of such a plateau on the frequency transfer functions of the photo-EMF (Fig. 2) and photoconductivity [Fig. 4(a)] can be explained as follows: the analysis of the effect shows that photocurrent signal is proportional to $(\omega\tau')^{-1}$ for modulation frequencies $\omega > \omega_2$. In the frequency range $\beta < \omega < \tau_T^{-1}$ the value $|\tau'|$ depends on frequency as ω^{-1} [see Eq. (30)] so the photocurrent and photoconductivity amplitudes remain approximately constant here.

The frequency dependencies of complex amplitudes N and M are shown in Fig. 5. As seen dependencies $|M^{++}(\omega)|$ and $|n^{++}(\omega)|$ are similar in the frequency range $\omega < \omega_3$ (in fact, $M^{++} \approx n^{++}T_T/\tau_T$). It means that space charge waves $en^{++}\exp(iKx+i\omega t)$ and $eM^{++}\exp(iKx+i\omega t)$ are summed for these frequencies and this sum can be considered as an effective wave of space charge traveling in the conduction band of the crystal characterized with the effective parameters μ', γ_D'. For frequencies $\omega > \omega_4$ we have $n^{++} \approx N^{++}$.

For the frequency range $\omega \ll \beta$ we can use the results obtained before, since expression (28) with $\tau' = \tau'(0)$ has the form similar to Eq. (19) in Ref. [41]. There are still, however, minor differences associated with definition of L_S and Θ. The photo-EMF signal in this frequency range can be characterized by maximum amplitude j_1 and corresponding cut-off frequencies ω_1, ω_2. [41] At these frequencies the photocurrent amplitude decreases by a factor of $1/\sqrt{2}$ (Fig. 2) and its phase equals to $-3\pi/4$ and $3\pi/4$ (Fig. 3).

The maximum amplitude j_1 and the cut-off frequencies ω_1, ω_2 are the values of interest since having been measured they allow to estimate the Maxwell relaxation time τ_M, effective lifetime $\tau'(0)$, diffusion length L_D, parameter Θ and the effective mobility $\mu' = eL_D^2/[k_BT\tau'(0)]$. Parameters τ_M, $\tau'(0)$, L_D can be obtained from the spatial frequency dependencies of characteristic frequencies $\omega_1(K)$ and $\omega_2(K)$. For the case of

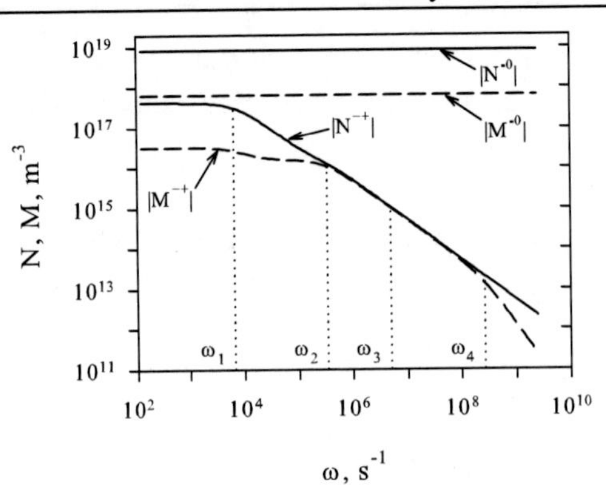

Figure 5. Amplitudes of densities N and M versus frequency of phase modulation ω ($I_0 = 10^4$ W/m^2, $K = 10^6$ m^{-1}, $m = 0.2, \Delta = 0.1$).

"relaxation type" photoconductor ($\tau'(0) \ll \tau_M$), linear photoconductivity ($\Theta \ll 1$) and low spatial frequencies ($KL_D, KL_S < 1$) we have $\omega_1 \approx \tau_M^{-1}$, $\omega_2 \approx \tau'(0)^{-1}$, and at spatial frequency $K = L_D^{-1}$ these cut-off frequencies change by a factor of 2: $\omega_1(K = L_D^{-1}) = 0.5\omega_1(K \approx 0)$, $\omega_2(K = L_D^{-1}) = 2\omega_2(K \approx 0)$. The value of diffusion length L_D can be also directly estimated from the dependence $j_1(K)$. As it follows from Eqs. (13), (14), (23), (32), (39) the value Θ can be expressed as the following ratio $\Theta = \tau'(0)/T_D$. So, if the photoconductivity σ_0 depends on light intensity as $\sigma_0(I_0) \propto I_0^x$ (in the finite region near I_0) then $\Theta(I_0) = 1/x - 1$. This means that $\Theta(I_0)$ can be estimated from the intensity dependence $\omega_1(I_0)$ measured for low spatial frequency ($KL_D, KL_S < 1$) in the crystal with $\tau'(0) \ll \tau_M$. Intensity dependence $\omega_2(I_0)$ can provide information about Θ as well. Finally, for crystal with $\tau'(0) \ll \tau_M$ and low spatial frequency ($KL_D < 1$) the value Θ can be estimated as follows: $\Theta(I_0) = [\omega_2(I_0) - \omega_2(0)]/[\omega_2(I_0) + \omega_2(0)]$.

As seen from Figs. 2, 3 and Eq. (41) there is a possibility for the experimental measurement of the thermal excitation rate from the shallow level β: at frequency $\omega = \beta$ the photocurrent equals $|j^\omega| = \sqrt{2}|j_2|$ and its phase is $3\pi/4$. Furthermore, from the measured values of corresponding cut-off frequencies ω_3, ω_4 and using Eqs. (24), (39), (41), (42) one can easily estimate the value of the electron lifetime with respect to the deep donor level

$$\tau_D \approx [1 + \Theta + K^2L_D^2 + \omega_3\tau'(0)]/(\omega_3 + \omega_4) \tag{44}$$

and the lifetime of electrons in the conduction band

$$\tau = (\tau_D^{-1} + \tau_T^{-1})^{-1} = [\omega_4 - (\Theta + K^2L_D^2)/\tau_D]^{-1}. \tag{45}$$

Besides that, from the estimated values of the diffusion length L_D and lifetime τ_D, we can easily obtain the real (band) mobility of photocarriers: $\mu = eL_D^2/[k_BT\tau_D]$.

Let us analyze Equation (28) for "relaxation type" photoconductor ($\tau'(0) \ll \tau_M$), for the case of linear recombination ($\Theta \ll 1$) and for low frequency range $\omega \ll \beta$. The most interesting dependencies $j^\omega(K)$ are those obtained for $\omega \ll \omega_1$ and for $\omega = \omega_{m1}$ (i.e. $j_1(K)$).

In the frequency range $\omega \ll \omega_1$ the signal amplitude can be written as:

$$j^\omega(K) \propto K(1 + K^2L_S^2)^{-2}. \tag{46}$$

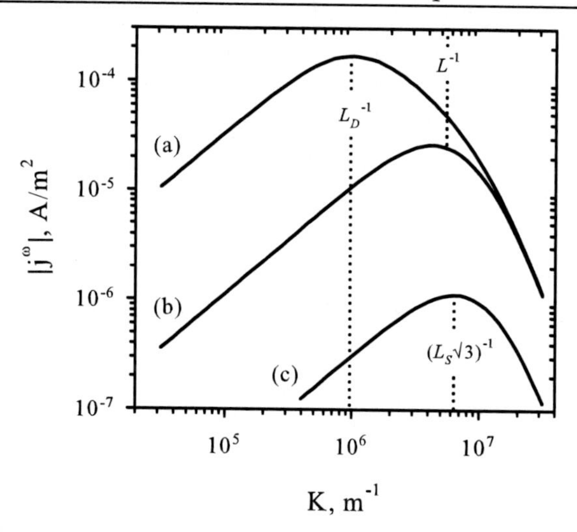

Figure 6. Dependence of the photocurrent amplitude j^{ω} on spatial frequency K calculated for characteristic temporal frequencies: (a) $\omega_1 < \omega < \omega_2$ [Eq. (37)], (b) $\omega_3 < \omega < \omega_4$ [Eq. (43)], (c) $\omega \ll \omega_1$ [Eq. (28)] for $\omega/2\pi = 2$ Hz; $I_0 = 10^4$ W/m^2, $m = 0.2, \Delta = 0.1$.

The signal amplitude grows linearly versus spatial frequency for $KL_S < 1/\sqrt{3}$ (Fig. 6, line c) which is due to increase of diffusion field amplitude E_d and decreases as $\sim K^{-3}$ for $KL_S > 1/\sqrt{3}$, behavior corresponding to saturation of deep traps. The maximum on this dependence is observed for $K = (\sqrt{3}L_S)^{-1}$.

For $\omega = \omega_{m1}$ as seen from Eq. (37) we have the following expression for the signal amplitude

$$j^{\omega}(K) \propto K(1 + K^2L_S^2)^{-1}(1 + K^2L_D^2)^{-1}. \tag{47}$$

In photorefractive sillenite crystals the value of the diffusion length of electrons is usually larger than the Debye screening length, i.e. $L_D > L_S$. In this case the decrease of the signal amplitude as $\sim K^{-1}$ (see dependence $j_1(K)$ in Fig. 6, line a) is due to smoothing of electron density distribution in the conduction band. The photocurrent amplitude peaks at spatial frequency $K \approx L_D^{-1}$ [20].

Let us analyze the spatial frequency dependence $j_2(K)$ of the photo-EMF signal for high excitation frequencies $\omega_3 < \omega < \omega_4$. For $\Theta \ll 1$, $\tau_D \ll \tau'(0) \ll \tau_M$ Equation (43) can be written in the following form:

$$j^{\omega}(K) \propto K(1 + K^2L_S^2)^{-1}(1 + K^2L^2)^{-1}. \tag{48}$$

Here $L = \sqrt{\mu\tau k_BT/e}$ is the actual diffusion length of electrons. If the condition $L > L_S$ is satisfied the maximum on this dependence is observed at $K \approx L^{-1}$ (Fig. 6, line b). This means that real parameters can be determined from the spatial frequency dependence of the photo-EMF signal measured in the frequency range $\omega_3 < \omega < \omega_4$. Such practical estimation of real diffusion length for sillenites may be complicated because of approaching L to L_S (for the chosen material parameters (33) $L = 1.8 \times 10^{-7}$ m and $L_S = 9.1 \times 10^{-8}$ m).

Let us first consider the intensity dependencies of the photocurrent amplitude and corresponding cut-off frequencies for low illumination levels. The shallow level is almost empty in this case [see Eq. (19)] and all intensity dependencies of the effect are determined by behavior of the deep level. It follows from the fact that the intensity dependencies of values $n_0, \sigma_0, \tau_M, \tau_D, \tau'(0), L_D, L_S, \Theta$ are expressed via the intensity dependence of the density N_0 [see Eqs. (14), (23), (29) - (32)]. The dependence $N_0(I_0)$ (22) in its turn is almost identical to the analogous dependence for the standard model of semiconductor with one deep level [41]. The difference is described by the presence of the intensity independent factor $(1 + M_T/N_{CM})$ and can be eliminated by using effective parameter γ'_D (25).

For the case of linear recombination of photocarriers ($N_0 \ll N_A$) and for $\tau'(0) \ll \tau_M$ corresponding values j_1, j_2 and ω_1 [Eqs. (37), (43), (35)] are proportional to the average light intensity I_0, other quantities, i.e. ω_2, ω_3, ω_4 [Eqs. (36), (41), (42)], are intensity independent. It follows from the linear character of the photoconductivity $\tau_M^{-1} \propto \sigma_0 \propto I_0$ and intensity independence of the other material parameters: $\tau'(0), \tau_D, \tau_T, L_D, L_S, \Theta \approx \mathrm{const}(I_0)$.

The situation of quadratic recombination ($N_0 \gg N_A$) and for $\tau'(0) \sim \tau_M$ is more complicated: all values except ω_3 are intensity dependent. This peculiarity appears due to the intensity dependencies of the material parameters: $\sigma_0, \tau_M^{-1}, \tau_D^{-1}, \tau'(0)^{-1}, L_D^{-2}, L_S^{-2} \propto I_0^{0.5}$. For low spatial frequencies ($KL_D, KL_S < 1$) we have $j_1, \omega_1, \omega_2 \propto I_0^{0.5}$, the character of dependencies $j_2(I_0), \omega_4(I_0)$ can vary according to the relation between τ_D and τ_T. For the sillenite crystals sublinear photoconductivity dependence is observed [46], $\tau_T \ll \tau_D$ ($\tau_T = 4 \times 10^{-9}$ s [2], $\tau_D \sim (3 - 160) \times 10^{-6}$ s) and hence $j_2 \propto I_0, \omega_4 \approx \mathrm{const}(I_0)$. The value τ_D is calculated using Eq. (39) and experimental estimations of the photoconductivity relaxation time $\tau'(0) = (0.15 - 8) \times 10^{-3}$ s obtained for the sillenite crystals grown in an oxygen-free atmosphere [45,46].

2.3. Strong Filling of Shallow Levels

Let us consider the signal dependence for the case of sufficient filling of the shallow level. In order to perform numerical calculations of photovoltage we have to estimate stationary and spatially uniform densities N_0, M_0 and n_0. Let us assume that the conditions of linear generation (20) and linear recombination

$$N_0 \ll N_A \tag{49}$$

are satisfied for the deep level. Then the necessary concentrations are easily calculated from (1), (2), (18) and equal to

$$n_0 = S_D I_0 (N_D - N_A)/(\gamma_D N_A), \tag{50}$$

$$M_0 = \frac{M_T}{1 + (\beta + S_T I_0)/(\gamma_T n_0)}, \tag{51}$$

and the quantity N_0 is defined as their sum (18). As seen from Eq. (50) the concentration of electrons in the conduction band (average photoconductivity) is proportional to the light intensity while M_0 tends to the finite value for high light intensity (compare to Eq. (15) in Ref. [40]):

$$M_\infty = \lim_{I_0 \to \infty} M_0 = \frac{M_T}{1 + [S_T N_A \gamma_D]/[S_D (N_D - N_A)\gamma_T]}. \tag{52}$$

For the case

$$S_D(N_D - N_A)/S_T M_T \gg \gamma_D N_A/\gamma_T M_T \tag{53}$$

this value is close to the total density of shallow traps M_T, so the strong filling of shallow level is possible for high light intensities.

In this section we analyse the effect of photo-EMF for high light intensity when the shallow level is filled substantially:

$$M_0 \approx M_T. \tag{54}$$

For arbitrary material parameters and occupancy degrees of both levels one should use the general expression [Eq. (17)] to calculate the photocurrent amplitude. Here we shall consider two interesting situations when Eq. (17) reduces to the form similar to the one-level model approximation.

Let us assume that the light excitation from the deep level strongly exceeds the light excitation from the shallow one [Eqs. (26), (53)]. In this case the shallow level is practically totally filled with electrons. For this reason we can neglect in Eq. (17) all terms containing parameters of shallow traps and we obtain the standard expression for photocurrent amplitude:

$$j^{\omega} = \frac{-0.5m^2 \Delta\sigma_0 E_d (1 + \Theta + K^2 L_S^2)^{-1} i\omega\tau_M}{1 + \Theta + K^2 L_S^2 - \omega^2 \tau_M \tau_D + i\omega[\tau_D + \tau_M(1 + \Theta + K^2 L_D^2)]}. \tag{55}$$

Here Θ is given by

$$\Theta = n_0/[(N_A + N_0)(1 - N_A/N_D)]. \tag{56}$$

Equation (55) is identical to the photocurrent amplitude calculated for one-level model of semiconductor [41]. In this case, however, investigation of frequency transfer function $j^{\omega}(\omega)$ and dependence $j^{\omega}(K)$ can provide estimations of real (but not effective) values τ_D, L_D and μ.

We have estimated the minimal light intensity needed for sufficient filling of shallow level so that Eq. (55) is the consequence of common expression (17) in whole frequency range: $I_0 \sim 3 \times 10^6$ W/m^2 (this estimation was obtained using Eqs. (18), (50), (51) for stationary concentrations, material parameters (33) and assuming $M_T = 1 \times 10^{20}$ m^{-3}, $S_T = 0$). For this intensity the shallow level is filled by 95%.

Let us consider the situation when significant light generation of electrons from the shallow level takes place:

$$S_T M_0 \sim S_D(N_D - N_A). \tag{57}$$

If $\gamma_T M_T \gg \gamma_D N_A$ the shallow level is practically filled for high light intensities [see Eq. (52)] and we can neglect the items containing τ_T in Eq. (17). The expression for the complex amplitude of photo-EMF can be written then as follows:

$$j^{\omega} = \frac{-0.5m^2 \Delta\sigma_0 E_d (1 + \Theta + K^2 L_S^2)^{-1} i\omega\tau_M A(0) A(\omega)}{1 + \Theta + K^2 L_S^2 - \omega^2 \tau_M \tau_D + i\omega[\tau_D + \tau_M(1 + \Theta + K^2 L_D^2)]}, \tag{58}$$

where

$$A(\omega) = 1 + \frac{S_T M_0 T_T}{S_D(N_D - N_A) T_D} \frac{1 + i\omega T_D}{1 + i\omega T_T}. \tag{59}$$

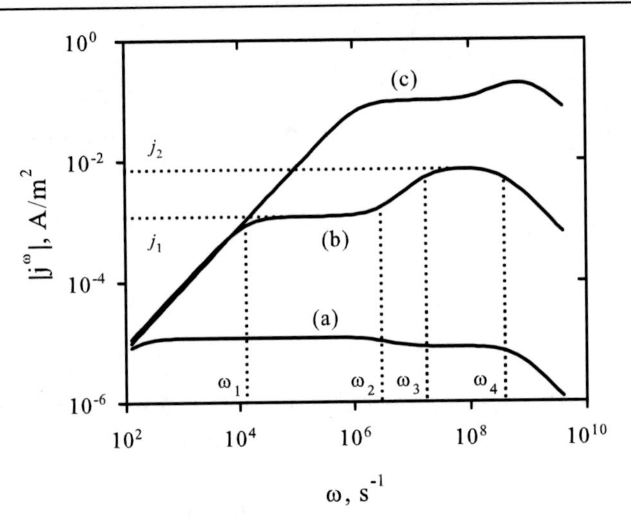

Figure 7. Frequency transfer function of the photo-EMF for different light intensities $I_0 = 3 \times 10^3, 3 \times 10^5, 3 \times 10^7$ W/m^2 (curves (a), (b) and (c) respectively). The calculations are performed using Eq. (17) for $K = 6.3 \times 10^6$ m^{-1}, $m = 0.2$, $\Delta = 0.1$.

The frequency transfer function described by Eqs. (58), (59) is presented in Fig. 7 (curves b, c). The calculations were performed using Eqs. (18), (50), (51) for stationary concentrations and material parameters (33). We also assumed that $M_T = 1 \times 10^{20}$ m^{-3}, $S_T = 10$ m^{2}J^{-1} (choosing such S_T we satisfy condition (57)). The characteristic cut-off frequencies are given by the following expressions:

$$\omega_1 \approx \frac{1 + \Theta + K^2 L_S^2}{\tau_D + \tau_M(1 + \Theta + K^2 L_D^2)}, \tag{60}$$

$$\omega_2 \approx \frac{1}{T_T[1 + S_T M_0 / S_D(N_D - N_A)]}, \tag{61}$$

$$\omega_3 \approx \frac{1}{T_T + \tau_D/(1 + \Theta + \tau_D/\tau_M + K^2 L_D^2)}, \tag{62}$$

$$\omega_4 \approx \frac{1}{\tau_M} + \frac{1 + \Theta + K^2 L_D^2}{\tau_D} + \frac{1}{T_T} + \frac{1}{T_T + \tau_D/(1 + \Theta + \tau_D/\tau_M + K^2 L_D^2)}. \tag{63}$$

Calculating ω_2, ω_3, ω_4 we have supposed $T_T \ll \tau_D + \tau_M(1 + \Theta + K^2 L_D^2)$. For the considered range of light intensities $I_0 = 0 - 10^8$ W/m^2 the following relation is satisfied: $\tau_M \geq 30 T_T$. The amplitude of the photo-EMF signal for the frequency range $\omega_3 < \omega < \omega_4$ can be written as:

$$j_2 \approx \frac{-0.5 m^2 \Delta \sigma_0 E_d A(0)[1 + S_T M_0 / S_D(N_D - N_A)]}{(1 + \Theta + K^2 L_S^2)(1 + \Theta + \tau_D/\tau_M + \tau_D/T_T + K^2 L_D^2)}. \tag{64}$$

For low frequencies of phase modulation ($\omega < \omega_2$) the effect of photo-EMF is described by standard expression (55) since $A(0) \approx 1$. As seen from Fig. 7 the presence of shallow traps reveals itself only for high excitation frequencies ($\omega > \omega_2$). In contrast to the low intensity approximation ($M_0 \ll M_T, S_T M_0 \ll S_D(N_D - N_A)$) the influence of the shallow level for the case of high light intensities ($M_0 \approx M_T$, $S_T M_0 \sim S_D(N_D - N_A)$) leads to the increase of photocurrent at high frequencies of phase modulation (compare Figs. 2 and 7).

This fact can be explained as follows: when the shallow level is filled slightly it play a role of an electron reservoir leading to the increase of the photoconductivity relaxation time and reducing the effective electron mobility. However, when this level is filled strongly the characteristic time τ_D tends to its actual value [see Eqs. (55), (58)] and sufficient light generation from this level leads to photocurrent increase with respect to standard estimation of (55).

In the first part of the Chapter we have analyzed the influence of shallow traps on the effect of non-steady-state photo-EMF. We have derived the general expression for amplitude of the photocurrent for apparent filling degrees of deep and shallow levels. We analyze the photocurrent behavior with respect to temporal and spatial frequencies and discuss the case of high illumination levels. The general calculations were performed for the crystal's parameters typical for photorefractive $Bi_{12}SiO_{20}$. The influence of shallow traps is not restricted to considered cases. The effect of non-steady-state photo-EMF in other photoconductive crystals (e.g. semi-insulating GaAs) with different material constants, may reveal peculiarities at other characteristic temporal and spatial frequencies.

3. Non-steady-State Photo-EMF in an External DC Electric Field

The subject of space charge waves covers a wide range of physics from plasmas, acoustics and semiconductors to the relatively new area of photorefractive optics [47,48]. Space charge waves can be regarded as quasiparticles in solids and are of great interest not only in the case of photorefractive materials but also in the case of many semiconductors. They determine the dynamic behavior of the space charge distribution in the crystal and can be excited optically by various methods [44,49–51]. A powerful technique is to use the spatially oscillating interference pattern created by two coherent laser beams. Then, if the frequency of oscillation and the grating spacing of the interference pattern coincide with the frequency and the wavelength of the corresponding space charge wave, a resonance excitation of the space charge wave occurs. Space charge and photoconductivity waves play important role in inducing non-steady-state photocurrents in photorefractive crystals [20,37]. This effect provide an unique opportunity for direct measurement of the photoelectron's drift mobility from the position of the corresponding resonance maxima in frequency transfer function of the photocurrent [37,52].

In this part of the Chapter we shall analyze the resonant excitation of space charge and photoconductivity waves and their detection using the non-steady-state photoelectromotive force effect. We consider the model of semiconductor with shallow energy level and present numerical calculations for the commonly used $Bi_{12}SiO_{20}$ crystal. It is important to point out that application of the external electric field completely changes the physics of the phenomenon and here we deal with the *resonant* excitation of the *eigen-waves* in the material

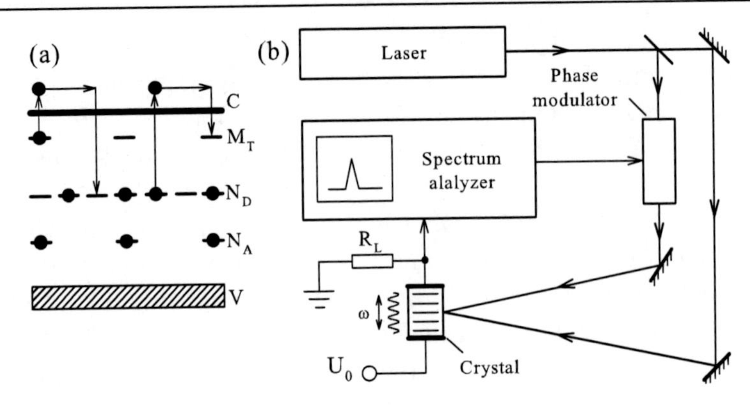

Figure 8. Model of semiconductor with shallow levels (a) and experimental setup for measurements of the non-steady-state photocurrent (b).

and not with the *forced* excitation of the effect as in Ref. [20].

Let us consider the model of semiconductor with two levels from which photoelectrons are generated [Fig. 8]. We confine ourselves to the photoexcitation from the deep donor level (with the total density N_D) and both light and thermal excitation from the shallow level (with the total density M_T). The shallow level is assumed to be empty in the dark and the deep one is partially filled (with the density of vacant states N_A).

Suppose a crystal is illuminated with the oscillating interference pattern $I(x,t) = I_0\{1 + m\cos[Kx + \Delta\cos(\omega t)]\}$. Theoretical analysis of the non-steady-state photocurrent generation is performed for the following simplifying assumptions: a negligible dark conductivity, a low contrast light pattern ($m \ll 1$), and low-amplitude vibrations of the interference pattern ($\Delta \ll 1$). No assumptions were made on the occupancy of both levels and the prevalence of any activation mechanism for the shallow level.

We have obtained the general expression for the photocurrent amplitude. It is too bulky, however, so we shall discuss the most interesting case of the linear generation from the deep level, low filling degree and negligible light excitation from the shallow level:

$$\begin{aligned} N_0 \ll N_D - N_A, \quad M_0 \ll M_T, \quad S_T I_0 \ll \beta, \\ S_T M_0 \ll S_D(N_D - N_A), \quad \beta T_D \gg 1. \end{aligned} \tag{65}$$

Here N_0, M_0, n_0 are the stationary and spatially uniform densities (defined from Eqs. (1), (2) and neutrality condition $N_0 = M_0 + n_0$), $T_D = (S_D I_0 + \gamma_D n_0)^{-1} \approx (\gamma_D n_0)^{-1}$ is the inverse of the sum of the generation and ion recombination rates for the deep level. The fifth condition ($\beta T_D \gg 1$) states that the number of free carriers excited from shallow level during lifetime of ionized deep level is larger than the number of carriers captured at this level.

Under these assumptions the expression for the photocurrent amplitude is simplified to:

$$\begin{aligned} j^\omega = & \frac{0.25m^2\Delta\sigma_0(Z^*)^{-1}[i2E_0 - \omega\tau_M(E_0 + iE_d)]}{Z - \omega^2\tau'\tau_M + i\omega[\tau' + \tau_M(1 + \Theta + K^2L_D^2 + iKL_0)]} \\ & - \frac{0.25m^2\Delta\sigma_0 Z^{-1}[i2E_0 - \omega\tau_M(E_0 - iE_d)]}{Z^* - \omega^2\tau'\tau_M + i\omega[\tau' + \tau_M(1 + \Theta + K^2L_D^2 - iKL_0)]}. \end{aligned} \tag{66}$$

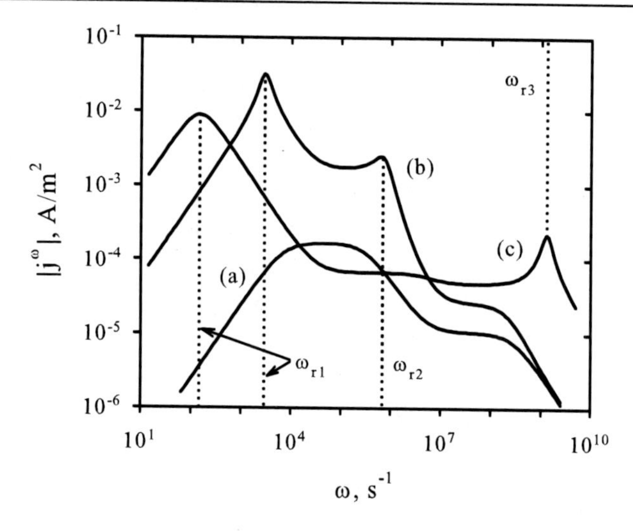

Figure 9. Theoretical dependencies of the photocurrent amplitude $|j^{\omega}|$ on the oscillation frequency ω ($I_0 = 10^4$ W/m^2, $m = 0.2$, $\Delta = 0.1$): (a) $E_0 = 0$, $K = 1.00 \times 10^6$ m^{-1}, (b) $E_0 = 6 \times 10^5$ V/m, $K = 1.89 \times 10^5$ m^{-1}, (c) $E_0 = 2 \times 10^6$ V/m, $K = 1.89 \times 10^6$ m^{-1}.

Here $\sigma_0 = e\mu n_0$ is the average photoconductivity of the crystal, E_0 is the external electric field, $E_d = Kk_BT/e$ is the diffusion field, $\tau_M = \epsilon\epsilon_0/\sigma_0$ is the Maxwell relaxation time,

$$\tau' = \tau'(\omega) = \tau_D \left(1 + \frac{\tau_T^{-1}\beta^{-1}}{1 + i\omega\beta^{-1}}\right) \tag{67}$$

is the complex and frequency-dependent effective electron lifetime, $\tau_D = [\gamma_D(N_A + N_0)]^{-1}$, $\tau_T = [\gamma_T(M_T - M_0)]^{-1}$ are the recombination times of electrons to the deep and shallow traps, $L_D = \sqrt{D\tau_D}$, $L_0 = \mu\tau_D E_0$ are the average diffusion and drift lengths of photocarriers, $L_S = [\epsilon\epsilon_0 k_B T/e^2(N_A + N_0)(1 - N_A/N_D)]^{1/2}$ is the Debye screening length [40], $Z = 1 + \Theta + K^2L_S^2(1 + iE_0/E_d)$ and $\Theta = [(1 + N_A/N_0)(1 - N_A/N_D)]^{-1}$ are the dimensionless parameters ($0 < \Theta < 1$).

The frequency dependence of the photovoltage amplitude is shown in Fig. 9. The material parameters used for numerical calculations are taken from Refs. [32,44]. Depending on the amplitude of the applied electric field E_0, spatial frequency K and material parameters the frequency response of the photocurrent can have from one to three maxima.

Let us first consider the low-frequency range of photocurrent excitation when $\omega \ll \beta$. For such modulation frequencies the effective electron lifetime is the real value: $\tau' \approx \tau'(0) = \tau_D(1 + \tau_T^{-1}\beta^{-1})$. It can be shown that for $KL_0 > 1 + \Theta + K^2L_D^2$ there is a maximum on the frequency dependence of the non-steady-state photocurrent at $\omega_{r1} \approx |Z|/|\tau'(0) + \tau_M(1 + \Theta + K^2L_D^2 + iKL_0)|$. The existence of such a peak is associated with the excitation of space charge waves [49,50]. For $L_S < L_D$ the resonant frequency ω_{r1} is a decreasing function of the applied electric field E_0 [37,49]. However, the finite screening length L_S taken into account makes ω_{r1} tend to a non-zero limit for high electric fields $E_0 \to \infty$: $\omega_{r1} \to T_D^{-1}$ (compare to Eq. (16) in Ref. [37]).

For the intermediate range of excitation frequencies when $\omega_{r1} \ll \omega \ll \beta$ and for the case of relaxation type semiconductor [$\tau'(0) \ll |\tau_M(1+\Theta+K^2L_D^2+iKL_0)|$] the second maximum is observed on the frequency transfer function. Two types of this maximum can be considered in this case. The first one corresponds to the resonant excitation of a photoconductivity wave and is detected at $\omega_{r2} \approx KL_0/\tau'(0) = \mu' E_0 K$, where $\mu' = \mu/(1+\tau_T^{-1}\beta^{-1})$ is the effective mobility of electrons. Such a resonant maximum exists in the frequency range $(1+\Theta+K^2L_D^2)/\tau'(0) < \omega < \beta$. This characteristic frequency ω_{r2} is an increasing function of values E_0 and K, but when it approaches the thermal excitation rate β, the corresponding peak disappears. Thus there is an optimum value of the external electric field that maximizes the amplitude of the resonant peak:

$$E_0 \approx \frac{\beta}{\mu' K}\sqrt{\frac{\tau_T}{\tau_D}(1+\Theta+K^2L_D^2)}, \tag{68}$$

$$j^\omega \approx \frac{-im^2\Delta\sigma_0\beta}{8Z\mu' K}\sqrt{\frac{\tau_T}{\tau_D(1+\Theta+K^2L_D^2)}}. \tag{69}$$

The second type of the maximum can be observed at another characteristic frequency: $\omega_{m2} \approx (1+\Theta+K^2L_D^2)/\tau'(0)$. This peak on the frequency response of the photocurrent is detected only for specific relations between the amplitude of the applied electric field E_0 and spatial frequency of the interference pattern K: $KL_0, KL_D, KL_S \leq 1$, $L_0 \approx [(1+\Theta)(L_D^2+L_S^2)]^{0.5}$. It is important to point out that the origin of this maximum is not associated with the resonant excitation of photoconductivity waves. The characteristic frequency ω_{m2} weakly depends on spatial frequency K, and the amplitude of this maximum increases with increasing K value.

For high excitation frequencies ($\omega > \tau_T^{-1}$) the effective lifetime of electrons is the real value as well: $\tau' \approx \tau_D$. The third resonant peak is observed in this case at the following frequency:

$$\omega_{r3} \approx KL_0/\tau_D = \mu E_0 K. \tag{70}$$

The presence of such a peak is the characteristic feature of the discussed model. This maximum can be observed in following frequency region $\omega_{r3} > (1+\Theta+K^2L_D^2)\tau_D^{-1} + \tau_T^{-1}$. In other words, this condition means that the photoconductivity waves are resonantly excited if an electron in conduction band is neither captured by deep nor by shallow traps during the period of interference pattern oscillation. The values of effective μ' and actual μ mobilities can be easily obtained from characteristic resonant frequencies ω_{r2} and ω_{r3} on the frequency dependence of the non-steady-state photocurrent.

The non-steady-state photocurrent amplitude for the second and third resonances can be estimated from the following expression

$$j^\omega_{r2,r3} \approx \frac{-i0.25m^2\Delta\sigma_0 E_0 Z^{-1}}{1+\Theta+K^2L_D^2+\tau_D/\tau_T(1+\beta^2/\omega_{r2,r3}^2)}. \tag{71}$$

We have assumed here that $\omega_{r2}, \omega_{r3} \gg \omega_{r1}$, $|\tau'| \ll |\tau_M(1+\Theta+K^2L_D^2+iKL_0)|$, $E_0 \gg E_d$.

The origin of the resonance maxima on the frequency transfer function can be explained as follows. The light intensity distribution $I(x,t)$, which is external periodic driven force in our case, contains the waves running in the opposite direction with respect to applied electric field ($E_0 > 0$): $I^{\pm\pm}\exp[\pm i(Kx+\omega t)]$. These light waves give rise the analogous waves of free carriers concentration: $n^{\pm\pm}\exp[\pm i(Kx+\omega t)]$. The phase velocities of these waves equal $v_{ph} = -\omega/K$. For $E_0 \gg E_d$ and $m \ll 1$ each separate electron migrates in semiconductor crystal with the average drift velocity $v_{dr} = -\mu E_0$ [during the actual electron lifetime $(\tau_D^{-1}+\tau_T^{-1})^{-1}$] and $v_{dr} = -\mu' E_0$ [during the effective electron lifetime $\tau'(0)$]. If the phase and drift velocities are not matched, i.e. $v_{ph} \neq v_{dr}$, this mentally selected electron outruns or lags behind the packet which phase velocity v_{ph} is determined by external influence $I(x,t)$. As the result, the concentration density distribution $n^{\pm\pm}\exp[\pm i(Kx+\omega t)]$ become blurred. At resonant frequencies ω_{r2}, ω_{r3} the photoconductivity wave and all the photocarriers migrate synchronously ($v_{ph} = v_{dr}$) giving rise to the appearance of corresponding resonant peaks. For the waves traveling in the direction of applied electric field, i.e. for $n^{\pm\mp}\exp[\pm i(Kx-\omega t)]$, the phase and drift velocities have the opposite signs. This means, in particular, that "smoothing" of the photoconductivity wave takes place for any modulation frequency ω and this component does not contribute to the resonant excitation of the external photocurrent.

Let us write down corresponding complex amplitude of the space-charge electric field formed in the crystal bulk for $E_0 \gg E_d$, $\Theta \ll 1$:

$$E^{++} = \frac{-i0.25m\Delta E_0}{Z^* - \omega^2\tau_M\tau' + i\omega[\tau' + \tau_M(1-iKL_0)]}. \tag{72}$$

It is seen from Eq. (72) that the time-dependent electric field grating amplitude is strongly decreased for high excitation frequencies $\omega \gg \omega_{r1}$. For example, this amplitude calculated at ω_{r3} and for the parameters used in Fig. 9, line (c) is only 3×10^{-3} V/m. This fact may essentially complicate the observation of the discussed effects using conventional holographic techniques.

The conventional experimental setup for observation of the non-steady-state photocurrents is presented in Fig. 8(b). A standard single-mode He-Ne laser with an average power of $P_0 = 40$ mW ($\lambda = 633$ nm) was used as the basic source of coherent radiation for formation of a recording interference pattern. To form the interference pattern with a specified fringe spacing and to cause its sinusoidal vibrations within the sample volume, we used a conventional Twyman-Green interferometer. An electro-optic phase modulator ML-102A produced phase modulation of the laser beam. The non-steady-state photocurrent was measured using standard SK4-56 spectrum analyzer. Figure 10 presents the frequency dependence of the non-steady-state photocurrent flowing in n-type photorefractive $Bi_{12}SiO_{20}$ grown in argon atmosphere. These crystals have several distinct features with respect to sillenite crystals grown in an air atmosphere [46]. First, these crystals have rather high photoconductivity in the red region of spectrum comparable with the photoconductivity of the usual sillenite crystals in the blue-green region of the spectrum [20]. Second, in all investigated crystals for moderate light intensity of the red light we observed unusual relation between the lifetime of photocarriers τ' and Maxwell relaxation time $\tau_M < \tau'(0)$. This means, in particular, that photocurrent measurements are done in the lifetime regime without external electric field. Nevertheless, the appropriate choice of the external electric field, makes the observation of the second resonance possible and photocurrent measurements are done in relaxation time regime $[\tau'(0) < \tau_M|1+\Theta+K^2L_D^2+iKL_0|]$. The drift mobility value was found to be $\mu' = (4.3 \pm 0.4) \times 10^{-6}$ m^2/V s ($I_0 = 60$ W/m^2) and corresponds

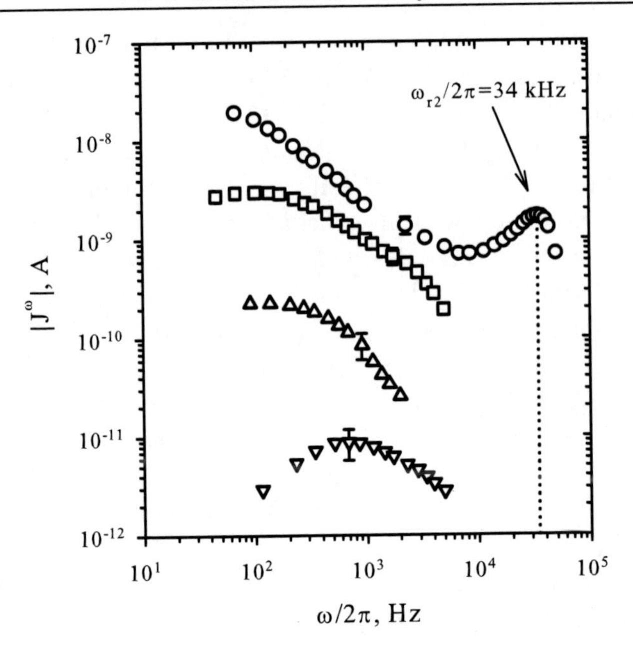

Figure 10. Frequency dependencies of photocurrent measured in $Bi_{12}SiO_{20}$ grown in argon atmosphere ($\lambda = 633$ nm, $K = 5.0 \times 10^4$ m^{-1}, $I_0 = 60$ W/m^2, $m = 0.2$, $\Delta = 0.2$): $E_0 = 0$ (∇), 2.0×10^5 ($\triangle$), 4.0×10^5 ($\square$), 1.0×10^6 V/m ($\circ$).

to the values of effective mobilities observed in conventional sillenites [37]. Thus we think that the crystal growth in oxygen free atmosphere affect the deep levels mainly, changing the properties of shallow levels slightly. The mobility turned out to be dependent on the light intensity which can be associated with both filling of shallow levels and heating of the crystal in the external electric field.

In conclusion, we have reported resonant excitation of space-charge and photoconductivity waves in semiconductors with shallow energy levels. We predicted the appearance of several resonance maxima on the frequency transfer function of the photocurrent and showed that observation of photoconductivity wave with conventional holographic technique is unlikely.

4. Non-steady-State Photo-EMF in an External AC Electric Field: Nonresonant Case

Let us present investigations of the non-steady-state photocurrent generation in an external AC field. We show both experimentally and theoretically that the frequency transfer function of the effect is similar to the response of the effect in the diffusion regime of excitation, i.e. linear growth of the signal for low modulation frequencies up to a certain characteristic cut-off frequency and frequency-independent region for high modulation frequencies. The new technique for the signal enhancement and determination of the $\mu\tau$-product is proposed.

4.1. Theoretical Analysis

This section presents calculations of the non-steady-state photocurrent amplitude excited in an alternating electric field. Suppose a crystal is illuminated with the oscillating interference pattern formed by two plane light waves one of which is phase modulated with frequency ω and amplitude Δ. We consider the widely used model of semiconductor with one type of partially compensated donor levels [2]. The balance equation for the concentration of electrons in the conduction band n can be written as follows:

$$\frac{\partial n}{\partial t} = g - \frac{n}{\tau} + \frac{1}{e}\frac{\partial j_e}{\partial x}. \tag{73}$$

Here $g(x,t) = (\alpha\beta/h\nu)I(x,t)$ is the electron generation rate (α is the light absorption coefficient, β is the quantum efficiency, and $h\nu$ is the photon energy). We assume low illumination levels when the generation and recombination rates are linear functions of the light intensity and electron concentration respectively, i.e. for the considered model conditions $N_A + N \ll N_D$, $N_A + N \gg n$ are fulfilled [N_D is the total concentration of donors, N is the concentration of donor centers emptied by light, N_A is the concentration of the compensating acceptor states. In this case the electron lifetime is independent on the light intensity and determined by the concentration N_A only. The effects of trap saturation are neglected as well (screening lengths are small compared to the inverse value of the spatial frequency). In order to calculate the distributions of the electron concentration, electric field and resulting photocurrent Eq. (73) should be added by the continuity and Poisson equations and by the expressions for the charge ρ and electron current j_e densities:

$$\frac{\partial \rho}{\partial x} + \frac{\partial j_e}{\partial x} = 0, \tag{74}$$

$$\epsilon\epsilon_0\frac{\partial E}{\partial x} = \rho, \tag{75}$$

$$\rho = e(N - n), \tag{76}$$

$$j_e = e\mu n E + k_B T \mu \frac{\partial n}{\partial x}. \tag{77}$$

where E is the electric field, μ is the electron mobility, ϵ is the dielectric constant, ϵ_0 is the free space permittivity, k_B is the Boltzmann constant, T is the temperature. Equations (73)-(77) can be represented as follows:

$$\frac{\partial n}{\partial t} = g - \frac{n}{\tau} - \frac{\epsilon\epsilon_0}{e}\frac{\partial^2 E}{\partial x \partial t}, \tag{78}$$

$$\frac{\partial}{\partial x}\left(e\mu n E + k_B T \mu \frac{\partial n}{\partial x} + \epsilon\epsilon_0 \frac{\partial E}{\partial t}\right) = 0. \tag{79}$$

The sinusoidal electric field with amplitude E_{ext} and frequency Ω is applied to the crystal: $E_{ext}(t) = E_{ext}\cos\Omega t$.

The calculation procedure simplifies substantially if we suppose the contrast m and amplitude of phase modulation Δ to be small: $m, \Delta \ll 1$. Then we can look for the solution of equations (78)-(79) to the lowest order with respect to the small parameters presenting $n(x,t)$, $E(x,t)$ as the sums of stationary and running gratings:

$$\begin{aligned} n &= n_0 + \sum_{p,q=-1}^{1} n^{+pq} \exp\{i[Kx + (p\omega + q\Omega)t]\} \\ &+ \sum_{p,q=-1}^{1} n^{-pq} \exp\{i[-Kx + (p\omega + q\Omega)t]\}, \end{aligned} \tag{80}$$

$$\begin{aligned} E &= E^{00+} \exp(i\Omega t) + E^{00-} \exp(-i\Omega t) \\ &+ \sum_{p,q=-1}^{1} E^{+pq} \exp\{i[Kx + (p\omega + q\Omega)t]\} \\ &+ \sum_{p,q=-1}^{1} E^{-pq} \exp\{i[-Kx + (p\omega + q\Omega)t]\}. \end{aligned} \tag{81}$$

It is important to point out that for the solution of similar tasks two basic approaches can be used: the utilization of truncated Fourier series [53] (as was also done in this paper) and the time averaging technique [54]. The former covers all time scales [55], the latter is convenient for arbitrary waveforms of the external field [56].

For the case of cyclic boundary conditions [20] the total current averaged over the interelectrode spacing L contains the drift and displacement components. The photocurrent signal is defined by the drift one: [20]

$$j(t) = \frac{1}{L} \int_0^L e\mu n(x,t) E(x,t) dx. \tag{82}$$

We confine ourselves to the analysis of the first harmonic of photocurrent (with frequency ω) which can be written in the form of the following combination of coefficients introduced in Eqs. (80), (81):

$$\begin{aligned} j^{\omega} &= 2e\mu \sum_{q=-1}^{1} \Big(n^{+1q} E^{-0(-q)} + n^{+0q} E^{-1(-q)} \\ &+ n^{-1q} E^{+0(-q)} + n^{-0q} E^{+1(-q)} \Big). \end{aligned} \tag{83}$$

The higher spatial and temporal harmonics (and even subharmonics) of values $n(x,t)$, $E(x,t)$ and resulting photocurrent $j(t)$ become substantial only for the case of large contrast and amplitude of phase modulation: $m, \Delta \sim 1$. [2,5,57] Although these problems are rather interesting and widely discussed in the literature the most valuable applications of the holographic photocurrents for the semiconductor material characterization are based on the measurements of the first harmonic of the signal.

Let us consider the case of "relaxation type" semiconductor , i.e. $\tau \ll \tau_M$ [$\tau_M = \epsilon\epsilon_0 / e\mu n_0$ is the Maxwell relaxation time]. We suppose that the frequency of phase modulation ω is much smaller than the frequency of the external electric field Ω; the period of the AC field is larger than the carrier lifetime τ and shorter than grating relaxation time τ_M:

$$\omega \ll \Omega, \tag{84}$$

$$\tau_M^{-1} \ll \Omega \ll \tau^{-1}. \tag{85}$$

We have obtained all complex amplitudes of the electron concentration and electric field necessary for calculation of the first harmonic of the photocurrent signal. Some of them are given below:

$$\begin{aligned}
n^{+00} &= \frac{m}{2}n_0, \quad n^{+01} = \frac{imKL_0}{2\sqrt{2}(1+K^2L_D^2)}n_0, \\
n^{+10} &= \frac{im\Delta(1+i\omega\tau_M)(1+K^2L_D^2)n_0/4}{1+K^2L_D^2+i\omega\tau_M[(1+K^2L_D^2)^2+K^2L_0^2]}, \\
n^{+11} &= \frac{-m\Delta(1+i\omega\tau_M)KL_0n_0/4\sqrt{2}}{1+K^2L_D^2+i\omega\tau_M[(1+K^2L_D^2)^2+K^2L_0^2]}, \\
E^{+00} &= -i\frac{m}{2}\left(E_D+E_0\frac{KL_0}{1+K^2L_D^2}\right), \\
E^{+01} &= \frac{imE_0}{2\sqrt{2}\Omega\tau_M(1+K^2L_D^2)}, \\
E^{+10} &= \frac{m\Delta[E_D(1+K^2L_D^2)+E_0KL_0]/4}{1+K^2L_D^2+i\omega\tau_M[(1+K^2L_D^2)^2+K^2L_0^2]}, \\
E^{+11} &= \frac{-m\Delta(1+i\omega\tau_M)E_0/4\sqrt{2}\Omega\tau_M}{1+K^2L_D^2+i\omega\tau_M[(1+K^2L_D^2)^2+K^2L_0^2]}.
\end{aligned} \tag{86}$$

Here $E_0 = E_{ext}/\sqrt{2}$ is the effective value of the external electric field, $E_D = (k_BT/e)K$ is the diffusion field, $\sigma_0 = e\mu n_0$ is the average conductivity, $L_0 = \mu\tau E_0$ is the drift length of photocarriers, $L_D = (k_BT\mu\tau/e)^{1/2}$ is the diffusion length.

Finally the expression for the complex amplitude of the photocurrent can be written as:

$$j^\omega = \frac{-0.5m^2\Delta\sigma_0[E_D+E_0KL_0/(1+K^2L_D^2)]i\omega\tau_M}{1+i\omega\tau_M[1+K^2L_D^2+K^2L_0^2/(1+K^2L_D^2)]}. \tag{87}$$

Let us list the main features of the nonstationary holographic photocurrents excited in the AC external field. As follows from Eq. (87) the frequency transfer function of the effect has the traditional form typical for the diffusion mechanism of photocurrent excitation, namely, the linear growth of the signal for low modulation frequencies and the frequency-independent region for high frequencies of phase modulation [20]. These two regions are separated by corresponding cut-off frequency:

$$\omega_0 = \frac{1}{\tau_M[1+K^2L_D^2+K^2L_0^2/(1+K^2L_D^2)]}. \tag{88}$$

For low spatial frequencies of the interference pattern (i.e. for $K^2L_D^2 \ll 1$) the most striking peculiarities can be observed. In this case the expressions for the photocurrent amplitude and corresponding cut-off frequency can be written as follows:

$$\begin{aligned}
j^\omega &= \frac{m^2\Delta}{2}\sigma_0E_D\frac{-i\omega\tau_M(1+L_0^2/L_D^2)}{1+i\omega\tau_M(1+K^2L_0^2)} \\
&= \frac{m^2\Delta}{2}\sigma_0E_D\frac{-i\omega\tau_M(1+E_0^2/E_L^2)}{1+i\omega\tau_M(1+E_0^2/E_M^2)},
\end{aligned} \tag{89}$$

$$\begin{aligned}
\omega_0 &= [\tau_M(1+K^2L_0^2)]^{-1} \\
&= [\tau_M(1+E_0^2/E_M^2)]^{-1}.
\end{aligned} \tag{90}$$

Here $E_L = k_B T/eL_D$, $E_M = (K\mu\tau)^{-1}$ are the characteristic values of the electric field [44]. As seen from Eq. (89) the dependence of the photocurrent amplitude on electric field has a S-like shape. The signal grows as E_0^2 between E_L and E_M due to the corresponding increase of the electric field grating amplitudes E^{+00}, E^{+10} [Eq. (86)]. Beyond this region ($E_0 < E_L$, $E_0 > E_M$) the signal is independent on the electric field. The external field does not affect the photocurrent amplitude while the effective drift length is smaller than the diffusion one ($L_0 \leq L_D$). For high values of the applied electric field (the so-called long drift length regime) an enhanced transfer of electrons characterized by the large effective drift length L_0 takes place. Under these conditions the nonstationary photoconductivity grating is blurred leading to the saturation of the signal amplitude. The corresponding cut-off frequency ω_0 [Eq. (90)] falls down when drift length L_0 becomes comparable to the spatial period of the recorded grating ($L_0 \geq K^{-1}$ or $E_0 \geq E_M$).

The dependencies of the photocurrent amplitude j^ω and cut-off frequency ω_0 versus the AC field amplitude E_0 can be used for the determination of transport parameters of photoexcited carriers in semiconductor materials. Indeed, the characteristic breakpoints in these dependencies give the values of the diffusion and drift lengths of photocarriers and consequently – the $\mu\tau$-product value.

We have also calculated the maximum gain of the photocurrent signal $j^\omega(\omega > \omega_0)$ which can be obtained in the sinusoidal AC field:

$$G_m = \frac{\lim\limits_{E_0\to\infty} j^\omega}{\lim\limits_{E_0\to 0} j^\omega} = \frac{1 + K^2L_D^2 + L_0^2/L_D^2}{1 + K^2L_D^2 + K^2L_0^2/(1+K^2L_D^2)}. \tag{91}$$

It depends on the spatial frequency of the interference pattern: $G_m = 1/K^2L_D^2$ for $K^2L_D^2 \ll 1$ and $G_m = 1$ for $K^2L_D^2 \gg 1$.

4.2. Experimental Setup and Results

The experimental setup used for the measurements of the nonstationary holographic currents under applied alternating electric field is shown in Fig. 11. A standard He-Cd laser with an average power of $P_{out} \approx 1$ mW ($\lambda = 442$ nm) was used as the basic source of coherent radiation for formation of a recording interference pattern. To form the interference pattern with a specified fringe spacing and to cause its sinusoidal vibrations within the sample volume, we used a conventional Twyman-Green interferometer. An electrooptic phase modulator ML-102A produced phase modulation of the laser beam (with frequency ω and amplitude $\Delta = 0.16$). The voltage from generator was amplified by conventional transformer and then applied to the crystal. The photocurrent signal was filtered and then measured by spectrum analyzer SK4-56 ($f = 0.01 - 50$ kHz, $\Delta f = 3$ Hz). In our experiments we used n-type $Bi_{12}SiO_{20}$ crystal with characteristic dimensions $1\times3\times10$ mm^3, the front and back surfaces of the crystal (1×10 mm^2) were polished, the silver paste electrodes (3×4 mm^2) were painted on the lateral surfaces. The interelectrode spacing was 1 mm. In order to change the parameters of the material (concentration of the trapping centers, characteristic recombination times, etc. [43]) the crystal was simultaneously illuminated by the light from the microscope lamp focused by lens and passed through the infrared filter KS-15 ($\lambda_{IR} = 650 - 2700$ nm, $I_{IR} = 13$ mW/mm^2).

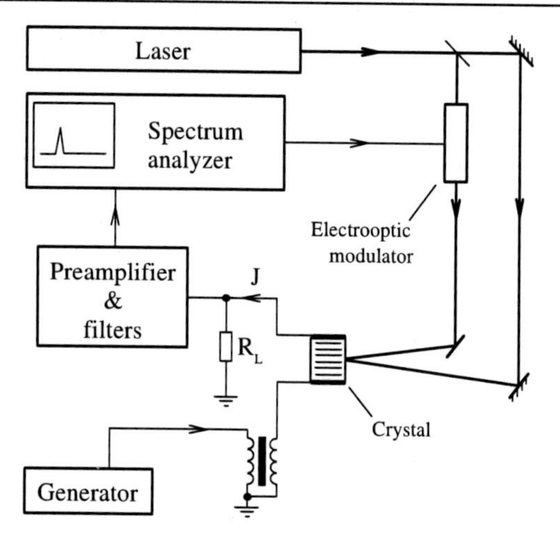

Figure 11. Experimental setup for measurements of the nonstationary holographic photocurrents in the presence of alternating electric field.

Let us present the results of the photocurrent measurements carried out in the external electric field and compare them with the data obtained for the diffusion excitation regime. Fig. 12 presents the frequency transfer function of the holographic photocurrent excited in $Bi_{12}SiO_{20}$ crystal in the absence of an external electric field. Recall that the frequency response of the effect is usually characterized by the corresponding cut-off frequencies [20]:

$$\omega_0 = [\tau_M(1+K^2L_D^2)]^{-1}, \tag{92}$$

$$\omega_0' = (1+K^2L_D^2)\tau^{-1}. \tag{93}$$

From the dependencies of the photocurrent amplitude measured at low spatial frequency of the interference pattern ($K^2L_D^2 \ll 1$) and without additional illumination we estimated corresponding cut-off frequencies: $\omega_0/2\pi \simeq 330$ Hz and $\omega_0'/2\pi \simeq 2.6$ kHz [Fig. 12 (circles)]. The present and earlier experiments [58] carried out in sillenite crystals (illumination wavelength $\lambda = 442$ nm) have revealed that the values of cut-off frequencies ω_0, ω_0' are close to each other. This means that the lifetime τ and Maxwell relaxation time τ_M are comparable: $\tau \sim \tau_M \sim 100\ \mu$s [41,58]. In this case the cut-off frequencies are defined by the more complicated expressions than Eqs. (92), (93) [41]:

$$\omega_0 = \frac{1+\tau/\tau_I+K^2l_D^2}{\tau+\tau_M(1+\tau/\tau_I+K^2L_D^2)}, \tag{94}$$

$$\omega_0' = \frac{1}{\tau_M}+\frac{1+\tau/\tau_I+K^2L_D^2}{\tau} \tag{95}$$

$$+\frac{1+\tau/\tau_I+K^2l_D^2}{\tau+\tau_M(1+\tau/\tau_I+K^2L_D^2)}, \tag{96}$$

where τ_I is the lifetime of the ionized donor state, l_D is the Debye screening length.

The condition $\tau \ll \tau_M$ is typical for the most of widegap semiconductor materials. To fulfill this condition and to conduct photocurrent measurements in the relaxation time

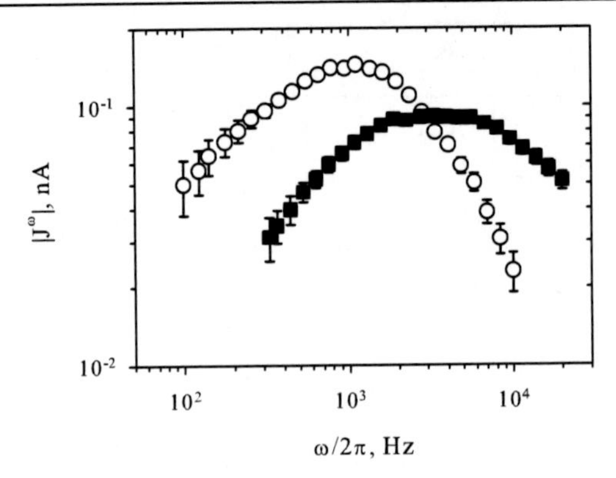

Figure 12. Frequency transfer functions of the nonstationary photocurrent measured with (squares) and without (circles) additional infrared illumination ($Bi_{12}SiO_{20}$, $\lambda = 442$ nm, $E_0 = 0$, $P_0 = 0.32$ mW, $m = 0.65$, $K = 5 \times 10^4$ m^{-1}).

regime additional infrared illumination of the crystal ($\lambda_{IR} = 650 - 2700$ nm, $I_{IR} = 13$ mW/mm^2) was used. For the first and second cut-off frequencies the following values were obtained: $\omega_0/2\pi \simeq 800$ Hz ($\tau_M \simeq 0.25$ ms) and $\omega_0'/2\pi \simeq 13$ kHz ($\tau \simeq 12$ μs) [Fig. 12 (squares)]. As seen from Fig. 12 the second cut-off frequency ω_0' is remarkably shifted to higher excitation frequencies what can be associated with the decrease of the electron lifetime. The growth of the first cut-off frequency ω_0 was clearly observed. This fact was unexpected since infrared illumination should not influence *stationary* photoconductivity of the crystal (and correspondingly the value of the Maxwell relaxation time). Note, that even 20 – 25% *decay* of the sample's conductance in the presence of the infrared light was observed. For this reason the behavior of ω_0 cannot be attributed to the growth of the conductivity. The observed behavior can be explained by the fact that $\tau \sim \tau_M$. In this case the cut-off frequency depends both on the electron lifetime and the Maxwell relaxation time. The former vanishes with the additional illumination resulting in the slight change of the first cut-off frequency.

The dependencies of the corresponding cut-off frequencies versus spatial frequency of the interference pattern K are presented in Fig. 13. The experimental dependencies were fitted using Eqs. (35), (36) providing the following estimations for the diffusion length of photocarriers: $L_D \simeq 1.2$ μm ($\mu\tau \simeq 0.6 \times 10^{-10}$ m^2/V) [from $\omega_1(K)$], $L_D \simeq 1.6$ μm ($\mu\tau \simeq 1.0 \times 10^{-10}$ m^2/V) [from $\omega_2(K)$].

Investigation of the spatial frequency dependence of the holographic photocurrent gives another possibility for the diffusion length determination. For the diffusion mechanism of recording ($E_0 = 0$) and for modulation frequencies higher than the first characteristic cut-off frequency ($\omega > \omega_0$) the simplest theory of the effect predicts the following dependence of the photocurrent amplitude versus spatial frequency [20]:

$$j^\omega(K) \propto \frac{K}{1 + K^2 L_D^2}. \tag{97}$$

We performed these measurements (Fig. 14) and obtained the following estimation for the diffusion length of electrons: $L_D = 1.4$ μm ($\mu\tau \simeq 0.7 \times 10^{-10}$ m^2/V).

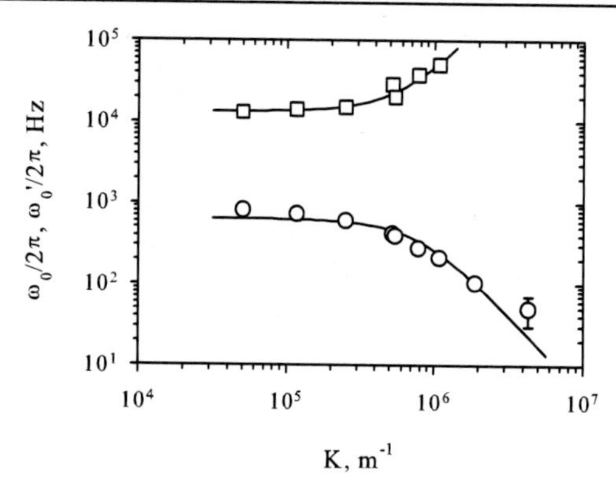

Figure 13. Dependence of the cut-off frequencies ω_0 (circles), ω_0' (squares) on the spatial frequency K of the interference pattern ($Bi_{12}SiO_{20}$, $\lambda = 442$ nm, $E_0 = 0$, $P_0 = 0.32$ mW, $m = 0.65$, infrared illumination is switched on). The theoretical curves are calculated using Eqs. (92), (93).

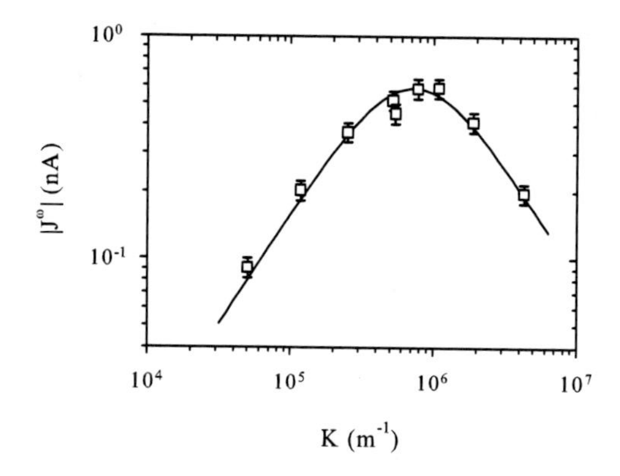

Figure 14. Dependence of the photocurrent amplitude $|J^{\omega}|$ on the spatial frequency K ($Bi_{12}SiO_{20}$, $\lambda = 442$ nm, $E_0 = 0$, $P_0 = 0.32$ mW, $m = 0.65$, infrared illumination is switched on). The solid line shows the theoretical curve calculated using Eq. (97).

Let us proceed to the case of non-zero alternating electric field applied to the crystal. The frequency dependence of the photocurrent amplitude measured for different values of electric field applied to the crystal is presented in Fig. 15. As seen from this Figure all dependencies keep the form typical for the diffusion mechanism of the photocurrent excitation. No resonant peaks on the frequency response were observed even for the highest voltage applied. The giant enhancement of the signal amplitude is another feature of the nonstationary photocurrent excitation in an external AC electric field (Figs. 15, 16). The characteristic growth of the signal amplitude as $\propto E_0^2$ was observed in the region $E_0 = (0.2 - 2) \times 10^5$ V/m. The photocurrent amplification by a factor of $G_m \simeq 350$ (~ 50 dB) has been achieved for low spatial frequencies of the interference pattern ($K^2 L_D^2 \ll 1$). This value is even higher than the theoretical estimation of the maximum gain factor by Eq. (91): $G_m = 140 - 280$ ($K = 5 \times 10^4$ m^{-1}, $L_D = 1.2 - 1.7$ μm).

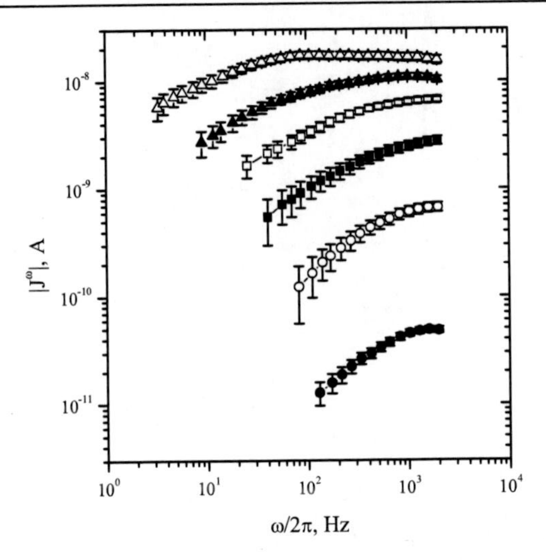

Figure 15. Frequency transfer functions of the photocurrent measured for $E_0 = 0$ (filled circles), 53 V/mm (open circles), 130 V/mm (filled squares), 210 V/mm (open squares), 410 V/mm (filled triangles), and 1000 V/mm (open triangles). $Bi_{12}SiO_{20}$, $\lambda = 442$ nm, $P_0 = 0.19$ mW, $m = 0.65$, $K = 5 \times 10^4$ m^{-1}, infrared illumination is switched on.

The $\mu\tau$-product or diffusion length of photocarriers can be estimated from the both characteristic breakpoints on the dependence of the photocurrent amplitude versus AC field amplitude (Fig. 16). We have approximated this dependence by Eq. (89) and obtained the following values: $E_L \simeq 1.5 \times 10^4$ V/m ($L_D \simeq 1.7$ μm, $\mu\tau \simeq 1.2 \times 10^{-10}$ m^2/V) for the first breakpoint and $E_M \simeq 2.6 \times 10^5$ V/m ($L_D \simeq 1.4$ μm, $\mu\tau \simeq 0.8 \times 10^{-10}$ m^2/V) for the second breakpoint.

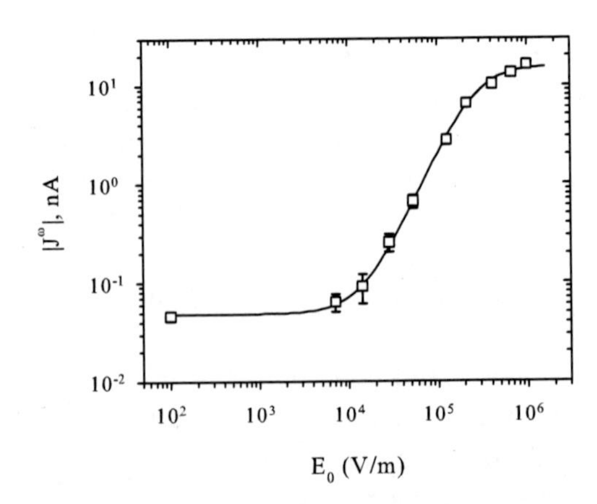

Figure 16. Photocurrent amplitude $|J^\omega|$ ($\omega > \omega_0$) versus effective value of the external AC field E_0 ($Bi_{12}SiO_{20}$, $\lambda = 442$ nm, $P_0 = 0.19$ mW, $m = 0.65$, $K = 5 \times 10^4$ m^{-1}, $\omega/2\pi = 2.0$ kHz, infrared illumination is switched on). The measurement performed at zero field is shown at $E_0 = 10^2$ V/m. The solid line shows the theoretical curve calculated using Eq. (89).

The behavior of the the first cut-off frequency ω_0 versus applied electric field is similar to the drift mechanism of recording – it shifts towards low excitation frequencies (Fig. 17). Approximation of this dependence by Eq. (90) provides the value of $\mu\tau$-product: $\mu\tau \simeq 1.2 \times 10^{-10}$ m^2/V ($E_M \simeq 1.6 \times 10^5$ V/m).

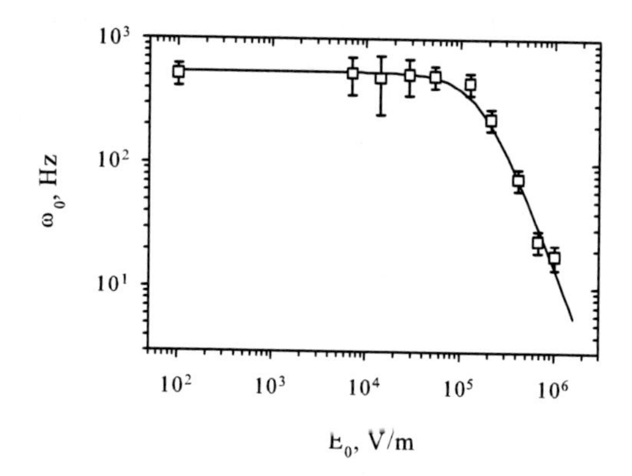

Figure 17. Dependence of the cut-off frequency ω_0 on the effective value of the external AC field E_0 ($Bi_{12}SiO_{20}$, $\lambda = 442$ nm, $P_0 = 0.19$ mW, $m = 0.65$, $K = 5 \times 10^4$ m^{-1}, infrared illumination is switched on). The measurement performed at zero field is shown at $E_0 = 10^2$ V/m. The approximation by Eq. (90) is shown by solid line.

The utilization of AC field for the photocurrent signal enhancement provides several important advantages over the application of DC field. First, the flat frequency response of the effect seems to be more convenient for measurements of phase modulated optical signals of small amplitude. The high frequency alternating electric field ($\Omega \gg \tau_M^{-1}$) applied to the crystal is more homogeneous than the DC field (no screening effects are presented) [43]. For this reason the measurements performed in fast AC fields are expected to be more plausible.

Determination of $\mu\tau$-product from the dependence of photocurrent amplitude versus applied AC voltage is much easier than the standard technique based on the measurements of the photocurrent amplitude on the spatial frequency of the interference pattern which is time consuming operation and includes numerous mechanical movements and readjustments of optical components (see, for example, Ref. [20]). The position of the first breakpoint on the dependence $J^\omega(E_0)$, i.e. E_L (see Sec. 4.1.) does not depend both on spatial frequency of the interference pattern K (for $K^2L_D^2 \ll 1$) and average light intensity I_0, what minimizes corresponding experimental errors.

5. Non-steady-State Photo-EMF in an External AC Electric Field: Resonant Case

Let us consider the resonant high-frequency excitation of the non-steady-state photocurrents in photoconductive crystals placed in a sinusoidal electric field ($\omega > \Omega$, $\tau_M^{-1} < \Omega < \tau$). Such combination of frequencies can give an unique opportunity for the signal magnification. Indeed the frequency of external field is higher than the inverse formation time of the space charge grating with period $\Lambda = 2\pi/K$: $\Omega > |\tau_M(1 + K^2L_D^2 + iKL_0)|^{-1}$, so the space charge amplitude will be increased as it was predicted in Ref. [54]. At the same time the frequency of the external field is lower than the inverse relaxation time of conductivity grating: $\Omega < |(1 + K^2L_D^2 + iKL_0)\tau^{-1}|$. For this reason the amplitude of photoconductivity grating will be increased as well via resonant mechanism analogous to discussed in Ref. [37]. As a result the resonant signal enhancement in an ac electric field is expected to be more pronounced than for the case of the dc field.

The crystal is illuminated with the oscillating interference pattern formed by two plane light waves one of which is phase modulated with frequency ω and amplitude Δ. The sinusoidal electric field with amplitude E_{ext} and frequency Ω is applied to the crystal: $E_{ext}(t) = E_{ext}\cos\Omega t$. For the external field we use the effective value $E_0 = E_{ext}/\sqrt{2}$ rather than amplitude E_{ext}. We consider the case of "relaxation type" semiconductor , i.e. $\tau \ll \tau_M$. The opposite situation (i.e. $\tau \geq \tau_M$) is slightly more complicated for the analysis, moreover the application of strong electric fields decreases the time constant of the conductivity relaxation and increase the buildup time for the space charge, so we obtain conditions analogous to the "relaxation type" regime.

We suppose also that the frequency of phase modulation ω is much higher than the frequency of the external electric field Ω; the period of the ac field is larger than the carrier lifetime τ and shorter than grating relaxation time τ_M:

$$\omega \gg \Omega, \tag{98}$$

$$\tau_M^{-1} \ll \Omega \ll \tau^{-1}. \tag{99}$$

The calculation procedure includes the solution of the balance, continuity, and Poisson equations. Since the main stages of this procedure are similar to the ones described above we omit them and write the final expression for the complex amplitude of the photocurrent with frequency ω:

$$\begin{aligned} j^\omega &= -\frac{m^2\Delta}{2}\sigma_0\left(E_D + E_0\frac{KL_0}{1+K^2L_D^2}\right) \\ &\times \frac{1+K^2L_D^2+i\omega\tau}{(1+K^2L_D^2+i\omega\tau)^2+K^2L_0^2}. \end{aligned} \tag{100}$$

Here $E_D = (k_BT/e)K$ is the diffusion field and $L_0 = \mu\tau E_0$ is the drift length of photocarriers, k_B is the Boltzmann constant, T is the temperature. The resonant maximum appears in the frequency transfer function of the photocurrent when the drift item in this expression dominates over the diffusion one: $E_0 \gg E_D$, $KL_0 \gg 1 + K^2L_D^2$. Then the resonant frequency and corresponding signal amplitude are equal to

$$\omega_r = \frac{KL_0}{\tau} = K\mu E_0, \tag{101}$$

$$j^\omega(\omega_r) = -\frac{m^2\Delta\sigma_0 E_0 KL_0}{4(1+K^2L_D^2)^2}. \tag{102}$$

Let us compare the photocurrent excitation in the dc and ac fields for the considered frequency region: $\omega \gg \tau_M^{-1}$. The photocurrent amplitude in dc electric field was found to be [37]:

$$j_{dc}^{\omega} = \frac{m^2\Delta}{2}\sigma_0 \frac{E_0 K L_0 - E_D(1 + K^2L_D^2 + i\omega\tau)}{(1 + K^2L_D^2 + i\omega\tau)^2 + K^2L_0^2}. \tag{103}$$

As seen the frequency transfer functions of the photocurrent for ac and dc fields are similar: the resonant peaks appear at the same frequency [Eq. (101)]. However there are distinctive features which differ these approaches. The resonant amplitude of the photocurrent excited in dc field equals

$$j_{dc}^{\omega}(\omega_r) = -i\frac{m^2\Delta\sigma_0 E_0}{4(1 + K^2L_D^2)}. \tag{104}$$

It is $KL_0/(1 + K^2L_D^2)$ times lower than those for ac field [Eq. (102)]. Moreover this difference grows when the amplitude of applied field increases. Such advantage of the ac voltage over dc one is associated with the fact that ac field enhances the stationary space charge grating proportionally to $\propto E_0^2$ (while dc fiels does $\propto E_0$) [54]. The phases of the photocurrent amplitude defined by Eqs. (102) and (104) differ by $\pi/2$ what is due to the fact that the spatial shift of the stationary electric field grating is equal to $-\pi$ for the dc field and $-\pi/2$ for the ac field applied. [2] Besides one can note the different shapes of the resonant peak for the cases of ac and dc external field [compare Eqs. (100) and (103)].

The measurements of the non-steady-state photocurrents under applied alternating electric field were carried out using the conventional experimental setup discussed above. The second harmonic of a Nd:YAG laser ($\lambda = 532$ nm) with an average power of $P_{out} \approx 50$ mW was used as the basic source of coherent radiation. A conventional Twyman-Green interferometer forms the interference pattern with a specified fringe spacing. The contrast of the interference pattern was settled to be small enough ($m = 0.12$) to avoid the excitation of higher spatial harmonics of the electric field grating. An electrooptic phase modulator ML-102A produced phase modulation of the laser beam with frequency ω and amplitude $\Delta = 0.81$. The additional low frequency phase modulation with frequency 100 Hz and amplitude 2.4 rad was utilized for separation of the volume and contact signals. This technique allows to transfer the contact component of the signal to the side frequencies which are rejected by the selective receiver. As a result the signal detected at carrier frequency contains the volume component only. The photocurrent signal arising in the crystal's volume was amplified, filtered and then measured by spectrum analyzer SK4-58 ($f = 0.4 - 600$ kHz, $\Delta f = 100$ Hz, with 10 Hz videofilter). In our experiments we used n-type $Bi_{12}SiO_{20}$ crystal with characteristic dimensions $1 \times 4 \times 10$ mm^3, the front and back surfaces of the crystal (1×10 mm^2) were polished, the silver paste electrodes (4×3 mm^2) were painted on the lateral surfaces. The interelectrode spacing was $L = 1$ mm. The crystal was placed in the teflon holder with styrofoam linings which damp the mechanical oscillations due to the piezoelectric effect. The sinusoidal voltage from generator ($\Omega/2\pi = 6$ kHz) was amplified by conventional transformer and then applied to the sample. In some experiments the crystal was simultaneously illuminated by the light from the microscope lamp focused by lens and passed through the different filters: IKS-5 ($\lambda_{IR} = 980 - 3000$ nm), KS-15 ($\lambda_R = 650 - 2700$ nm), SZS-22 ($\lambda_B = 380 - 530$ nm). The crystal's temperature was measured by the PtRh-Pt thermocouple.

Figure 18 presents the frequency transfer functions of the nonstationary holographic photocurrent measured for different amplitudes of external electric field $E_0 = U_0/L$ (U_0 is

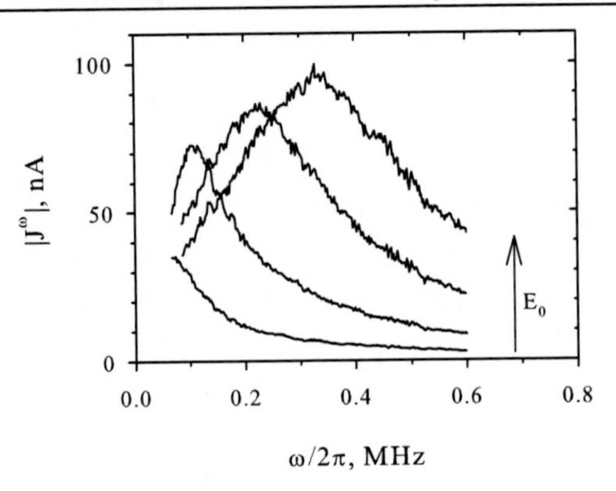

Figure 18. Frequency transfer functions of the holographic current measured in photorefractive $Bi_{12}SiO_{20}$ for different effective values of the sinusoidal electric field: $E_0 = 4.0$, 5.8, 8.0, 9.5 kV/cm ($\lambda = 532$ nm, $I_0 = 2.2$ W/cm^2, $K = 500$ cm^{-1}, $\Omega/2\pi = 6$ kHz).

the effective amplitude of applied voltage). The resonant maximum appears in the investigated frequency region ($\omega/2\pi = 60 - 600$ kHz) when the effective value of the sinusoidal field reaches ~ 5 kV/cm. The further increase of the applied voltage leads to the nonlinear growth of the resonant frequency and current amplitude (Figs. 18, 19).

The photocurrent amplitude grows slower than E_0^2 what is probably due to nonlinear effects. The stationary electric field grating has amplitude $\sim mE_0KL_0$ [59]. When this value approaches the amplitude of external field $E_{ext} = \sqrt{2}E_0$ we can expect the saturation of the photocurrent. For the chosen m, K and for $\mu\tau \sim 10^{-6}$ cm^2/V [59] the discussed effect should appear at $E_0 \sim 2$ kV/cm.

Nonlinear dependence $\omega_r(E_0)$ points out to the fact that strong electric fields influence the electron mobility varying in the range $\mu = 0.19 - 0.43$ cm^2/V s (Fig. 19). Dependence $\mu(E_0)$ was approximated with the power function: $\mu \propto E_0^{1.3}$. Such behavior is most probably explained by heating of the sample in the electric field. Indeed, the crystal's temperature raises by ~ 20 K in the external field of 9.5 kV/cm (Fig. 19).

The forbidden gap of sillenite type crystals is characterized by a rather complicated structure of local levels [2]. In particular, the shallow traps affect the charge transport significantly magnifying the conductivity relaxation time and reducing the effective drift mobility of electrons [43]. When the temperature grows electrons spend less time being captured on shallow traps so the effective mobility increases.

The dependencies of the resonant frequency and photocurrent amplitude versus spatial frequency of the interference pattern are presented in Fig. 20. Dependence $\omega_r(K)$ is well fitted with the linear function as predicted by Eq. (101). This approximation allows us to estimate electron mobility: $\mu = 0.42$ cm^2/V s ($T \simeq 313$ K). This value is more than one order of magnitude higher than the ones measured earlier by the non-steady-state photocurrent technique in an external dc electric field [37, 60]. In contrast to the electric field the spatial frequency of the interference pattern does not influence the photoelectric parameters (lifetime, mobility, etc.) so the behavior of the resonant frequency are in good agreement with the simplest theory described above and we can state that the maximum on the frequency transfer functions of the photocurrent is associated with the resonant excitation of

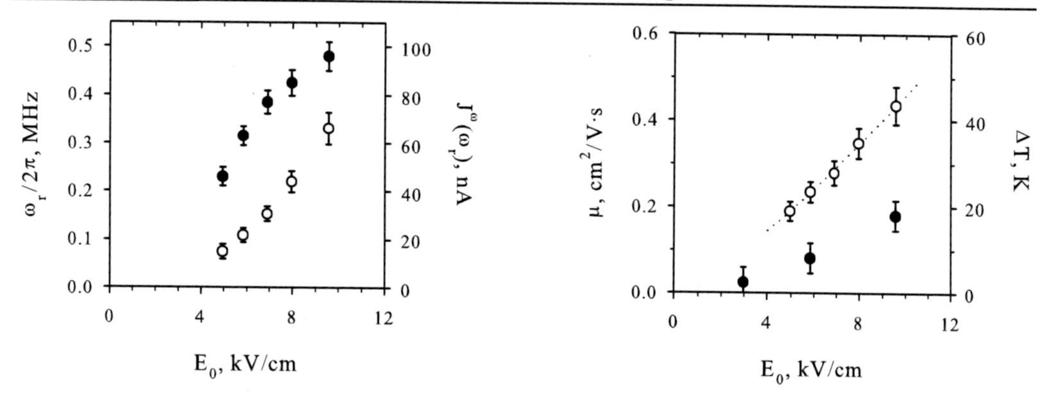

Figure 19. Resonant frequency ω_r (○) and resonant photocurrent amplitude $J^\omega(\omega_r)$ (●) versus effective value of the external field E_0 (left figure) and dependence of the drift mobility μ (○) versus effective value of the external field E_0 (right figure). The dotted line shows fitting function $\mu \propto E_0^{1.3}$. The increment of the crystal's temperature $\Delta T = T - 293$ K versus the external field amplitude E_0 is shown as well (●). $\lambda = 532$ nm, $I_0 = 2.2$ W/cm^2, $K = 500$ cm^{-1}, $\Omega/2\pi = 6$ kHz.

the photoconductivity gratings.

As seen from Fig. 20 the photocurrent amplitude is maximal for the lowest settled spatial frequency $K = 3\times10^2$ cm^{-1} though the theoretical estimation by Eq. (102) predicts maximum at $K = (\sqrt{3}L_D)^{-1} \sim 6 \times 10^3$ cm^{-1} ($L_D \sim 1$ μm). Perhaps this discrepancy is typical for sillenites and associated with the grating saturation effect discussed above.

Dependencies of the resonant frequency and current amplitude versus light intensity are shown in Fig. 21. According to the basic theory presented above no dependence of the resonant frequency ω_r should be observed [Eq. 101]. In practice, however, this statement is not valid since light intensity can influence the effective mobility via at least two mechanisms. First, the mobility increases due to Joule heating of the illuminated crystal placed in the electric field. Second, the increase of the light intensity changes the population of the shallow traps, the capture rate of electrons to these traps becomes weaker what consequently increases the effective mobility. As a result we obtained the wide range for electron mobility: $\mu = 0.13 - 0.43$ cm^2/V s. The similar behavior of the photocurrent was observed earlier for the case of dc external field [60].

In order to vary the total light intensity and spectrum the crystal was illuminated by the additional infrared, red, blue and "white" light from the microscope lamp (Fig. 21). The most powerful "white" light with the intensity $I_W = 2.9$ W/cm^2 shifts the resonant maximum beyond the studied frequency range ($\omega_r/2\pi > 600$ kHz) what correspond to the value of drift mobility $\mu > 0.8$ cm^2/V s ($T \simeq 370$ K).

The dependence of the resonant peak amplitude on the light intensity ($\lambda = 532$ nm) was found to be linear (Fig. 21) what is associated with the linear character of the photoconductivity. As seen from the same figure the additional illumination suppress the signal amplitude. Two reasons of this phenomena can be adduced. First, the incoherent backlight decreases the contrast of the interference pattern and photocurrent amplitude as a consequence. Second, the simultaneous infrared illumination redistributes the electrons between

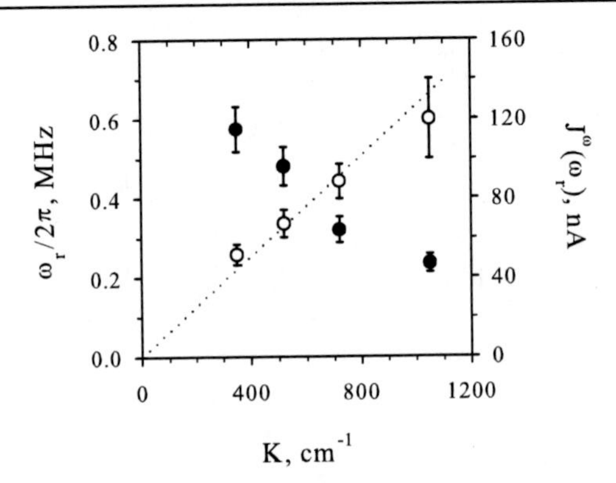

Figure 20. Dependencies of the resonant frequency ω_r (○) and resonant photocurrent amplitude $J^\omega(\omega_r)$ (●) on the spatial frequency of the interference pattern K ($\lambda = 532$ nm, $I_0 = 2.2$ W/cm^2, $E_0 = 9.5$ kV/cm, $\Omega/2\pi = 6$ kHz, $T \simeq 313$ K). The dotted line shows the approximation by Eq. (101) for $\mu = 0.42$ cm^2/V s.

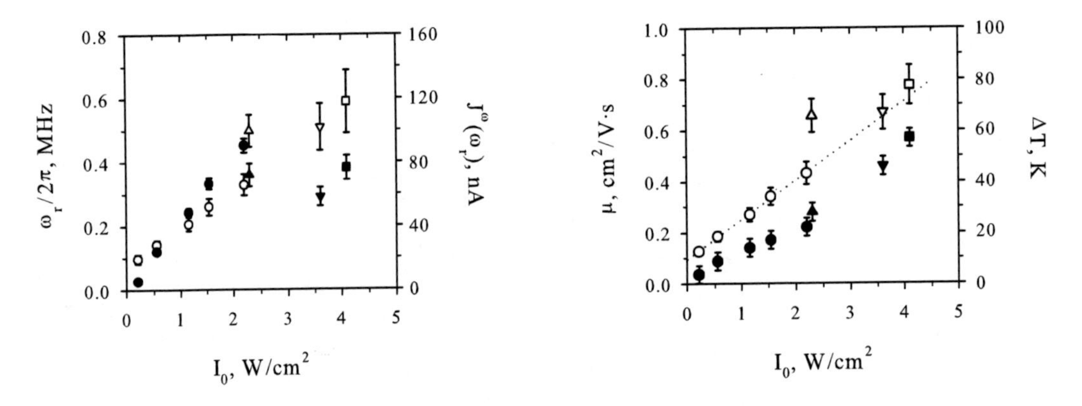

Figure 21. Resonant frequency ω_r (○, △, ▽, □) and resonant photocurrent amplitude $J^\omega(\omega_r)$ (●, ▲, ▼, ■) versus total light intensity I_0 (left figure) and dependence of the drift mobility μ (○, △, ▽, □) on the total light intensity I_0 (right figure). Dependence $\mu(I_0)$ measured without additional illumination (○) is fitted by the linear function $\mu \propto I_0$ with $\mu(0) = 0.95$ cm^2/V s. Dependence of the temperature increment $\Delta T = T - 293$ K on the light intensity I_0 is shown by the filled symbols (●, ▲, ▼, ■) . The circles (●, ○) demonstrate the measurements with the only laser illumination ($\lambda = 532$ nm), the other symbols correspond to the measurements with the additional illumination: ▲, △ – $\lambda_B = 380 - 530$ nm, $I_B = 0.10$ W/cm^2; ▼, ▽ – $\lambda_{IR} = 980 - 3000$ nm, $I_{IR} = 1.4$ W/cm^2; ■, □ – $\lambda_R = 650 - 2700$ nm, $I_R = 1.9$ W/cm^2. $K = 500$ cm^{-1}, $E_0 = 9.5$ kV/cm, $\Omega/2\pi = 6$ kHz.

different types of local levels reducing the effective lifetime, nonstationary response of the photoconductivity, and resulting photocurrent [43, 59].

We have compared two techniques for the drift mobility determination and carried out measurements of the frequency transfer functions of the nonstationary photocurrent excited in ac and dc fields with approximately equal effective values $E_0 \simeq 10$ kV/cm (Fig. 22). As

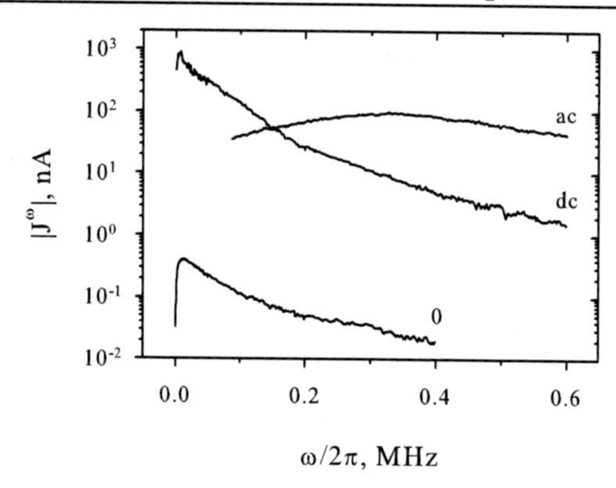

Figure 22. Frequency transfer functions of the holographic photocurrent measured in ac ($E_0 = 9.5$ kV/cm, $\Omega/2\pi = 6$ kHz, $T \simeq 313$ K), dc ($E_0 = 10$ kV/cm, $T \simeq 313$ K), and zero ($E_0 = 0, T \simeq 293$ K) external field. $\lambda = 532$ nm, $I_0 = 2.2$ W/cm^2, $K = 500$ cm^{-1}.

follows from this figure the photocurrent generation in ac field is more efficient in the high frequency region $\omega > 300$ kHz: the signal enhancement with respect to diffusion regime equals $\sim$ 70 dB for the case of ac field and $\sim$ 50 dB for dc one. This fact proves the theoretical estimations [Eqs. (102), (104)] that predicts the prevalence of ac field excitation over dc one by a factor $KL_0 \sim 10$.

One can notice the most interesting difference between these two approaches: the frequency transfer function of the nonstationary photocurrent excited in dc field contains no resonant maximum at high modulation frequencies $\omega > \omega_0'$ [where $\omega_0'/2\pi \simeq 20$ kHz is the second cut-off frequency of the nonstationary photocurrent excited without an external field (Fig. 22)].

It is quite difficult to point out the origin of this feature. (i) First the silver paste contacts to $Bi_{12}SiO_{20}$ crystals are usually rectifying so the 10-20% dc voltage drop in the contact area is expected. This decreases KL_0-product partially eliminating the resonant conditions for the dc field. The blocking contact gives rise to the inhomogeneous distribution of dc field over the interelectrode spacing as well. It makes the resonant frequency to be dependent on the local value of the electric field what can results in blurring of the resonant peak. In the case of high frequency ac field ($\Omega > \tau_M^{-1}$) such effects are negligible as the displacement current through the contacts decreases corresponding voltage drops and makes the electric field more homogeneous.

(ii) Another and perhaps the most essential reason for the discussed difference between photocurrent generation in ac and dc fields can be associated with the complicated processes of conductivity relaxation in $Bi_{12}SiO_{20}$. The frequency dependence of the photoconductivity response to the amplitude modulated light [43] does not adequately described by the simplest band transport model with one type of partially compensated donor level. Namely there are at least two regions where photoconductivity relaxation is characterized by different time constants (Fig. 23). In our experiments the frequency of the ac field ($\Omega/2\pi = 6$ kHz) was higher than the characteristic frequency separating these regions ($\omega_s/2\pi \sim 100$ Hz). We suppose that frequency transfer functions measured in ac and dc fields differ because the effective photoelectric parameters governing the charge transport does not co-

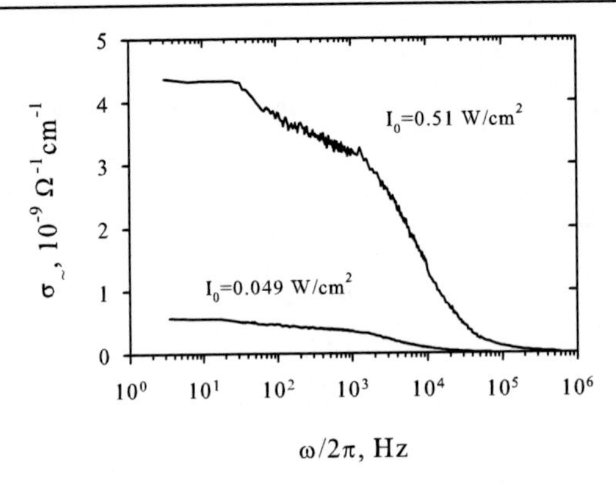

Figure 23. Frequency dependence of the nonstationary photoconductivity response to the amplitude modulated light with different average intensities $I_0 = 0.049, 0.51$ W/cm^2. $\lambda =$ 532 nm, $m = 0.018$.

incide for these cases: $\tau_{ac} < \tau_{dc}$ and $\mu_{ac} > \mu_{dc}$. This assumption is proved by the values of effective mobility measured by the discussed technique in ac field $\mu_{ac} = 0.13 - 0.8$ cm^2/V s and dc one $\mu_{dc} = 0.016 - 0.043$ cm^2/V s [37, 60]. So the larger mobility in ac field helps to fulfill resonant condition $\omega_r \gg \omega_0'$ and obtain the resonant maximum on the frequency dependence of the photocurrent.

The resonant excitation of the non-steady-state photocurrents in an ac field has demonstrated several distinctive features that can improve the techniques and devices based on the discussed effect. In particular, the proposed above resonant technique allows the direct measurements of the photocarrier's drift mobility. The utilization of an ac external voltage instead of the dc one minimizes the experimental errors of the mobility measurements since the voltage drop in the contact area is smaller and the electric field distribution in the sample volume is more homogeneous. Besides, variations of the voltage frequency may provide additional possibilities for the characterization of crystals with complicated conductivity relaxation processes.

The drift mobility of electrons in $Bi_{12}SiO_{20}$ was measured using this method. It lies in the range $\mu = 0.13 - 0.8$ cm^2/V s which is approximately an order lower than the range of actual mobility $\mu = 3.4 - 5.5$ cm^2/V s estimated by other techniques. On the other hand the measured values are $3 - 10$ times higher than the ones obtained by the non-steady-state photo-EMF method realized in dc fields [37, 60]. So the application of an ac electric field allowed us to partially overcome difficulties of the mobility measurements in widegap semiconductors with shallow energy levels.

The giant signal enhancement in the high frequency region is another attractive feature of the photocurrent generation in an alternating field. The enhancement factor reaches ~ 70 dB for low spatial frequencies of the interference pattern against the factor of ~ 50 dB typical for the case of dc field.

For practical applications it is more useful to calculate and compare the signal-to-noise ratio for zero and ac fields. Let us estimate it for the resonant frequency ($\omega = \omega_r \gg \omega_0'$), low spatial frequency ($K^2 L_D^2 \ll 1$), and low load resistance ($R_L \ll Z_{cr}$, Z_{cr} is the crystal's impedance). For the mentioned frequency region the signal amplitude can be calculated using Eq. (100). The thermal noise of the load resistor is dominant for the diffusion regime of the photocurrent generation: $J_{th} = (4k_B T R_L^{-1} \Delta f)^{1/2}$ (Δf is the detection bandwidth). If the external voltage is applied then generation-recombination noise can be prevalent: $J_{g/r} = 2e\mu\tau E_0[Sg_0\Delta f / L(1+\omega^2\tau^2)]^{1/2}$ (g_0 is the average generation rate of photoelectrons, S is the cross section of the illuminated volume) [61]. Finally the magnification of the signal-to-noise ratio equals

$$\frac{(S/N)_{ac}}{(S/N)_0} = \frac{E_0 K^2 L_0^2}{2}\sqrt{\frac{L}{Sg_0 k_B T R_L}} \quad \text{for } J_{g/r} \gg J_{th} \tag{105}$$

or

$$\frac{(S/N)_{ac}}{(S/N)_0} = \frac{eE_0 K L_0^2}{2k_B T} \quad \text{for } J_{g/r} \ll J_{th}. \tag{106}$$

This gain factor is very large: for the sensible experimental conditions and crystal parameters ($E_0 = 10$ kV/cm, $K = 500$ cm^{-1}, $L_0 = 0.02$ cm, $L = 0.1$ cm, $S = 10^{-2}$ cm^2, $g_0 = 10^{19}$ cm^{-3}s^{-1}, $R_L = 10^4$ Ω) it reaches 90 dB. In practice, however, some factors can reduce this striking value: nonlinear recording of the space charge grating, trap saturation, etc.

6. Combined Excitation of Running Space Charge and Conductivity Gratings

The evolution of photoinduced charge in semiconductor materials mainly occurs in two basic stages: formation of the photoconductivity distribution and creation of the space charge [2, 43]. For the crystal placed in an external electric field the relaxation of photoconductivity and space charge distributions has complex behavior (oscillation and running character) [5]. The existence of two eigenmodes (branches) of space charge oscillation was predicted in Ref. [47]. One of these eigenmodes describe oscillation of the space charge trapped on the local levels, while the other is associated with the oscillation of the density of electrons in the conduction band (conduction electrons). The former branch is historically called "space charge wave". To avoid the complexity we shall use the term "running grating of the conductivity" for the latter one. The dispersion laws of these waves, i.e. dependencies of the temporal oscillation frequency on the spatial frequency K can be written as follows:

$$\omega_{sc} = (\tau_M K L_0)^{-1}, \tag{107}$$

$$\omega_{pc} = K\mu E_0, \tag{108}$$

where τ_M is the Maxwell relaxation time, μ is the mobility, and L_0 is the drift length of electrons in dc electric field E_0. A number of approaches for investigations of the running space charge and conductivity gratings has been realized by the present time [20, 31, 36, 37, 49, 62–68].

In spite of the large number of publications dedicated to the subject of space charge waves the question of simultaneous excitation and interaction of the eigenmodes mentioned above is still unsolved. Most of the proposed optical techniques use the interference pattern with the specified spatial frequency K and temporal frequency ω for the excitation of either space charge wave or running photoconductivity grating. This is due to the fact that the eigenfrequency of space charge oscillations in high resistive crystals (such as $Bi_{12}SiO_{20}$) is in the range $\omega_{sc}/2\pi = 10 - 100$ Hz while the eigenfrequency of the photoconductivity grating with the same spatial period can reach considerably higher values: $\omega_{pc}/2\pi = 0.01-$ 1 MHz. The situation is partly similar to the light scattering on the acoustic waves where interaction of particles with comparable momenta (wave vectors) and completely different energies (frequencies) takes place.

As it is known from the oscillation theory, the behavior of systems with two degrees of freedom have distinctive features with respect to the one in systems with one degree of freedom. This features result in such effects as incoherent parametric amplification, generation, wideband parametric frequency division, etc. Considering two eigenmodes of space charge oscillations in photorefractive media one can expect some analogs in holographic recording.

For the realization of the simultaneous excitation and interaction of the space charge wave (ω_{sc}, K) travelling along the dc electric field with the velocity ω_{sc}/K and electron conductivity grating $(\omega_{pc}, -K)$ running in the opposite direction with velocity $-\omega_{pc}/K$ the third oscillation $(\Omega, 0)$ satisfying the conditions analogous to the energy and momentum conservation laws is required: $\omega_{sc}+\omega_{pc} = \Omega$, $K-K = 0$. The electric field with frequency Ω can play the role of such oscillation and provide necessary coupling.

In this section we consider one of the possible manifestation of this phenomenon, namely the excitation of the space charge wave provided by the interaction of the running photoconductivity grating with an applied ac electric field. The effect is observed and investigated using optical and electrical methods [69–71]. The first one is based on light diffraction on photorefractive volume holograms, the second one uses the non-steady-state photoelectromotive force effect.

6.1. Theoretical Analysis

Let us suppose the photorefractive crystal placed in external dc and ac electric fields is illuminated by the interference pattern formed by two plain light waves one of which is phase modulated with amplitude Δ and frequency ω [Fig. 24, Eq. (6)]. The external dc and sinusoidal ac voltages applied to the crystal can be presented as follows:

$$U_{ext} = U_0 + U_A \cos \Omega t. \tag{109}$$

We consider the widely used model of semiconductor with monopolar photoconductivity and one type of partially compensated donor levels (Fig. 8) [2,43]. The formulation of the problem and initial stages are similar to those presented in Section 4.: we have to solve Eqs. (78), (79) and find distributions of the electron density $n(x,t)$ and electric field $E(x,t)$ in the form of stationary and running sinusoidal gratings [see Eqs. (80), (81)]. The right hand of Eq. (81) should also contain the dc component of the electric field $E_0 = U_0/L$ in the present analysis.

Let us analyze the case of "relaxation type" semiconductor:

$$\tau \ll \tau_M. \tag{110}$$

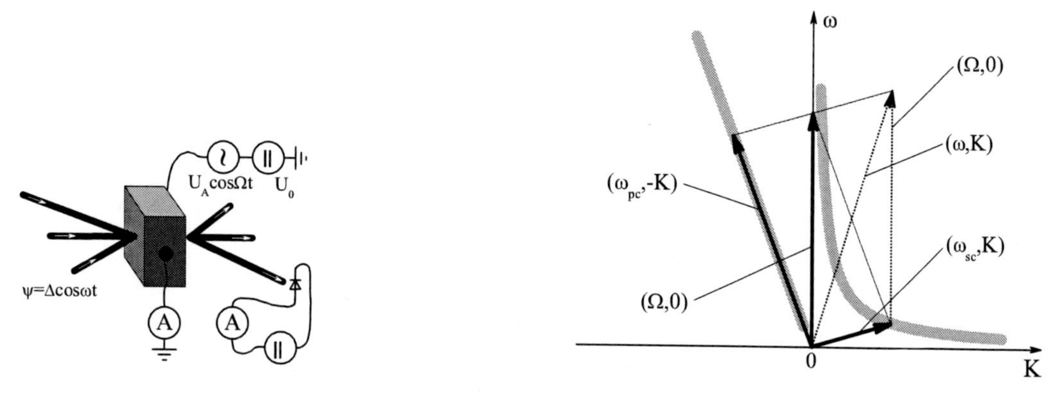

Figure 24. Scheme for observation of the combined excitation of running space charge and conductivity gratings (left figure). Diagram illustrating the interaction of running space charge and conductivity gratings with applied ac electric field. The arrows show the space charge grating (ω_{sc}, K) travelling with the velocity ω_{sc}/K along the dc field, the photoconductivity grating $(\omega_{pc}, -K)$ running with the velocity $-\omega_{pc}/K$ in the opposite direction, the photoconductivity grating (ω, K) driven with the velocity ω/K along the dc field, and the external ac field $(\Omega, 0)$ with frequency Ω. The half-tone lines show the dispersion laws for the space charge wave ω_{sc}(K) and running photoconductivity grating $\omega_{pc}(-K)$ (right figure).

Note, that the opposite case (i.e. $\tau \geq \tau_M$) is slightly more complicated for the analysis. However, the application of strong electric fields decreases the time constant of the conductivity relaxation and increases the buildup time of the space charge, so we obtain conditions analogous to the "relaxation type" regime.

Let us consider the "long-drift length" regime of holographic recording:

$$KL_0 \gg 1 + K^2 L_D^2, \tag{111}$$

where $L_0 = \mu\tau E_0$ and $L_D = (k_B T \mu\tau/e)^{1/2}$ are the drift and diffusion lengths of photocarriers. This assumption is rather typical for theoretical considerations of various phenomena in photorefractive media: it ensures the oscillatory relaxation of the space charge and photoconductivity for large applied dc fields (in contrast to the exponential one in the diffusion regime) [2].

In this section we shall develop two approaches. In the former one the frequency of an ac electric field is equal to the eigenfrequency of photoconductivity oscillation, and the difference frequency lies in the range of the space charge wave excitation:

$$\Omega = \omega_{pc}, \tag{112}$$

$$|\omega - \Omega| \sim \omega_{sc}. \tag{113}$$

The complex amplitudes of values $n(x,t)$ and $E(x,t)$ necessary for further analysis have been calculated. Below the most interesting terms presenting gratings with spatial frequency K and oscillation frequency $\pm(\omega - \Omega)$ are given:

$$n^{++-} = \frac{-im\Delta}{8D_1} n_0(\omega - \Omega)\tau_M K L_A, \tag{114}$$

$$E^{++-} = \frac{-im\Delta}{8D_1} E_A, \tag{115}$$

$$n^{+-+} = \frac{i3m\Delta}{8D_2} n_0(\omega - \Omega)\tau_M K L_A, \tag{116}$$

$$E^{+-+} = \frac{-i3m\Delta}{8D_2} E_A, \tag{117}$$

where

$$D_1 = (1 + K^2 L_D^2)[1 + (\omega - \Omega)\tau_M K L_0] - i[(1 + K^2 L_D^2)^2 + K^2 L_A^2/4]/K L_0, \tag{118}$$

$$D_2 = (1 + K^2 L_D^2)[5 - 11(\omega - \Omega)\tau_M K L_0] - i6K L_0[1 - (\omega - \Omega)\tau_M K L_0], \tag{119}$$

and $L_A = \mu\tau E_A$.

The latter approach implies that frequencies of phase modulation and external field vary in the region of the photoconductivity grating oscillation maintaining the difference frequency constant and equal to the frequency of the space charge oscillation:

$$\omega, \Omega \sim \omega_{pc}, \tag{120}$$

$$\omega - \Omega = -\omega_{sc}. \tag{121}$$

The choice of negative sign in right hand of Eq. (121) will be clear below when we shall analyze the frequency characteristics of the non-steady-state photo-EMF and diffraction efficiency.

The corresponding complex amplitudes of electron density and electric field are the following:

$$n^{++-} = -\frac{m\Delta}{8D_3} n_0 K L_A \left[1 + K^2 L_D^2 + i(2\omega\tau - K L_0)\right], \tag{122}$$

$$E^{++-} = \frac{m\Delta}{8D_3} E_0 K L_A \left[1 + K^2 L_D^2 + i(2\omega\tau - K L_0)\right], \tag{123}$$

where

$$D_3 = (1 + K^2 L_D^2)\left[(1 + K^2 L_D^2)^2 + K^2 L_A^2/2 - K^2 L_0^2 + 3\omega\tau K L_0 - 2\omega^2\tau^2\right] + i\left[(1 + K^2 L_D^2)^2(3\omega\tau - 2K L_0) + K^2 L_A^2(2\omega\tau - K L_0)/4\right]. \tag{124}$$

Let us now proceed with the calculation of the non-steady-state photo-EMF and diffraction efficiency amplitudes – values which will be measured in the experiment. As it was shown in Ref. [20] the non-steady-state photo-EMF in crystal biased by voltage source is defined by the averaged drift component of the photocurrent [Eq. (15)]. As seen from Eqs. (80), (81), (15) the signal with frequency $\omega - \Omega$ is produced by the following combination:

$$j^{\omega-\Omega}(t) = \mathrm{Re}\{2e\mu(n^{++-}E^{-00} + n^{-00}E^{++-} + n^{-+-}E^{+00} + n^{+00}E^{-+-} + n^{++0}E^{-0-} + n^{-0-}E^{++0} + n^{-+0}E^{+0-} + n^{+0-}E^{-+0})\exp[i(\omega - \Omega)t]\}. \tag{125}$$

The different nature of the items in this expression should be pointed out. The first four components are associated with the resonant excitation of the space charge waves with temporal frequency $\omega - \Omega$. The last four components in Eq. (125) are just mixing of the "fast" gratings with temporal frequencies ω and Ω. It is reasonable to expect that the former group produces larger contribution to the total current. This supposition is proved by the numerical calculation. Substituting the calculated above terms (114)-(117) into the Eq. (125) we can write down the expression for the complex amplitude of the nonstationary photocurrent as:

$$j^{\omega-\Omega} = -i\frac{m^2\Delta}{8}\sigma_0 E_A \left[\frac{1-(\omega-\Omega)\tau_M K L_0}{D_1} - 3\frac{1+(\omega-\Omega)\tau_M K L_0}{(D_2)^*}\right]. \tag{126}$$

The frequency transfer function of the non-steady-state photo-EMF, i.e. dependence $j^{\omega-\Omega}(\omega-\Omega)$ has two resonant maxima at $\omega - \Omega \simeq \pm\omega_{sc}$. The amplitudes of the resonant maxima at $\omega - \Omega = -\omega_{sc}$ and at $\omega - \Omega = \omega_{sc}$ are the following:

$$j_{r1}^{\omega-\Omega} = \frac{m^2\Delta}{4}\frac{\sigma_0 E_A K L_0}{(1+K^2L_D^2)^2 + K^2L_A^2/4}, \tag{127}$$

$$j_{r2}^{\omega-\Omega} = -i\frac{m^2\Delta}{8}\frac{\sigma_0 E_A}{1+K^2L_D^2}. \tag{128}$$

The resonant maximum for the negative frequency difference is approximately $2KL_0$ times larger than for positive frequency difference. This fact is associated with the principal distinction in the physical mechanism of grating's excitation in these frequency regions.

For the negative frequency difference range ($\omega - \Omega < 0$) the space charge wave is originated from the interaction of an external field with the photoconductivity grating propagating in the direction opposite to the dc field. When the oscillations of the interference pattern are synchronized with the movement of drifting electrons the resonant enhancement of the photoconductivity grating takes place. Thus the simultaneous generation of the running space charge and photoconductivity gratings, which are the eigenwaves in this case, should be observed.

For positive frequency difference ($\omega - \Omega > 0$) the space charge wave arises due to the interaction of the external field with the conductivity grating driven in the direction of the dc field. The photoconductivity grating is not an eigenmode of semiconductor in this case. It just follows the movements of the oscillating interference pattern in the direction of dc field.

The diagram illustrating the interaction of the space charge and photoconductivity gratings with the applied ac field is presented in Fig. 24. As seen from the figure in the former case the applied ac field couples two eigenmodes of semiconductor satisfying the equations $\omega_{sc} + \omega_{pc} = \Omega$, $K - K = 0$. In the latter case the only eigenmode, namely, space charge wave is excited. The fulfillment of the resonant conditions for both the space charge wave and photoconductivity grating ensures observation of larger signal's amplitudes.

Dependence $j^{\omega-\Omega}(\omega)$ with the fixed frequency difference $\omega - \Omega$ [Eqs. (120) and (121)] can be measured in the experiment as well. Direct substitution of Eqs. (122), (123) into Eq. (125) gives the following expression for this dependence:

$$j^{\omega-\Omega} = \frac{m^2\Delta}{4D_3}\sigma_0 E_A K L_0 \left[1 + K^2L_D^2 + i(2\omega\tau - KL_0)\right]. \tag{129}$$

The analysis of the denominator in this expression [see Eq. (124)] revealed that the resonant maximum can be observed at $\omega \simeq \omega_{pc}$ and its amplitude is expressed by Eq. (127). This resonant peculiarity can be explained as follows: since the excited space charge oscillations at combination frequency are due to interaction of the external ac electric field and the photoconductivity grating, the amplitude of the latter should affect the detected signal amplitude. The photoconductivity grating is resonantly excited at $\omega = \omega_{pc}$. So the resonant dependence of the low frequency signal $j^{\omega-\Omega}$ versus phase modulation frequency ω (or the frequency of an ac electric field Ω) can be expected as well.

The diffraction efficiency of thin phase hologram [2] recorded in photorefractive crystal equals

$$\eta(t) = |\pi n_p^3 r E^+(t) d/\lambda_p|^2, \tag{130}$$

where n_p and r are the refractive index and electrooptic coefficient of the crystal for wavelength λ_p of the probe readout light, $E^+(t)$ is the complex amplitude of $+K$ spatial component of the electric field, and d is the thickness of the crystal. The oscillations of the diffraction efficiency with frequency $\omega - \Omega$ are produced by the following components:

$$\begin{aligned} \eta^{\omega-\Omega}(t) &= \mathrm{Re}\{2q^2[(E^{+00})^* E^{++-} + E^{+00}(E^{+-+})^* \\ &\quad + (E^{+0+})^* E^{++0} + E^{+0-}(E^{+-0})^*] \exp[i(\omega-\Omega)t]\}, \end{aligned} \tag{131}$$

here $q = \pi n_p^3 r d/\lambda_p$. One can note the difference between two first and two last terms of the expression – similar to the ones mentioned above for the nonstationary photocurrent analysis. Direct substitution of Eqs. (115), (117) into Eq. (131) provides the expression for the complex amplitude of the diffraction efficiency:

$$\eta^{\omega-\Omega} = i\frac{m^2\Delta}{8} q^2 E_0 E_A \left[\frac{1}{D_1} - \frac{3}{(D_2)^*}\right]. \tag{132}$$

This frequency transfer function has two resonant maxima. The oscillation amplitudes of the diffraction efficiency at resonant frequencies $\omega - \Omega = -\omega_{sc}$ and $\omega - \Omega = \omega_{sc}$ are the following:

$$\eta_{r1}^{\omega-\Omega} = -\frac{m^2\Delta}{8}\frac{q^2 E_0 E_A K L_0}{(1+K^2L_D^2)^2 + K^2L_A^2/4}, \tag{133}$$

$$\eta_{r2}^{\omega-\Omega} = i\frac{m^2\Delta}{16}\frac{q^2 E_0 E_A}{1+K^2L_D^2}. \tag{134}$$

Like in Eqs. (127), (128) these resonant peaks differ by factor of $\sim 2KL_0$ due to the same reason: the simultaneous excitation of two eigenmodes is more efficient.

Dependence of the diffraction efficiency amplitude $\eta^{\omega-\Omega}$ on the frequency of phase modulation ω with fixed difference frequency $\omega - \Omega = -\omega_{sc}$ can be calculated substituting Eq. (123) into Eq. (131):

$$\eta^{\omega-\Omega} = -\frac{m^2\Delta}{8D_3} q^2 E_0^2 K L_A \left[1 + K^2L_D^2 + i(2\omega\tau - KL_0)\right]. \tag{135}$$

The resonant maximum with amplitude (133) is observed at frequency $\omega \simeq \omega_{pc}$.

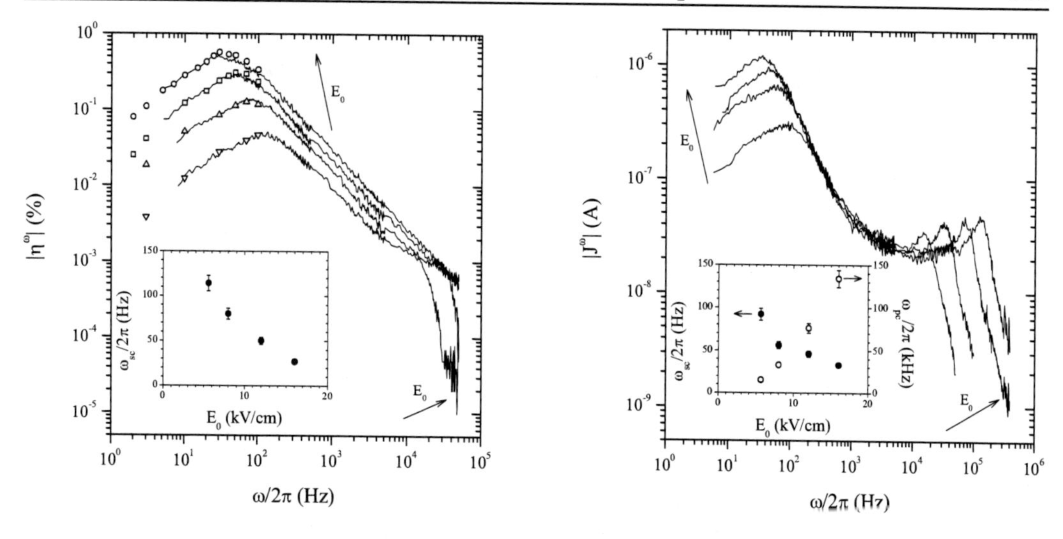

Figure 25. Frequency transfer functions of the diffraction efficiency oscillations (left figure) and non-steady-state photo-EMF (right figure) excited in $Bi_{12}SiO_{20}$ at the frequency of phase modulation ω (standard technique, $E_A = 0$). The dependencies are measured for four values of the applied dc electric field: $E_0 = 5.6$ kV/cm (▽), 8.0 kV/cm (△), 12 kV/cm (□), 16 kV/cm (○). The measurements performed using spectrum analyzer and lock-in voltmeter are presented by the solid lines and symbols respectively. The arrows shows the directions of curve evolution with increasing field amplitude E_0. Dependencies of the first (●) and second (○) resonant frequencies versus applied dc field are presented in the insets.

This frequency dependence along with the signal dependence of the non-steady-state photo-EMF effect [Eq. (129)] can be utilized for direct measurements of the photocarrier's drift mobility $\mu = \omega_{pc}/KE_0$.

We have considered here the simplest case of thin hologram recorded in slightly absorbing media. The generalization of the theory for the cases of thick holograms in absorbing and optically active materials can be done as required using the approaches known in literature [2,5,72].

6.2. Experimental Results

The experimental arrangement and procedure used for the investigation of the combined excitation of space charge and conductivity gratings in photorefractive crystals are described in Refs. [69–71].

For the determination of eigenfrequencies of space charge and conductivity oscillations let us consider the results of the diffraction measurements in photorefractive $Bi_{12}SiO_{20}$ using standard technique of signal excitation.

The diffraction efficiency of the hologram was defined as $\eta = I_{p1}/I_p$, where I_{p1} is the intensity of the probe beam for the first order of diffraction. The diffraction efficiency of the stationary hologram recorded in the electric field $E_0 = U_0/L = 16$ kV/cm was equal to $\eta_0 = 2.0\%$. Figure 25 presents the frequency transfer function of the diffraction efficiency oscillations η^ω measured at frequency of phase modulation ω. The first resonant maximum

Table 1. Drift mobility of electrons in $Bi_{12}SiO_{20}$ crystal estimated from the frequency transfer function of the non-steady-state photo-EMF $|J^{\omega}(\omega)|$.

E_0, kV/cm	T, K	μ, cm²/V s
5.6	294	0.89×10^{-2}
8.0	296	1.4×10^{-2}
12	298	2.1×10^{-2}
16	300	2.8×10^{-2}

is observed in the low frequency range ($\omega/2\pi = 27 - 110$ Hz). When the electric field is increased the signal amplitude increases as well and the first maximum shifts to lower excitation frequencies (Fig. 25, inset). The shoulder at frequencies $\omega/2\pi = 10 - 50$ kHz was observed as well. Perhaps, for larger applied electric fields this shoulder could transfer into the second maximum, but field amplitudes $E_0 > 8$ kV/cm shift this shoulder beyond the investigated frequency range. The signal amplitude at $\omega/2\pi > 50$ kHz was so small that utilization of spectrum analyzer with appropriate frequency range and larger detection bandwidth (SK4-58) was not successful.

For this reason we used another approach based on the non-steady-state photo-EMF measurements. The frequency transfer function of the non-steady-state photo-EMF signal excited in $Bi_{12}SiO_{20}$ crystal is presented in Fig. 25. Two resonant maxima on the frequency dependencies were observed. The increase of an applied field amplitude shifts the first and second maxima to the low and high frequency bands respectively (Fig. 25, inset). As it was shown in Ref. [37] such signal behavior is associated with the resonant excitation of running gratings of the space charge and conductivity described by the dispersion laws (107) and (108). Note, that the first resonant maximum is wide, this can be due to nonlinearity of the space charge formation for large contrast of the interference pattern [73].

The non-steady-state photo-EMF technique provides the unique opportunity for the direct determination of the photocarrier drift mobility. The value of the second resonant frequency measured for the specified K and E_0 is associated with the electron mobility value (Table 1). The obtained data are in good agreement with the values measured earlier [34, 37, 60].

The preliminary experiments allowed us to estimate corresponding resonant frequencies which can be used for further analysis of the simultaneous generation of the space charge and photoconductivity gratings.

For the realization of such regime the frequency of the external ac field was settled to be approximately equal to the frequency of the second resonant maximum. Small deviations ($0.3 - 1.0$ kHz) of frequency $\Omega/2\pi$ from the resonant frequency value $\omega_{pc}/2\pi$ are acceptable since the width of the second resonant maximum reaches $15 - 130$ kHz (see Fig. 25). The amplitude of an external ac electric field was chosen noticeably smaller than the value of the dc field ($E_A = U_A/L = 1.4 - 5.6$ kV/cm), so possible shifts of the resonant frequencies are expected to be negligible.

Figure 26 (left figure) presents the frequency transfer function of the diffraction efficiency oscillations excited at difference combination frequency $\omega - \Omega$. Two maxima

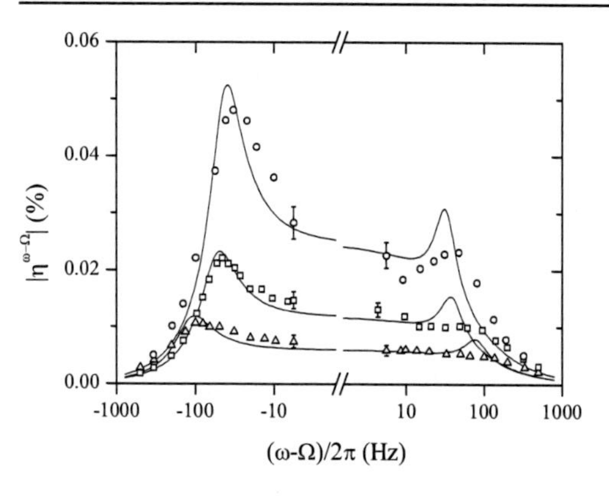

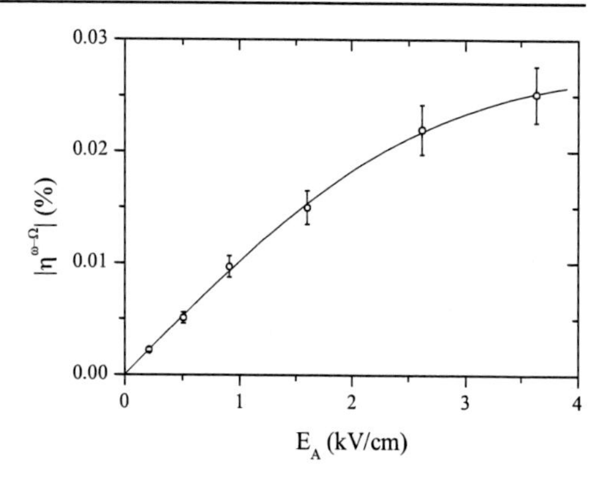

Figure 26. Frequency transfer function of the diffraction efficiency oscillations excited in $Bi_{12}SiO_{20}$ at the combination frequency $\omega - \Omega$ (left figure). The dependencies are obtained for three values of the applied dc field and for three corresponding frequencies of the applied ac field: $E_0 = 5.6$ kV/cm, $E_A = 1.4$ kV/cm, $\Omega/2\pi = 15$ kHz (△), $E_0 = 8.0$ kV/cm, $E_A = 2.6$ kV/cm, $\Omega/2\pi = 33$ kHz (□), and $E_0 = 16$ kV/cm, $E_A = 5.6$ kV/cm, $\Omega/2\pi = 130$ kHz (○). The approximation of the experimental dependencies by Eq. (132) is shown by the solid lines. Dependence of the diffraction efficiency amplitude $|\eta^{\omega-\Omega}|$ versus amplitude of an ac electric field E_A (right figure). $Bi_{12}SiO_{20}$, $E_0 = 8.0$ kV/cm, $\Omega/2\pi = 33$ kHz, $(\omega - \Omega)/2\pi = -54$ Hz. The solid line shows approximation by Eq. (133).

Table 2. Material parameters used for approximation of the experimental dependencies in Fig. 26 (left figure) by Eq. (132).

E_0, kV/cm	E_A, kV/cm	τ_M, ms	$\mu\tau$, cm²/V
5.6	1.4	0.83	1.8×10^{-7}
8.0	2.6	1.7	1.3×10^{-7}
16	5.6	1.8	7.2×10^{-8}

were observed at this dependence: the first one – at the negative difference frequency $(\omega - \Omega)/2\pi = -(100 - 33)$ Hz ($E_0 = 5.6 - 16$ kV/cm), and the second one – at the positive frequency $(\omega - \Omega)/2\pi = 40$ Hz ($E_0 = 16$ kV/cm). The increase of the external dc field leads to the growth of the signal amplitude and corresponding shift of the resonant maxima to lower frequencies. This behavior is the typical manifestation of the space charge waves [2]. The experimental curves are well fitted by Eq. (132) with the parameters presented in Table 2. Different values of the maxima in the region of negative and positive difference frequencies is an important peculiarity of the frequency transfer function (Fig. 26). One can notice another feature: the values of $\mu\tau$-product estimated from this frequency dependence noticeably decreases at large applied fields (Table 2). In fact, decrease of the quality factor of the space charge wave is observed. This decrease is most probably due to effects of trap saturation and it is not associated with any decay of $\mu\tau$-product. According to Ref. [74] trap saturation effects reduce quality factor at $E_0 > [(N_A/\epsilon\epsilon_0)(e/K^2\mu\tau + k_BT)]^{1/2}$. For $Bi_{12}SiO_{20}$ crystal with $N_A \simeq 10^{16}$ cm^{-3},

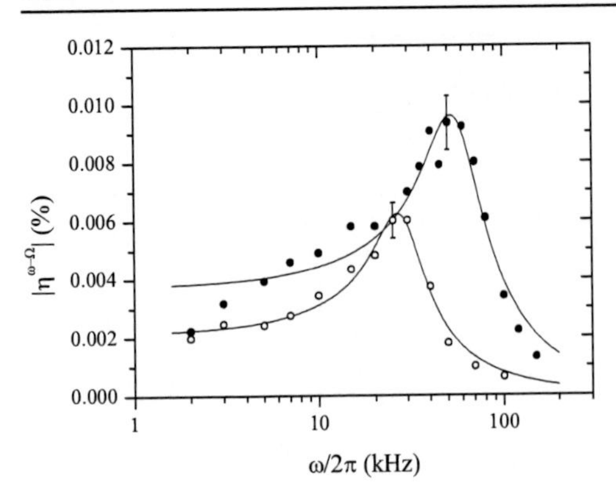

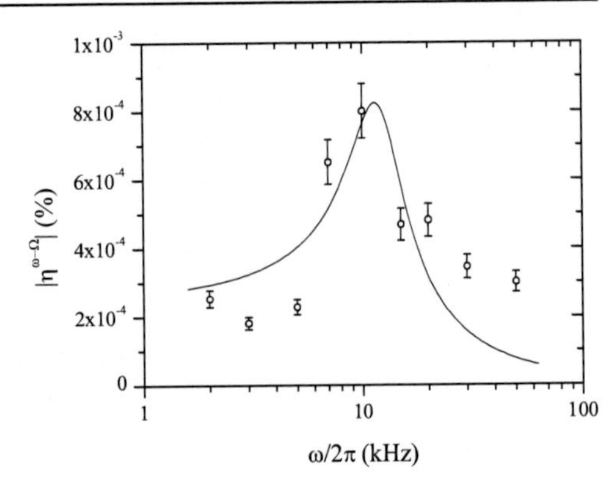

Figure 27. Dependencies of the diffraction efficiency amplitude $|\eta^{\omega-\Omega}|$ versus frequency of phase modulation ω measured in $Bi_{12}SiO_{20}$ (left figure) and $Bi_{12}TiO_{20}$ (right figure). $E_A = 0.92$ kV/cm. The measurements in $Bi_{12}SiO_{20}$ are carried out for two values of dc field and corresponding combination frequencies: $E_0 = 8.0$ kV/cm, $(\omega - \Omega)/2\pi = -48$ Hz (○) and $E_0 = 12$ kV/cm, $(\omega - \Omega)/2\pi = -38$ Hz (●). The measurements in $Bi_{12}TiO_{20}$ are carried out for $E_0 = 12$ kV/cm, $(\omega - \Omega)/2\pi = -10$ Hz. The approximations by Eq. (135) are shown by the solid lines.

$\epsilon = 56$, and $\mu\tau \simeq 10^{-6}$ cm²/V this reduction should appear at $E_0 \simeq 10$ kV/cm what well corresponds to the observed behavior. The proper description of the effect in trap saturation regime can be done by the appropriate modification of the basic equation set [Eqs. (73)-(77)] and subsequent calculations.

The dependence of the diffraction efficiency $|\eta^{\omega-\Omega}|$ measured at negative combination frequency versus amplitude of the applied ac field E_A is presented in Fig. 26 (right figure). This dependence seems to be linear in the region $E_A = 0 - 1.0$ kV/cm. The higher fields cause the noticeable saturation of the signal amplitude. This behavior can be explained as follows. The interaction of the ac field with the running photoconductivity grating is more efficient for higher amplitudes of an ac field, this also results in the growth of space charge wave. However, large ac fields perturb the propagation of the photoconductivity grating. As it is known, [37, 60] the excitation of running photoconductivity gratings occurs when the moving light pattern is synchronized with the propagation of photocarriers drifting in dc electric field. Higher values of the applied ac field comparable with the dc bias lead to considerable change of drift velocities, the synchronization is partially lost leading to the reduction of the running photoconductivity grating amplitude. This causes the saturation of the resulting space charge wave and corresponding detected signal. The obtained experimental dependence is approximated by Eq. (133) with $\mu\tau = 2.1 \times 10^{-7}$ cm²/V.

Dependencies of the diffraction efficiency versus frequency of phase modulation $|\eta^{\omega-\Omega}(\omega)|$ were measured in $Bi_{12}SiO_{20}$ and $Bi_{12}TiO_{20}$ crystals for the constant combination frequency $\omega - \Omega \approx -\omega_{sc}$ (Fig. 27). The experimental curves demonstrate the behavior typical for running photoconductivity gratings: the growth of the dc electric field increases the amplitude of the detected signal and shifts the resonant peak to higher modulation frequencies (108). The resonant frequency values are slightly smaller than those obtained for the non-steady-state photo-EMF experiments (Fig. 25). The approximation of the experi-

Table 3. Material parameters of the $Bi_{12}SiO_{20}$ crystal estimated from the frequency dependence of the diffraction efficiency amplitude $|\eta^{\omega-\Omega}(\omega)|$.

E_0, kV/cm	T, K	μ, cm²/V s	$\mu\tau$, cm²/V
8.0	296	1.1×10^{-2}	1.9×10^{-7}
12	298	1.4×10^{-2}	1.1×10^{-7}

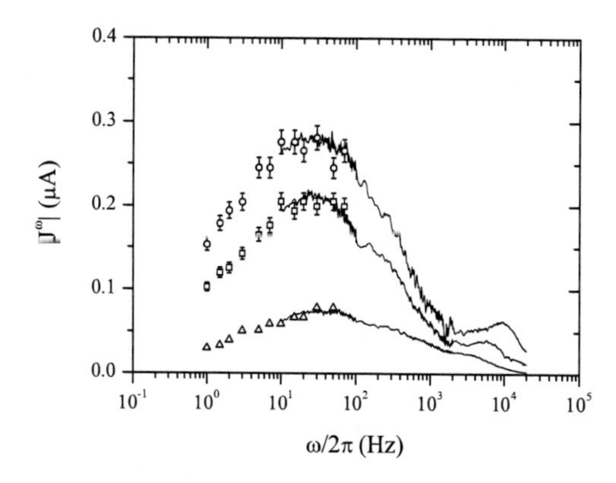

Figure 28. Frequency transfer functions of the non-steady-state photo-EMF excited in $Bi_{12}SiO_{20}$. Standard technique ($E_A = 0$). The dependencies are measured for three values of the applied dc electric field: $E_0 = 6$ kV/cm (△), 10 kV/cm (□), 14 kV/cm (o).

mental dependencies by Eq. (135) allowed determination of the drift mobility of electrons as well as $\mu\tau$-product for $Bi_{12}SiO_{20}$ crystal (Table 3). The mobility measurements were also carried out in photorefractive $Bi_{12}TiO_{20}$. This crystal has sufficiently lower photoconductivity and corresponding resonant frequency of space charge oscillations ω_{sc}. Thus the minimal possible combination frequency $(\omega - \Omega)/2\pi = -10$ Hz was settled close to the resonant frequency. The amplitude of the detected signal was significantly lower than the one in $Bi_{12}SiO_{20}$ crystal due to the same reason: the smaller conductivity grating excites weak space charge wave. The electron mobility and the $\mu\tau$-product were estimated from the frequency dependence (Fig. 27): $\mu = 3.1 \times 10^{-3}$ cm²/V s, $\mu\tau = 1.4 \times 10^{-7}$ cm²/V ($T = 293$ K).

The obtained drift mobility values in $Bi_{12}SiO_{20}$ and $Bi_{12}TiO_{20}$ crystals are lower by 2-3 orders of magnitude than the actual band [38] and Hall [75] mobilities. Such difference is usually attributed to the process of strong trapping of photoelectrons on the shallow levels [34] or large polarons influence [38].

In this section we also consider the application of another, namely, the non-steady-state photo-EMF technique for detection of the discussed processes in $Bi_{12}SiO_{20}$ at $\lambda = 442$ nm.

Figure 28 presents the frequency transfer functions of the non-steady-state photo-EMF measured in $Bi_{12}SiO_{20}$ crystal for three values of the applied dc field E_0. If the value of an applied electric field is increased the signal amplitude increases as well. Initially, the first resonant maximum appears in the low frequency region ($\omega/2\pi \sim 100$ Hz). When the dc field reaches $E_0 \simeq 6$ kV/cm the shoulder is observed on the frequency transfer function at $\omega/2\pi \sim 3$ kHz. The further increase of an applied field amplitude transforms this shoulder into the second resonant maximum. The growth of the applied field shifts the first and second maxima to lower and higher excitation frequency regions respectively. For example, the resonant frequencies equals $\omega_{sc}/2\pi \simeq 25$ Hz, $\omega_{pc}/2\pi \simeq 5.6$ kHz for the electric field $E_0 = 10$ kV/cm and $\omega_{sc}/2\pi \simeq 20$ Hz, $\omega_{pc}/2\pi \simeq 9.1$ kHz for the electric field $E_0 = 14$ kV/cm. It was shown [37] that such signal behavior is associated with the resonant excitation of the space charge and photoconductivity running gratings. The first resonance is very wide which can also be attributed to the nonlinear character of the space charge formation for large contrast of the interference pattern [73]. The light intensity inhomogeneity can be another reason for such behavior: strong light absorption ($\alpha \sim 30$ cm^{-1}) makes the space charge relaxation time in the surface layer smaller than in the crystal's bulk.

Direct measurements of the photocarrier's drift mobility were carried out using the non-steady-state photo-EMF technique. The mobility value was estimated from the position of the second resonant frequency for fixed K and E_0 values: $\mu \simeq 0.015$ cm^2/V s. This estimation is in good agreement with an electron mobility value $\mu = 0.016$ cm^2/V s measured earlier at the wavelength $\lambda = 458$ nm [37]. Nevertheless the large light absorption the parameters estimated from the high frequency resonant maximum seem to be rather correct since the resonant frequency and the shape of the resonant peak are not defined by parameters dependent on light intensity [Eqs. (129), (135)].

Let us proceed with the combined excitation of the space charge and photoconductivity gratings. The frequency of an external ac field was settled approximately equal to the frequency of the second resonant maximum: $\Omega/2\pi = 5.9$ kHz for $E_0 = 10$ kV/cm and $\Omega/2\pi = 8.8$ kHz for $E_0 = 14$ kV/cm. The amplitude of an external ac field ($E_A = 3.5$ kV/cm) was chosen to be considerably smaller than the dc field value.

Figure 29 (left figure) presents the frequency transfer function of the non-steady-state photo-EMF excited at difference combination frequency $\omega - \Omega$. Two maxima are clearly seen at this dependence: the first one is observed for the negative difference frequency $(\omega - \Omega)/2\pi = -(50 - 200)$ Hz ($E_0 = 10$ kV/cm), $(\omega - \Omega)/2\pi = -(20 - 200)$ Hz ($E_0 = 14$ kV/cm) and the second one – for the positive frequency $(\omega - \Omega)/2\pi = 50$ Hz ($E_0 = 10$ kV/cm), $(\omega - \Omega)/2\pi = 30$ Hz ($E_0 = 14$ kV/cm). The increase of the external dc field causes the growth of the signal amplitude as well as the shift of the resonant maxima to lower frequencies. This behavior is typical for the space charge waves [2]. The resonant maxima turned out to be wide. This fact can be attributed to the nonlinear limitation of the space charge field for large contrasts of the interference pattern. The inhomogeneity of the light distribution through the sample thickness can be another reason of such behavior.

Figure 29 (right figure) presents the dependencies of the non-steady-state photo-EMF amplitude at resonant frequencies versus amplitude of the applied ac field. These dependencies seems to be linear in the investigated range of ac field amplitudes. The saturation of the signal amplitude can be expected for such values of an ac field E_A when $KL_A \sim 2$ [see Eqs. (127), (128)]. For $\mu\tau \sim 10^{-6}$ cm^2/V (see, for example Ref. [59]) and $K = 25$ mm^{-1} this saturation effect should be observed for $E_A \sim 8$ kV/cm.

In this section we have considered the combined excitation of two eigenmodes in semiconductor: the running gratings of the space charge and photoconductivity. The novel technique based on the diffraction efficiency measurements allowed us to detect the space

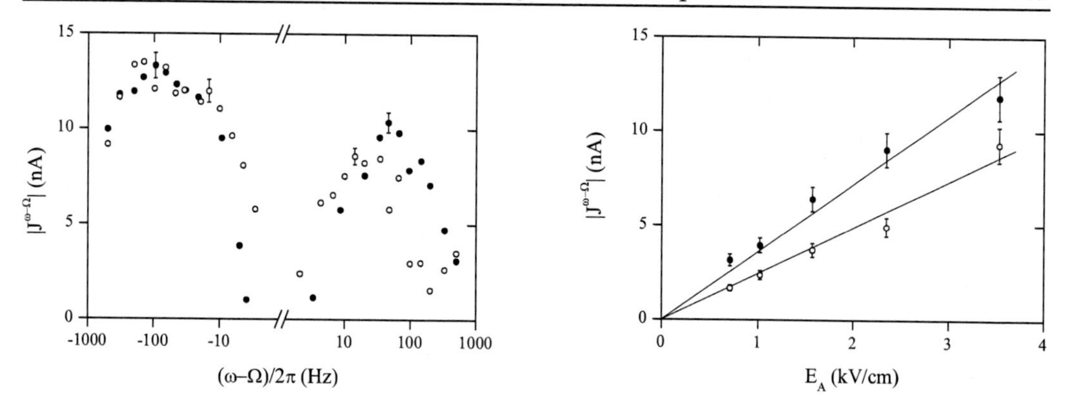

Figure 29. Frequency transfer function of the non-steady-state photo-EMF excited in $Bi_{12}SiO_{20}$ by the combination technique (left figure). The dependencies are obtained for two values of the applied dc field and for two corresponding frequencies of the applied ac field: $E_0 = 10$ kV/cm, $\Omega/2\pi = 5.9$ kHz (○) and $E_0 = 14$ kV/cm, $\Omega/2\pi = 8.8$ kHz (●). $E_A = 3.5$ kV/cm. Dependence of the non-steady-state photo-EMF amplitude $|J^{\omega-\Omega}|$ on the amplitude of the applied ac field (right figure). The measurements are carried out for two resonant frequencies: $(\omega - \Omega)/2\pi = -20$ Hz (●), $(\omega - \Omega)/2\pi = +20$ Hz (○). $E_0 = 14$ kV/cm, $\Omega/2\pi = 8.8$ kHz.

charge wave originated from the nonlinear interaction of the running photoconductivity grating with the applied ac field.

The most important feature of this approach is the difference between the spatiotemporal characteristics of the arising space charge wave and the spatiotemporal characteristics of the external driving forces (illumination and applied field). In this sense, the experiments presented in the paper is another demonstration of the eigen-nature of space charge wave – the oscillation which properties are defined by the material parameters rather than external forces.

The combined excitation of the photoconductivity and space charge waves is very promising for the investigation of fast processes in high resistive semiconductors. The measurement of the actual drift mobility of photocarriers in widegap semiconductors with complicated structure of recombination levels is an important example of the tasks, where detection of high frequency signals is necessary. Indeed, the resonant frequency of the photoconductivity grating can reach 100 MHz in $Bi_{12}SiO_{20}$ crystal [60]. Due to the large values of Maxwell relaxation time ($\tau_M \sim 1$ ms) oscillations of the space charge field and corresponding diffraction efficiency should be negligible in this frequency range: the reduction coefficient is $(\omega\tau_M)^{-1}$. This makes observation of the high frequency resonances practically impossible. In contrast to the standard techniques the developed approach of combined excitation allows the detection of fast processes in the low-frequency region.

The nonlinear interaction of applied ac field with running gratings of the photoconductivity and space charge is not restricted to the results obtained in this paper. Further experimental investigations in other widegap semiconductors should be carried out, and more detailed theoretical analysis including trap saturation, optical activity and absorption effects should be performed to complete description of the phenomena discussed.

7. Non-steady-state Photo-EMF in SnS_2 Crystals

Tin disulphide belongs to a class of compounds of the form MX_2, where M represents an element of group IV (Sn, Zr, Hf, Tl, etc.) and X an element of Group VIB (S, Se, Te) of the periodic table. These compounds crystallize in CdI_2, C-6 type crystal structure. The X atoms form densely packed hexagonal layers. The M atoms are arranged in layers with the same hexagonal structure to give a stacking sequence X-M-X, X-M-X, etc in the c-direction of the crystal. Each metal atom is surrounded by three X atoms of each layer so as to form octahedron. The bonds between these sandwich layers are weak and thought to be of Van der Waals type.

The optical properties of SnS_2 have been studied in Refs. [76–78]. The crystals of SnS_2 are highly sensitive in the visible region of spectrum, however, there are only several papers [79–81] dealing with the photoconductivity investigation in these crystals. Theoretical calculations of the energy band structure of SnS_2 crystals [82] showed that this material has several conductive sub-bands, that can have various energy with respect to the valence band (the energy interval depends on the crystal polytype). So the measurements at different light wavelengths should reveal this peculiarity. In this section we present the experimental results of the non-steady-state photocurrent measurements in SnS_2 crystals in the red (633 nm), blue (442 nm) and ultraviolet (325 nm) regions of spectrum [83, 84]. The material parameters such as type of photoconductivity, average diffusion length of photocarriers, dielectric relaxation time and the Debye screening length were determined using the non-steady-state photo-EMF technique [20, 41, 85–87]. Since this effect is based on the same principles as the dynamic holographic recording in photorefractive materials the results of this paper can also be useful for the expansion of our understanding of the space charge field formation process in these media.

7.1. Experimental Results

The observation of the non-steady-state photo-EMF in SnS_2 crystal was performed using the conventional experimental setup (Fig. 8, b). The detailed description of the experimental procedure is presented in Refs. [83, 84]. The SnS_2 crystal was grown by the vapor transport method using iodine as a transporting agent. The size of the sample was $6 \times 5 \times 0.042$ mm^3. The silver paste electrodes were painted on the front surface 6×5 mm^2. The distance between electrodes was about 1 mm. The band gap for SnS_2 is equal to 2.0-2.1 eV [77,82,88], the absorption coefficient for the illumination wavelengths is $\alpha \simeq 10$ cm^{-1} (for $\lambda = 633$ nm), $\alpha \sim 10^4$ cm^{-1} (for $\lambda = 442$, 325 nm) [77]. It means that the red light is absorbed slightly and practically uniformly within the crystal thickness whereas the near-surface absorption takes place for the blue and UV light. The optical axis of the crystal was perpendicular to the front surface so the grating vector laid in the plane where the crystal properties are uniform. The quoted value of the dielectric constant is $\epsilon = 6 - 20$, the electron mobility equals $\mu = 50$ cm^{2}V^{-1}s^{-1} [88]. We have measured the ϵ value for our crystal and it was found to be 7.0.

We observed the pronounced photocurrent signal J^ω in SnS_2 crystal for all illumination wavelengths. The signal to noise ratio (S/N) for the optimal spatial and temporal frequencies was about $S/N = 50$ dB for $\lambda = 633$ nm, $P_0 = 33$ mW, $\tau_{int} = 0.1$ s, $S/N = 43$

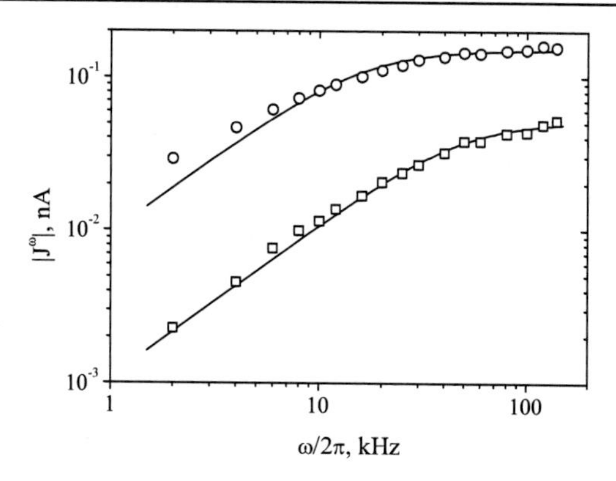

Figure 30. Frequency transfer functions of the non-steady-state photo-EMF in SnS_2 crystal. $\lambda = 633$ nm, $P_0 = 28$ mW, $K = 0.44\ \mu m^{-1}$ (○) and $\lambda = 442$ nm, $P_0 = 0.28$ mW, $K = 0.16\ \mu m^{-1}$ (□).

dB for $\lambda = 442$ nm, $P_0 = 0.28$ mW, $\tau_{int} = 0.3$ s, and $S/N = 40$ dB for $\lambda = 325$ nm, $P_0 = 3.8$ mW, $\tau_{int} = 1$ s (P_0 is the light power on the crystal surface, τ_{int} is the integration time).

It was shown in [20] that using non-steady-state photo-EMF technique one can define the type of photoconductivity. We compared the phase of the photocurrent signal in SnS_2 and $Bi_{12}SiO_{20}$ for all illumination wavelengths and spatial frequencies of the interference pattern $K = 0.1-20\ \mu m^{-1}$ and found out that SnS_2 has the same type of photoconductivity as $Bi_{12}SiO_{20}$, i.e. electron one [20]. Note that the sign of the dominant photocarriers can be obtained directly from the sign of the non-steady-state photocurrent signal, since J^ω has opposite signs for electrons and holes.

The experimental frequency dependencies of the photocurrent amplitude for wavelengths $\lambda = 633$, 442 nm are presented in Fig. 30. For low modulation frequencies the signal amplitude grows linearly with ω up to characteristic cut-off frequency ω_0; for higher frequencies of phase modulation the signal amplitude is frequency-independent. This behavior is typical for photorefractive materials and can be simply explained [20]. At high frequencies $\omega > \omega_0$ the photoconductivity distribution follows the movements of the interference pattern whereas the space charge field grating is nearly motionless. The condition for photocurrent excitation is optimal. For low excitation frequencies ($\omega < \omega_0$) the photocurrent decreases since both photoconductivity and electric field distributions can follow slow shifts of the interference pattern. The theory predicts the following frequency dependence of the photocurrent amplitude [20, 41, 85]:

$$J^\omega = \frac{Sm^2 \Delta\sigma_0 E_d}{2(1 + K^2 L_D^2)(1 + K^2 L_d^2)} \frac{-i\omega/\omega_0}{1 + i\omega/\omega_0}, \tag{136}$$

where S is the sample cross-section, σ_0 is the average photoconductivity of the material, $E_d = K k_B T/e$ is the diffusion field, k_B is the Boltzmann constant, T is the temperature, e is the electron charge, L_d is the diffusion length of photoelectrons and L_D is the Debye screening length. The cut-off frequency ω_0 is associated with the time of grating recording

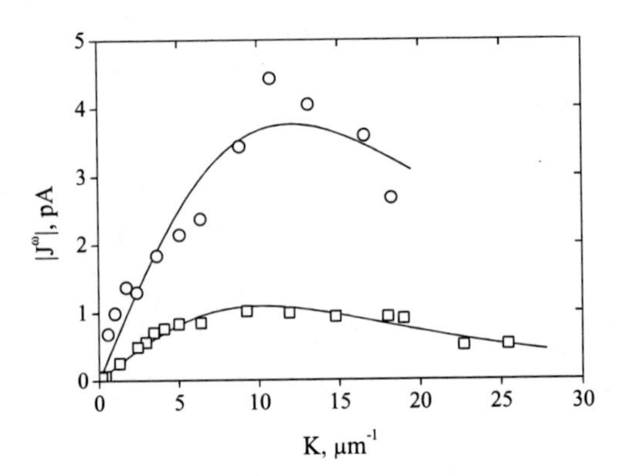

Figure 31. Dependencies of the non-steady-state photo-EMF on the spatial frequency measured for $\omega/2\pi = 30$ Hz $\ll \omega_0/2\pi$. $\lambda = 633$ nm, $P_0 = 34$ mW (○); $\lambda = 442$ nm, $P_0 = 0.42$ mW (□).

τ_{sc} as [41]:

$$\omega_0 = \frac{1}{\tau_{sc}} = \frac{1 + K^2 L_D^2}{\tau_M (1 + K^2 L_d^2)}, \tag{137}$$

here $\tau_M = \epsilon\epsilon_0/\sigma_0$ is the Maxwell relaxation time, ϵ_0 is the free space permittivity, ϵ is the static dielectric constant. According to (2) one can estimate the values of Maxwell relaxation time τ_M and photoconductivity σ_0 from the measurement of the cut-off frequency ω_0 for low spatial frequencies K ($K^2 L_d^2$, $K^2 L_D^2 \ll 1$). We performed such measurements and obtained for the cut-off frequency $\omega_0/2\pi$, Maxwell relaxation time τ_M, photoconductivity of the sample σ_0 and for average concentration of photocarriers n_0 the following: $\omega_0/2\pi = 16$ kHz, $\tau_M = 10$ μs, $\sigma_0 = 6.2 \times 10^{-6}$ Ω^{-1}m^{-1}, $n_0 = 7.8 \times 10^{15}$ m^{-3} ($\lambda = 633$ nm, $P_0 = 28$ mW) and $\omega_0/2\pi = 48$ kHz, $\tau_{SC} = 3.3$ μs ($\lambda = 442$ nm, $P_0 = 0.28$ mW, $K = 0.16$ μm^{-1}). The estimations were done for $\epsilon = 7.0$. The results of fitting are presented in the same figure. We observed linear dependence of the cut-off frequency ω_0 and photocurrent amplitude J^ω (for $\omega > \omega_0$) on the light power P_0 for the excitation wavelength $\lambda = 633$ nm and sub-linear behavior (ω_0, $J^\omega \propto P_0^{0.8}$) for $\lambda = 442$ nm. The photoconductivity σ_0 depends on the light intensity in the same way. We observed no reduction (for the modulation frequencies up to 150 kHz) of the photocurrent amplitude due to finiteness of the electron lifetime τ [41], so we can state that the lifetime of photocarriers τ is less than 1 μs.

We have performed the measurements of the non-steady-state photocurrent amplitude versus spatial frequency K (see Fig. 31) at low frequencies of phase modulation ($\omega \ll \omega_0$). According to Eq. (137) the cut-off frequency ω_0 is a decreasing function of K. Condition $\omega \ll \omega_0$ should be satisfied for whole range of spatial frequencies used. In our experiments we settled $K_{max} \simeq 20$ μm^{-1} and corresponding cut-off frequencies were $\omega_0/2\pi = 200$ Hz (for $\lambda = 633$ nm) and $\omega_0/2\pi = 2.8$ kHz (for $\lambda = 442$ nm). For both excitation wavelengths we fixed the frequency of phase modulation $\omega/2\pi = 30$ Hz so that condition $\omega \ll \omega_0$ was

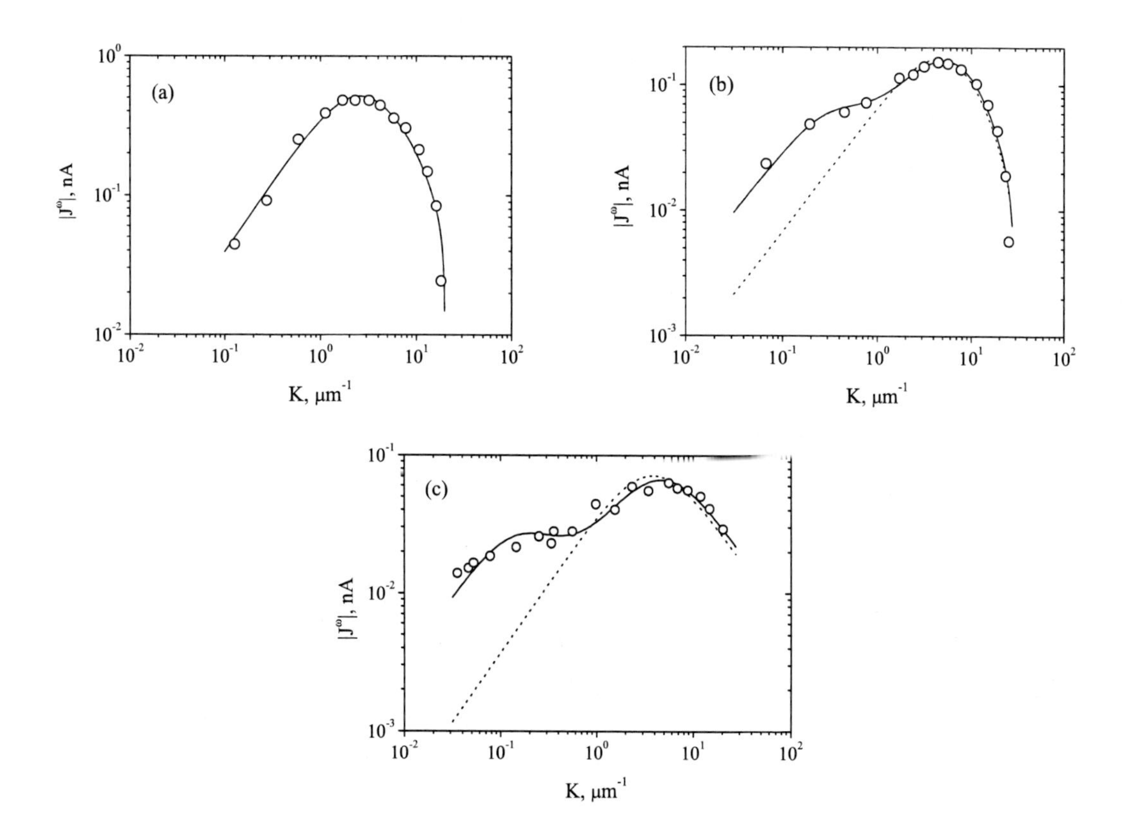

Figure 32. Dependencies of the non-steady-state photo-EMF on the spatial frequency measured for $\omega > \omega_0$. a – $\lambda = 633$ nm, $P_0 = 33$ mW, $\omega/2\pi = 100$ kHz; b – $\lambda = 442$ nm, $P_0 = 0.29$ mW, $\omega/2\pi = 140$ kHz; c – $\lambda = 325$ nm, $P_0 = 3.8$ mW, $\omega/2\pi = 94$ kHz. The dotted and solid lines show approximation by Eqs. (139) and (155), respectively.

valid. For low frequency of phase modulation the photocurrent signal amplitude depends on spatial frequency of the interference pattern K as [41]:

$$J^\omega \propto \frac{K}{(1 + K^2 L_D^2)^2}. \tag{138}$$

The signal amplitude J^ω peaks at spatial frequency $K = (\sqrt{3} L_D)^{-1}$. Approximation of the experimental results provides the following estimations for the Debye screening lengths: $L_D = 48$ nm (for $\lambda = 633$ nm) and $L_D = 61$ nm (for $\lambda = 442$ nm). Using the well-known expression for the Debye screening length [41] we calculated the density of capture centers: $N_A = 4.3 \times 10^{21}$ m^{-3} (for $\lambda = 633$ nm), $N_A = 2.6 \times 10^{21}$ m^{-3} (for $\lambda = 442$ nm).

The measurements of dependence $J^\omega(K)$ for $\omega > \omega_0$ were performed as well (Fig. 32). As it was shown in [20, 41] this dependence can also provide information about material parameters. The behavior of the nonstationary photocurrent is defined by the following

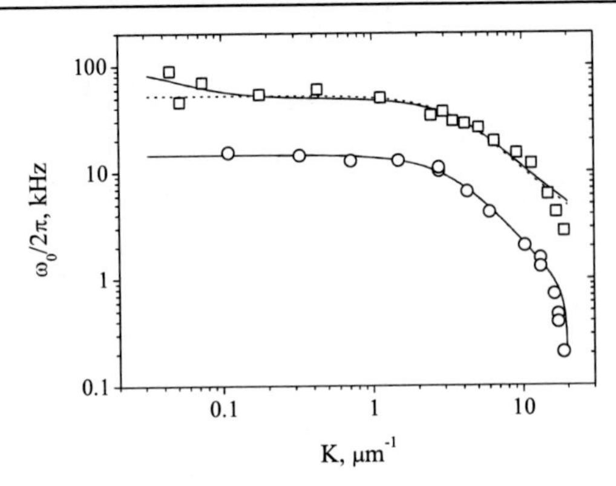

Figure 33. Dependence of the cut-off frequency ω_0 on the spatial frequency of the interference pattern. $\lambda = 633$ nm, $P_0 = 29$ mW (○); $\lambda = 442$ nm, $P_0 = 0.35$ mW (□).

expression [41]:

$$J^{\omega} \propto \frac{K}{(1 + K^2 L_D^2)(1 + K^2 L_d^2)}. \tag{139}$$

The linear growth of the signal J^{ω} for low spatial frequencies K is due to increasing of the diffusion field E_d [20, 41]. Smoothing of the photoconductivity distribution at $K > L_d^{-1}$ results in decreasing of the non-steady-state photocurrent flowing through the crystal. For $L_d > L_D$ the dependence $J^{\omega}(K)$ has a maximum at $K \simeq L_d^{-1}$. Using equation (139) we determined the value of the diffusion length of photocarriers: $L_d = 0.38$ μm ($\lambda = 633$ nm).

As seen from Fig. 32, b the photocurrent excited at $\lambda = 442$ nm is not adequately described by Eq. (139) at low spatial frequencies (dash line). We tried to approximate the obtained results using the theory of the effect based on the concept of bipolar conductivity [85–87], but we did not obtain satisfactory correspondence as well. We think that the observed experimental dependencies can be explained by the presence of two types of photocarriers – light and heavy electrons. Equation (155) obtained below for such model of semiconductor describes experimental results at low spatial frequencies more correctly (see solid line in Fig. 32, b). Corresponding diffusion lengths of photocarriers and photoconductivity ratio were found: $L_{d1} = 0.15$ μm, $L_{d2} = 2.0$ μm, $\sigma_2/\sigma_1 = 4.3$.

Even more noticeable difference ($\sim$ 18 dB) between the experimental results and approximation by standard expression (139) is observed for excitation wavelength $\lambda = 325$ nm (Fig. 32, c). Approximation by Eq. (155) seems more reliable and provides the following estimations of material parameters: $L_{d1} = 0.14$ μm, $L_{d2} = 5.0$ μm, $\sigma_2/\sigma_1 = 11$. Hypothetically there is a possibility for the excitation of three types of electrons and in this case we get the following parameters: $L_{d1} = 0.16$ μm, $L_{d2} = 0.85$ μm, $L_{d3} = 11$ μm, $\sigma_2/\sigma_1 = 2.1$ and $\sigma_3/\sigma_1 = 20$.

The dependence of cut-off frequency ω_0 versus spatial frequency of the interference pattern can also be used for estimation of the Maxwell relaxation time τ_M and diffusion length of photocarriers L_d. We measured this dependence (see Fig. 33) and approximated it using Eq. (137) (solid line for $\lambda = 633$ nm, dash line for $\lambda = 442$ nm) and Eq. (153) (solid line for $\lambda = 442$ nm). Values τ_M and L_d for $\lambda = 633$ nm were found: $\tau_M = 11$ μs,

$L_d = 0.25\ \mu$m ($P_0 = 29$ mW). For $\lambda = 442$ nm we got the following results: $\tau_M = 1.6$ μs, $L_{d1} = 0.22\ \mu$m, $L_{d2} = 20\ \mu$m, $\sigma_2/\sigma_1 \simeq 1$ ($P_0 = 0.35$ mW).

Using the obtained value of τ_M we also estimated the average photoconductivity and concentration of free electrons for $\lambda = 442$ nm: $\sigma_0 \simeq 3.9\times10^{-5}\ \Omega^{-1}\mathrm{m}^{-1}$, $n_0 \simeq 4.9\times10^{16}$ m^{-3}. Since there is no enough information about mobilities for each type of electrons it should be reminded that by using terms "light" and "heavy" electrons we imply that these carriers have different diffusion lengths.

7.2. Theoretical Analysis and Discussion

It is well known that there are several minima (valleys) in the conduction band of many semiconductors (e.g. GaAs). The electrons in these valleys are characterized with different effective masses, lifetimes and mobilities. In this section we analyze the effect of the non-steady-state photo-EMF for the model of semiconductor with two channels of charge migration characterized by different transport parameters.

Let us suppose that the crystal is illuminated by an oscillating interference pattern:

$$I(x,t) = I_0\left[1 + \frac{m(t)}{2}\exp(iKx) + \frac{m^*(t)}{2}\exp(-iKx)\right], \tag{140}$$

where

$$m(t) = m\exp(i\Delta\cos\omega t) \tag{141}$$

is the contrast of the interference pattern.

Let us consider the model of semiconductor with two types of charge carriers (electrons) characterized with mobilities μ_1, μ_2 and lifetimes τ_1, τ_2. The excitation and recombination of electrons are assumed to be linear. The transitions of electrons between valleys are neglected. Then the balance equations for the corresponding concentrations of electrons n_1, n_2 are as follows:

$$\frac{\partial n_{1,2}}{\partial t} = g_{1,2} - \frac{n_{1,2}}{\tau_{1,2}} + \mu_{1,2}\frac{\partial}{\partial x}(n_{1,2}E) + D_{1,2}\frac{\partial^2 n_{1,2}}{\partial x^2}, \tag{142}$$

where g_1, g_2 are the generation rates of electrons and $D_{1,2} = (k_BT/e)\mu_{1,2}$ are their diffusion coefficients.

In order to calculate the distributions of the carrier density and space charge field the continuity and Poisson equations should be added [2]. Differentiating the Poisson equation with respect to time and using the continuity equation we get

$$\epsilon\epsilon_0\frac{\partial^2 E}{\partial x\partial t} = -\frac{\partial j_e}{\partial x}. \tag{143}$$

The current density j_e created by the transport of electrons of both types equals

$$j_e = e\mu_1 n_1 E + eD_1\frac{\partial n_1}{\partial x} + e\mu_2 n_2 E + eD_2\frac{\partial n_2}{\partial x}. \tag{144}$$

The calculations of the photocurrent and electric field are performed using the technique proposed in [2, 19]. For low contrast of the interference pattern $m \ll 1$ values $n(x,t)$ and

$E(x,t)$ can be written using their complex amplitudes [2]:

$$n_{1,2}(x,t) = n^0_{1,2}\left[1+\frac{a_{1,2}(t)}{2}\exp(iKx)+\frac{a^*_{1,2}(t)}{2}\exp(-iKx)\right], \tag{145}$$

$$E(x,t) = \frac{E_{sc}(t)}{2}\exp(iKx)+\frac{E^*_{sc}(t)}{2}\exp(-iKx). \tag{146}$$

Here $n^0_{1,2} = g^0_{1,2}\tau_{1,2}$ are the average concentrations of electrons of both types. Then the nonstationary photocurrent equals

$$j(t) = \sum_{q=1,2} e\mu_q n^0_q[a_q(t)E^*_{sc}(t)+a^*_q(t)E_{sc}(t)]. \tag{147}$$

Substituting Eqs. (145), (146) into Eqs. (142)-(144) we can calculate complex amplitudes $a_{1,2}(t)$ and $E_{sc}(t)$. For the simplicity let us suppose the electron densities are quasi-stationary: $\partial n_{1,2}/\partial t \simeq 0$ [2]. Then complex amplitudes $a_{1,2}(t)$ are easily obtained [2, 19]:

$$a_{1,2}(t) = \frac{m(t)+iK\mu_{1,2}\tau_{1,2}E_{sc}}{1+K^2L^2_{d1,d2}}. \tag{148}$$

Here $L_{d1,d2} = (D_{1,2}\tau_{1,2})^{1/2}$ are the diffusion lengths of electrons of both types. Further substitution of Eqs. (145), (146) into(143), (144) with (148) leads to the equation describing the process of space charge grating formation:

$$\frac{dE_{sc}}{dt} = -\left[\frac{\sigma_1}{1+K^2L^2_{d1}}+\frac{\sigma_2}{1+K^2L^2_{d2}}\right]\frac{imE_d+E_{sc}}{\epsilon\epsilon_0}. \tag{149}$$

Here $\sigma_{1,2} = e\mu_{1,2}n^0_{1,2}$ are the average values of the conductivity corresponding to the two types of electrons. The obtained equation differs from the classic equation of hologram recording [2] by the presence of two components in the brackets.

Let us suppose the amplitude of phase modulation to be small: $\Delta \ll 1$, then $m(t) \simeq m[1+i0.5\Delta\exp(i\omega t)+i0.5\Delta\exp(-i\omega t)]$. Amplitudes $a_{1,2}(t)$ and $E_{sc}(t)$ can be presented as

$$a_{1,2}(t) = a^0_{1,2}+a^+_{1,2}\exp(i\omega t)+a^-_{1,2}\exp(-i\omega t), \tag{150}$$

$$E_{sc}(t) = E^0_{sc}+E^+_{sc}\exp(i\omega t)+E^-_{sc}\exp(-i\omega t), \tag{151}$$

Substituting Eq. (151) into Eq. (149) we obtain

$$E^0_{sc} = -imE_d, \qquad E^{\pm}_{sc} = \frac{m\Delta}{2}\frac{E_d}{1\pm i\omega/\omega_0}. \tag{152}$$

Here we introduced the cut-off frequency of the non-steady-state photo-EMF:

$$\omega_0 = \frac{1}{\epsilon\epsilon_0}\left[\frac{\sigma_1}{1+K^2L^2_{d1}}+\frac{\sigma_2}{1+K^2L^2_{d2}}\right]. \tag{153}$$

It is seen from the expression for E^0_{sc} that amplitude of the stationary space charge field grating recorded in the semiconductor with two types of carriers (with the same sign) is exactly equal to the standard value $-imE_d$ [2]. It is worth to point out that the amplitude of the grating recorded in semiconductor with bipolar conductivity decreases due to partial compensation of the space charges created by electrons and holes [2]. After substitution of Eq. (152) into Eq. (148) one can obtain expressions for the complex amplitudes of the electron density:

$$a_{1,2}^0 = m, \qquad a_{1,2}^{\pm} = \frac{im\Delta/2}{1 \pm i\omega/\omega_0}\left[1 \pm \frac{i\omega/\omega_0}{1 + K^2 L_{d1,d2}^2}\right]. \tag{154}$$

At last using the obtained expressions for $a_{1,2}$ and E_{sc} we can calculate the amplitude of the non-steady-state photo-EMF:

$$j^{\omega} = \frac{m^2 \Delta E_d}{2} \frac{-i\omega/\omega_0}{1 + i\omega/\omega_0}\left[\frac{\sigma_1}{1 + K^2 L_{d1}^2} + \frac{\sigma_2}{1 + K^2 L_{d2}^2}\right]. \tag{155}$$

Let us specify the main features of the non-steady-state photo-EMF in semiconductors with two types of charge carriers. The frequency transfer function is standard, i.e. there are linear growth of the signal at low frequencies of phase modulation ($\omega \ll \omega_0$) and frequency-independent region at higher frequencies ($\omega \gg \omega_0$). The dependence of the photocurrent on the spatial frequency (for $\omega \gg \omega_0$) can have different shapes for various relations between parameters of the carriers (L_{d1}, L_{d2}, σ_1, σ_2). We just point out that for low spatial frequencies ($K^2 L_{d1,d2}^2 \ll 1$) the amplitude of the photocurrent j^{ω} ($\omega \gg \omega_0$) and cut-off frequency ω_0 are defined by electron type producing the dominant contribution in the total photoconductivity, whereas for high spatial frequencies ($K^2 L_{d1,d2}^2 \gg 1$) it is defined by electron type with higher generation rate (compare with Ref. [85]). In contrast to the case of bipolar conductivity [85], the photo-EMF signal does not change its sign for any spatial frequency.

It is important to point out that crystal SnS_2 has a high photosensitivity: corresponding cut-off frequencies are in kilohertz range for all illumination wavelengths and powers used. For the well-known photorefractive sillenite-type crystals $Bi_{12}Si(Ge,Ti)O_{20}$ this value is in the range of hundreds of hertz for the same light intensities.

The Debye screening length value is rather small so the space charge field formation without saturation is possible in wide range of spatial frequencies $K = 0 - 10\ \mu m^{-1}$. In spite of the high photoconductivity this material has small diffusion length, especially for $\lambda = 442$ nm (for comparison: typical diffusion length of photoelectrons in photorefractive $Bi_{12}SiO_{20}$ at $\lambda = 442$ nm is about 2 μm).

The non-steady-state photo-EMF excited in SnS_2 at $\lambda = 633$ nm is well described in the frames of the standard model of semiconductor. Indeed, such illumination generates conduction band electrons from the impurity centers since the energy of photons ($h\nu = 1.96$ eV) is not enough for the interband transitions (forbidden band is of $2.07 - 2.18$ eV).

For the light with $\lambda = 442$ nm the energy of photons is of 2.81 eV. Such energy is enough for the generation of electrons not only to the minimum with lowest energy (it corresponds to the indirect transition in SnS_2) but also to the minimum which energy is higher by ~ 0.8 eV (direct transitions [77, 89]). Although there in no data concerning effective masses or mobilities of electrons in these valleys, we can suppose at least that diffusion lengths of electrons are different, so the utilization of the considered model is reasonable.

The dependence of the photocurrent on the spatial frequency measured at shorter wavelength $\lambda = 325$ nm ($h\nu = 3.82$ eV) differs from the standard one even more noticeably. In this case the carriers in layers with other polytype modification can also contribute to the total photocurrent. For example, layers of polytype $9R$ with forbidden band of 3.38 eV [90] can be present between layers of basic polytype modification $1H$.

The assumption of independent recombination and transport processes used in the theoretical analysis implies that the inter-sub-band relaxation time τ_{21} should exceed the electron lifetimes for the both sub-bands, i.e. $\tau_{21} \gg \tau_1, \tau_2$. We think that such relation is possible because the minimums of these sub-bands are shifted in the k-space [82] and the presence of the phonon is required for the inter-sub-band transition.

8. Non-steady-State Photo-EMF in Pyrolytic BN Crystal

Let us present experimental investigations of the non-steady-state photocurrents excited in pyrolytic crystal of boron nitride at the wavelength $\lambda = 532$ nm. The experiments are carried out for the diffusion (zero electric field) and drift mechanisms of recording (in an external ac electric field). The average photoconductivity and diffusion length of photoholes are estimated from the frequency transfer functions and dependence of the signal on electric field amplitude.

Modern optoelectronics requires new materials to manufacture devices operating at different wavelengths, electric fields, temperatures, etc. Wide-gap semiconductors such as GaN, ZnSe with band gap $E_g > 2.5$ eV are of particular importance for applications in semiconductor lasers and light emitting diodes for blue and ultraviolet spectral regions. The same materials are to be utilized for production of photoresistors and photodiodes. Boron nitride crystals are characterized by an unique combination of physicochemical properties that advance its wide application in various areas of science and technology. Due to its high thermal stability boron nitride is a widely used material in vacuum technology. It has been employed in microelectronic devices, for x-ray lithography masks, and as a wear-resistant lubrificant. The hexagonal phase is also the underlying structure of BN nanotubes, which are systems of growing interest at present. [91]

The chemical vapor deposition method [92] provides bulk samples of anisotropic modifications of boron nitride with high order degree of crystal lattice and nanostructure. Polarization infrared spectroscopy of well-ordered samples shows the existence of interlayer ion-covalent bonds which define the one-dimensional order-disorder phase transition between hexagonal and rhombohedral modifications of BN. The most perfect samples of pyrolytic BN crystallize in the form of large monocrystal layers of rhombohedral modification with transverse dimensions of 1 cm^2, thickness of 2000 Å and translation period $C_R = 3d$, where d is an interlayer distance. The crystal axes of the adjacent crystallites are correlated and the deviation of their c-axes does not exceed $1°$. The samples also contain about 20% of hexagonal phase with unordered interlayer bonds with the typical translation period of $C_H \simeq 2d$. The dimensions of hexagonal crystallites are $\sim$ 100 Å with the off-orientation degree of their hexagonal axes $15°$. Typical microstructure of an as grown pyrolytic BN is presented in Fig. 34.

Optical and electrical properties of boron nitride were investigated in a number of papers (see, for example, Refs. [93–98]). However, this does not mean that layered BN is a well

Figure 34. The microstructure of the investigated pyrolytic boron nitride. The c-axis is in the plane of the figure (parallel to the scale line of 1 μm).

investigated material: the scattering of the experimental data is significant. For example the estimated values of the band gap vary in the range 3.6-5.9 eV. [94] The absorption spectrum *of an as grown* rhombohedral BN measured in wide energy range corresponds to direct transitions and provides estimation of the forbidden gap to be $E_g = 5.67$ eV.

The measurements of the nonstationary holographic currents in BN crystal were carried out using standard experimental setup (Fig. 8, b). The detailed description of the arrangements is presented in Ref. [99].

First let us consider the diffusion regime of photocurrent excitation when the external electric field is not applied to the sample. The sign of the detected signal in the investigated frequency range ($\omega/2\pi = 1.5 - 200$ Hz) corresponds to the hole photoconductivity of the crystal.

The frequency transfer functions of the nonstationary photocurrent are shown in Fig. 35. These dependencies demonstrate typical behavior of the signal excitation, i.e. the linear growth for low modulation frequencies and the plateau for higher frequencies. Such behavior is an important manifestation of adaptive nature of space charge formation in photoconductive materials. As seen from Fig. 35 experimental points are in satisfactory agreement with the theoretical dependencies calculated using following expression [20]:

$$J^{\omega} = \frac{Sm^2 \Delta\sigma_0 E_D}{2(1 + K^2 L_D^2)} \frac{-i\omega/\omega_0}{1 + i\omega/\omega_0}, \tag{156}$$

where

$$\omega_0 = [\tau_M(1 + K^2 L_D^2)]^{-1} \tag{157}$$

is the cut-off frequency.

For high light intensities we observed the growth of the photocurrent amplitude and corresponding shift of the cut-off frequency (Fig. 36), what is associated with the increase of the crystal's photoconductivity. These dependencies are slightly nonlinear in the investigated range of light intensities: $|J^{\omega}| \propto I_0^{0.75}$ and $\omega_0 \propto I_0^{0.64}$. The non-steady-state photo-EMF technique provides the possibility for determination of crystal's photoconductivity from the position of corresponding cut-off frequency ω_0 [Eq. (157)] measured at low spatial frequency $KL_D \ll 1$. For the layered BN crystals the dielectric constant equals $\epsilon_{\perp} = 6.85 - 7.04$ (Ref. [93]), the diffusion length is measured below ($L_D = 35 - 50$ nm, $KL_D = 0.03 - 0.05$) so we obtain for the photoconductivity value of $\sigma_0 = (0.9 - 2.4) \times 10^{-10}\ \Omega^{-1}\text{cm}^{-1}$ ($I_0 = 0.22 - 0.86$ W/cm^2). This estimate is about

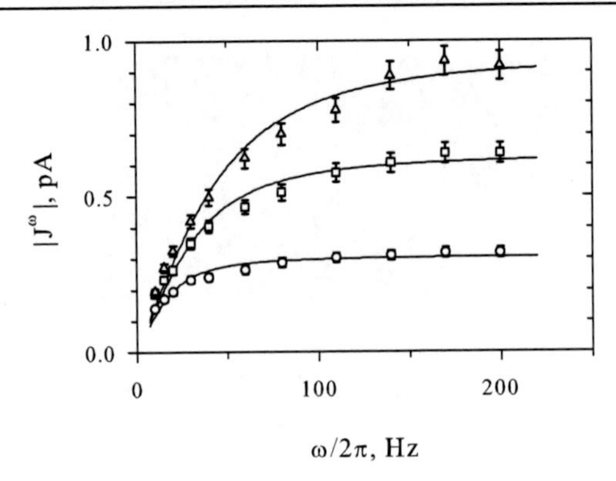

Figure 35. Frequency transfer functions of the nonstationary photocurrent measured for different intensity levels of the incident light: $I_0 = 0.22$ W/cm^2 (○), 0.44 W/cm^2 (□), 0.86 W/cm^2 (△). BN, $\lambda = 532$ nm, $K = 0.90\ \mu\text{m}^{-1}$, $E_0 = 0$. Solid lines show the approximation by Eq. (156).

two orders of magnitude lower than the one for $Bi_{12}SiO_{20}$ crystal which is a model object for non-steady-state photo-EMF technique [20]. We should note here that Eqs. (156), (157) were obtained for case when the crystal is illuminated uniformly. In our experiments the light penetration into the crystal volume is prohibited by the layered structure of the material which operates like a multilayer dielectric mirror (light penetration depth $\sim$1 μm). The influence of the layer geometry on the holographic photocurrent excitation can be noticeable only for low spatial frequencies of the interference pattern and reveals as an additional coefficient $[1 + (\epsilon K d/2)^{-1}]$ after τ_M in the expression for cut-off frequency (d is the thickness of the layer where photocurrent is excited) [100]. Our estimations shows that this coefficient should not exceed 1.3 for $K \geq 0.90\ \mu\text{m}^{-1}$. Another peculiarity of the photocurrent excitation in layered crystals can reveals as follows: the crystal is similar to a multilayered dielectric mirror which reflectance and transmittance are sensitive to the wavelength and incidence angle. So the dependencies of the nonstationary photocurrent and cut-off frequency versus spatial frequency (incidence angle) can differ from the theoretical ones. Although such effects are smoothed away in crystals with varying thickness of layers an additional experimental error can occur.

The dependence of the cut-off frequency versus spatial frequency of the interference pattern is presented in Fig. 37. As follows from Eq. (157) such a dependence has a breakpoint at spatial frequency corresponding to the inverse value of diffusion length. We have approximated the experimental dependence by Eq. (157) and found the diffusion length of holes to be $L_D \simeq 35$ nm. The fast decay of the cut-off frequency at high K is most probably resulted from the light reflectance increasing at large incident angles. The slight decrease of light intensity and corresponding cut-off frequency is typical for low spatial frequencies as well: when the crystal is moved far from the mirrors the light divergence become noticeable. So the estimation of the diffusion length obtained by this method is rather rough and can be considered as the lower bound for this parameter.

A new approach for investigation of materials with low conductivity and for determination of photocarrier transport parameters was developed in Ref. [59]. It is based on

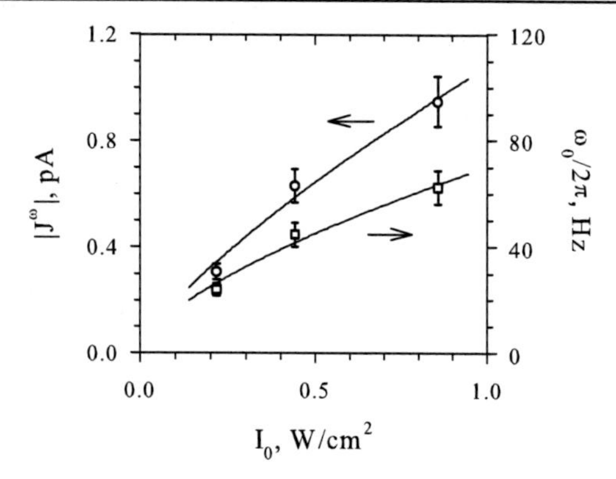

Figure 36. Dependencies of the signal amplitude (o) and cut-off frequency (□) on the light intensity. BN, $\lambda = 532$ nm, $K = 0.90\ \mu\text{m}^{-1}$, $\omega/2\pi = 200$ Hz, $E_0 = 0$. Solid lines show the approximation by power functions: $J^{\omega} \propto I_0^{0.75}$, $\omega_0 \propto I_0^{0.64}$.

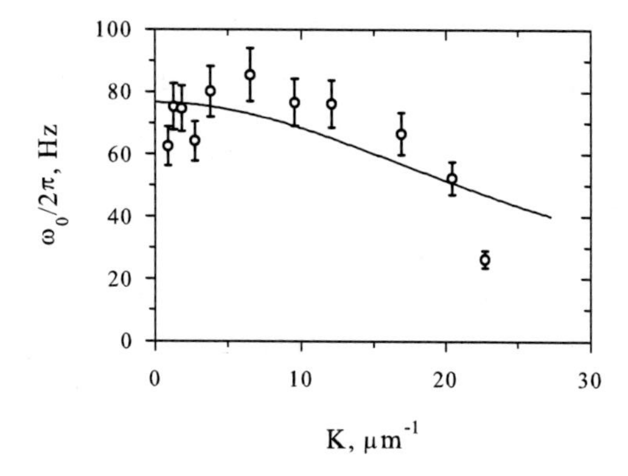

Figure 37. Dependence of the cut-off frequency on the spatial frequency of the interference pattern. BN, $\lambda = 532$ nm, $I_0 = 0.86$ W/cm^2, $E_0 = 0$. Solid line shows the approximation by Eq. (157).

the excitation of the non-steady-state photocurrents in an external alternating electric field. This novel technique allow us to avoid variations of the spatial frequency in wide range accompanied with the time consuming readjustments of optical set-up and hardly controlled experimental errors. The dependence of the holographic photocurrent on the effective value of the external field is presented in Fig. 38. This dependence can be approximated by the following expression obtained for $\omega > \omega_0$ [59]:

$$J^{\omega}(E_0) = -\frac{Sm^2\Delta\sigma_0 E_D}{2(1+K^2L_D^2)} \frac{1+E_0^2/(E_D^2+E_L^2)}{1+E_0^2/(E_D+E_M)^2}, \tag{158}$$

here $E_L = (k_BT/e\mu\tau)^{1/2}$, $E_M = (K\mu\tau)^{-1}$ are the characteristic values of the electric field (μ is the mobility and τ is the lifetime of photocarriers). For the investigated BN crystal they equal $E_L = 5.0$ kV/cm ($\mu\tau = 1.0 \times 10^{-9}$ cm^2/V, $L_D = 50$ nm), $E_M = 5.2$ kV/cm. Characteristic fields E_L, E_M are not independent parameters by definition. In

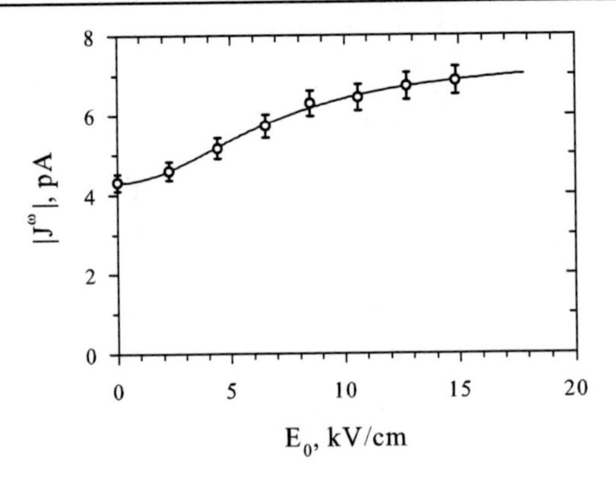

Figure 38. Dependence of the photocurrent amplitude on the effective value of applied ac electric field. BN, $\lambda = 532$ nm, $K = 6.9\ \mu\text{m}^{-1}$, $I_0 = 1.1$ W/cm^2, $\omega/2\pi = 200$ Hz. Solid line shows the approximation by Eq. (158).

practice, however, it is reasonable to consider them separately since saturation effects can influence the position of the second breakpoint on dependence $|J^\omega(E_0)|$ and add error in measurement of E_M.

We have also investigated the influence of a dc electric field on non-steady-state photocurrent amplitude in BN crystal. No considerable effect was observed up to the breakdown values of electric field amplitudes ($\sim$ 15 kV/cm). The similar difference between application of ac and dc external fields is reported above where ac electric field contributed to resonant excitation of photoconductivity gratings while dc one did not. Several reasons of such behavior can be pointed out: first there are voltage drop on the blocking contact and inhomogeneity of electric field distribution in the crystal placed in dc electric field. If ac field has rather high frequency ($\Omega > \tau_M^{-1}$) these effects are imperceptible. This difference can be associated with the complicated processes of conductivity relaxation. The photoconductivity response to the amplitude modulated light is shown in Fig. 39. As seen there are two frequency independent regions $\omega/2\pi = 1.5 - 7$ Hz and $\omega/2\pi = 150 - 3000$ Hz. The frequency of external field $\Omega/2\pi = 560$ Hz corresponds to the second region so the effective photoelectric parameters governing the charge transport may not coincide for the cases of ac and dc fields applied. It should be noted that the origin of the extrinsic hole photoconductivity for $\lambda = 532$ nm is unknown. Perhaps it can be associated with carbon impurities which serve as luminescent centers and create local levels located $1.2 - 1.6$ eV above the valence band [101] or with nitrogen vacancies creating several donor and acceptor levels [102].

Besides the nonresonant excitation of photocurrent presented in this paper the resonant technique exists (see, for example, Ref. [20]). The latter approach provides the possibility of direct determination of photocarrier's mobility. The resonant excitation of the non-steady-state photocurrent is accompanied by a significant growth of the photocurrent amplitude and can be observed in the long drift length regime when the product $KL_0 \gg 1$ (here L_0 is the drift length of photocarriers). The application of an external electric field did not significantly increase the photocurrent amplitude in BN as was in our previous experiments with $Bi_{12}SiO_{20}$ crystals. The influence of the dc field was not observed at all. This fact can

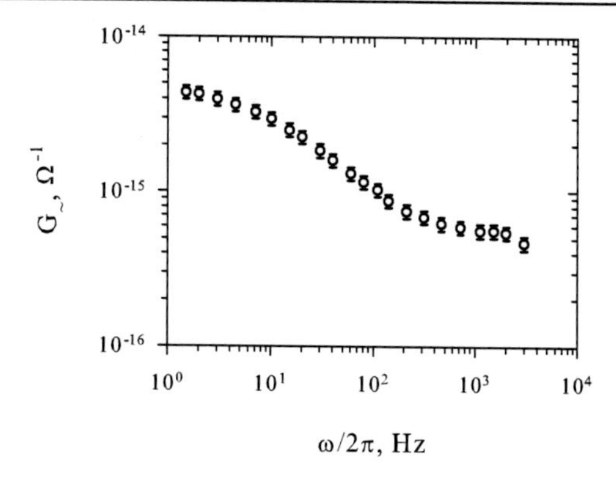

Figure 39. Frequency transfer function of the photoconductivity response. BN, $\lambda = 532$ nm, $I_0 = 0.32$ W/cm^2.

be attributed to the small value of the $\mu\tau$-product of BN. The measured transport parameters ($L_D = 35 - 50$ nm) allow estimation of the characteristic value of an ac or dc field amplitudes providing resonant excitation of the non-steady-state photocurrent ($KL_0 = 10$, $K = L_D^{-1}$): $E_0 = 50 - 70$ kV/cm. This value is considerably larger than the breakdown field of the investigated BN sample (~ 15 kV/cm). The resonant excitation technique can be used for investigation of BN crystals in future when bulk single crystals of boron nitride will be available.

To summarize, we have studied the non-steady-state photocurrent generation in new widegap crystal of boron nitride. The photoconductivity and diffusion length of photoholes were estimated: $\sigma_0 = (0.9 - 2.4) \times 10^{-10}\ \Omega^{-1}\text{cm}^{-1}$ ($I_0 = 0.22 - 0.86$ W/cm^2), $L_D = 35 - 50$ nm [99].

9. Non-steady-State Photo-EMF in Nanostructured GaN within Porous Glass Matrix

The utilization of noncrystalline materials instead of single crystal analogs is the new trend in modern optics and optoelectronics. Such materials are practically feasible and cheap, they can be chemically stable, and their photoelectric properties can be easily varied.

The porous media can be considered as new objects attracting great interest of the specialists in the solid state physics, optics, and semiconductors. Asbestoses, ceramics, and porous glasses are typical examples of this class of materials. The nanopores in these materials with the average diameter ranging from 10 to 100 nm may be doped with almost any gas, liquid or a liquid solution of a solid. The physical properties of fabricated structures can be different. As the result the material may be insulating or conductive, transparent, absorbing or reflecting. The nanoscale of the filling component advances observation of such effects as charge localization, quantum size effects, Peierls phase transition.

There are applications and devices where semiconductors with low conductivity are required (wafers for high-power microwave transistors, X-ray detection, holographic recording including optical data storage and processing, nondestructive testing). Gallium nitride

(GaN) is considered as one of the most perspective material for such areas. Undoped GaN samples are characterized with rather high conductivity $\sigma = 1.2 \times 10^1 - 3.2 \times 10^2$ $\Omega^{-1}\text{cm}^{-1}$ [103, 104]. A certain success has been achieved in fabrication of insulating GaN films using hydride vapor phase epitaxy (HVPE) methods. The conductivity of that samples reaches $\sigma \sim 10^{-12}$ $\Omega^{-1}\text{cm}^{-1}$ [105]. The production of a composite material filled with nanoparticles of GaN can be considered as another way to obtain insulating samples (especially bulk ones).

The application of the standard methods for semiconductor characterization (see, for example, Ref. [43]) is often restricted by the space charge formation near the blocking contacts. This problem is of vital importance for investigation of high-resistive materials where space charge formation time can be very large and introduce inadmissible errors. The novel techniques based on holographic recording are more suitable for semi-insulating samples with low conductivity and mobility of photocarriers [2]. The non-steady-state photoelectromotive force (photo-EMF) technique based on the detection of the alternating electric current arising in the sample illuminated by an oscillating interference pattern is the most versatile and advanced method [20]. Since the current is resulted from the periodic relative shifts of the photoconductivity and space charge gratings the technique based on this effect allows determination of the number of photoelectric parameters (conductivity, carrier sign, lifetime, diffusion length, and drift mobility) and can be applied for the investigation of both non-centrosymmetric and centrosymmetric media.

The investigation of the new class of the semi-insulating materials, namely, porous glasses filled with nanoscaled components using the non-steady-state photo-EMF technique is the main goal of this section.

We used the conventional experimental arrangement for investigations of the non-steady-state photo-EMF (Fig. 8, b). One can find details of the setup in Refs. [106, 107]. A porous silica glass is a random structure of two interpenetrating phases, namely the solid and pore networks [108]. The pores in the glasses are connected to each other and the pore size distribution is narrow. The structure is obtained due to spinodal decomposition of the two phases SiO_2 and B_2O_3+Na_2O. After the glass is heat-treated and annealed, the B_2O_3-rich phase is removed by leaching with acid, leaving an almost pure SiO_2 skeleton. The pore sizes were tested by the mercury intrusion porosimetry and the average pore diameter was found to be 7 ± 1 nm. Gallium nitride has been synthesized directly inside the pores by chemical bath deposition technique similar to the described elsewhere [109]. At first, solid precursor (Ga_2O_3) was embedded into the pores of vycor template. We used alcohol solutions of gallium salts (gallium nitrate or gallium formate) as a soluble precursors to infiltrate Ga_2O_3. The filling procedure was repeated periodically. To produce GaN the vycor sample filled with gallium oxide was annealed in presence of ammonia (1000 Torr) at 750 °C for 50 h. The fill factor of pores with GaN was measured by gravimetric method and has been found to be 25%.

The structural state of the composite was determined by X-ray diffraction (XRD) measurements. The XRD patterns were obtained by using CuK_α radiation (Ni filter). Polycrystalline germanium was used as a standard. Figure 40 (curve 3) shows XRD patterns of the nanocomposite. As evident from these angular dependencies of the intensities scattered the material synthesized in the pores is the hexagonal GaN. Small amounts of crystalline phases such as α-Ga_2O_3 (JCPDS 6-0503), β-Ga_2O_3 (JCPDS 11-370), and α-tridymite (JCPDS 18-

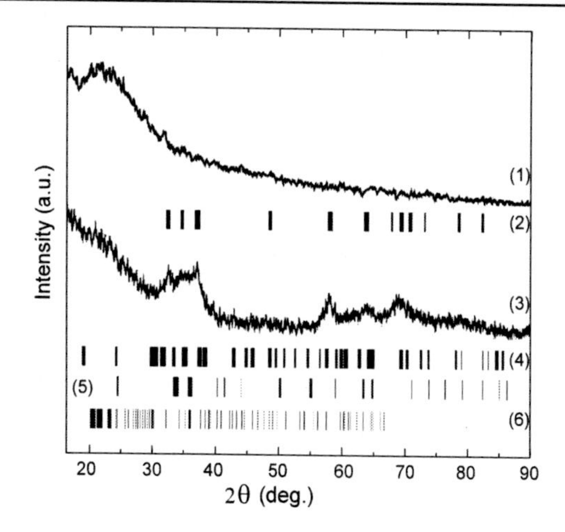

Figure 40. X-ray diffraction patterns. 1 – bare porous glass, 2 – bulk hexagonal GaN (JCPDS 2-1078), 3 – nanocomposite porous glass-GaN, 4 – α-Ga_2O_3 (JCPDS 6-0503), 5 – β-Ga_2O_3(JCPDS 11-370), 6 – α-tridymite (JCPDS 18-1170).

1170) also present in the composite. The analysis of Fig. 40 has also provided the typical size of GaN particles: 15 ± 5 nm. Note that this value is about 2 times larger than average pore dimension. Similar feature was observed earlier in the samples of Pb within porous glass matrix [110] and it was attributed to the elongated form of the nanoparticles.

9.1. $\lambda = 442$ nm

The presence of the effect of the non-steady-state photo-EMF in new class of materials – porous glasses – is the first result that should be pointed out [106, 107]. In spite of the small signal amplitudes we can obtain reliable results by the use of high sensitive lock-in nanovoltmeter with integration time up to 100 s. We have determined the type of the photoconductivity at $\lambda = 442$ nm. The sign of the detected signal corresponds to the electron one.

The frequency transfer function of the non-steady-state photo-EMF signal amplitude is shown in Fig. 41. The signal demonstrates typical behavior, namely, there is a linear growth of the amplitude at low frequencies of the phase modulation ω and frequency independent region at high frequencies. This regions are separated by the so-called cut-off frequency $\omega_0 = 48 - 73$ Hz. Such a behavior is an important manifestation of the adaptive nature of space charge formation in photoconductive materials, and it is described by the following expression obtained earlier for the simplest model of the semiconductor crystal with one type of partially compensated donor centers [20]:

$$J^\omega = \frac{Sm^2 \Delta\sigma_0 E_D}{2(1 + K^2 L_D^2)} \frac{-i\omega/\omega_0}{1 + i\omega/\omega_0}, \tag{159}$$

where

$$\omega_0 = [\tau_M (1 + K^2 L_D^2)]^{-1}, \tag{160}$$

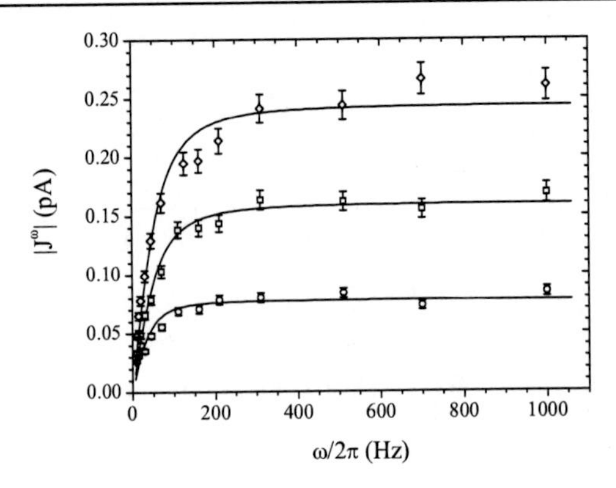

Figure 41. Frequency transfer functions of the non-steady-state photo-EMF measured in GaN nanostructure for different light intensities: $I_0 = 0.045$ W/cm^2 (○), 0.12 W/cm^2 (□), 0.19 W/cm^2 (◇). $\lambda = 442$ nm, $K = 7.4 \times 10^4$ cm^{-1}, $m = 0.87$, $\Delta = 0.82$. The solid lines show the approximation by Eq. (159).

S is the electrode area, σ_0 is the average photoconductivity of the crystal, $E_D = Kk_BT/e$ is the diffusion field, L_D is the diffusion length of photocarriers, $\tau_M = \epsilon\epsilon_0/\sigma_0$ is the Maxwell relaxation time, T is the crystal's temperature, ϵ is the dielectric constant, ϵ_0 is the free space permittivity, k_B is the Boltzmann constant, and e is the elementary charge.

If the light intensity increases we observe the corresponding growth of the signal amplitude (at $\omega > \omega_0$) and cut-off frequency (Fig. 42, left figure). The both dependencies demonstrate approximately linear behavior in the investigated intensity range: $J^\omega, \omega_0 \propto I_0$, what is due to increase of the sample's photoconductivity σ_0 [see Eqs. (159), (160)]. The method of the non-steady-state photo-EMF provides possibility for determination of the conductivity in high resistive materials: on measuring the cut-off frequency we can easily calculate the Maxwell relaxation time and conductivity of the sample [Eq. (160)]. The investigated glass has $\epsilon \simeq 4$, so the conductivity equals $\sigma = (1.1 - 1.6) \times 10^{-10}\ \Omega^{-1}\text{cm}^{-1}$ for $I_0 = 0.045 - 0.19$ W/cm^2. One can note that cut-off frequency tends to nonzero limit at $I_0 = 0$. It is most probably associated with the finite dark conductivity of the material, which is estimated as $\sigma_d = 9 \times 10^{-11}\ \Omega^{-1}\text{cm}^{-1}$.

The dependence of the signal amplitude and cut-off frequency versus spatial frequency of the interference pattern is presented in Fig. 42, right figure. The dependence of the signal amplitude $|J^\omega(K)|$ seems to be linear in the range $K = 0 - 14\ \mu\text{m}^{-1}$ what is due to increase of the diffusion field E_D. The sufficient deviation of the last point from the claimed linear dependence is probably caused by the increase of the light reflection and shadowing of the near contact area but not by the blurring of the oscillating photoconductivity distribution observed at $K > L_D^{-1}$. The scattering of the cut-off frequency values is noticeable as well, and no decrease of this parameter at high spatial frequencies is guessed [see Eq. (160)]. So we can state that diffusion length of electrons is too small to be measured using this technique: $L_D < 40$ nm.

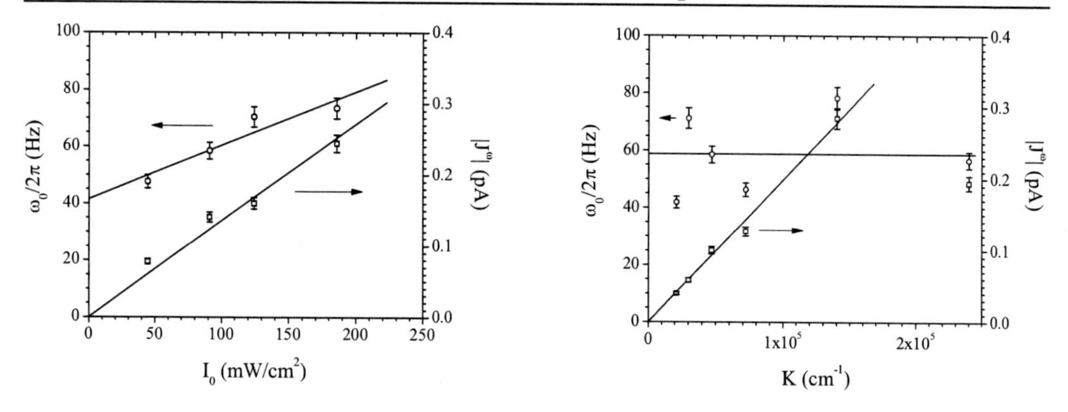

Figure 42. Dependencies of the non-steady-state photo-EMF amplitude (□) and cut-off frequency (○) on the intensity of incident light (left figure). $\lambda = 442$ nm, $K = 7.4 \times 10^4$ cm^{-1}. Both dependencies are approximated by linear function: $J^\omega, \omega_0 \propto I_0$. Dependencies of the non-steady-state photo-EMF amplitude (□) and cut-off frequency (○) on the spatial frequency of the interference pattern (right figure). $\lambda = 442$ nm, $I_0 = 0.19$ W/cm^2. The solid lines represent approximation by Eqs. (159) and (160) for the case of small diffusion lengths $KL_D \ll 1$.

9.2. $\lambda = 532$ nm

In this subsection we consider the excitation of the non-steady-state photo-EMF signal at $\lambda = 532$ nm and consider its temperature dependence. The sign of the signal corresponds to the electron type of the conductivity.

The frequency transfer functions of the non-steady-state photo-EMF measured at different temperatures are shown in Fig. 43, left figure. The signal demonstrates the behavior similar to the one at $\lambda = 442$ nm. The conductivity values derived from the cut-off frequencies are the following: $\sigma = (3.5 - 4.6) \times 10^{-10}\ \Omega^{-1}\text{cm}^{-1}$ ($T = 249 - 388$ K). The signal amplitude and cut-off frequency are slightly higher than the ones at $\lambda = 442$ nm but it is resulted from the much higher intensities of light so we can state that excitation of the signal at $\lambda = 532$ nm is less effective. The following reasons of this difference can be pointed out: the material have larger absorption coefficient at shorter wavelength ($\alpha = 440$ cm^{-1} at $\lambda = 442$ nm versus $\alpha = 280$ cm^{-1} at $\lambda = 532$ nm) and, perhaps, larger quantum efficiency of the conductivity.

The temperature dependence of the cut-off frequency is rather linear in the investigated range of temperatures $T = 249 - 388$ K (Fig. 43, right figure):

$$\omega_0 = A(T + T'), \tag{161}$$

where A and T' are some coefficients describing the slope and shift of the line. The theoretical dependence is unknown. We can only suppose the conductivity of the material is hopping. The temperature dependence of the conductivity in glasses is described by the conventional exponential law: $\sigma \propto \exp(-E/k_B T)$ [111]. The investigated range of temperatures is not wide enough to reveal the nonlinear growth of the cut-off frequency. So the used linear approximation can be considered as a truncated series of the unknown dependence.

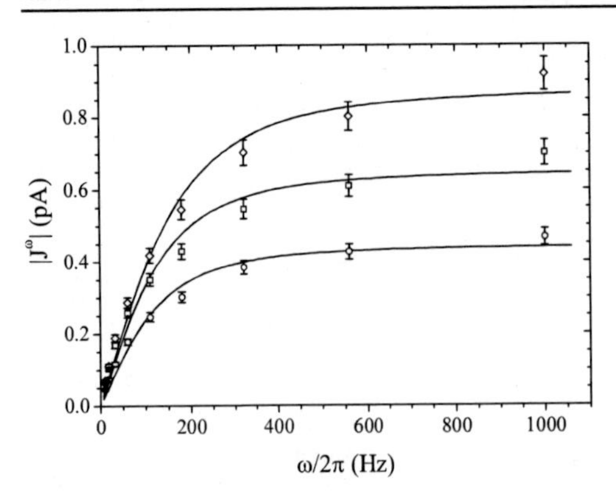

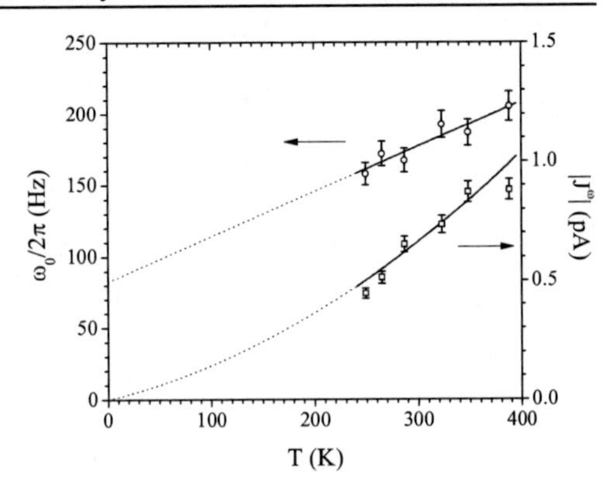

Figure 43. Frequency transfer functions of the non-steady-state photo-EMF measured in GaN nanostructure for different temperatures: $T = 249$ K (○), 287 K (□), 388 K (◇) (left figure). The best fit of the experimental data by Eq. (159) is shown by the solid lines. Dependencies of the non-steady-state photo-EMF amplitude (□) and cut-off frequency (○) on the sample temperature (right figure). The experimental dependencies $J^{\omega}(T)$ and $\omega_0(T)$ are approximated by Eqs. (162) and (161). $\lambda = 532$ nm, $I_0 = 2.3$ W/cm^2, $K = 7.1 \times 10^4$ cm^{-1}, $m = 0.91, \Delta = 0.61$.

Let us consider the temperature dependence of the signal amplitude measured at $\omega > \omega_0$ (Fig. 43, right figure). As follows from Eq. (159) there is an explicit temperature dependence of the signal arising from the temperature dependence of the diffusion field: $E_D \propto T$. There can be also an implicit function of the temperature resulted from the temperature dependence of material parameters, e.g. photoconductivity σ_0. The linear temperature dependence of σ_0 was assumed above, so the expected dependence of the signal amplitude becomes quadratic:

$$J^{\omega} = B(T + T')T, \tag{162}$$

where B is a coefficient. One can notice the saturation of the photo-EMF signal at $T > 350$ K. The nature of this saturation is not clear. Perhaps, it is due to temperature dependence of the hole conductivity: in the case of bipolar conductivity the detected signal must decrease since the stationary space charge gratings formed by electrons and holes are shifted by a half of spatial period and compensate each other [85].

Utilization of the modified technique for characterization of the as-grown structure in an external ac electric field was not successful [59]. We tried to measure a dependence of the signal amplitude on the electric field strength. No dependencies were detected in the fields up to $E_0 = 14$ kV/cm ($\Omega/2\pi \simeq 30$ kHz). This fact is explained by the small value of diffusion length which can be estimated as $L_D < k_B T/eE_0 = 18$ nm ($\mu\tau < 1.3 \times 10^{-10}$ cm^2/V). The higher amplitudes of the external field are beyond the breakdown limit which is about 14 kV/cm for our sample. If the mentioned dependence of the signal on applied field was observed, we could detect the well known dependence of the conductivity (cut-off frequency) on the frequency of applied field typical for hopping transport: $\omega_0 \propto \sigma \propto \Omega^s$. Such experiments can be considered as the task for further investigation of the effect in other nanostructured materials where transport parameters are more appropriate.

To summarize, we have investigated the effect of the non-steady-state photo-EMF in porous glass filled with nanoscaled GaN semiconductor. The detected signal revealed the behavior similar to the one typical for widegap semiconductor crystals. The conductivity of the material was estimated from the frequency transfer functions of the effect and temperature dependence of the detected signal and conductivity was observed and discussed.

10. Non-steady-State Photo-EMF in Nanostructured Selenium within Chrysotile Asbestos

Nanowires, and particularly semiconductor nanowires, are expected to be key elements of future generation of devices used in electronics, optics, biology and medicine. Nowadays electronic circuits are close to the limit when further shrinking of elements leads to a large spread of their characteristics due to size fluctuations while the improvements of the resolution in technological equipment becomes an economical challenge [112, 113]. The utilization of nanowires as active components and interconnects seems to be perspective way to continue progress in the integration of electronic circuits. The concept of nanowire crossbar, in which active elements are placed at each crosspoint of two nanowire arrays, has been proposed [113]. It advances new types of resistive memory, reconfigurable logic circuits and new hardware realization of neuromorphic networks. The capability to assemble a wide range of different direct band gap nanowires enables the facile creation of multicolor light emitting diodes on a single substrate. Coherent laser emission has been detected from a number of nanowire cavities [114, 115]. The effect of quantum confinement inherent to nanowires should result in lower lasing thresholds, reduced temperature dependence, and a narrower gain spectrum of future devices [115]. In addition to nanoscale light sources, crossed nanowire junctions can also be configured as photodetectors critical for integrated photonics [114].

To fully explore the potential of nanowire building blocks for integrated devices and circuits, efficient and scalable strategies are needed to assemble large amount of nanowires into increasingly complex architectures. Several methods have been developed to date. They include the electric field assembly technique [116], which implies that nanowires tend to align with the direction of an applied field, fluidic flow-directed assembly technique based on passing a suspension of nanowires through a microfluidic channel formed between a mould and a flat substrate [112], and Langmuir-Blodgett assembly technique where the uniaxial compression of nanowire-surfactant monolayer on an aqueous subphase produces aligned nanowires with controlled spacing [112]. Filling dielectric matrices with desired semiconductors is one of the possible ways to get aligned nanoscaled structure [117]. The utilization of natural asbestoses allows to produce nanowires with diameter of 2-10 nm and length of 1 mm. In this paper we study composite of selenium within chrysotile asbestos for the first time.

Selenium exhibits a number of interesting and useful properties which contribute this material in investigation and production of semiconductor nanowires. Synthesis of selenium nanowires can be performed without any physical template what is due to anisotropic characteristic of the target material: the extended spiral chains of selenium atoms in the trigonal phase provide a natural template to define and guide the growth along c-axis. Several methods for selenium nanowires synthesis were proposed, they include refluxing processes [118], hydrothermal routes [119, 120], cellulose-directed growth [121], chemical vapor deposition [122], sonochemical approach [123]. Methods using physical templates have also been reported [124]: selenium nanowires can be produced by injecting chalcogene into matrix from melt under pressure or by absorption from the vapor phase.

Generally, t-Se is a p-type extrinsic semiconductor, with an indirect band gap of 1.6 eV. It has a relatively low melting point ($\sim$ 217 °C), a high photoconductivity ($\sim 0.8 \times 10^5$ Ω^{-1}cm^1), catalytic activity toward hydration and oxidation reactions, and high piezoelectric, thermoelectric, and nonlinear optical responses. Selenium also demonstrates a high reactivity towards a wealth of chemicals that can be potentially exploited to convert selenium into other functional materials such as Bi_2Se_3, ZnSe, and CdSe.

The composition of the chrysotile asbestos is $Mg_3Si_2O_5(OH)_4$. It is a regular array of closely packed, parallel, ultrathin dielectric nanotubes. The mineral structure is described as a layer of partially hydrated MgO, which is bounded to a corresponding SiO_2. Since the lattice constant in the MgO monolayer differs from that of the SiO_2 layer, this double layer rolls up in a very thin tubes, with the outermost magnesium oxide layer. The fibers are therefore hollow cylindrical tubes with the outer diameter of 20-40 nm and the inner diameter of 2-10 nm [125, 126] that is convenient for nanowire preparation.

The investigation of the optical properties and electronic transport in nanowires and their structures is usually confined by the measurements of the light absorption [118, 120] and current-voltage characteristics [120, 127]. This approach seems to be not completely informative in the case of dissipative (non-ballistic) transport at room temperatures when scattering and even trapping of carriers occurs. The application of the standard methods for semiconductor characterization (see, for example, Ref. [43]) is often restricted by the space charge formation near the blocking contacts. This problem is of vital importance for investigation of high-resistive materials where space charge formation time can be very large and introduce inadmissible errors. The novel techniques based on holographic recording are more suitable for semi-insulating samples with low conductivity and mobility of photocarriers [2]. The non-steady-state photoelectromotive force (photo-EMF) technique based on the detection of the alternating electric current arising in the sample illuminated by an oscillating interference pattern is the most versatile and advanced method [20]. Since the current is resulted from the periodic relative shifts of the photoconductivity and space charge gratings the technique based on this effect allows determination of the number of photoelectric parameters (conductivity, carrier sign, lifetime, diffusion length, and drift mobility) and can be applied for the investigation of both non-centrosymmetric and centrosymmetric media. So the expansion of the non-steady-state photo-EMF technique to the characterization of semiconductor nanowires within dielectric matrices is the main goal of this section.

10.1. Experimental Setup and Results

We used the rather conventional experimental arrangement for the investigation of the non-steady-state photo-EMF (Fig. 8, b). The light from the conventional He-Ne ($\lambda = 633$ nm, $P_{out} \simeq 1$ mW) laser was split into two beams forming the interference pattern with spatial frequency K and contrast m on the sample's surface. In the most experiments the light polarization was perpendicular to the incidence plane. The signal beam was phase modulated with frequency ω and amplitude $\delta = 0.49$ by the electrooptic modulator ML-

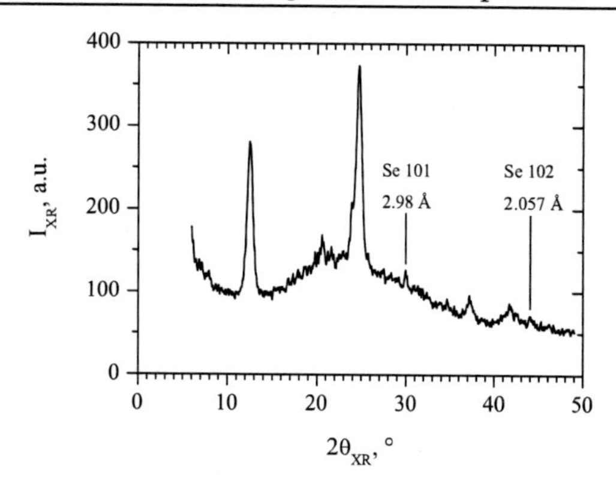

Figure 44. X-ray diffraction pattern of selenium within chrysotile asbestos.

102A. Additional low frequency modulation ($\Omega/2\pi = 3$ Hz, $\Delta = 2.4$) was created by the mirror attached to the phone membrane. We studied two sample orientations: the grating vector **K** and the nanowires were parallel in the former case ("longitudinal" geometry) and they were perpendicular in the latter one ("transversal" geometry). The photocurrent arising in the sample generated the corresponding voltage on the load resistor ($R_L = 1$ MΩ for "longitudinal" geometry and $R_L = 10$ MΩ for the "transversal" one). Then this voltage was amplified and measured with the lock-in nanovoltmeter Unipan-232B or spectrum analyzers SK4-56, SK4-58.

Selenium was embedded into chrysotile asbestos matrix from the melt under pressure. Three samples were studied in this paper. Samples S1 and S2 were prepared for the measurements in the "longitudinal" geometry and sample S3 for the "transversal" one. The characteristic dimensions of the samples were $10 \times 1 \times 0.5$ mm^3. The silver paste electrodes (1×1 mm^2) were painted on the front surface (10×1 mm^2) in the "longitudinal" geometry. The electrodes (2×0.5 mm^2) were painted on the lateral surfaces in the "transversal" geometry.

The most of experiments were carried out at room temperature ($T = 16$ °C). In certain experiments the temperature of studied sample was changed using the Peltier thermal converter and controlled using the analog temperature-to-voltage converter.

Figure 44 shows XRD pattern of the studied nanocomposite. Marked weak peaks at this diagram indicate the presence of the hexagonal phase of selenium.

The presence of the non-steady-state photo-EMF effect in the new class of light-sensitive materials – quasi-1D semiconductor nanowire arrays – is the first results that should be pointed out. The amplitude of the signal detected in the experiments is of order of $0.1 - 10$ pA which is noticeable lower than that in the model objects – sillenite crystals $Bi_{12}Si(Ti,Ge)O_{20}$, where it reaches 1 nA [20]. Nevertheless this amplitude is quite enough for consistent detection of the signal with signal-to-noise ratio of $1 - 10$.

The phase of the detected signal indicates that holes prevail in the process of photo-EMF excitation. We should mention here that a nontrivial behavior of the signal amplitude was observed. In the standard experimental procedure, when there is no additional low frequency modulation, the signal amplitude is very unstable: its deviation reaches almost 100% of the mean value. These deviations are very slow: their characteristic "period" is of

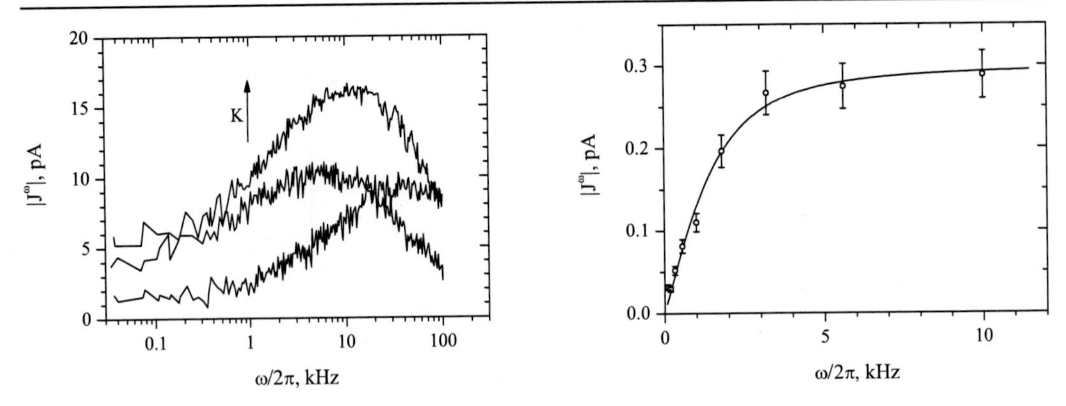

Figure 45. Frequency transfer functions of the non-steady-state photo-EMF signal in Se nanowires within chrysotile asbestos. The dependencies are measured for two geometries: $\mathbf{K} \parallel$ nanowires (S1), $K = 0.087\ \mu\mathrm{m}^{-1}$, $0.26\ \mu\mathrm{m}^{-1}$, $2.6\ \mu\mathrm{m}^{-1}$ (left figure), and $\mathbf{K} \perp$ nanowires (S3), $K = 0.58\ \mu\mathrm{m}^{-1}$ (right figure). $\lambda = 633$ nm, $I_0 = 370$ W/m^2, light polarization is perpendicular to the incidence plane. Solid line shows the approximation by Eq. (163).

about 10 minutes. When additional phase modulation with frequency $\Omega/2\pi = 3$ Hz and amplitude $\Delta > 1$ is switched on the signal increases by ~ 2 times, and its amplitude keeps stable within $\pm 10\%$ for hours. Such unstable behavior is most probably originated from the slow drift of the light phase in the interferometric setup. There can be two mechanisms transferring this slow drift into deviations of the detected signal. The former implies the presence of the contact photo-EMF which amplitude is comparable with that of the volume photo-EMF [129]. The latter is associated with the formation of competitive space charge gratings with different relaxation times. Since the introduction of additional modulation increases the signal amplitude regularly and for all spatial frequencies we are inclined to consider the mechanism with competitive gratings to be dominating (the contact component can both increase and decrease total signal). Anyway we settled the frequency and amplitude of additional modulation so that the influence of both factors is eliminated ($\Omega/2\pi = 3$ Hz, $\Delta = 2.4$).

We have measured the frequency transfer functions of the non-steady-state photo-EMF signal (Fig. 45) in the "longitudinal" and "transversal" geometries of the experiments, i.e. for the cases when direction of the space charge field (grating vector) is parallel and perpendicular to the nanowires. The signal demonstrates typical behavior, namely, there is a linear growth of the amplitude at low frequencies of the phase modulation $\omega < \omega_1$, frequency independent region at higher frequencies $\omega_1 < \omega < \omega_2$, and decay at frequencies $\omega > \omega_2$. This regions are separated by the so-called cut-off frequencies $\omega_1/2\pi = 0.42 - 12$ kHz and $\omega_2/2\pi = 32 - 210$ kHz. These dependencies are well described by the following expression which is equivalent to that obtained earlier for the simplest model of the semiconductor crystal with one type of partially compensated donor centers [37]:

$$J^{\omega} = J^{\omega}_{m} \frac{i\omega(\omega_2 - \omega_1)}{\omega_1\omega_2 - \omega^2 + i\omega(\omega_2 - \omega_1)}, \tag{163}$$

where J^{ω}_{m} is the current amplitude at the maximum of frequency dependence. The linear

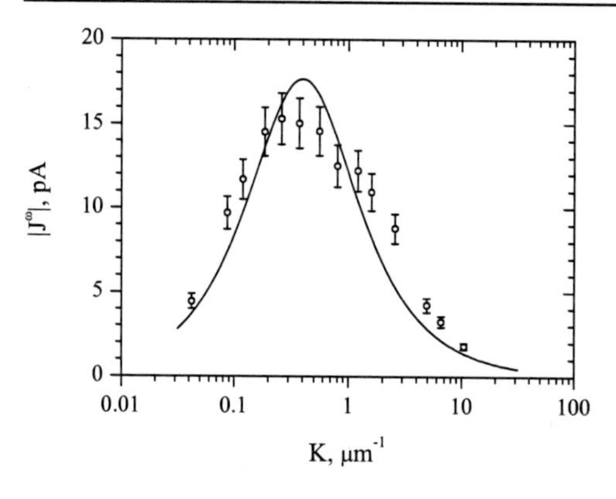

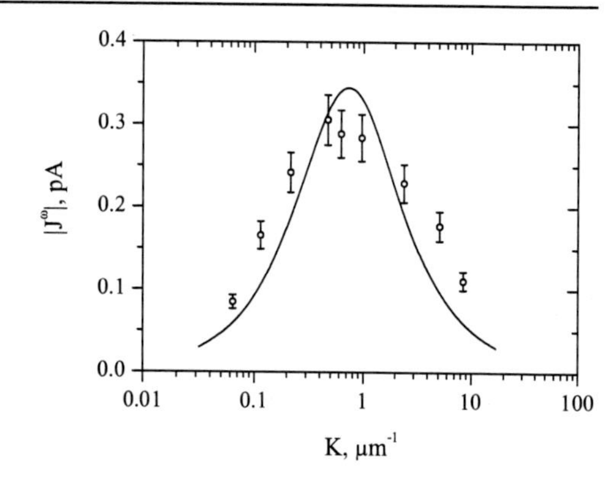

Figure 46. Dependence of the maximal non-steady-state photo-EMF amplitude on the spatial frequency of the interference pattern. The dependencies are measured for two geometries: **K** $\|$ nanowires (left figure), and **K** $\perp$ nanowires (right figure). $\lambda = 633$ nm, $I_0 = 370$ W/m^2, light polarization is perpendicular to the incidence plane. Solid line shows the approximation by Eq. (164).

growth at low frequencies is an important manifestation of the adaptive nature of space charge formation in photoconductive materials. The signal in this frequency region is small since both the space charge field grating and grating of free carriers (photoconductivity grating) follow the movement of the interference pattern. The spatial shift between the gratings maintain to be nearly equal to $\pi/2$ what results in small amplitude of current. At higher frequencies (frequency independent region) the grating with larger relaxation time becomes "frozen in", the periodic spatial shifts increase, and the photo-EMF signal reaches maximum. At very high frequencies the amplitude of oscillations for both gratings and corresponding relative shifts between them become negligible, and resulting current vanishes. Sample S2 behaves similarly to sample S1 in this and further experiments except for the temperature dependence of the effect.

The frequency transfer function measured in the "transversal" geometry looks analogously except the decay region which was not observed because of the signal smallness and corresponding frequency limits. The first cut-off frequency in this dependence equals $\omega_1 = 2.1$ kHz and it is almost the same value as that for "longitudinal" geometry (2.4 kHz).

The dependence of the signal amplitude versus spatial frequency of the interference pattern is another important characteristic that is usually measured in the non-steady-state photo-EMF experiments (Fig. 46). The behavior of the signal can be easily explained: the signal increases at low K due to the growth of space charge field amplitude which is proportional to the so-called diffusion field $E_d = Kk_BT/e$, the signal decrease at high K is resulted from either diffusion blurring of the conductivity grating or Debye screening of the stationary space charge grating. As the first cut-off frequency increases with spatial frequency (it will be shown below), the latter reason for the signal suppression is dominating. The dependencies are well fitted by the following simple expression known from the non-steady-state photo-EMF theory [20]:

$$J^{\omega}(K) \propto \frac{K}{1 + Q + K^2 L_S^2}. \tag{164}$$

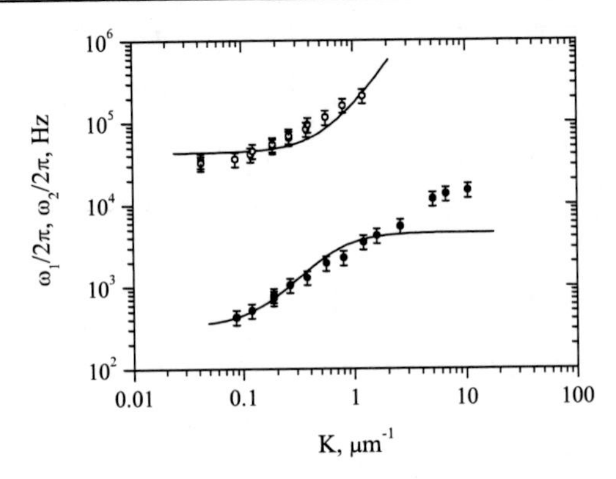

Figure 47. Dependencies of the first (•) and second (∘) cut-off frequencies on the spatial frequency of the interference pattern. $\lambda = 633$ nm, $I_0 = 370$ W/m^2, $\mathbf{K} \parallel$ nanowires, light polarization is perpendicular to the incidence plane. Solid lines show the approximation by Eqs. (165) and (166).

Here Q is the dimensionless parameter characterizing the recombination process ($0 < Q < 1$ in general case, $Q = 0$ for linear recombination, $Q = 1$ for quadratic recombination) [39, 41]. The photoconductance of the studied material demonstrated nearly square root dependence on the light intensity, i.e. quadratic recombination is present and $Q \simeq 1$. The screening length was estimated from the measured dependence: $L_S = 3.6$ μm for the "longitudinal" geometry and $L_S = 1.9$ μm for the "transversal" one.

We have also measured the dependencies of the cut-off frequencies on the spatial frequency of the interference pattern (Fig. 47). These dependencies noticeably differ from those measured earlier in our model objects – $Bi_{12}Si(Ti)O_{20}$ crystals. Dependence $\omega_1(K)$ is expected to be decreasing function in particular. The expressions for the cut-off frequencies obtained earlier for the simplest model of semiconductor with one type of compensated donor centers are as follows [37]:

$$\omega_1 = \frac{1 + Q + K^2 L_S^2}{\tau + \tau_M(1 + Q + K^2 L_d^2)}, \tag{165}$$

$$\omega_2 = \omega_1 + \frac{1}{\tau_M} + \frac{1 + Q + K^2 L_d^2}{\tau}. \tag{166}$$

The only explanation in the frames of existing theory can be done for the observed growing function $\omega_1(K)$, if we do not neglect the screening effects: $L_S^2 > L_d^2 \tau_M / (\tau + \tau_M)$. The following parameters were determined from the experimental dependencies: $\tau = 0.96$ ms, $\tau_M = 3.7$ μs, $L_d = 27$ μm, $L_S = 8.9$ μm. As seen from Fig. 47 the correspondence of the experimental and theoretical dependencies $\omega_1(K)$ is just qualitative: they noticeably diverge at high spatial frequencies.

The anisotropy of the studied material reveals not only in the difference of the current amplitude along and across nanowires (Fig. 45) but also in the experiments with varied light polarization (Fig. 48). The higher signal amplitudes are observed in the case when the light is polarized perpendicular to nanowires. The influence of mobility anisotropy is

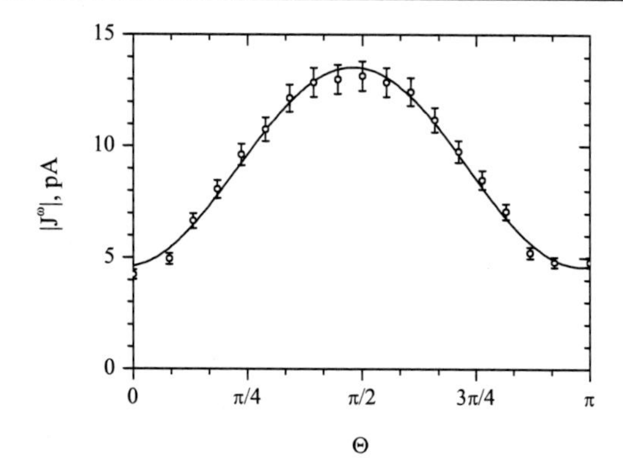

Figure 48. Dependence of the non-steady-state photo-EMF amplitude on the angle of light polarization. Angle $\Theta = 0$ corresponds to the case when light is polarized in the incidence plane (polarization is nearly parallel to the nanowires). $\lambda = 633$ nm, $I_0 = 370$ W/m^2, $K = 0.20\ \mu$m^{-1}, $\omega/2\pi = 35$ kHz, $\mathbf{K} \parallel$ nanowires.

eliminated in this part of experiments since the current always flow along the nanowires, so the difference in signal amplitude should be attributed to the photoexcitation processes. Such photoexcitation anisotropy is probably associated with the orientation of selenium crystallites with respect to the matrix: dipole momentum of Se is perpendicular to asbestos nanotubes.

We have mentioned above that the the signal nontrivially behave in the presence of additional low-frequency modulation. The dependencies of the non-steady-state photo-EMF signal versus amplitude of additional modulation is presented in Fig. 49. The monotonous growth of the signal amplitude in this dependence is rather unexpected. The basic non-steady-state photo-EMF theory predicts the signal decay as $(1 + \Omega\Delta/\omega_1)^{-0.5}$ [128]. It was pronounced in the beginning of the section that the grating competition is responsible for such behavior. At the first glance two mechanisms for grating competition can be suggested: influence of shallow traps and bipolar conductivity. The analysis of the effect for the model with shallow traps is presented in the next section. In this case the charge gratings with the opposite signs are formed at deep donor and shallow levels. The solid lines in Fig. 49 represent approximation by Eq. (187) with fitting parameter $\beta T = 0.91$ for $K = 0.099\ \mu$m^{-1}, $\omega/2\pi = 9.0$ kHz and $\beta T = 0.49$ for $K = 12\ \mu$m^{-1}, $\omega/2\pi = 59$ kHz. The consideration of the effect in the presence of bipolar conductivity was not successful: the desired theoretical dependence is not obtained (at least for the simplest model).

The relaxation of the signal amplitude between the regimes with and without additional phase modulation occurs during considerable time amount (Fig. 49). We have approximated this dependence by the simplest exponential decay function and found the time constant $\tau_{am} = 35$ s as a fitting parameter.

The temperature dependence of the non-steady-state photo-EMF was also observed (Fig. 50). The two samples prepared for measurements in the "longitudinal" geometry demonstrate different behavior in the range of room temperatures. The non-steady-state photo-EMF signal in sample S1 linearly grows with the temperature. The linear growth also follows from the theoretical approach which implies that the signal amplitude is pro-

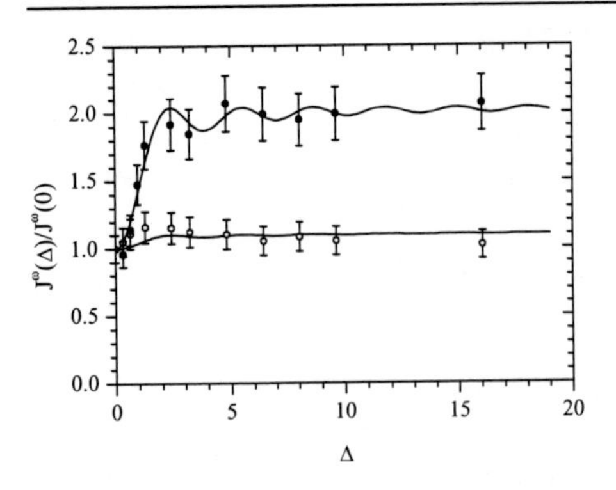

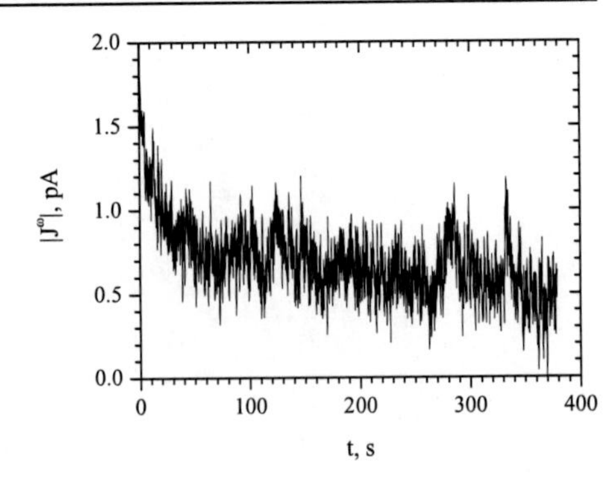

Figure 49. Dependencies of the non-steady-state photo-EMF amplitude on the amplitude of additional low-frequency phase modulation ($\Omega/2\pi = 3$ Hz) (left figure). (○) – $K = 0.099$ μm^{-1}, $\omega/2\pi = 9.0$ kHz; (•) – $K = 12$ μm^{-1}, $\omega/2\pi = 59$ kHz. Time evolution of the signal amplitude after the moment when additional phase modulation ($\Omega/2\pi = 3$ Hz, $\Delta = 16$) is switched off (right figure). $K = 12$ μm^{-1}, $\omega/2\pi = 59$ kHz, integration time $\tau_{int} = 0.3$ s. $\lambda = 633$ nm, $I_0 = 370$ W/m^2, $\mathbf{K} \parallel$ nanowires, light polarization is perpendicular to the incidence plane.

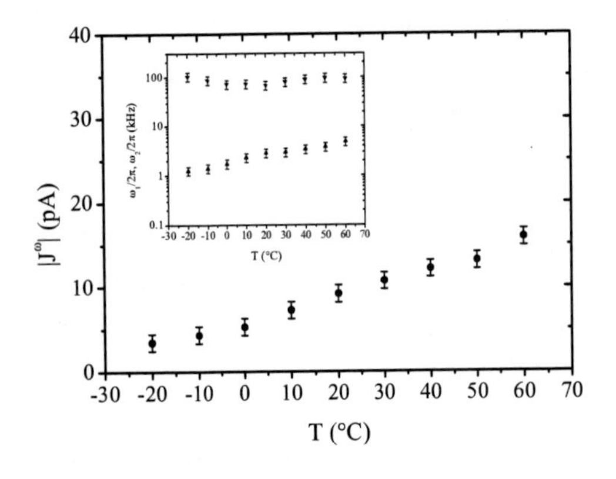

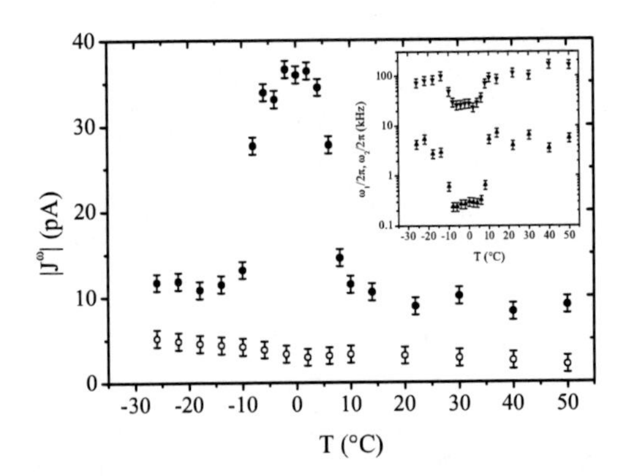

Figure 50. Temperature dependencies of the non-steady-state photo-EMF amplitude measured in sample S1 (left figure) at $K = 0.36$ μm^{-1} and in sample S2 (right figure) at $K = 0.11$ μm^{-1} (•) and $K = 7.2$ μm^{-1} (○). The insets present temperature dependencies of the first (▲) and the second (▼) cut-off frequencies. $\lambda = 633$ nm, $I_0 = 370$ W/m^2, $\mathbf{K} \parallel$ nanowires, light polarization is perpendicular to the incidence plane.

portional to the diffusion field $E_d = Kk_BT/e$. The slope of the experimental dependence is larger however: being extrapolated to the lower temperatures it crosses the axis of abscissae at $T = -38$ °C (235 K) but not at $T = 0$ K as follows from the simplest theory. This difference can be explained by the temperature of the material parameters (e.g. conductivity). The temperature dependence of the cut-off frequencies proves this assumption. So the observed linear dependence in studied region should be considered as a linear approximation of more complicated temperature dependence. An unexpected temperature dependence of the effect was revealed in sample S2. There are an pronounced peak in the dependence of the signal amplitude and wells in the dependencies of the both cut-off frequencies measured at low K. These peculiarities appear at $T \simeq 0$ °C. In this case we can hardly attribute such behavior to the change of material parameters: such assumption requires simultaneous growth of the photoconductivity σ_0, Maxwell relaxation time τ_M, and conductivity relaxation time τ, but $\tau_M \propto \sigma_0^{-1}$. The more realistic explanation implies that dc electric field appears when temperature approaches to 0 °C. It is known from the basic theory of the non-steady-state photo-EMF effect that the amplitude of photocurrent grows in the presence of external dc field and its maximum at the frequency transfer function shifts to the lower frequencies [37]. We performed measurements of the dc electric current arising in the samples under uniform illumination and without it. The temperature dependencies of this dc current demonstrate peak in the vicinity of 0 ° in both samples S1 and S2. The peak in sample S1 is of 10 pA only, and in sample S2 it reaches 200 pA. We should note here that this dc current is almost independent on the illumination level, so it is rather pyroelectric than photogalvanic. The origin of this pyroelectricity is not clear, but evidently it is associated with the crystallization of atmosphere water accumulated in the sample pores unfilled with selenium.

10.2. Theoretical Analysis and Discussion

The studied material belongs to the class of quasi one-dimensional non-crystalline media. The most of obtained experimental results, however, are well described by the theory developed for bulk crystals. The dependence of signal amplitude versus amplitude of additional phase modulation is the only result that seriously contradicts with the theory developed earlier for the simplest one-level model of semiconductor crystal [128, 129]. It was mentioned that the unexpected behavior can be associated with the competition of space charge gratings characterized by different relaxation times. Grating competition is one of the feature of the two-level model of semiconductor [40]. There is no theory of the non-steady-state photo-EMF in the presence of additional modulation for this model. So we shall try to explain this certain feature of the signal behavior using the approach based on band transport two-level model of semiconductor [40, 43]. For the simplicity the material is assumed to be isotropic as well.

The initial stage of the analysis is similar to that described above in Section 2.1. Let us suppose the sample is illuminated by the interference pattern formed by two plain light waves one of which is phase modulated:

$$I(x,t) = I_0\{1 + m\cos[Kx + \varphi(t)]\}, \tag{167}$$

$$\varphi(t) = \delta\cos(\omega t) + \Delta\cos(\Omega t). \tag{168}$$

Here I_0, m, and K are the average light intensity, contrast, and spatial frequency of the interference pattern, Δ and ω are the amplitude and frequency of phase modulation. The non-steady-state photo-EMF signal is defined by the drift component of the electric current averaged over interelectrode spacing [Eq. (15)]. Necessary distributions $n(x,t)$ and $E(x,t)$ are derived from Eqs. (1)-(5).

The calculation procedure simplifies substantially if we suppose the contrast m to be small: $m \ll 1$. Then we can look for the solution of equation set to the lowest order with respect to small parameter m presenting $n(x,t)$, $E(x,t)$ as the sums of sinusoidal gratings:

$$n(x,t) = n_0 + n_+(t)\exp(iKx) + n_-(t)\exp(-iKx), \tag{169}$$

$$E(x,t) = E_+(t)\exp(iKx) + E_-(t)\exp(-iKx). \quad (170)$$

Furthermore, we suppose the amplitude of primary phase modulation to be small $\delta \ll 1$, and thus we can present complex amplitudes n_+, n_-, E_+, E_- like the following series:

$$n_+(t) = \sum_{p=-1}^{1} \sum_{q=-\infty}^{\infty} n_{+pq} \exp[i(p\omega + q\Omega)t]. \quad (171)$$

Then the complex amplitude of the nonstationary photocurrent with frequency ω can be rewritten as the following combination of coefficients:

$$j^\omega = 2e\mu \sum_{q=-\infty}^{\infty} \left(n_{+1q}E_{-0-q} + n_{-0-q}E_{+1q} + n_{-1q}E_{+0-q} + n_{+0-q}E_{-1q}\right). \quad (172)$$

After substitution of Eqs. (167)-(172) into Eqs. (1)-(5), linearization and some routine calculations we obtain rather common expression for the amplitude of the non-steady-state photo-EMF signal:

$$j^\omega = \frac{m^2\delta}{2}\sigma_0 E_d \sum_{q=-\infty}^{\infty} \mathrm{J}_q^2(\Delta)\frac{A(\omega, q\Omega)}{B(-q\Omega)B(\omega+q\Omega)}, \quad (173)$$

where

$$\begin{aligned} A(\omega, q\Omega) &= -\tau_M\tau_T^2[iq\Omega(\omega^2 + 3\omega q\Omega + 2q^2\Omega^2) \\ &\quad + i(\omega + 2q\Omega)(\beta + \tau_{TI}^{-1})(i\omega + \beta + \tau_{TI}^{-1})], \quad (174) \\ B(\omega) &= -i\omega^3\tau_M\tau_D\tau_T - \omega^2[\tau_M\tau_D\tau_T(S_D I_0 + \tau_{DI}^{-1} + \beta + \tau_{TI}^{-1}) \\ &\quad + \tau_M(\tau_D + \tau_T) + \tau_D\tau_T(1 + K^2L_M^2)] + i\omega[\tau_D + \tau_T \\ &\quad + \tau_D(S_D I_0 + \tau_{DI}^{-1})(\tau_T + \tau_M(1 + K^2L_T^2)) + \tau_T(\beta + \tau_{TI}^{-1}) \\ &\quad \times (\tau_D + \tau_M(1 + K^2L_D^2)) + \tau_M\tau_D\tau_T(S_D I_0 + \tau_{DI}^{-1})(\beta + \tau_{TI}^{-1})] \\ &\quad + \tau_D(S_D I_0 + \tau_{DI}^{-1}) + \tau_T(\beta + \tau_{TI}^{-1}) \\ &\quad + \tau_D\tau_T(S_D I_0 + \tau_{DI}^{-1})(\beta + \tau_{TI}^{-1})(1 + K^2L_M^2), \quad (175) \end{aligned}$$

$E_d = Kk_BT_s/e$ is the diffusion field, k_B is the Boltzmann constant, T_s is the sample's temperature, $L_D = \sqrt{D\tau_D}$, $L_T = \sqrt{D\tau_T}$ and $L_M = \sqrt{D\tau_M}$ correspond to the average diffusion shifts of electrons for times τ_D, τ_T and τ_M, respectively, $\mathrm{J}_q(\Delta)$ is the Bessel function of the first kind of the q-th order.

Equations (173)-(175) are evidently too complicated for the further analysis, so let us simplify them using the following assumptions. We suppose the light intensity to be not very high so that the deep donors keep almost filled:

$$N_D \gg N_A + N_0. \quad (176)$$

The shallow traps are assumed to be almost empty on the contrary:

$$M_0 \ll M_T. \quad (177)$$

The low illumination level does not allow to produce high density of electrons in the conduction band:

$$n_0 \ll N_0, M_0. \tag{178}$$

Other set of conditions concern the characteristic timescales and frequencies of phase modulation. First let us suppose that the frequency of primary modulation is of order of τ_M^{-1}, while the frequency of additional modulation is much lower: it is chosen with respect to the rate of slow relaxation processes at shallow traps. So the following conditions are valid:

$$\omega \gg \Omega\Delta, \tag{179}$$

$$\omega \gg \beta, \tag{180}$$

$$\tau_M \ll \beta^{-1}. \tag{181}$$

Another condition establish the relation between the Maxwell relaxation time and lifetime of ionized donors and it is often present in semiinsulating materials:

$$\tau_M \ll \tau_{DI}. \tag{182}$$

At last, the spatial frequency of interference pattern is assumed to be not very high:

$$K^2 L_M^2 \ll M_0/n_0. \tag{183}$$

If conditions (176)-(183) are fulfilled, the expression for the complex amplitude of the non-steady-state photo-EMF signal reduces to the form very similar to that known from basic theory of the effect:

$$j^\omega = \frac{-i\omega\tau_M(m^2\delta/2)\sigma_0 E_d}{1-\omega^2\tau\tau_M + i\omega[\tau+\tau_M(1+K^2L_d^2)]}\frac{N(\Delta,\Omega)}{(1+\tau_D/\tau_T)(1+\tau_D/\beta\tau_T\tau_{DI})} \tag{184}$$

where $\tau = \left(\tau_D^{-1}+\tau_T^{-1}\right)^{-1}$, $L_d = \sqrt{D\tau}$ are the lifetime and diffusion length of electrons, respectively, and

$$N(\Delta,\Omega) = \mathrm{J}_0^2(\Delta) + 2\sum_{q=1}^{\Delta_m} \mathrm{J}_q^2(\Delta)\frac{1+(q\Omega)^2\beta^{-1}T}{1+(q\Omega T)^2} \tag{185}$$

is the normalized non-steady-state photo-EMF amplitude: $N = j^\omega(\Delta)/j^\omega(0)$ – the main goal of this analysis, with characteristic time

$$T = \frac{\tau_D+\tau_T}{\tau_D\tau_{DI}^{-1}+\beta\tau_T}. \tag{186}$$

The upper limit of sum Δ_m is chosen to be equal to the nearest integer greater than amplitude Δ.

Function $N(\Delta,\Omega)$ is real and it can be both growing and decaying depending on relation between material parameters β, T and frequency Ω. It should be noted that characteristic time T is rather large: the denominator in Eq. (186) can be rewritten as $n_0/(N_A+N_0)+n_0/M_0$, and according to Eq. (178) it makes time T much larger than sum $\tau_D+\tau_T$. So frequency of additional modulation Ω remaining low in sense of Eq. (179) can

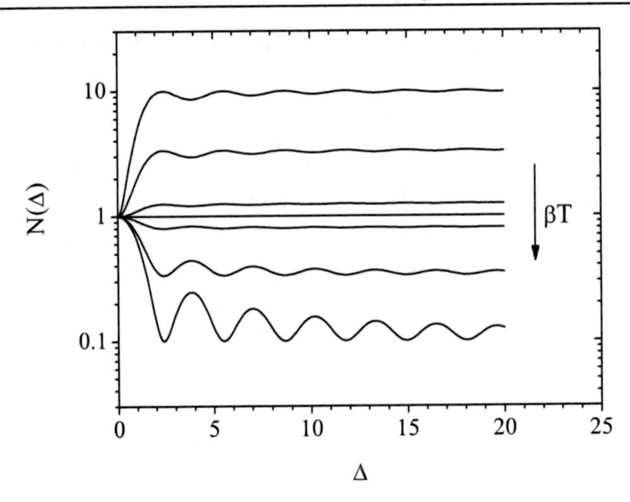

Figure 51. Normalized amplitude of the non-steady-state photo-EMF N versus amplitude of additional low-frequency phase modulation Δ. The dependencies are calculated using Eq. (187) for $\beta T = 0.1, 0.3, 0.8, 1.0, 1.25, 3.0, 10$.

be chosen to satisfy conditions $(\Omega T)^2 \gg 1$ and $\Omega^2\beta^{-1}T \gg 1$. In this case the expression for normalized amplitude N can be further simplified:

$$N(\Delta) = \mathrm{J}_0^2(\Delta) + \frac{2}{\beta T}\sum_{q=1}^{\Delta_m} \mathrm{J}_q^2(\Delta). \tag{187}$$

As seen, this function has the only argument – the amplitude of additional phase modulation Δ, and it contains the only parameter – product βT. Dependence $N(\Delta)$ calculated for different values of parameter βT is presented in Fig. 51. The dependence acquires the growing character at $\beta T < 1$ and decaying one at $\beta T > 1$. The growth (decay) at low amplitudes is nearly quadratic: $N \propto \Delta^2$. The saturation level is reached at $\Delta \simeq 2$ and equals $(\beta T)^{-1}$.

To summarize we studied the effect of non-steady-state photo-EMF in new class of composite materials – nanostructured selenium within chrysotile asbestos. The average photoconductivity of the material was estimated from the measurements of the signal amplitude versus frequency of phase modulation. The dependence of the signal on spatial frequency provided the estimation of the diffusion and Debye screening lengths. The polarization dependencies revealed the anisotropy of photoexcitation while the temperature dependence of the detected signal pointed out to the presence of water in the porous matrix. The increase of the signal amplitude was observed on application of additional low-frequency phase modulation. To explain this peculiarity the theoretical analysis has been performed for the model of semiconductor with shallow traps.

The obtained results have demonstrated that the non-steady-state photo-EMF technique can be considered as a powerful tool in investigation of nanostructured media. A number of photoelectric parameters are estimated using the only arrangement which can be easily automatized in the case of need. The technological problem of ohmic contacts to widegap semiconductors, especially nanosized ones, is not so significant here as that for standard methods, since the effect of the non-steady-state photo-EMF is based on the excitation of an alternating current. The developed theoretical approach is not restricted to considered material and can be applied to any semiconductor material (bulk or nanostructured) where electronic transport is accompanied by the charge trapping. The behavior of the non-steady-state photo-EMF in semiconductor nanowires at low temperatures, where ballistic transport and Coulomb blockade effects can be present, as well as the application of such materials in adaptive detectors of phase-modulated optical signals seem to be interesting future tasks.

11. Conclusion

In this Chapter we have discussed the space-and-time current spectroscopy for characterization of wide-gap semiconductors (photorefractive sillenites, BN, SnS_2). Since the non-steady-state photocurrent is originated from the interaction of both the photoconductivity and space charge gratings a lot of photoelectric parameters of wide-gap semiconductors were measured.

Acknowledgements

The financial support from the Russian Foundation for Basic Research and from the Ministry of Knowledge Economy of Korea is gratefully acknowledged.

References

[1] Gunter P.; Huignard J.-P. *Photorefractive effects and materials*. Topics in applied physics: Photorefractive materials and their applications Volume 61 (Eds. P. Gunter and J.-P. Huignard) Berlin, Heidelberg: Springer-Verlag, p. 7-73, 1988.

[2] Petrov M.P.; Stepanov S.I.; Khomenko A.V. *Photorefractive crystals in coherent optical systems;* Springer-Verlag: Berlin, 1991.

[3] Agullo-Lopez F.; Cabrera J.M.; Agullo-Rueda F. *Electrooptics: phenomena, materials and applications;* Academic Press: London, San Diego, New York, Boston, Sydney, Tokyo, Toronto, 1994.

[4] Nolte D.D. *Photorefractive transport and multi-wave mixing;* The Kluwer International Series in Engineering and Computer Science Photorefractive effects and materials (Edited by D. Nolte) Kluwer Academic Publishers: Boston, Dordrecht, London, 1995; pp 1-66.

[5] Solymar L.; Webb D.J.; Grunnet-Jepsen A. *The physics and applications of photorefractive materials;* Oxford: Clarendon Press, 1996.

[6] Deferrari H.A.; Darby R.A.; Andrews F.A. *J. Acost. Soc. Am.* 1967, 42, 982-990.

[7] Forward R.L. *Phys. Rev. D.* 1978, 17, 379-390.

[8] Gorecki C. *J. Optics (Paris)* 1995, 26, 29-34.

[9] Pernick B.J. *Appl. Opt.* 1973, 12, 607-610.

[10] Nokes M.A.; Hill B.C.; Barelli A.E. *Rev. Sci. Instrum.* 1978, 49, 722-728.

[11] Eberhardt F.J.; Andrews F.A. *J. Acoust. Soc. Am.* 1970, 48, 603-609.

[12] Dragsten P.R.; Webb W.W. *J. Acoust. Soc. Am.* 1976, 60, 665-671.

[13] Martinussen H.; Aksnes A.; Engan H. *Optics Express* 2007, 18, 11370-11384.

[14] Paul M.; Betz B.; Arnold W. *Appl. Phys. Lett.* 1987, 50, 1569-1571.

[15] Stepanov S.I.; Sokolov I.A. In *IEEE Proc. of Second International Conference on Holographic Systems, Components and Applications;* Institution of Electrical Engineers: London, U.K., 1989; Vol. 311, pp 95-100.

[16] Stepanov S.I., In *International Trends in Optics;* Goodman J.; Ed.; Academic Press: Boston, 1991; pp 124-140.

[17] Pouet B.; Krishnaswamy S. *Appl. Opt.* 1996, 35, 787-793.

[18] Stepanov S. In *Handbook of Advanced Electronic and Photonic Materials;* Nalwa H.S.; Ed.; Academic Press, 2000; Vol. 2, pp 205-272.

[19] Trofimov G.S.; Stepanov S.I. *Fiz. Tverd. Tela (Leningrad)* 1986, 28 2785-2789 [*Sov. Phys. Solid State* 1986, 28 1559-1562].

[20] Petrov M.P.; Sokolov I.A.; Stepanov S.I.; Trofimov G.S. *J. Appl. Phys.* 1990, 68, 2216-2225.

[21] Stepanov S.I.; Sokolov I.A.; Trofimov G.S.; Vlad V.I.; Popa D.; Apostol I. *Opt. Lett.* 1990, 15, 1239-1241.

[22] Sochava S.L.; Buse K.; Krätzig E. *Opt. Commun.* 1993, 105, 315-319.

[23] Sochava S.L.; Buse K.; Krätzig E. *Opt. Commun.* 1993, 98, 265-268.

[24] Gerwens A.; Buse K.; Krätzig E.; et al. *J. Opt. Soc. Am. B* 1998, 15, 2143-2146.

[25] Korneev N.; Mayorga D.; Stepanov S.; Gerwens A.; Buse K.; Kratzig E. *Opt. Commun.* 1998, 146, 215-219.

[26] Bittner R.; Meerholz K.; Stepanov S. *Appl. Phys. Lett.* 1999, 74, 3723-3725.

[27] Korneev N.; Mayorga D.; Stepanov S.; Veenhuis H.; Buse K.; Kuper C.; Hesse H.; Krätzig E. *Opt. Commun.* 1999, 160, 98-102.

[28] Seres I.; Stepanov S.; Mansurova S. *J. Opt. Soc. Am. B* 2000, 17, 1986 -1991.

[29] Trofimov G.S.; Kosarev A.I.; Kovrov A.G.; LeComber P.G. *J. Non-Cryst. Solids* 1991, 137,138, 483-486.

[30] Kosarev A.I.; Trofimov G.S. *Int. J. Electronics* 1994, 56, 1023-1028.

[31] Haken U.; Hundhausen M.; Ley L. *Phys. Rev. B* 1995, 51, 10579-10590.

[32] Pauliat G.; Roosen G. *J. Opt. Soc. Am. B* 1990, 7, 2259-2267.

[33] Strohkendl F.P. *J. Appl. Phys.* 1989, 65, 3773-3780.

[34] Hou S.L.; Lauer R.B.; Aldrich R.E. *J. Appl. Phys.* 1973, 44, 2652-2658.

[35] Kostyuk B.Kh.; Kudzin A.Yu.; Sokolyanskii G.Kh. *Sov. Phys. Solid State* 1980, 22, 1429-1432.

[36] Partanen J.P.; Jonathan J.M.C.; Hellwarth R.W. *Appl. Phys. Lett.* 1990, 57, 2404-2406.

[37] Sokolov I.A.; Stepanov S.I. *J. Opt. Soc. Am. B* 1993, 10, 1483-1488.

[38] Biaggio I.; Hellwarth R.W.; Partanen J.P. *Phys. Rev. Lett.* 1997, 78, 891-894.

[39] Bryushinin M.A.; Sokolov I.A. *Phys. Rev. B* 2000, 62, 7186-7194.

[40] Tayebati P.; Mahgerefteh D. *J. Opt. Soc. Am. B* 1991, 8, 1053-1064.

[41] Sokolov I.A.; Stepanov S.I. *Optik* 1993, 93, 175-182.

[42] G.C. Valley, *IEEE J. Quantum Electron.* 1983, QE-19, 1637-1645.

[43] Ryvkin S.M. *Photoelectric effects in semiconductors;* Consultants Bureau: New York, 1964.

[44] Refregier Ph.; Solymar L.; Rajbenbach H.; Huignard J.P. *J. Appl. Phys.* 1985, 58, 45-57.

[45] Bryushinin M.A.; Petrov A.A.; Sokolov I.A. *Appl. Phys. Lett.* 1999, 75, 445-447.

[46] Mokrushina E.V.; Bryushinin M.A.; Kulikov V.V.; Petrov A.A.; Sokolov I.A. *J. Opt. Soc. Am. B* 1999, 16, 57-62.

[47] Kazarinov R.F.; Suris R.A.; Fuks B.I. *Sov. Phys. Semicond.* 1973, 7, 102-107.

[48] Hutson A.R.; McFee J.H.; White D.L. *Phys. Rev. Lett.* 1961, 7, 237-239.

[49] Stepanov S.I.; Kulikov V.V.; Petrov M.P. *Opt. Commun.* 1982, 44, 19-23.

[50] Lyuksyutov S.F.; Buchhave P.; Vasnetsov M.V. *Phys. Rev. Lett.* 1997, 79, 67-70.

[51] Segev M.; Collings B.; Abraham D. *Phys. Rev. Lett.* 1996, 76, 3798-3801.

[52] Hundhausen M. *J. Non-Cryst. Solids* 1996, 198-200, 146-152.

[53] Kumar J.; Albanese G.; Steier W.H. *J. Opt. Soc. Am. B* 1987, 4, 1079-1082.

[54] Stepanov S.I.; Petrov M.P. *Opt. Commun.* 1985, 53, 292.

[55] Johansen P.M.; Pedersen H.C. *J. Opt. Soc. Am. B* 1998, 15, 1366-1374.

[56] Walsh K.; Powell A.K.; Stace C.; Hall T.J. *J. Opt. Soc. Am. B* 1990, 7, 288-303.

[57] Sokolov I.A.; Stepanov S.I. *Appl. Opt.* 1993, 31, 1958-1964.

[58] Sokolov I.A.; Stepanov S.I. *Electron. Lett.* 1990, 26, 1275-1277.

[59] Bryushinin M.; Kulikov V.; Sokolov I. *Phys. Rev. B* 2002, 65, 245204.

[60] Bryushinin M.A.; Sokolov I.A. *Phys. Rev. B* 2001, 63, 153203.

[61] van der Ziel A. *Noise. Sources, characterization, measurement;* Prentice-Hall. Inc.: Englewood Cliffs, N. J., 1970.

[62] Huignard J.P.; Marrakchi A. *Opt. Commun.* 1981, 38, 249-254.

[63] Hamel de Montchenault G.; Loiseaux B.; Huignard J.P. *Electron. Lett.* 1986, 22, 1030-1032.

[64] Heaton J.M.; Solymar L. *IEEE J. Quantum Electron.* 1988, 24, 558-567.

[65] Petrov M.P.; Petrov V.M.; Bryksin V.V.; Gerwens A.; Wevering S.; Krätzig E. *J. Opt. Soc. Am. B* 1998, 15, 1880-1888.

[66] Mallick S.; Imbert B.; Ducollet H.; Herriau J.P.; Huignard J.P. *J. Appl. Phys.* 1988, 63, 5660-5663.

[67] Pedersen H.C.; Johansen P.M. *Opt. Lett.* 1995, 20, 689-691.

[68] Petrov M.P.; Bryksin V.V.; Vogt H.; Rahe F.; Krätzig E. *Phys. Rev. B* 2002, 66, 085107.

[69] Bryushinin M. *Appl. Phys. B* 2004, 79, 851-856.

[70] Bryushinin M.A. *Zhurn. Tekh. Fiz. (St. Petersburg)* 2004, 74, 62-66 [*Tech. Phys.* 2004, 49, 1016-1020].

[71] Bryushinin M.; Kulikov V.; Sokolov I. *Phys. Rev. B* 2005, 71, 165208.

[72] Kogelnik H. *Bell. Syst. Tech. J.* 1969, 48, 2909-2947.

[73] Mansurova S.; Stepanov S.; Korneev N.; Dibon C. *Opt. Commun.* 1998, 152, 207-214.

[74] Sturman B.I.; Mann M.; Otten J.; Ringhofer K.H. *J. Opt. Soc. Am. B* 1993, 10, 1919-1932.

[75] Sochava S.L.; Buse K.; Krätzig E. *Phys. Rev. B* 1995, 51, 4684-4686.

[76] Greenaway D.L.; Nitsche R. *J. Phys. Chem. Solids* 1965, 26, 1445-1458.

[77] Domingo G.; Itoga R.S.; Kannewurf C.R. *Phys. Rev.* 1966, 143, 536-541.

[78] Au-Yang M.Y.; Cohen M.L. *Phys. Rev.* 1969, 178, 1279-1283.

[79] Patil S.G.; Tredgold R.H. *J. Phys. D* 1971, 4, 718-722.

[80] Nakata R.; Yamaguchi M.; Zembutsu S.; Sumita M. *J. Phys. Soc. Jap.* 1972, 32, 1153-1153.

[81] Shibata T.; Makanushi Y.; Miura T.; Kishi T. *J. Phys. Chem. Sol.* 1990, 51, 1297-1300.

[82] Dubrovskii G.B. *Fiz. Tverd. Tela (St. Petersburg)* 1998, 40, 1712-1718 [*Phys. Solid State* 1998, 40, 1557-1562].

[83] Bryushinin M.A.; Dubrovsky G.B.; Sokolov I.A. *Appl. Phys. B* 1999, 68, 871-875.

[84] Bryushinin M.A.; Dubrovskii G.B.; Petrov A.A.; Sokolov I.A.; Vainos N.A.; Kalpouzos C. *Fiz. Tverd. Tela (St. Petersburg)* 2002, 44, 1203-1205 [*Phys. Solid State* 2002, 44, 1254-1256].

[85] Stepanov S.I.; Trofimov G.S. *Fiz. Tverd. Tela (Leningrad)* 1989, 31, 89-92 [*Sov. Phys. Solid State* 1989, 31, 49-50].

[86] Korneev N.A.; Stepanov S.I. *J. Appl. Phys.* 1993, 74, 2736-2741.

[87] Korneev N.; Mansurova S.; Rodriguez P.; Stepanov S. *J. Opt. Soc. Am B* 1997, 14, 396-399.

[88] Landolt-Börnstein *Numerical data and functional relationships in science and technology;* Madelung O.; Ed.; Springer-Verlag: Berlin, 1983; Vol. 17, Subvol. f.

[89] Dubrovskii G.B.; Shelykh A.I. *Fiz. Tverd. Tela (St. Petersburg)* 1998, 40, 1426-1427 [*Phys. Solid State* 1998, 40, 1295-1296].

[90] Acharya S.; Srivastava O.N. *Phys. Stat. Sol. (a)* 1981, 65, 717-723.

[91] Chopra N.G.; Luyken R.J.; Cherrey K.; Crespi V.H.; Cohen M.L.; Louie S.G.; Zettl A. *Science* 1995, 269, 966-967.

[92] Sharupin B.N. Structure and properties of boron pyronitride. In *Chemical vapor deposition of refractory inorganic materials;* Shpak V.S.; Avarbe R.G.; Ed.; Leningrad, 1976.

[93] Geick R.; Perry C.H.; Rupprecht G. *Phys. Rev.* 1966, 146, 543-547.

[94] Zunger A.; Katzir A.; Halperin A. *Phys. Rev. B* 1976, 13, 5560-5573.

[95] Xu Y.-N.; Ching W.Y. *Phys. Rev. B* 1991, 44, 7787-7798.

[96] El-Yadouni A.; Soltani A.; Boudrioua A.; Thevenin P.; Bath A.; Loulergue J.C. *Opt. Mater.* 2001, 17, 319-322.

[97] Cappellini G.; Satta G.; Palummo M.; Onida G. *Phys. Rev. B* 2001, 64, 035104-1-11.

[98] Plaksin O.A.; Stepanov V.A.; Stepanov P.A.; Demenkov P.V.; Chernov V.M.; Krutskikh A.O., *Nucl. Inst. Met. Phys. Res. B* 2002, 193, 265-270.

[99] Sokolov I.A.; Bryushinin M.A.; Ordin S.V.; Kulikov V.V.; Petrov A.A. *J. Phys. D* 2006, 39, 1063-1068.

[100] Korneev N.A.; Stepanov S.I. *IEEE J. Quantum Electron.* 1994, 30, 2721-2725.

[101] Katzir A.; Suss J.T.; Zunger A.; Halperin A. *Phys. Rev. B* 1975, 11, 2370-2377.

[102] Grinyaev S.N.; Konusov F.V.; Lopatin V.V. *Phys. Solid State* 2002, 44, 286-293.

[103] Ilegems M. *J. Cryst. Growth* 1972, 13/14, 360-364.

[104] Crouch R.K.; Debnam W.J.; Fripp A.L. *J. Mater. Sci.* 1978, 13, 2358-2364.

[105] Kuznetsov N.I.; Nikolaev A.E.; Zubrilov A.S.; Melnik Yu.V.; Dmitriev V.A. *Appl. Phys. Lett.* 1999, 75, 3138-3140.

[106] Bryushinin M.; Golubev V.; Kumzerov Yu.; Kurdyukov D.; Sokolov I. *Physica B* 2009, 404 12511254.

[107] Bryushinin M.; Golubev V.; Kumzerov Yu.; Kurdyukov D.; Sokolov I. *Appl. Phys. B* 2009, 95 489495.

[108] Levitz P.; Ehret G.; Sinha S.K.; Drake J.M. *J. Chem. Phys.* 1991, 95, 6151-6161.

[109] Davydov V.Yu.; Dunin-Borkovski R.E.; Golubev V.G.; Hutchison J.L.; Kartenko N.F.; Kurdyukov D.A.; Pevtsov A.B.; Sharenkova N.V.; Sloan J.; Sorokin L.M. *Semicond. Sci. Technol.* 2001, 16, L5.

[110] Golosovsky I.V.; Delaplane R.G.; Naberezhnov A.A.; Kumzerov Y.A. *Phys. Rev. B* 2004, 69, 132301.

[111] Mott N.F.; Davis E.A. *Electron processes in non-crystalline materials;* Clarendon Press: Oxford, 1979.

[112] Lu W.; Lieber C. *J. Phys. D: Appl. Phys.* 2006, 39, R387-R406.

[113] Likharev K.K. *J. Nanoelectron. Optoelectron.* 2008, 3, 203-230.

[114] Li Y.; Qian F.; Xiang J.; Lieber C.M. *Mater. Today* 2006, 9, 18-27.

[115] Pauzauskie P.J.; Yang P. *Mater. Today* 2006, 9, 36-45.

[116] Duan X.; Huang Y.; Cui Y.; Wang J.; Lieber C.M. *Nature* 2001, 409 66-69.

[117] Kumzerov Y.; Vakhrushev S. *in Encyclopedia of nanoscience and nanotechnology;* Nalva H.S.; Ed.; American Scientific Publishers: Los Angeles 2004; Vol. 7, pp 811-849.

[118] Gates B.; Mayers B.; Cattle B.; Xia Y. *Adv. Funct. Mater.* 2002, 12, 219-227.

[119] Xiong S.; Xi B.; Wang W.; Wang C.; Fei L.; Zhou H.; Qian Y. *Cryst. Growth Des.* 2006, 6, 1711-1716.

[120] Cheng L.; Shao M.; Chen D.; Wei X.; Wang F.; Hua J. *J. Mater. Sci.: Mater. Electron.* 2008, 19, 1209-1213.

[121] Lu Q.; Gao F.; Komarneni S. *Chem. Mater.* 2006, 18, 159-163.

[122] Cao X.; Xie Y.; Zhang S.; Li F.; *Adv. Mater.* 2004, 16, 649-653.

[123] Gates B.; Mayers B.; Grossman A.; Xia Y. *Adv. Mater.* 2002, 14 1749-1752.

[124] Bogomolov V.N.; Kholodkevich S.V.; Romanov S.G.; Agroskin L.S. *Solid State Commun.* 1983, 47, 181-182.

[125] Bragg L.; Claringbull G.F. *Crystal structure of minerals;* G. Bell and Sons: London, 1965.

[126] Sobolev N.D. *Introduction to asbestos science (in Russian);* Nedra: Moscow, 1971.

[127] Qin D.; Tao H.; Zhao Y.; Lan L.; Chan K.; Cao Y. *Nanotechnology* 2008, 19, 355201.

[128] Sokolov I.A.; Stepanov S.I.; Trofimov G.S. *Zhurn. Tekhn. Fiz. (Leningrad)* 1989, 59, 126-129 [*Sov. Tech. Phys.* 1989, 34, 1165-1167].

[129] Bryushinin M.A.; Kulikov V.V.; Sokolov I.A. *Zhurn. Tekhn. Fiz. (St. Petersburg)* 2002, 72, 79-86 [*Tech. Phys.* 2002, 47, 1283-1290].

In: Advances in Condensed Matter ... Volume 10
Editors: H. Geelvinck and S. Reynst
ISBN: 978-1-61209-533-2

Chapter 6

HYDRODYNAMICAL AND OPTICAL PHENOMENA IN CONFINED NEMATIC AND SMECTIC LIQUID CRYSTALS

Juan Adrian Reyes[1,*] ***and Laura O. Palomares***[2]
[1]Departamento de Física
Universidad Autónoma Metropolitana,
Ixtapalapa, Apartado Postal 55 534 09340,
México D. F., México
[2]Instituto de Física,
Universidad Nacional Autónoma de México,
Apartado Postal 20 364 01000,
México D. F., México

Abstract

Liquid crystals are signature systems of soft condensed matter media. Their unique features make them very susceptible to the action of external agents. We focus here in the study of their hydrodynamical and optical properties so we first review the formalism to describe these properties. Nematic liquid crystals are fluids that exhibit long-range orientational order over distances many times larger than the dimensions of the molecules of which they are composed. Their intrinsic anisotropy is responsible for considering liquid crystals as interesting fluids for electrorheological and optical applications. They offer obvious advantages over the more conventional fluids such as avoiding the problems associated with the settling of the dispersed phase; these complications are nonexistent for liquid crystals since they are homogeneous phases. We study various viscometric properties of a nematic submitted to distinct external imposed fields and confined under different geometries including the planar and cylindrical ones. We show the existence of an electrorheological effect for which the effective viscosity increased various magnitude orders its value upon the application of an electric field. We discuss also the appearance of various non newtonian and directional behavior as a function of an external applied electric field. We develop an asymptotic geometrical formalism to describe the propagation or light in liquid crystals. We use

*On leave from Instituto de Física, Universidad Nacional Autónoma de México

this to study the propagation of monochromatic rays in nematic and smectic bent core liquid crystals confined in planar cells, cylindrical pores and droplets. We show how the photo refractive effect can be controlled by externally imposed agents like electric fields and imposed flows. Various phenomena related like beam steering, dispersion, total internal reflection and scattering of light are also discussed.

1. Theoretical Frame

1.1. Bulk and Surface Elasticity of Nematics

Liquid crystal systems [?] are well defined and specific phase of matter (mesophases) characterized by a noticeable anisotropy in many of their physical properties as solid crystals do, although they are able to flow. Liquid crystal phases who undergo a phase transition as a function of temperature (thermotropics), exist in relatively small intervals of temperature lying between solid crystals and isotropic liquids.

Liquid crystals are synthesized from organic molecules, some of which are elongated and uniaxial, so they can be represented as rigid rods; others are formed by disc-like molecules [?]. This molecular anisotropy in shape is manifested macroscopically through the anisotropy of the mechanical, optical and transport properties of theses substances.

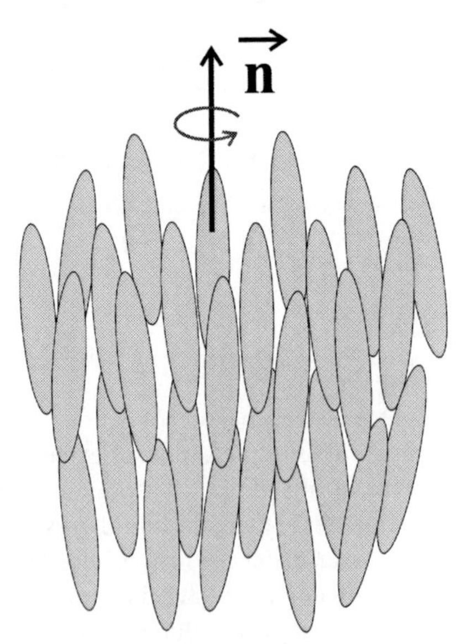

Figure 1. Schematic of the nematic phase.

Liquid crystals are classified by symmetry. As it is well known, isotropic liquids with spherically symmetric molecules are invariant under rotational, $O(3)$, and translational, $T(3)$, transformations. Thus, the group of symmetries of an isotropic liquid is $O(3) \times T(3)$. However, by decreasing the temperature of these liquids, the translational symmetry $T(3)$ is usually broken corresponding to the isotropic liquid-solid transition. In contrast, for a liquid formed by anisotropic molecules, by diminishing the temperature the rotational symmetry is broken $O(3)$ instead, which leads to the appearance of a liquid crystal. The mesophase for which only the rotational invariance has been broken is called nematics (see Fig. 1). As shown, the centers of mass of the molecules of a nematic have arbitrary positions whereas the principal axes of their molecules are spontaneously oriented along a preferred direction $\hat{n}$. If the temperature decreases even more, the symmetry $T(3)$ is also partially broken. The mesophases exhibiting the translational symmetry $T(2)$ are called smectics, Fig. 2a, and those having the symmetry $T(1)$ are called columnar phases, Fig. 2b.

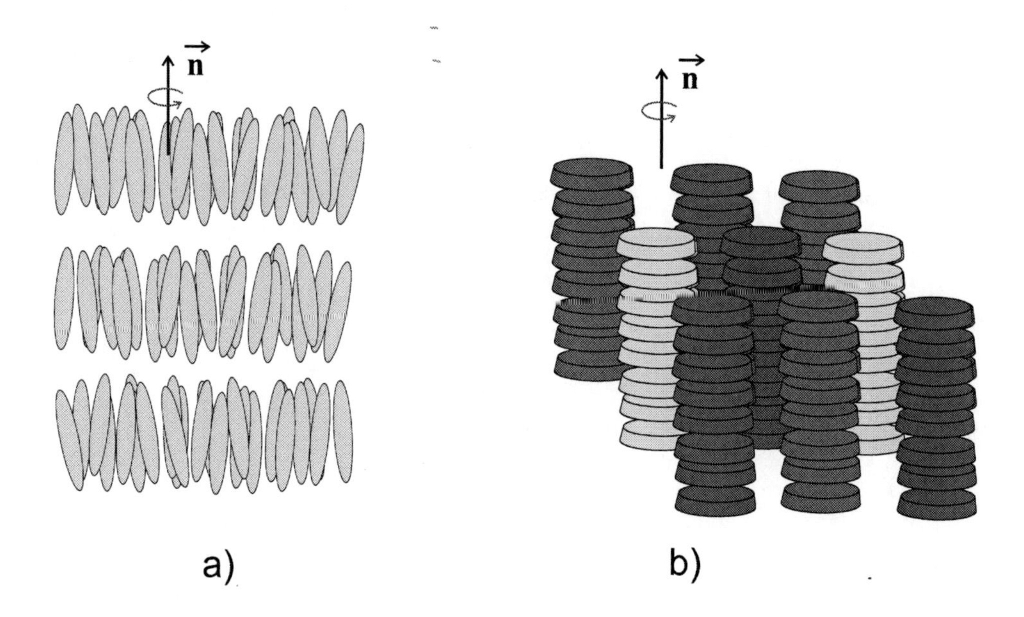

Figure 2. a) Molecular arrangement in the a) smectic A phase and b) columnar phase.

The elastic properties of liquid crystals determine their behavior in the presence of external fields and play an essential role in characterizing many of the electro-optical and magneto-optical effects occurring in them. In this work we shall adopt a phenomenological approach to describe these elastic and viscous properties. A liquid crystal will be considered as a continuum, so that its detailed molecular structure will be ignored. This approach is feasible because all the deformations observed experimentally have a minimum spatial extent that greatly exceed the dimensions of a nematic molecule. The macroscopic description of the Van der Waals forces between the liquid crystal molecules is given in terms of the following formula [?, ?] for the elastic contribution to the free-energy density:

$$\begin{aligned} \mathcal{F}_{el} &= (1/2)\int_V dV \left[K_{11}(\nabla\cdot\hat{n})^2 + K_{22}(\hat{n}\cdot\nabla\times\hat{n})^2 + \right. \\ &+ \left. K_{33}(\hat{n}\times\nabla\times\hat{n})^2 - K_{24}\nabla\cdot[\hat{n}\nabla\cdot\hat{n} + \hat{n}\times\nabla\times\hat{n}]\right] \\ & +(1/2)\int W_0\cos^2\theta dS. \end{aligned} \tag{1}$$

Here the unit vector $\hat{n}$ is the director, the elastic moduli K_{11}, K_{22}, and K_{33} describe, respectively, transverse bending (splay), torsion (twist), and longitudinal bending (bend) deformations. K_{24} is called the surface elastic constant because it is the coefficient of a divergence term which can be transformed to a surface integral by using Green's theorem. The last term provides the interaction between the NLC and the confining surface. There θ is the angle between $\hat{n}$ and the easy direction evaluated at the frontier of the nematic, and W_0 denotes the strength of interaction, in units of energy per area, between a nematic molecule and the surface.

The free energy of the LC cylinder has, in addition to the above elastic part, also an electromagnetic part due to the applied electrostatic field. As we have already discussed, the first contribution is given by Eq. (1). The electromagnetic free energy density, in MKS units,

$$\mathcal{F}_{em} = -(1/2) \int_V \text{Re}\left[\bar{E} \cdot \bar{D}^* + \bar{B} \cdot \bar{H}^*\right] dV \tag{2}$$

where the displacement field $\bar{D}$ and the magnetic flux vector $\bar{B}$ are related to the electric field $\bar{E}$ and magnetic field $\bar{H}$ by means of the constitutive relations

$$\begin{aligned} \overline{D} &= \epsilon_0 \boldsymbol{\epsilon} \cdot \overline{E}, \\ \overline{B} &= \mu_0 \boldsymbol{\mu} \cdot \overline{H}, \end{aligned} \tag{3}$$

characterized specifically by dielectric and magnetic tensors [?]

$$\boldsymbol{\epsilon} = \epsilon_\perp \delta_{ij} + \epsilon_a \widehat{n}\widehat{n}, \tag{4}$$

$$\boldsymbol{\mu} = \mu_\perp \delta_{ij} + \mu_a \widehat{n}\widehat{n}. \tag{5}$$

Here δ_{ij} is the identity tensor, $\epsilon_a = \epsilon_\parallel - \epsilon_\perp$ and $\mu_a = \mu_\parallel - \mu_\perp$ are, respectively, the dielectric and magnetic anisotropies of medium. $\epsilon_\perp$, $\mu_\perp$ ($\epsilon_\parallel$, $\mu_\parallel$) represent the dielectric permittivity and magnetic susceptibility perpendicular (parallel) to the director $\hat{n}$. Also, ϵ_0 and μ_0 are the dielectric permittivity and magnetic permeability constants in vacuum.

1.2. Nematodynamics

The hydrodynamic description of complex condensed matter systems like super fluids, ferromagnets, polymeric solutions, etc., has been possible thanks to the deep understanding of the role played by the symmetries and thermodynamics properties of the system [?], [?], [?], [?]. The extension of this linear hydrodynamic of the liquid crystals has been started in seventies, [?], [?], [?] and in recent years it has been generalized to the nonlinear case and for liquid crystal phases more complex[?], [?].

The key idea of the hydrodynamical formalism is based on the observation that for most of the complex condensed matter systems in the limit of very large temporal and spatial scales, only survive a very small number of slow processes compared with the enormous number microscopic freedom degrees. The evolution of these processes is described by the evolution of the corresponding hydrodynamical variables which describe cooperative phenomena which are not to be relaxed in a finite time for a spatially homogeneous limit. That is to say, the hydrodynamical variables are such that their Fourier transform satisfy the relation

$$\omega(k \longrightarrow 0) \longrightarrow 0. \tag{6}$$

Moreover the hydrodynamical variables can be identified uniquely by utilizing conservation laws (global symmetries) and breaking of symmetries, for spatio-temporal scales such that the microscopic freedom degrees have already been relaxed. for these scales the description of the systems is exact. When the microscopic freedom degrees reach their thermodynamical equilibrium (local equilibrium), we can use locally the thermodynamics to follow the evolution of the slow variables. This implies to consider a thermodynamical potential, for instance the internal energy as a function of the system variables [?], [?], [?]. In a second step we obtain the dynamics of the system by expressing the currents or thermodynamical fluxes in terms of their corresponding thermodynamical forces, which are the gradients of the conjugated thermodynamical variables, and performing and a series expansion of the fluxes in powers of the forces. this expansion will be expressed in terms of dynamical phenomenological coefficients (transport coefficients) which can be only determined from an experiment or a microscopic theory. Then, we separate the fluxes in those for which the entropy is conserved (reversible part) and those that make the entropy increase (irreversible part) and use the classical thermodynamics laws to find the evolution equations for the hydrodynamical variables. After obtaining these equations for the liquid crystal is possible to include the effects of external fields like: electromagnetic fields, stresses, thermal gradient, etc.

In what follows we sketch the steps of this theoretical formalism for nematics. The first class of hydrodynamical variables is associated with the local conservation laws which express the fact that quantities mass, momentum or energy cannot be locally destroyed and only can be transported. If $\rho(\overrightarrow{r},t)$, $\overrightarrow{g} = \rho\overrightarrow{\mathrm{v}}(\overrightarrow{r},t)$ and $\epsilon(\overrightarrow{r},t)$, where $\overrightarrow{\mathrm{v}}$ is the hydrodymical velocity, denote respectively, the density of these quantities, the corresponding conservation equations are [?]

$$\frac{d}{dt}\rho + \rho\, div\, \overrightarrow{\mathrm{v}} = 0, \tag{7}$$

$$\frac{d}{dt}\mathrm{v}_i + \frac{1}{\rho}\,\nabla_j\sigma_{ij} = 0, \tag{8}$$

$$\frac{d}{dt}\left(\frac{\epsilon}{\rho}\right) + \frac{1}{\rho}\, div\, \overrightarrow{\mathrm{j}^{\epsilon}} = 0. \tag{9}$$

Here $\frac{d}{dt} \equiv \frac{\partial}{\partial t} + \mathrm{v}_i\nabla_i$ denotes the hydrodynamical velocity, σ_{ij} is the nematic's stress and $\overrightarrow{\mathrm{j}^{\epsilon}}$ is the energy flow.

When a phase transition to the liquid crystal state occurs after reducing the temperature, the rotational symmetry $O(3)$ is broken spontaneously and the number of hydrodynamical variables increase. Any rotation around an axis different from $\overrightarrow{n}$ transforms the system to a different and distinguishable state form the state without rotation. This rotational symmetry broken is called spontaneous since the energy is a rotational invariant and there no exists energy which favors one orientation of $\overrightarrow{n}$ with respect any other. Thus, the state of the system becomes infinitely degenerate. Under these conditions, one slow variation of the degeneracy parameter is related takes to a slow relation of the system. which augments as long as $q \rightarrow 0$. This type of behavior is the basic content of the Goldstone Theorem [?]. Therefore, the degeneracy parameter is related with the order parameter of the liquid crystal which adopts different structures for the different mesophases. For a nematic the order parameter has the following form

$$Q_{ij} = S(n_i n_j - \frac{1}{3}\delta_{ij}), \tag{10}$$

where S the degree of order, i. e., $S = 0$ for the isotropic phase and $S = 1$ for a nematic phase having the molecules completely aligned. In agreement with this statement the dynamics of Q_{ij} is determined by that of $\vec{n}$. In summary, the macroscopic state of a nematic can be described by means of two scalar variables that can be chosen as $\rho(\vec{r}, t)$, $\epsilon(\vec{r}, t)$, one vectorial variable, $\vec{g} = \rho\vec{v}(\vec{r}, t)$ and one tensorial variable Q_{ij}, that can be selected, for instance, as the anisotropic part of the dielectric tensor.

Since **n** is related with a conservation law, its balance equation is a dynamical equation of the form

$$\left(\frac{\partial}{\partial t} + v_j \nabla_j\right) n_i + Y_i = 0, \tag{11}$$

where Y_i is not a current, since its surface integral is not a flux, but a quasi-current . This quantity must be orthogonal to $\vec{n}$ for fulfilling nematic symmetry $\vec{n} \rightarrow -\vec{n}$; however, there are other contributions to Y_i which does not come from the symmetries but from the thermodynamics requirements.

If a specific physical situation is given, the state of the system can be described in terms of an appropriate thermodynamical potential. This can be chosen, for example, as the total free energy$E$, [?]

$$E = \epsilon V = E(V, \rho V, \vec{g}V, \rho V \nabla_j n_i, \rho V \vec{n}, \sigma V), \tag{12}$$

where V denotes the volume of the system and σ is the entropy per unit of volume. From this assumption and using the Euler relation, we can derive the Gibbs relation

$$d\epsilon = \mu d\rho + T d\sigma + \vec{v} \cdot d\vec{g} + \Phi_{ij} d\nabla_j n_i + h_i' dn_i \tag{13}$$

and the Gibbs-Duhem relation

$$p = -\epsilon + \mu\rho + T\sigma + \vec{v} \cdot \vec{g}. \tag{14}$$

Here μ is the chemical , Φ_{ij} y h_i' are called the molecular fields, which are defined as the partial derivatives of the thermodynamical potential with respect to the corresponding conjugated variable. Since in equilibrium the state variables are constants, any inhomogeneous distribution of theses variables takes outside of equilibrium the system. For this reason the gradients of theses quantities are taken as thermodynamical forces. Hence, the presence of $\nabla\mu$, ∇T, $\nabla_j v_i$ y $\nabla_j \Phi_{ij}$ give place to irreversible processes in the system. The dynamical part of the hydrodynamical equations is obtained by expressing the currents σ_{ij}, $\vec{j^{\epsilon}}$, Y_i in terms of the thermodynamical variables T, μ, v_i, Φ_{ij}. If, additionally,we separate in these expressions the reversible part, which does not generate entropy augment and is invariant under temporal inversion, from the irreversible part which increase the entropy and is no invariant under the transformation: $t \rightarrow -t$, we obtain the following expressions for the fluxes [?], [?],

$$\sigma_{ij} = \sigma_{ij}^R + \sigma_{ij}^D = p\delta_{ij} + \Phi_{lj}\nabla_i n_l - \frac{1}{2}\lambda_{kji}h_k - \nu_{ijkl}\nabla_l v_k, \tag{15}$$

$$Y_i = Y_i^R + Y_i^D = -\frac{1}{2}\lambda_{kji}\nabla_j v_k + \frac{1}{\gamma_1}\delta_{ik}^{\perp}h_k, \tag{16}$$

$$\overrightarrow{\mathrm{j}^{\epsilon}} = T\, j_i^{\sigma D} + v_j\sigma_{ij}^D. \tag{17}$$

In these equations the superscript indexes R an D denote, respectively, the reversible and irreversible or dissipative parts, *and c v cv*

$$\Phi_{li} = K_{\lim n}\nabla_n n_m \tag{18}$$

where $K_{\lim n}$ is given by

$$K_{\lim n} = K_1\delta_{li}^{\perp}\delta_{mn}^{\perp} + K_2 n_p\epsilon_{pli}n_q\epsilon_{qmn} + K_3 n_i n_n\delta_{lm}^{\perp}. \tag{19}$$

Here K_1, K_2, and K_3 are the elastic constants of the nematic and ϵ_{ijk} is the totally anti-symmetric tensor of Levy-Civitta. The Kronecker delta tensor $\delta_{ik}^{\perp}$ is given by

$$\delta_{ik}^{\perp} = \delta_{ik} - n_i n_k \tag{20}$$

and λ_{kji} can be expressed as

$$\lambda_{kji} = (\lambda - 1)\delta_{kj}^{\perp}n_i + (\lambda + 1)\delta_{ki}^{\perp}n_j. \tag{21}$$

In this expression $\lambda = \frac{v_1}{v_2}$, is the reversible parameter so called flux alignment parameter, being v_1, v_2 two of the five independent viscosities of the nematic. The molecular field h_k, that we have already defined as $h_k \equiv h_i' - \nabla_j\Phi_{ij}$, turns out to be explicitly

$$h_k = -K_{kjnl}\nabla_j\nabla_l n_n + \delta_{kq}^{\perp}\left(\frac{\partial}{2\partial n_q}K_{pjkl} - \frac{\partial}{\partial n_q}K_{qjkl}\right)\nabla_l n_k\nabla_j n_p. \tag{22}$$

Finally, the stress viscous tensor ν_{ijkl} contains five independent viscosities for the nematic, $\nu_i, i = 1, 2...5$,

$$\begin{aligned}\nu_{ijkl} &= \nu_2(\delta_{jl}\delta_{ik}+\delta_{il}\delta_{jk}) + 2(\nu_1 + \nu_2 - 2\nu_3)n_i n_j n_k n_l \\ &\quad +(\nu_3 - \nu_2)(n_j n_l\delta_{ik} + n_j n_k\delta_{il} + n_i n_k\delta_{jl} \\ &\quad +n_i n_l\delta_{jk}) + (\nu_4 - \nu_2)\delta_{ij}\delta_{kl} \\ &\quad +(\nu_5 - \nu_4 + \nu_2)(\delta_{ij}n_k n_l + \delta_{kl}n_i n_j).\end{aligned} \tag{23}$$

The second law of thermodynamics establishes that any irreversible process that occurs in the system should increase the entropy. Thus, the entropy obeys the following balance equation

$$\frac{\partial}{\partial t}\sigma + \nabla\cdot(\overrightarrow{\mathrm{v}}\sigma) + div\,(j_i^{\sigma R} + j_i^{\sigma D}) = \frac{R}{T}, \tag{24}$$

where R is the dissipation function *for irreversible processes.* This quantity can be interpreted as the energy per *unit of volume dissipated by* the microscopic freedom degrees, and divided by the temperature $\frac{R}{T}$, represents the entropy *production* of the nematic. If as we did previously, we relate Eq. (24) with Eqs. (7), (8), (9), (11), by using the Gibbs relation (13) ant the expressions (15)-(23), we obtain an explicit formula for R, that is,

$$\begin{aligned} R &= -\nabla_i(j_i^{\epsilon D} - T\, j_i^{\sigma D} - \mathrm{v}_j\sigma_{ij}^D) - j_i^{\sigma D}\nabla_i T - \sigma_{ij}^D \nabla_j v_i + h_i\delta_{ij}^{\perp} Y_j^D \\ &= \frac{1}{2\gamma_1} h_i\delta_{ij}^{\perp} h_j + \frac{1}{2}\nu_{ijkl}(\nabla_j v_i)(\nabla_l v_k) + \frac{1}{2}\kappa_{ij}(\nabla_i T)(\nabla_j T), \end{aligned} \tag{25}$$

where γ_1^{-1} is the rotational viscosity and the tensor κ_{ij} describes the heat conduction (thermal conductivity). The second law of thermodynamics requires for R to be a definite positive form, which implies in turns that every single coefficient of the previous expression is positive. Notice that Eq. (25) implies as well that the dissipative currents and quasi-currents are given by the partial derivatives of the dissipation function, that is

$$j_i^{\sigma D} \equiv -\frac{\partial R}{\partial \nabla_i T} = -\,\kappa_{ij}\,\nabla_j T, \tag{26}$$

$$\sigma_{ij}^D \equiv -\frac{\partial R}{\partial \nabla_j v_i} = -\,\nu_{ijkl}\,\nabla_l v_k, \tag{27}$$

$$Y_k^D = \frac{\partial R}{\partial h_k} = \frac{1}{\gamma_1}\delta_{ik}^{\perp} h_i. \tag{28}$$

Equations (7), (8), (9), (11) and (25) constitute a complete set to describe the irreversible dynamics of a nematic of low molecular weigh (thermotropic) in absence of external fields

1.3. Constitutive Equations

Is is usual to applied external fields like electric and magnetic field, gravity temperature gradients, pressure and concentration, shear and vortex flows to carry out a nematic to a new equilibrium state so that these fields must be included in the hydrodynamical equations.

It is well known that for any polarizable medium an electric field $\mathbf{E}$ induces a polarization $\mathbf{P} = \mathbf{D} - \epsilon_0\mathbf{E}$, which is oriented with the field, where $\mathbf{D}$ is the displacement electric vector. Now, in a nematic the molecular dipolar moments are oriented approximately parallel with respect to to the long axis of the molecules. Thus,the induced polarization gives rise to a director orientation. In contrast with the electric fields, the influence of the magnetic fields in a nematic is much more weaker and in general, the induced magnetization can be neglected. A very well known result based on conventional thermodynamical arguments establishes that the work associated to an electric field $\mathbf{E} = -\nabla\Phi$,is given by

$$dw_{el} = -\frac{1}{2}\,\mathbf{E}\cdot\mathbf{D}, \tag{29}$$

which should be added to the Gibbs relation (13) and to the Gibbs-Duhem relation (14). By modifying these expression and using a procedure completely analogous to the one follows in the last section, it is possible to show that in the presence of the electric field the equation (8) converts into

$$\frac{d}{dt}\mathrm{v}_i + \frac{1}{\rho}\,\nabla_j\sigma_{ij} = \rho_e E_i + P_j\nabla_j E_i, \tag{30}$$

whose charge density is given by $\rho_e = \epsilon_0 div\, \overrightarrow{D}$. Up to linear order, the expression for σ_{ij} in terms of the thermodynamical forces is not longer given by (15), but by the expression

$$\sigma_{ij} = \sigma_{ij}^R + \sigma_{ij}^D = \widetilde{p}\,\delta_{ij} + \Phi_{lj}\nabla_i n_l - \frac{1}{2}\lambda_{kji}h_k - \nu_{ijkl}\nabla_l v_k \tag{31}$$

with

$$\widetilde{p} = p - \frac{1}{2}\overrightarrow{E}^2. \tag{32}$$

Analogously, the currents (16), now are given by

$$Y_i = Y_i^R + Y_i^D = -\frac{1}{2}\lambda_{kji}\nabla_j v_k + \frac{1}{\gamma_1}\delta_{ik}^{\perp}h_k - \zeta_{ijk}^E\nabla_j E_k, \tag{33}$$

$$j_i^e = \sigma_{ij}^E\, E_j + \kappa_{ij}^E\, \nabla_j T + \nabla_j(\zeta_{kji}^E h_k), \tag{34}$$

where σ_{ij}^E is the electric conductivity; and in consequence the entropy current is

$$j_i^\sigma = -\kappa_{ij}\nabla_j T - \kappa_{ij}^E E_j. \tag{35}$$

Here the material tensors of second rank κ_{ij}, κ_{ij}^E and σ_{ij}^E have the uniaxial form and each one should be expressed in terms of two dissipative transport coefficient, that is

$$\alpha_{ij} = \alpha_\perp \delta_{ij}^\perp + \alpha_\parallel n_i n_j. \tag{36}$$

On the other hand, the third order tensor ζ_{kji}^E is irreversible and contains and dynamical coefficient, the flexoelectric coefficient ζ^E,

$$\zeta_{ijk}^E = \zeta^E(\delta_{ij}^\perp n_k + \delta_{ik}^\perp n_j); \tag{37}$$

whereas the fourth order tensor ν_{ijkl} is defined by (23).

Following the same steps we used to obtain Eq. (25), the dissipation function R show in this case additional terms which involves the electric field,

$$\begin{aligned} R \;=\; & \frac{1}{2\gamma_1}h_i\delta_{ij}^\perp h_j + \frac{1}{2}\nu_{ijkl}(\nabla_j v_i)(\nabla_l v_k) + \frac{1}{2}\kappa_{ij}(\nabla_i T)(\nabla_j T) \\ & + \frac{1}{2}\sigma_{ij}^E E_i E_j + \kappa_{ij}^E E_i \nabla_j T - \zeta_{ijk}^E h_i \nabla_j E_k. \end{aligned} \tag{38}$$

Most of the parameter involved in the hydrodynamical and electrodynamics equations for a nematic have been measured for different substances that show a uniaxial nematic phase. Among these are included the elastic constants [?]; specific heat, the flux alignment parameter λ and the viscosities ν_i, $i = 1, 2...5$, [?]; the inverse of the diffusion constant γ_1 [?]; the thermal conductivity [?], the dielectric tensor anisotropy $\epsilon_a = \epsilon_\perp$-$\epsilon_\parallel$, and the electric conductivity σ_{ij}^E, [?], [?], [?].

Finally the dynamical equations for a nematic in an isothermal process can be obtained by inserting Eqs. (15) and (16) in Eqs. (8) and (11). this leads

$$\frac{dv_i}{dt} + \frac{1}{\rho}\nabla_j\left(p\delta_{ij} + \Phi_{ij}\nabla_i n_l - \lambda_{kjl}h_k - \nu_{ijkl}\nabla_l v_k\right) = 0 \tag{39}$$

$$\frac{dn_i}{dt} + \frac{1}{\gamma}\delta_{ik}^\perp h_k - \lambda_{kji}\nabla_j v_k = 0 \tag{40}$$

1.4. Viscometric Functions

The viscosity function or apparent viscosity connects the[?] force per unit area and the magnitude of the local shear. It depends on the orientation of the director through the expression

$$\eta\left(\theta(\zeta)\right) = (2\alpha_1 \sin^2\theta \cos^2\theta + (\alpha_5 - \alpha_2)\sin^2\theta + (\alpha_6 + \alpha_3)\cos^2\theta + \alpha_4)/2, \qquad (41)$$

where α_1, α_2, α_3,α_5 and α_6 are the Leslie coefficients[?] and η_c is the transverse Miesowicz viscosity[?]. Since the orientation angle θ is given by Eq. (74), from the above equation it follows that the dependence of η on θ indicates that the system is non-Newtonian in its behavior, in the sense that η is strongly dependent on the driving force. If we integrate the result over the cross section area of the flow to obtain the averaged apparent viscosity

$$\overline{\eta}(q,\Lambda) \equiv \frac{1}{A_T}\int \eta(\theta\,(x))dA. \qquad (42)$$

where A_T is the total area of the cross section.

1.5. First Normal Stress Difference

One of the distinctive phenomena observed in the flow of liquid crystal polymers in the nematic state is that of a negative steady-state first normal stress difference, N_1, in shear flow over a range of shear rates. N_1 is zero or positive for isotropic fluids at rest over all rates of shear, which means that the force developed due to the normal stresses, tends to push apart the two surfaces between which the material is sheared. In liquid crystalline solutions, positive normal stress differences are found at low and high shear rates, with negative values occurring at intermediate shear rates [?].

On the other hand, Marrucci et al [?] have solved a two dimensional version of the Doi model for nematics [?], in which the molecules are assumed to lie in the plane perpendicular to the vorticity axis, that is, in the plane parallel to both, the direction of the velocity and the direction of the velocity gradient. Despite this simplification, the predicted range of shear rates over which N_1 is negative, is in excellent agreement with observations. This result opens up the possibility that negative first normal stress differences may be predicted in a two dimensional flow. Indeed, in this section we shall show that over a range of Reynolds' numbers, negative values of N_1 are predicted for the plane Poiseuille flow, taking into account the effect of an external electric field.

We shall now examine the effects produced by the stresses generated during the reorientation process by calculating the viscometric functions which relate the shear and normal stress differences. For the geometry under consideration and using the convention in [?], the first normal stress difference is defined by

$$N_1 \equiv \sigma_{xx} - \sigma_{zz}, \qquad (43)$$

where σ_{ij} are the components of the stress tensor of the nematic used by De Gennes [?].

$$\begin{aligned} \sigma_{ij} &= \alpha_1 n_i n_j n_\mu n_\rho A_{\mu\rho} + \alpha_2 n_i \Omega_j + \alpha_3 n_j \Omega_i \\ &\quad \alpha_4 A_{ij} + \alpha_5 n_i n_\mu A_{\mu j} + \alpha_6 n_i n_\mu A_{\mu j}. \end{aligned} \qquad (44)$$

Here $A_{ij} \equiv (1/2)(\partial v_j/\partial x_i + \partial v_i/\partial x_j)$ is the symmetric part of the velocity gradient $\partial v_i/\partial x_j$ and $\mathbf{\Omega} \equiv d\mathbf{n}/dt - (1/2)\nabla \times \mathbf{v} \times \mathbf{n}$ represents the rate of change of the director with respect to the background fluid. The α_i for $i = 1, ..6$, denote the Leslie coefficients of the nematic.

The integration of the first normal stress difference, Eq. (43) over the whole cell and along the velocity gradient direction renders the net force between the plates as a function of the Reynolds number, which is proportional to N.

$$f \equiv \frac{1}{A_T} \int N_1 \left(\theta(\zeta)\right) dA \tag{45}$$

A positive force exerted by the fluid motion tends to push the plates apart, or otherwise, if the force is negative, the fluid tends to pull the plates close together.

2. Orientational Transition for Nematics Confined in Coaxial Cylinders

Let us consider a nematic liquid crystal confined between two coaxial cylindrical surfaces whose internal and external radii are R_1 and R_2, respectively. The z-axis coincides with the cylinder axis and the director is weakly anchored to the surfaces of the cylinders with preferential directions parallel and perpendicular to the internal and external cylinders, respectively.

We shall assume that under these arbitrary boundary conditions the initial nematic's orientation occurs in the plane $r - z$, and that is only function of r, that is, the director $\hat{\mathbf{n}}$ is given by

$$\mathbf{n} = \left[\sin\theta(r), 0, \cos\theta(r)\right], \tag{46}$$

where θ is defined in Fig. 3.

Then by expressing $\nabla \cdot \hat{\mathbf{n}}$ and $\nabla \times \hat{\mathbf{n}}$ in cylindrical coordinates we obtain inserting this in Eq. (1) the free energy, $\mathcal{F}_{elast}$, per unit length:

$$\begin{aligned} \mathcal{F}_{elast} &= \pi K_1 \int_{R_1/R_2}^{1} \left[\left(\frac{d\theta}{dx}\right)^2 \left(\cos^2\theta + \kappa \sin^2\theta\right) + \frac{\sin^2\theta}{x^2} \right] x dx \\ &\quad + \pi K_1 \left[\sigma_2 \cos^2\theta\left(R_2\right) - \sigma_1 \cos^2\theta\left(R_1\right) + \frac{R_1 W_\theta}{K_1} \right] \end{aligned} \tag{47}$$

where $\kappa = K_3/K_1$, $x = r/R_2$, and $\sigma_1 = R_1 W_\theta/K_1 + K_4/K_1 - 1$ and $\sigma_2 = R_2 W_\theta/K_1 + K_4/K_1 - 1$ are the dimensionless surface anchoring parameters.

2.1. Axial Electric Field

The contribution to the total free energy taking into account the interaction of NLC with an external electric field is expressed by the electric free energy, F_{elect}. By taking $\mathbf{E}_0$ along the z-axis, substitution in Eq. (2) leads to

$$F_{elect} = -(1/2) \int_V \epsilon_{zz}(r) E_0^2 dv. \tag{48}$$

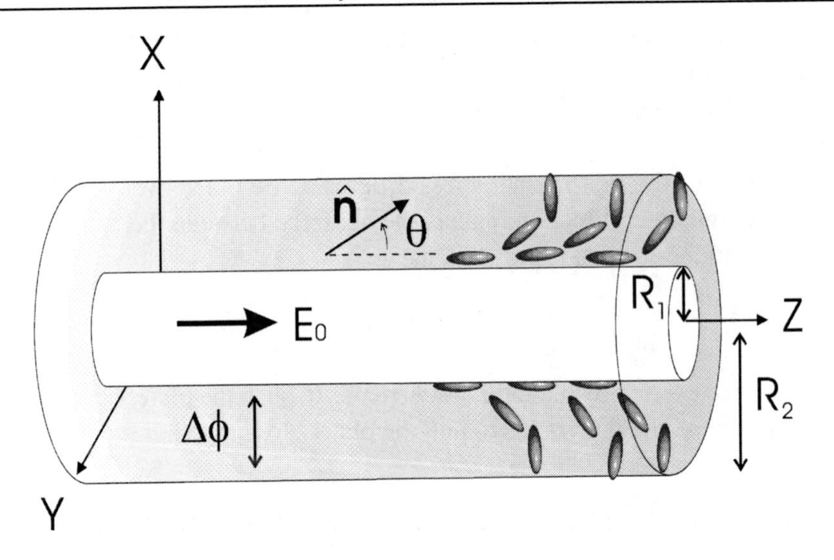

Figure 3. Coaxial cylinders inÞtreated by a nematic liquid crystal and submitted to a low-frequency electric field.

Now, Eq. (46) can be written as $\hat{\mathbf{n}} = \sin\theta(r)\mathbf{e}_r + \cos\theta(r)\mathbf{e}_z$ where $\mathbf{e}_r$ and $\mathbf{e}_z$ are the cylindrical unit vectors along the r and z directions, respectively. For infinite circular cylinders the symmetry implies that θ only depends on the radial distance r. Using the Eqs. (48) and (4), the electric energy per unit length, $\mathcal{F}_{elect}$, becomes

$$\mathcal{F}_{elect} = -\pi K_1 q_a \int_{R_1/R_2}^{1} \left(\frac{\epsilon_\perp^s}{\epsilon_a^s} + \cos^2\theta \right) x dx \tag{49}$$

where ϵ_a^s is the low frequency dielectric constant and q_a is an important dimensionless parameter defined as

$$q_a \equiv \epsilon_0 \frac{\epsilon_a^s E_0^2 R_2^2}{K_1} \tag{50}$$

which represents the ratio of the electric and elastic energies; for $q_a << 1$ the influence of the applied field is weak, while for $q_a >> 1$ the field essentially overcomes the elastic forces between the molecules.

Using the Eqs. (47) and (49), the total free energy per unit length, $\mathcal{F} = \mathcal{F}_{elast} + \mathcal{F}_{elect}$, becomes

$$\begin{aligned} \mathcal{F} &= \pi K_1 \int_{R_1/R_2}^{1} \left[\left(\frac{d\theta}{dx}\right)^2 (\cos^2\theta + \kappa\sin^2\theta) + \frac{1}{x^2}\sin^2\theta - q_a\cos^2\theta \right] x dx \\ &+ \pi K_1 \left\{ \sigma_2\cos^2\theta(R_2) - \sigma_1\cos^2\theta(R_1) + \frac{R_1 W_0}{K_1} - \frac{1}{2} q_a \frac{\epsilon_\perp^s}{\epsilon_a^s} \left[1 - \left(\frac{R_1}{R_2}\right)^2 \right] \right\} \end{aligned} \tag{51}$$

The stationary orientational configuration $\theta(x)$ is determined by minimizing the free energy, Eq. (51). This minimization leads to the Euler-Lagrange equation [?] in the bulk

$$x^2 \frac{d^2\theta}{dx^2}\left(\cos^2\theta + \kappa\sin^2\theta\right) + \frac{1}{2}x^2\left(\frac{d\theta}{dx}\right)^2(\kappa - 1)\sin 2\theta$$
$$+ x\frac{d\theta}{dx}\left(\cos^2\theta + \kappa\sin^2\theta\right) - \frac{1}{2}\left(1 + q_a x^2\right)\sin 2\theta = 0, \quad (52)$$

with $\theta(x)$ satisfying the following weak anchoring boundary conditions on the cylinders walls at R_1 and R_2, respectively

$$\left.\frac{d\theta}{dx}\right|_{x=R_1/R_2} = \left.\frac{\frac{1}{2}\left(\frac{R_2}{R_1}\right)\sigma_1 \sin 2\theta}{\cos^2\theta + \kappa\sin^2\theta}\right|_{x=R_1/R_2}, \quad (53)$$

$$\left.\frac{d\theta}{dx}\right|_{x=1} = \left.\frac{\frac{1}{2}\sigma_2 \sin 2\theta}{\cos^2\theta + \kappa\sin^2\theta}\right|_{x=1}. \quad (54)$$

The quasi-escaped radial configuration can be obtained by solving Eq. (52) subjected to the conditions given by Eqs. (53) and (54).

In what follows, we will present results for the specific case of 4'-n-pentyl-4-cyanobiphenyl (5CB) liquid crystal. The material parameters used are [?] $T_{IN} - T = 10^0C$ with $T_{IN} = 35^0C$, $\kappa = 1.316$, $K_1 = 1.2\times10^{-11}N$, $W_\theta/K_1 = 40\mu m^{-1}$, and $K_4/K_1 = 1$. The later relation between surface and bulk constants of the nematic results in the following relation between the surface parameters σ_1 and σ_2

$$\sigma_1 = \frac{R_1}{R_2}\sigma_2 \quad (55)$$

and in this case the boundary conditions given by Eqs. (53) and (54) take the same form.

Figure 4 shows a surface defined by the critical values of the applied axial electric field necessary to align the director parallel to z-axis, as a function of the parameters x_0 and σ_2. This plot defines two regions separated by the critical values of the electric field for which the NLC molecules are completely aligned with the axial direction (AX configuration), this phase transition has been described in Ref. [?], and' for fields smaller than the critical field the inhomogeneous configurations can take place. When the field is zero, the liquid crystal configuration is produced only by the elastic action of the walls on the molecules. This influence is larger when the interior cylinder radius grows and the field intensity grows too. In this case the strength of interaction with the boundary, σ_2, presents a threshold, such that after certain value, the walls help to the alignment of the liquid crystal; this is the reason because a lower applied field aligns the molecules.

2.2. Radial Electric Field

Now we shall consider the case of an applied electric field $\mathbf{E}$ acting along the radial direction from the inner cylindrical surface to the outer. In this case the electric free energy is given by

$$F_{elect} = -(1/2)\int_V \epsilon_{rr}(r) E^2 dv \quad (56)$$

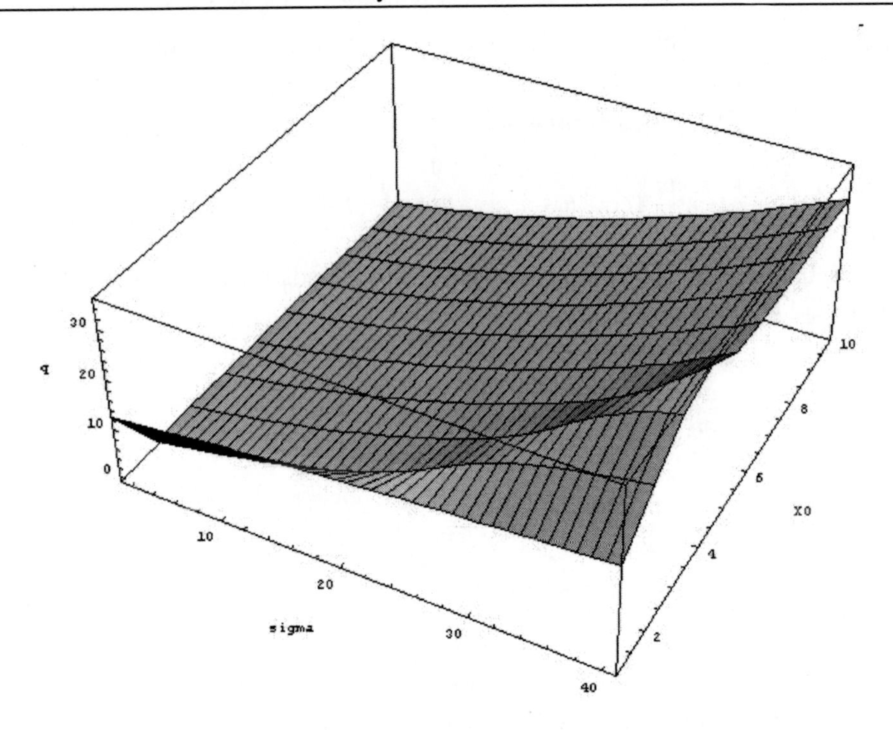

Figure 4. a) q_a against x_0 and σ_2 for 5CB. b) EquiÞfield curves are represented by contour lines. Dotted lines sepa- rate three regions of different behavior.

where $\epsilon_{rr}(r)$ is an element of the dielectric tensor ϵ_{ij} defined by Eq. (4). Here we have employed the electrostatic field $\mathbf{E} \equiv -\mathbf{e}_r \Delta\phi / r \ln(R_2/R_1)$ generated between the two coaxial cylinders subjected to a potential difference $\Delta\phi$. Using Eqs. (56) and (4), the electric energy per unit length, $\mathcal{F}_{elect}$, becomes

$$\mathcal{F}_{elect} = -\pi K_1 q_r \int_{R_1/R_2}^{1} \left(\frac{\epsilon_\perp^s}{\epsilon_a^s} + \sin^2\theta \right) \frac{dx}{x} \tag{57}$$

where the dimensionless parameter q_r is defined as

$$q_r \equiv \frac{\epsilon_a^s \Delta\phi^2}{K_1 \ln^2(R_2/R_1)}, \tag{58}$$

Using the Eqs. (47) and (57), the total free energy per unit length, $\mathcal{F}$, becomes

$$\begin{aligned} \mathcal{F} &= \pi K_1 \int_{R_1/R_2}^{1} \left[\left(\frac{d\theta}{dx} \right)^2 (\cos^2\theta + \kappa \sin^2\theta) + \frac{1}{x^2}(1 - q_r)\sin^2\theta \right] x dx \\ &+ \pi K_1 \left\{ \sigma_2 \cos^2\theta(R_2) - \sigma_1 \cos^2\theta(R_1) + \frac{R_1 W_\theta}{K_1} \right. \\ &\left. + q_r \frac{\epsilon_\perp^s}{\epsilon_a^s} \ln\left(\frac{R_1}{R_2} \right) \right\} \end{aligned} \tag{59}$$

The stationary orientational configuration $\theta(x)$ is determined by minimizing the free energy, Eq. (59). This minimization leads to the Euler-Lagrange equation in the bulk

$$x^2\frac{d^2\theta}{dx^2}\left(\cos^2\theta+\kappa\sin^2\theta\right)+\frac{1}{2}x^2\left(\frac{d\theta}{dx}\right)^2(\kappa-1)\sin 2\theta$$
$$+x\frac{d\theta}{dx}\left(\cos^2\theta+\kappa\sin^2\theta\right)+\frac{1}{2}(q_r-1)\sin 2\theta=0, \quad (60)$$

with $\theta\,(x)$ satisfying the weak anchoring boundary conditions on the cylinders walls at R_1 and R_2, given respectively by Eqs. (53) and (54).

Numerical solutions of Eq. (60) were calculated by using the same material parameters that those used for the axial electric field, for different values of $x_0 = R_1/R_2$, σ_2 and q_r. Figures 5a and 5b show threshold field surface and equifield curves, respectively, versus x_0 and σ_2. The radial threshold field is defined similarly as in the axial case.

This is also a configurational transition since for values smaller that this critical value the nematic acquires radially dependent textures. In Figure 4b the equifield curves are represented as contour lines to show that the threshold field are essentially hyperbola-like curves which implies that q_r increases by enlarging either, x_0 or σ_2. By contrast to the axial case discussed in last section, here the dependence on both parameters is completely monotonous. This is so because the spatial dependence of the radial electric field $(1/r)$, balances the effect of the line defect resulting from the escaped radial configuration, obtained in the limit case when $x_0 \to 0$.

We have discussed that for an axially orient a nematic confined between two cylinders there exist three regions in the $\sigma_2 - x_0$ space showing different behaviors, which are essentially the result of the competition between the surface elastic forces at both cylinders and the electric force [**?**]. In contrast, the behavior of the system under the action of a radial field is quite monotonous because its inversely proportional dependence compensate the augment of elastic bulk energy around the inner cylinder, as its radius decreases. Our results suggest to use this coaxial nematic cored fiber to design a device for electrically switch the propagation of a desired number of TM modes of optical fields.

3. Plane Poiseuille Flow

Consider a nematic layer of thickness l constrained between two parallel conducting slides, as depicted in *Fig. 6*. We shall assume that the orientation of the director is planar and, therefore, when an external *dc* electric field E is applied along the z direction, the director $\mathbf{n}$ will reorient inside the cell for values of $E > E_c$, where E_c is the critical field that has to be exceeded to start the reorientation. Owing to the low aspect ratio of the cell, we shall assume that the nematic's director and the the hydrodynamic flow belongs to the $x - z$ plane. That is to say

$$\mathbf{n} = \left[\sin\theta\,(z,t)\,,0,\cos\theta\,(z,t)\right]. \quad (61)$$
$$\mathbf{v} = (v_x,0,0) \quad (62)$$

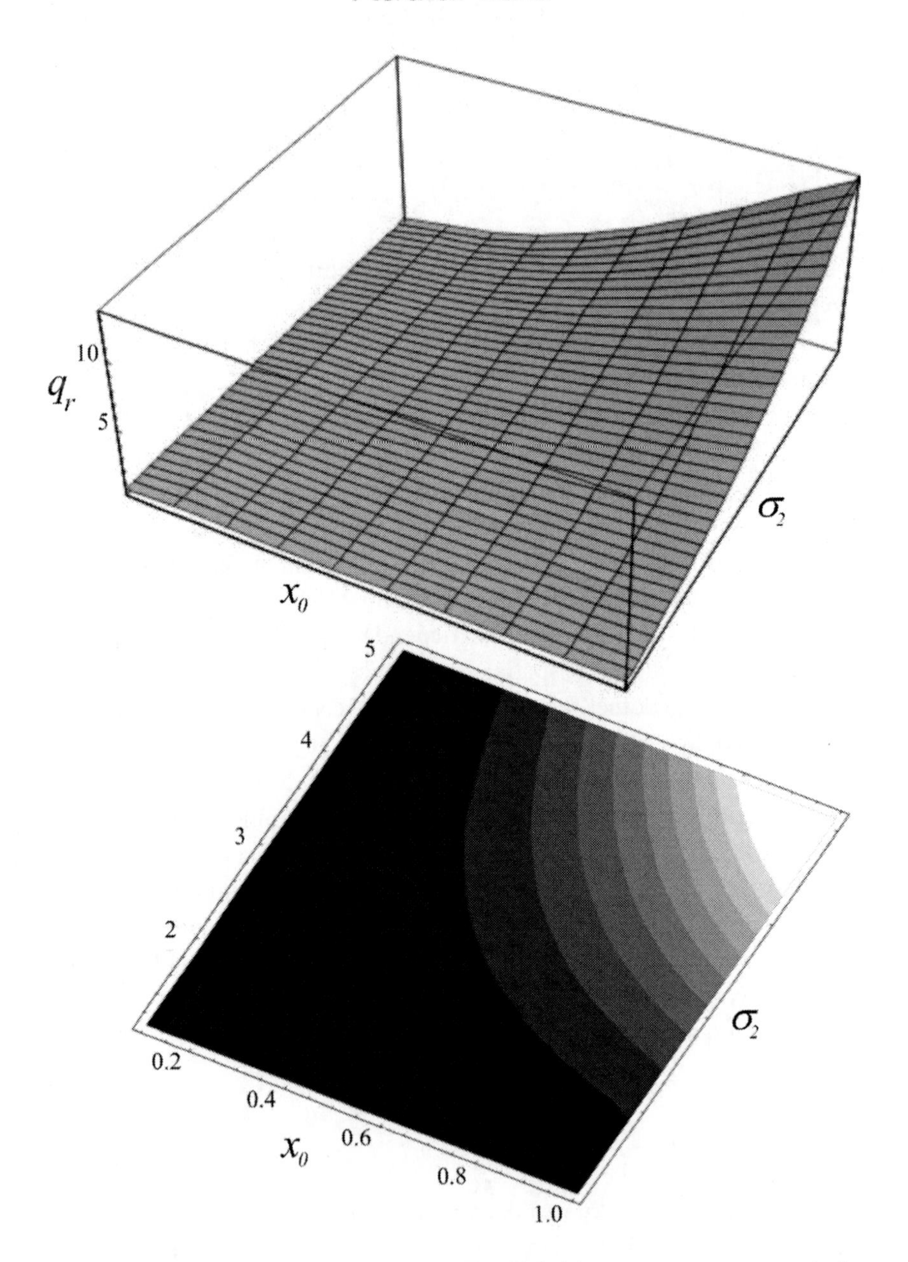

Figure 5. a) q_r versus x_0 and σ_2 for 5CB. b) EquiÞfield curves are represented as contour lines.

We also assume that the reorientation angle θ satisfies strong anchoring conditions at the plates and that the only relevant component of the velocity field is v_x which will be assumed to satisfy non slip boundary conditions,

$$\theta\left(z = \pm\frac{l}{2}\right) = \pm\frac{\pi}{2}, v_x(z = \pm\frac{l}{2}) = 0. \tag{63}$$

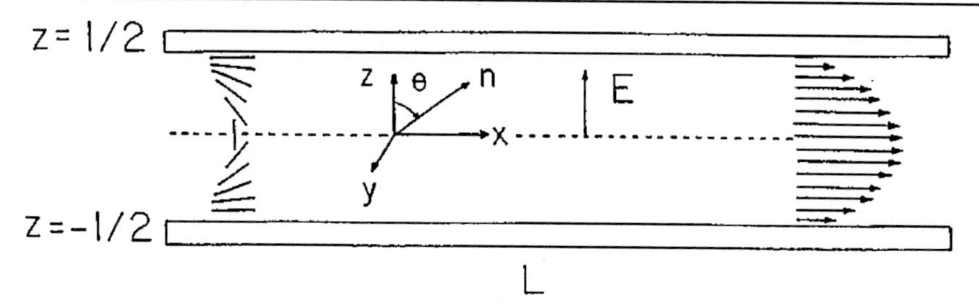

Figure 6. Schematics of a planarly aligned liquid crystal Þlm in the presence of a constant electric Þeld. The velocity proÞle of a plane Poiseuille flow is also shown.

If the reorientation of the director is considered to be an isothermal process, its equilibrium states may be described in terms of the associated Helmholtz free energy functional given Eq. (1), which for this model takes the form

$$F = \int_V dV \left\{ \frac{K}{2}\left(\frac{d\theta}{dz}\right)^2 - \frac{E^2}{2}\epsilon_0\left(\epsilon_\perp + \epsilon_a \cos^2\theta\right) + \frac{1}{2}\rho v_x^2(z) \right\}. \quad (64)$$

Here $\rho(z,t)$ denotes its local mass density. The last term in this equation represents the contribution to F due to the external pressure gradient along the x direction. It should be remarked that in writing this expression we have neglected the surface contributions K_{24} and W_0 since we are taking strong anchoring conditions. We have also made the assumption of equal elastic constants for the splay, bend and twist elastic deformations,$K \equiv K_1 = K_2 = K_3$.

If we substitute in Eqs. (39) and (40) the explicit form for $\mathbf{n}$ and the one component of velocity v_x we arrive at the following set of coupled dynamical equations for θ and v_x

$$\frac{\partial\theta}{\partial t} = -\frac{2}{\gamma_1}\frac{\delta F}{\delta\theta} - (1-\lambda)\cos\theta\frac{\partial v_x}{\partial z}, \quad (65)$$

$$\frac{\partial v_x}{\partial t} = \frac{\nu_3}{\rho}\frac{\partial^2 v_x}{\partial z^2} + \frac{\lambda-1}{2\rho}\frac{\partial}{\partial z}\left[\frac{1}{\cos\theta}\frac{\delta F}{\delta\theta}\right] - \frac{1}{\rho}\frac{\partial p}{\partial x}. \quad (66)$$

where the variational derivative $\delta F/\delta\theta$ is given explicitly by

$$\frac{\delta F}{\delta\theta} = K\frac{d^2\theta}{dz^2} - \frac{\epsilon_0\epsilon_a}{2}E^2\sin 2\theta \quad (67)$$

and $\gamma_1, \gamma_2, \nu_3$, with $\lambda \equiv \gamma_1/\gamma_2$, denote the various viscosity coefficients of the nematic. Note that the last term in (66) represents the externally applied pressure gradient along the x direction.

It is worth to note that, strictly speaking, equations (65)-(67) provide a closed set of hydrodynamic equations for a low molecular weight nematic (thermotropic), since in this case the director is the only additional hydrodynamic variable.

However, the corresponding description for a polymeric nematic (lyotropic) is much more complicated due to the large number of degrees of freedom that may contribute to the dynamics in the hydrodynamic limit. But since a complete formulation of a hydrodynamic description for arbitrary lyotropics is not yet known, as a first approximation we shall use the above formalism to describe the hydrodynamic behavior of a polymeric nematic solution of *PHIC [?]*. This is a strong approximation; but it is expected to be realistic because of the fact that molecular weight of *PHIC* ($\sim 10^5$) is quite low.

We consider the final stationary state for which the reorientation has already occurred and is defined by setting the left hand sides of (65) and (66) equal to zero. Hence, from (66) we arrive at,

$$\frac{d^2 v_x(\zeta)}{d\zeta^2} = \frac{l}{\nu_3}(\nabla p)_{ef}, \tag{68}$$

where $(\nabla p)_{ef}$ is an effcctive pressure gradient defined by

$$(\nabla p)_{ef} \equiv \frac{Q}{\rho l L} = \frac{\nu_3}{2}\frac{dp(\chi)}{d\chi}\left[\nu_3 + \frac{(\lambda-1)^2\gamma_1}{4}\right]^{-1}; \tag{69}$$

Q is the mass flow rate. Here we have defined the dimensionless variables $\zeta \equiv z/l$ and $\chi \equiv x/l$.

3.1. Orientational Configuration

We shall assume that the pressure gradient is constant, $dp(x)/dx \equiv \Delta p/L = const.$ Hence, the solution of Eq. (68) is given by

$$v_x(\zeta) = \frac{1}{\nu_3}(\nabla p)_{ef}(1/4 - \zeta^2). \tag{70}$$

This defines the parabolic velocity profile known as the plane Poiseuille flow. If we now substitute this solution into Eq. (40) with $d\theta/dt = 0$, we arrive at the following closed, nonlinear and dimensionless equation for the final orientational state

$$\frac{d^2\theta}{d\zeta^2} + q\overline{E}^2 \sin 2\theta + 6N\zeta\cos\theta = 0. \tag{71}$$

Here $\overline{E} \equiv E/E_c$, where $E_c = \frac{\pi}{l}\sqrt{2K/\epsilon_a\epsilon_0}$ is the critical field, such that reorientation occurs if $\overline{E} > 1$. It is essential to point out that there are two important physical parameters in (71). On the one hand, q was defined in Eq. (50) and on the other hand, the parameter

$$N \equiv 6R\nu_3\gamma_1(\lambda-1)/K\rho \tag{72}$$

contains the effects due to the hydrodynamic flow through the Reynolds number

$$R \equiv \rho l^2(\nabla p)_{ef}/\nu_3^2. \tag{73}$$

Eq. (71) presents a singular behavior is the result of the competition between the strong external field and the hard-anchoring boundary condition. Since exact solutions of Eq. (71)

when $N \neq 0$ are not analytical, the presence of the dimensionless parameters q and N in Eq. (71) allows for the possibility of carrying out systematic expansions in powers of them.

As mentioned for the case $N = 0$, it is to be expected that in the bulk of the cell the electric energy would be much larger than the nematic's elastic energy, so that $q >> 1$. In contrast, near the solid boundaries, the opposite behavior is expected, $q < 1$. These different physical situations and behaviors of the solutions of Eq. (71) may be modelled by using the boundary layer technique. The application of this formalism gives rise to the expression

$$\begin{aligned} \theta(\zeta; N, q\overline{E}^2) &= \pi/2 + sign\zeta \arcsin \frac{3N\zeta}{2q\overline{E}^2} - \arcsin \frac{3N}{q\overline{E}^2} \\ &-2 \arctan \left(e^{-\zeta} \tan \left\{ \pi/4 + \frac{1}{2} sign\,(\zeta) \arcsin \frac{3N\zeta}{2q\overline{E}^2} \right\} \right), \end{aligned} \quad (74)$$

where $sign\,(\zeta)$ is the sign function. Notice that even though θ is defined in terms of this discontinuous functions, it is easy to show that θ and $d\theta/d\zeta$ are continuous functions, while $d^2\theta/d\zeta^2$ has a finite discontinuity at $\theta = 0$[**?**].

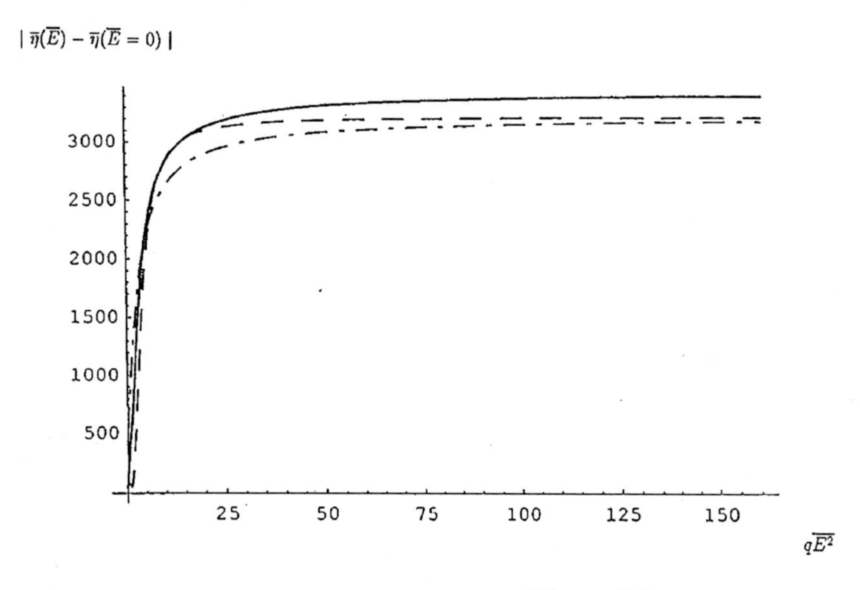

Figure 7. Apparent viscosity (Pa-s) for PHIC versus $q\bar{E}^2$ for (ÑÑ) N=50; (- ¥ -)N=100; (- Đ -) N=160.

3.2. Electrorheological Effect for Polymeric Nematic

We take $\alpha_1 = -1100\,Pa\,s$, $\alpha_2 = -3700\,Pa\,s$, $\alpha_3 = 320\,Pa\,s$, $\eta_c = 2100\,Pa\,s$ and $\epsilon_a\epsilon_0 = 6.19 \times 10^{-9} N/V^2$. Here we have also taken $K \sim 10^{-12} N$ for the elastic constant and $l = 4 \times 10^{-6} m$ for the separation distance between the plates. In this case it turns out that $E_c = 4490 V/m$ [**?**].

We shall substitute Eqs. (74) into (42) to get the averaged apparent viscosity and plot it versus $\overline{E}$ in *Fig.* 7.These curves show that the $PHIC$ nematic solution exhibits a significant

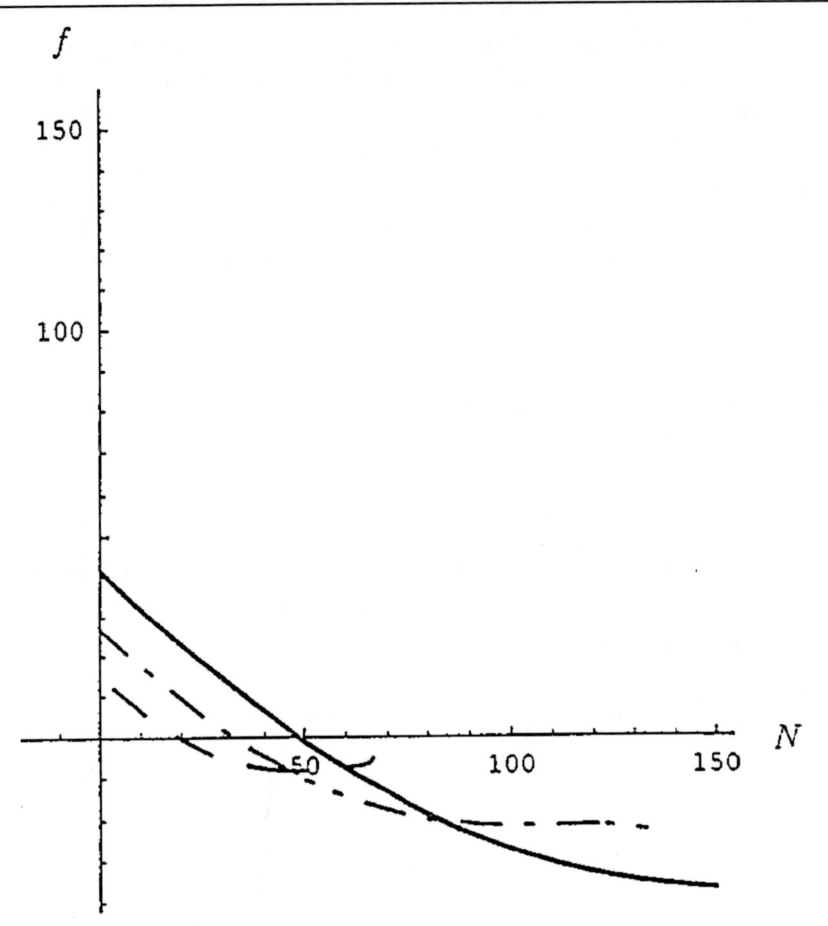

Figure 8. The dimensionless force f in units of the elastic constant $K = 10^{-}-12$N as a function of N for different values of $q\bar{E}^2$. (ÑÑ) $q\bar{E}^2 = 50$; (Ð ¥ Ð) $q\bar{E}^2 = 100$; (Ð Ð Ð Ð) $q\bar{E}^2 = 160$.

electrorheological effect, which is evidenced by the sharp increase of almost three orders of magnitude in its apparent viscosity. To estimate the corresponding value of the applied field, take a saturation value $q\overline{E}^2 = 10$ in *Fig.* 7. If we take $q = 10^{-1}$, which corresponds to a strong interaction between the electric field and the nematic, this implies that $\overline{E} = 10$ or $E = 10E_c$. This shows that for rather small applied fields, $\sim 45 \times 10^3 V/m$, a strong electrorheological effect may be produced in the cell.

To find the first normal stress difference we substitute Eq. (74) into Eqs. (43) and (44) to get

$$\begin{aligned} N_1\left(\theta(\zeta)\right) &= \nu_3(\nabla p)_{ef}\zeta(\nu_1+\nu_2-2\nu_3)\sin 4\theta \\ &\quad -\lambda\frac{K}{d^2}\left[\sin 2\theta\frac{d^2\theta}{d\zeta^2}+\cos 2\theta\left(\frac{d\theta}{d\zeta}\right)^2-q\cos^2\theta\right]. \end{aligned} \tag{75}$$

In *Figs.* 8 and 9, As observed in *Fig.* 8, as the Reynolds number increases, the net force changes sign from positive to negative, and the effect is more pronounced as the electric field increases in magnitude. In *Fig.* 8, the force is plotted as a function of the electric field strength for several values of N (or R). For sufficiently slow flows, the force is always positive, but at high fluid velocities the force is always negative over all the electric field strengths considered. The results shown in these figures do not exclude the possibility that a further increase in the fluid velocity will induce another change of sign in the force, becoming positive at very high velocity gradients. These prediction would be in accordance with the observed and predicted behavior of liquid crystal solutions in simple shear flow[**?**]. It should be remarked that in *Figs.* 8 and 9 we have plotted the dimensionless force f in units of the elastic constant $K = 10^{-12} Newtons$.

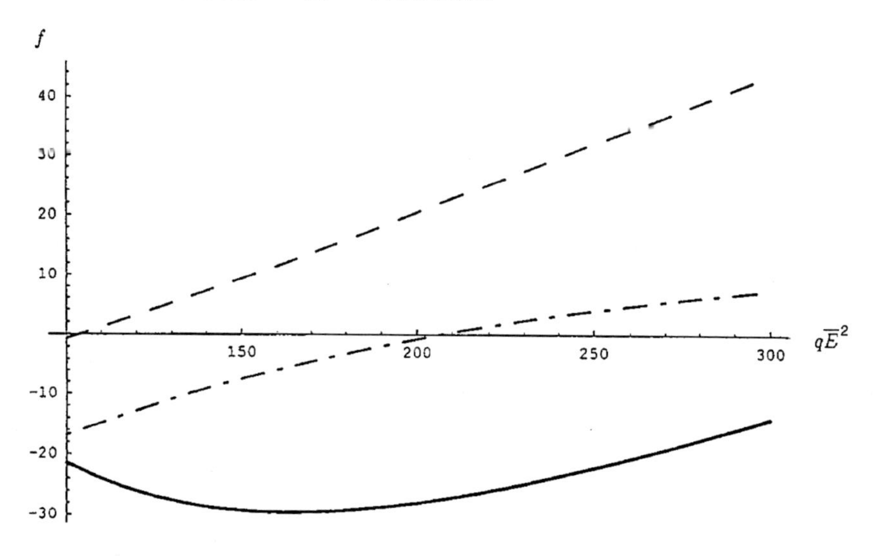

Figure 9. f vs $q\bar{E}^2$ for different values of N. (ÑÑ) N =50; (Ð ¥ Ð) N = 100; (Ð Ð Ð Ð) N = 160.

4. Planar Couette Flow

A nematic layer of thickness l is contained between two parallel conducting plates (*Fig. 10*). T If the initial orientation of the director is planar, when an external *d.c.* electric field E is applied along the z direction, the director $\vec{n}$ will reorient inside the cell for values of $E > E_c$, where E_c is the critical field that has to be exceeded to initiate the reorientation. If the reorientation is only transverse dependent and occurs in a plane and the velocity profile has only one shear component

$$\mathbf{n} = [\sin\theta\,(z,t)\,, 0, \cos\theta\,(z,t)] \tag{76}$$

$$\mathbf{v} = [v_x(z), 0, 0] \tag{77}$$

where it is also assumed that the reorientation angle θ satisfies strong anchoring conditions at the plates and $v_x(z)$ fulfils no-slip boundary conditions,

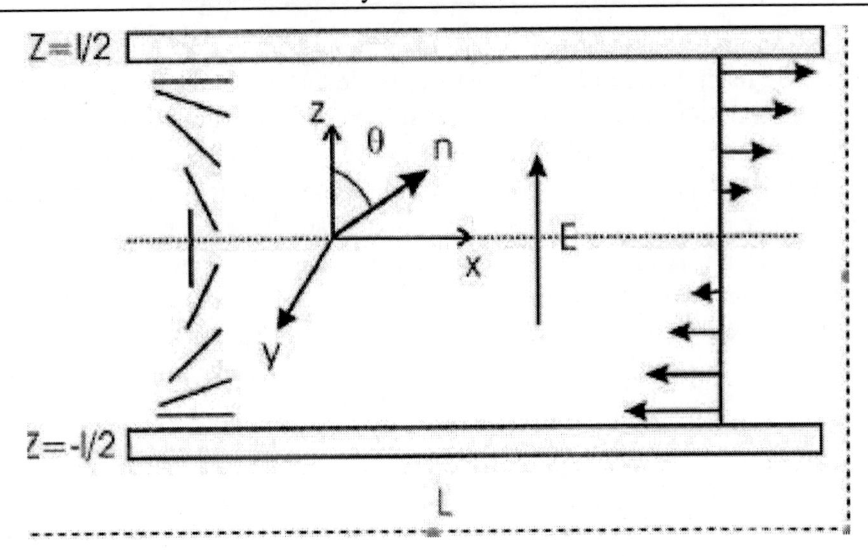

Figure 10. Schematics of a planary aligned liquid crystal film in the presence of a constant electric field. The velocity profile of a uniform shear flow is also shown.

$$\theta\left(z=\pm\frac{l}{2}\right)=\pm\frac{\pi}{2},\ v_x(z=\pm\frac{l}{2})=0. \tag{78}$$

As usual, the reorientation of the director will be considered to be an isothermal process and, therefore, its equilibrium states may be described in terms of the associated Helmholtz free energy Eq. (1), which for this geometry takes the form [**?**]

$$F=\int_V dV\left\{\frac{K}{2}\left(\frac{d\theta}{dz}\right)^2-\frac{E^2}{2}\epsilon_0\left(\epsilon_\perp+\epsilon_a\cos^2\theta\right)+\frac{1}{2}\rho v_x^2(z)\right\}, \tag{79}$$

where in writing this expression the assumption of equal elastic constants for the splay, bend and twist elastic deformations, $K\equiv K_1=K_2=K_3$ has been made. The last term in this equation represents the contribution to F due to the uniform shear flow along the x direction. Inserting Eqs. (76) and (77) into Eqs. (39) ad (40) we arrive at the following set of coupled dynamical equations for θ and v_x

$$\frac{\partial\theta}{\partial t}=\frac{2}{\gamma_1}\frac{\delta F}{\delta\theta}-(1-\lambda)\cos\theta\frac{\partial v_x}{\partial z}, \tag{80}$$

$$\frac{\partial v_x}{\partial t}=\frac{\nu_3}{\rho}\frac{\partial^2 v_x}{\partial z^2}+\frac{\lambda-1}{2\rho}\frac{\partial}{\partial z}\left[\frac{1}{\cos\theta}\frac{\delta F}{\delta\theta}\right], \tag{81}$$

where the variational derivative $\delta F/\delta\theta$ is given explicitly by

$$\frac{\delta F}{\delta\theta}=K\frac{d^2\theta}{dz^2}-\frac{\epsilon_0\epsilon_a}{2}E^2\sin 2\theta. \tag{82}$$

Strictly speaking, equations (80)-(82) provide for a closed set of hydrodynamic equations for a low molecular weight nematic (thermotropic), since in this case the director is the only additional hydrodynamic variable, apart from the usual conserved variables of mass, specific entropy and momentum densities [?]. Since for a polymeric nematic (lyotropic) the corresponding description is much more involved owing to the large number of degrees of freedom that may contribute to the dynamics in the hydrodynamic limit, its complete formulation is still an open issue. Therefore, as a first approximation we shall use the above formalism to describe the hydrodynamic behavior of a polymeric nematic solution of *PHIC,* as described by [?]. It is important to stress that this is, indeed, a strong approximation, however, it is expected to be a reasonable one owing to the fact that molecular weight of *PHIC* ($\sim 10^5$) is not too large. If in addition, the system is always away from a critical point, it is not necessary to account for the dynamics of an order parameter, which is not a hydrodynamic variable, and therefore its behavior can be described by considering the director field as the only additional hydrodynamic variable. This approximation has the great advantage of keeping the description simple enough so that an analytical treatment is possible and specific calculations can be carried out. In this work we only consider the final stationary state characterized by the fact that the reorientation has already occurred, but flow effects are still present. For this case the calculation of some rheological properties such as the viscometric functions, can be carried out explicitly [?], as we shall see below. The final stationary state is defined by setting the left hand sides of (80) and (81) equal to zero, which yields a closed equation for the final stationary orientational configuration. We shall carry out this procedure explicitly for a particular flow, namely, the uniform shear flow.

4.1. Uniform Shear Flow

This flow occurs when the plates move in opposite directions with constant velocity $\pm v_0$ in such a way that the velocity gradient $\frac{dv_x(\zeta)}{d\zeta} \equiv \dot{\gamma} = const.$ Eq. (66) reduces to

$$\frac{d^2 v_x}{d\zeta^2} = 0, \tag{83}$$

where $\zeta \equiv z/l$ and whose solution is satisfying the boundary conditions $v_x(\zeta = \pm\, 1/2) = \pm\, v_0$ is given by $v_x = 2v_0\zeta/l$. By inserting this solution into the orientational Eq. (65) we arrive at the following closed equation for θ

$$\frac{d^2\theta}{d\zeta^2} + p\sin 2\theta - m\cos\theta = 0, \tag{84}$$

with the dimensionless field strength

$$p \equiv q\overline{E}^2 \tag{85}$$

where

$$q \equiv \epsilon_0\epsilon_a E_c^2 l^2/2K, \tag{86}$$

and a dimensionless shear rate given by

$$m \equiv R\gamma_1\nu_3(1-\lambda)/\rho K. \tag{87}$$

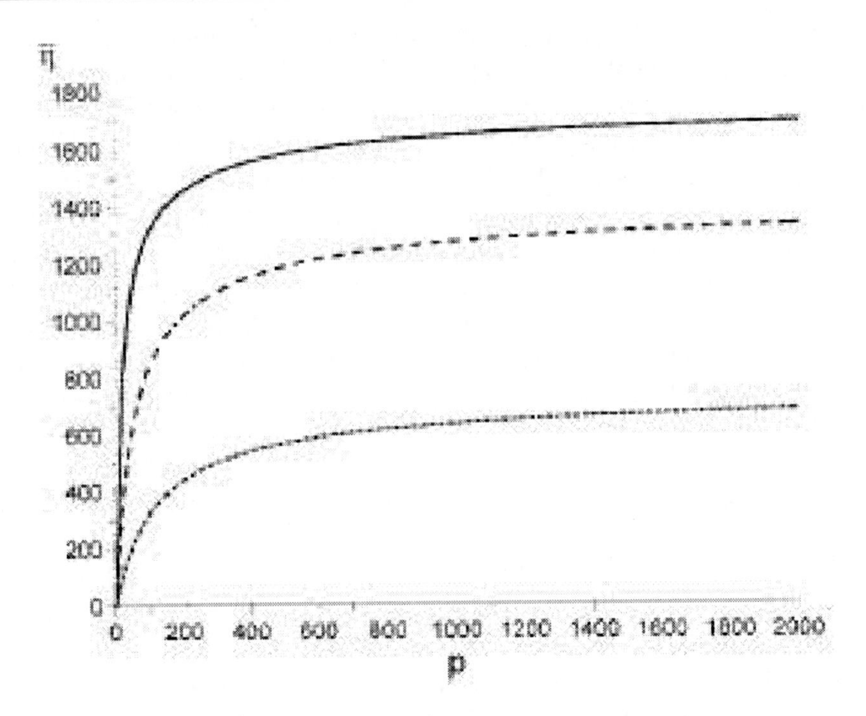

Figure 11. Averaged viscosity $\bar{\eta}$ vs p parametrized by m/p.( ) $m/p = 0.1$; ( ) $m/p = 1$; ( ) $m/p = 1.5$.

Here $\overline{E} \equiv \frac{E}{E_c}$, where $E_c = \frac{\pi}{l}\sqrt{2K/\epsilon_a\epsilon_0}$ is the critical field [?] and is such that reorientation occurs if $\overline{E} > 1$. The asymptotic solution of Eq. (84) for large values of p and m is obtained by using the boundary layer formalism which leads to the expression:

$$\begin{aligned}\theta(\zeta; m, p) &= \arcsin\frac{m}{2p} + \left(Sign\zeta\frac{\pi}{2} - \arcsin\frac{m}{2p}\right) \\ &\quad \frac{\sinh[\sqrt{2p - Sign[\zeta]m}\,|\zeta|]}{\sinh[\sqrt{2p - Sign[\zeta]m}/2]}. \end{aligned} \tag{88}$$

Notice from this equation that θ is continuous in the asymptotic limit of strong fields and flows, even though its first derivative is discontinuous[?].

4.2. Electrorheological Effect for Polymeric Nematic

The required viscosity coefficients and other material parameters for $PHIC$ are taken from the predictions of Doi's theory [?][?], $\alpha_1 = -2590\, Pa\, s$, $\alpha_2 = -3770\, Pa\, s$, $\alpha_3 = 169\, Pa\, s$, $\eta_c = 3640\, Pa\, s$ and $\epsilon_a\epsilon_0 = 6.19 \times 10^{-9} N/V^2$. Here we have also taken $K \sim 10^{-12} N$ for the elastic constant and $l = 4 \times 10^{-6} m$ for the separation distance between the plates. The ER in simple shear flow may be exhibited by showing the variation of the spatially-averaged viscosity given by Eq. (42) as a function of the electric p, parametrized by m. The variation of this quantity as a function of the electric field is shown in Fig. 11. The viscosity tends to the zero shear rate asymptote as m tends to zero.

The increase in the electric field strength gives rise to higher viscosity, since the director tends to be aligned with the electric field positioned normal to the flow direction, inducing higher resistance to flow. As in the case of Poiseuille flow [?], the nematic solution exhibits a significant electrorheological effect, which is evidenced by the sharp increase in the viscosity $\overline{\eta}(\overline{E})$. The magnitude of the electrorheological effect diminishes as the flow strength rises, i.e., for larger values of m/p . A very interesting effect is observed if we plot $\overline{\eta}$ as a function of m in a logarithmic scale. To estimate a typical value of the applied field, take, for instance, a saturation value $p = 800$ in the curve corresponding to $m/p = 0.1$ in Fig. 11. Then from the above given definitions of p, q, E_c and using the values of K, ϵ_a corresponding to $PHIC$, we find that the critical field $E_c = 1.41 \times 10^3 V/m$ and that the applied field is $E = 1.274 \; x \; 10^5 V/m \sim 90 E_c$. This shows that for rather small applied fields, a strong electrorheological effect may be produced in the cell.

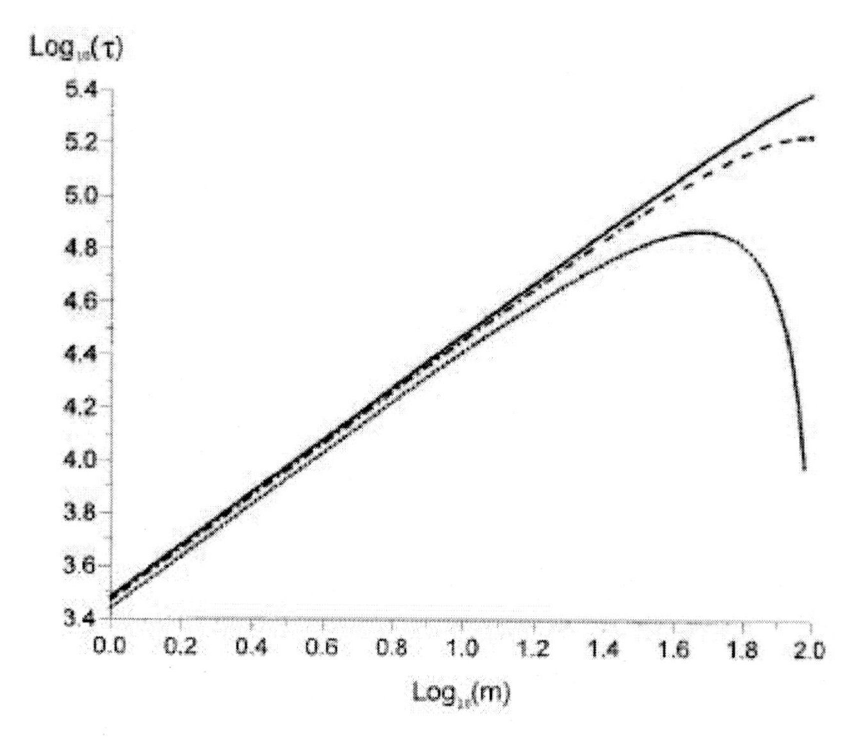

Figure 12. Logarithmic plot of the stress τ vs m for different values of the electric field.( ) $p = 160$; ( ) $p = 100$; ( ) $p = 50$.

4.3. Rheological Properties

For flows that are simple variations of uniform shear flow, the stresses are uniform throughout the fluid and the material functions depend on the shear rate alone. In steady simple shear flows all transient stresses have died out and the steady stress depends on the shear rate $\dot{\gamma}$. The behavior of the stress τ vs. m for different values of the electric field is shown in Fig. 12. First note that the initial slope of the curves is one, since the viscosity is constant for low values of m. At relatively high velocity gradients and for low electric fields, the stress reaches a maximum and shows a sharp drop at a critical velocity gradient. It is well known that fluids which exhibit a maximum in the stress may present a mechanical instability, since for a given stress there are two available values of the velocity gradient. In other words, regions of low and high viscosity co-exist for a given stress, implying that two average orientations also co-exist for the same stress. The hydrodynamic treatment by [?] in the absence of electric fields, also allows for multiple values of the velocity gradient for a given stress. It is interesting that this effect induced by both flow and electric fields, has been predicted by this model analytically; however, to our knowledge, there are no experimental data available to compare with. In simple shear flow, the effects produced by the stresses generated during the reorientation process are qualitatively different to those obtained in pressure driven flows [?]. The first normal stress difference is given explicitly by

$$N_1[\theta(\zeta)] = -\lambda\frac{K}{l^2}\left[\sin 2\theta\frac{d^2\theta}{d\zeta^2} + \cos 2\theta\left(\frac{d\theta}{d\zeta}\right)^2 - q\cos^2\theta\right]. \tag{89}$$

In Fig. 13, the integration of the first normal stress difference profile over the whole cell renders the net force between the plates,Eq. (45), as a function of the velocity gradient (m) for varying electric field strength (p). The net force increases with the electric field and remains almost constant with the velocity gradient for high values of p. The external field forces the average director orientation to be aligned perpendicular to the flow direction, producing a larger orientation gradient in the velocity gradient direction, and hence increasing the magnitude of the first normal stress difference. It is interesting to note that when $p = 50$, the magnitude of the force, f, decreases as the velocity gradient increases, and vanishes at $m = 100$. A vanishing N_1 is ascribed to a complete alignment of the director, on the average, with the flow direction. This situation agrees with results presented in Fig. 13, where for $p = 50$, the shear stress presents a steep drop at around $m = 100$.

Carlsson and Sharp [?] have also analyzed the stabilizing effect of an electric field on the shear flow of nematic liquid crystals. They showed that the presence of an ac field across the sample produces a stable boundary layer type of flow. They obtained explicit expressions for the flow alignment angle, the boundary layer and the relaxation time as functions of the electric field. Although their physical situation is essentially the same as the one considered in this work, there are important differences in the goals of both works. The solutions of the respective orientational equations for the final stationary state, our Eq. (84) and their Eq. (10), yield different orientational configurations because they neglect the influence of the walls. Therefore, a direct comparison is not feasible.

5. Bistability for Couette Flow

Consider a pure thermotropic nematic liquid crystal (LC) confined between two infinitely long coaxial cylinders with radii R_1 and R_2 $(R_1 < R_2)$ rotating with angular velocities Ω_1 and Ω_2, respectively. The system is under the action of a radial low-frequency electric field, as depicted in Fig. 14. Under the action of the flow and the electric field, the nematic's director $\hat{\mathbf{n}}$ adopts a stationary configuration which depends only on the radial coordinate r. If θ is the angle that the director forms with respect to the ϕ-axis, then the components of the director and the velocity of the LC are given by

$$\hat{\mathbf{n}} = \sin\theta(r)\mathbf{e}_r + \cos\theta(r)\mathbf{e}_\phi, \tag{90}$$

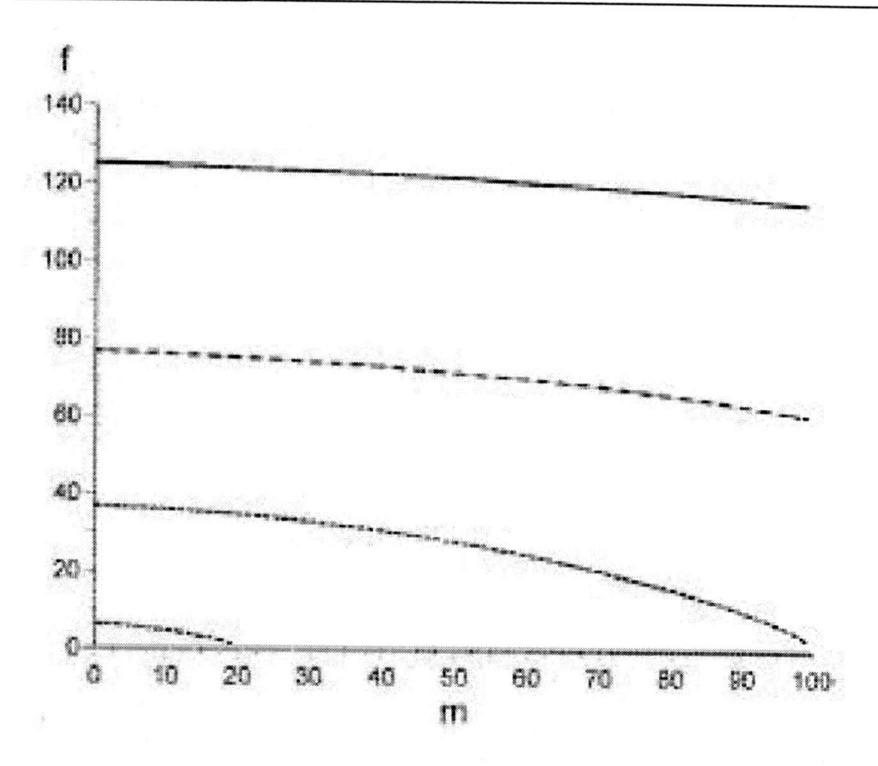

Figure 13. First normal stress difference N_1 vs f for different values of m and p. ( ) $m = 50, p = 160$; ( ) $m = 100, p = 160$; ( ) $m = 160$,$p = 160$; ( ) $m = 10, p = 10$.

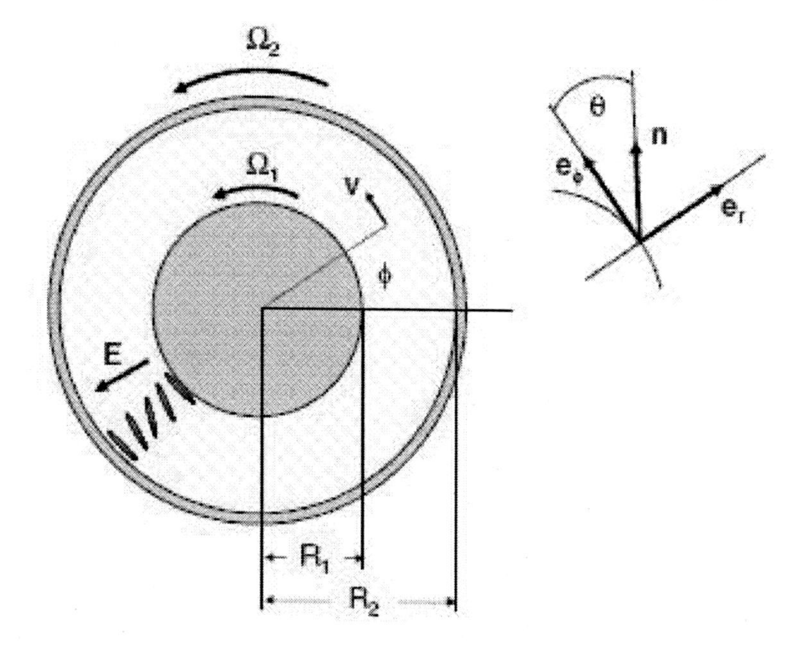

Figure 14. Sketch of a nematic liquid crystal conÞned by two rotating coaxial cylinders and subjected to a radial electric Þfield.

and

$$\mathbf{v} = \omega(r) r \mathbf{e}_\phi, \tag{91}$$

where $\mathbf{e}_r$ and $\mathbf{e}_\phi$ are the cylindrical unit vectors along the r and ϕ directions, respectively. We will assume homogeneous hard anchoring and non-slip boundary conditions of the nematic molecules at the surface of both cylinders:

$$\theta(R_1) = 0, \quad \theta(R_2) = 0, \tag{92}$$

and

$$\omega(R_1) = \Omega_1, \quad \omega(R_2) = \Omega_2. \tag{93}$$

In order to obtain the equilibrium configuration of the director, one has to consider the free energy of the nematic

$$\mathcal{F} = \mathcal{F}_{el} + \mathcal{F}_{em} + \mathcal{F}_h. \tag{94}$$

The elastic part of the free energy is given by the Frank-Oseen expression Eq. (1)

Since the electric field $\mathbf{E}_0$ points along the $r-$axis, the electromagnetic part of the free energy due to the applied electric field, $\mathcal{F}_{em}$ is,

$$\mathcal{F}_{em} = -\frac{1}{2} \int_V \epsilon_{rr}(r) E_0^2 dV. \tag{95}$$

Finally, there is a contribution arising from the motion of the fluid and is given by

$$\mathcal{F}_h = \frac{1}{2} \rho \int_V v^2 dV. \tag{96}$$

The total free energy of the LC is obtained by expressing the integrals of Eqs. (1), (2) and (96) in cylindrical coordinates to obtain the free energy per unit length:

$$\begin{aligned} F &= F_{el} + F_{em} + F_h \\ &= \pi K_1 \int_{R_1/R_2}^{1} \left[\left(\frac{d\theta}{dx} \right)^2 \left(\cos^2\theta + \kappa \sin^2\theta \right) + \frac{\sin^2\theta + \kappa\cos^2\theta}{x^2} \right] x dx \\ &\quad - \pi q K_1 \int_{R_1/R_2}^{1} (\sin^2\theta + \epsilon_\perp^s/\epsilon_a^s) \frac{dx}{x} \\ &\quad + \pi\rho \int_{R_1/R_2}^{1} v_\phi^2 x dx, \end{aligned} \tag{97}$$

where $\kappa = K_3/K_1$, $x = r/R_2$, and we have employed the electrostatic field $\mathbf{E}_0 \equiv -\mathbf{e}_r \Delta\phi/(r \ln[R_2/R_1])$ generated between the two coaxial cylinders subjected to a potential difference $\Delta\phi$. We define the important parameter q as

$$q_p \equiv \epsilon_a^s \Delta\phi^2 / \left(K_1 \ln^2 [R_2/R_1] \right) \tag{98}$$

where ϵ_a^s is the low-frequency dielectric anisotropy.

Equations (8) and (11) can be expressed as

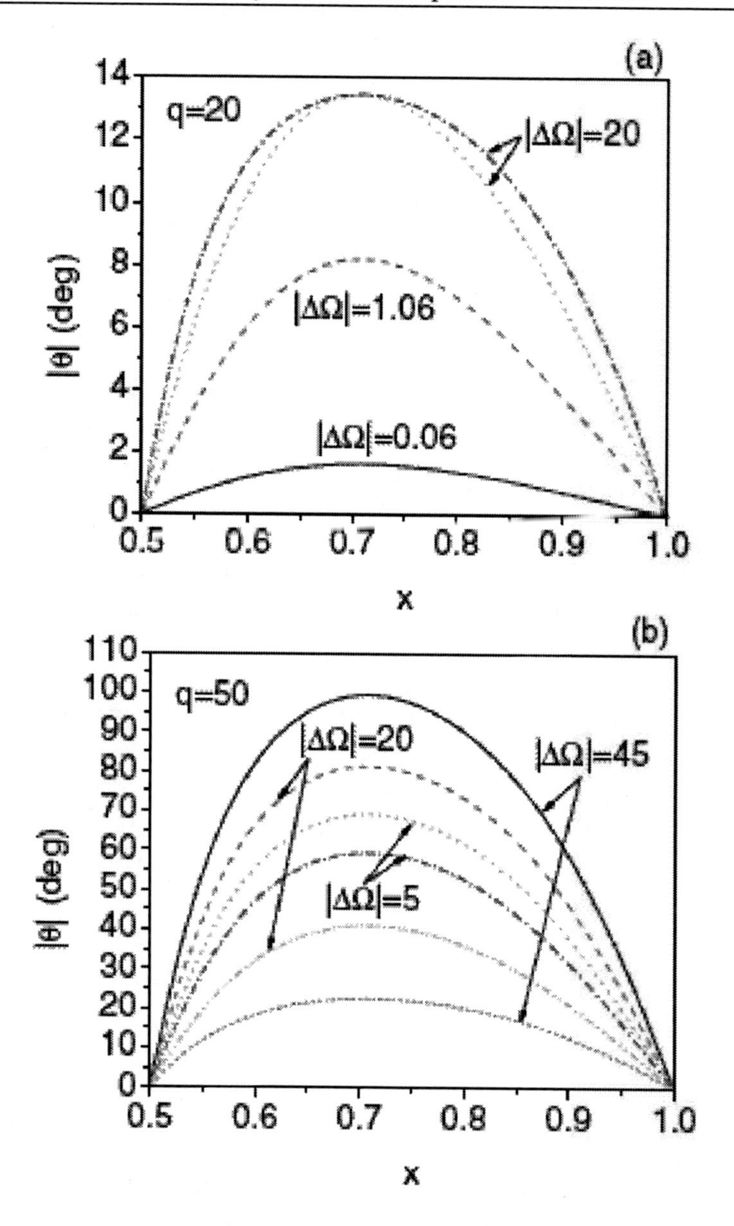

Figure 15. Nematicos configuration $\theta(x)$ as a function of x for 5CB and $R_1/R_2 = 0.5$ a) $q = 20$ and b) $q = 50$. The units of $\Delta\Omega$ are (rad/s).

$$
\begin{aligned}
0 = & \frac{d^2\theta}{dx^2}x^2\left(\cos^2\theta + \kappa\sin^2\theta\right) + x\frac{d\theta}{dx}\left(\cos^2\theta + \kappa\sin^2\theta\right) \\
& + \left(\frac{d\theta}{dx}\right)^2 \frac{x^2}{2}(\kappa - 1)\sin 2\theta - \frac{c_1}{2\eta(\theta)}\left[\gamma_1 + \gamma_2\cos 2\theta\right] \\
& + \frac{(\kappa - 1 + q_p)}{2}\sin 2\theta,
\end{aligned} \tag{99}
$$

and the relative angular velocity is determined by using

$$\omega(r) = \Omega_1 + \frac{c_1 K_1}{R_2^2} \int_{R_1/R_2}^{x} \frac{ds}{s^3 \eta(\theta(s))}. \tag{100}$$

where is the local viscosity of the nematic given by Eq. (41). The resulting configuration is radially dependent and can be obtained by solving Eq. (99) subjected to the conditions given by Eqs. (92).

5.1. Nematic's Director Configuration

In what follows, the numerical calculations are performed for 4'-n-pentyl-4-cyanobiphenyl (5CB). The Leslie coefficients for this LC at $T = 25$ °C are [?]: $\alpha_1 = -0.0060$ Pa-s, $\alpha_2 = -0.0812$ Pa-s, $\alpha_3 = -0.0036$ Pa-s, $\alpha_4 = 0.0652$ Pa-s, $\alpha_5 = 0.0640$ Pa-s, and $\alpha_6 = -0.0208$ Pa-s, $\kappa = 1.318$, and $K_1 = 12$ pN. Numerical solutions of Eq. (99) for 5CB were calculated by using the shooting method [?]. In Fig. 15a we present the director's orientation for $q_p = 20$ and different values of $\Delta\Omega$. We can observe that for low values of the electric field, the angle grows from zero up to a maximum value around $x = 0.7$, then, it decreases again to zero at the outer cylinder. The value of the maximum increases as we increase the value of $\Delta\Omega$. This means simply that the nematic's molecules tend to adopt a more inclined position as the flow increases. In Fig. 15b we plot the same as in Fig. 15a but for $q_p = 50$. For the values of $\Delta\Omega$ shown the system may adopt multiple steady-state solutions. In the figure, we have plotted different possible solutions for the case $\Delta\Omega > 0$ and $\Delta\Omega < 0$. Notice that the selected solutions for $\Delta\Omega < 0$ are not the negative of the solutions for $\Delta\Omega > 0$. Indeed it is possible to obtain a phase diagram in the q_p vs. $\Delta\Omega$ space for which a transition "curve" separates a region in which the system adopts a single steady-state solution from a region in which the system may adopt several stationary solutions. The phase diagram is shown in Fig. 16. In the case of zero flow ($\Delta\Omega = 0$), the critical value of the field which correspond to the transition between these two types of behaviors is given by the Fredericks field [?]

$$q_F = \frac{\pi^2}{\ln^2 [R_1/R_2]} + 1 - \kappa. \tag{101}$$

For fields larger than q_F the system may adopt two possible equivalent configurations each one being the specular image of the other. When the flow is different from zero ($\Delta\Omega \neq 0$) there is also a critical field q_c from which there are multiple configurations but this configurations are not the specular image of the other, as we have already shown in Fig. 15b. Moreover, unlike the Frederiks transition occurring at zero flow which is always of second order, the transition for the rest of the points on the curve of Fig. 16 corresponds to a first order transition. In other words, at zero shear flow the director's orientation continuously changes from $\theta = 0$ below the Frederiks field to two symmetric configurations that gradually separate from each other as one increases the electric field. On the other hand, if there is a finite shear flow, the director's orientation is distorted even for small electric field. As we increase the value of the electric field the director's orientation changes continuously even if one overpasses a critical value given by the transition line of Fig.16. However, once this critical value is surpassed, a new solution appears representing a totally different director's configuration.

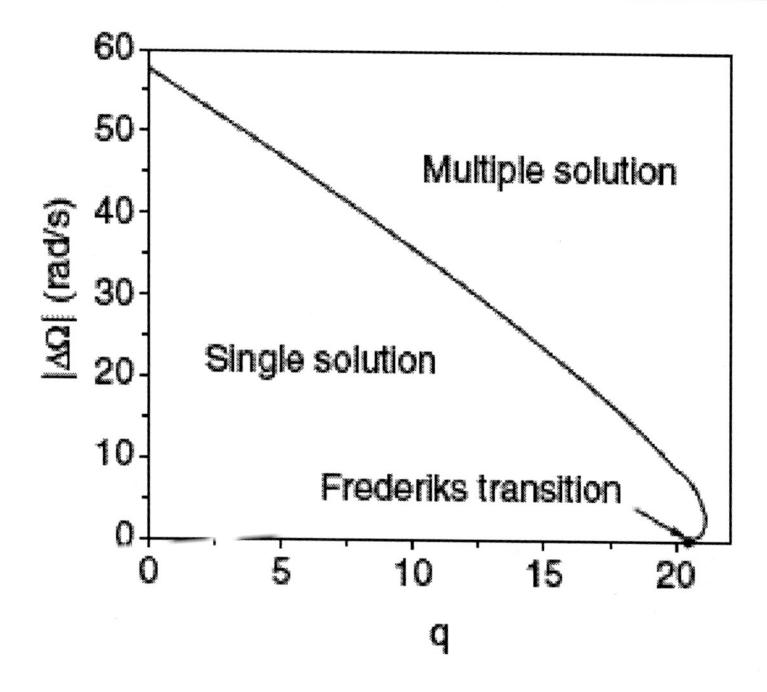

Figure 16. Phase diagram in the $\Delta\Omega$ vs q space showing a region with single valuated steady-state solutions and a region with multiple solutions. In this plot $R_1/R_2 = 0.5$.

5.2. Apparent Viscosity

If we now plot $\overline{\eta}(q_p, \Delta\Omega)$ given by Eq. (42) as a function of q_p and $\Delta\Omega$, we get the curves displayed in Fig.17. The plots show that $\overline{\eta}$ presents different kinds of behavior depending on the value of q_p. When the electric field q_p is small ($q \lesssim 21$), $\overline{\eta}$ has two solutions that increases as a function of $\Delta\Omega$ (see Fig. 17a).One of the solutions gives values for the viscosity considerably larger than the other one and they start at values of $\Delta\Omega$ different from zero. On the other hand, for large values of q_p ($\gtrsim 21$) the viscosity has solutions that decrease as $\Delta\Omega$ increases but the other solution first increases and then diminishes with an augment of $\Delta\Omega$. This is shown in Fig. 17b, where, for clarity, we have plotted the solution showing a shear thickening region in the negative part of the $\Delta\Omega$ axis.

6. Directional Dependent Response in a Nematic Capillary

Consider a pure thermotropic nematic confined between two coaxial pipes with radii R_1 and R_2, under the action of a radial low frequency electric field, as depicted in Fig. 18. Under these conditions, the director's configuration is spatially homogeneous along the axis of the

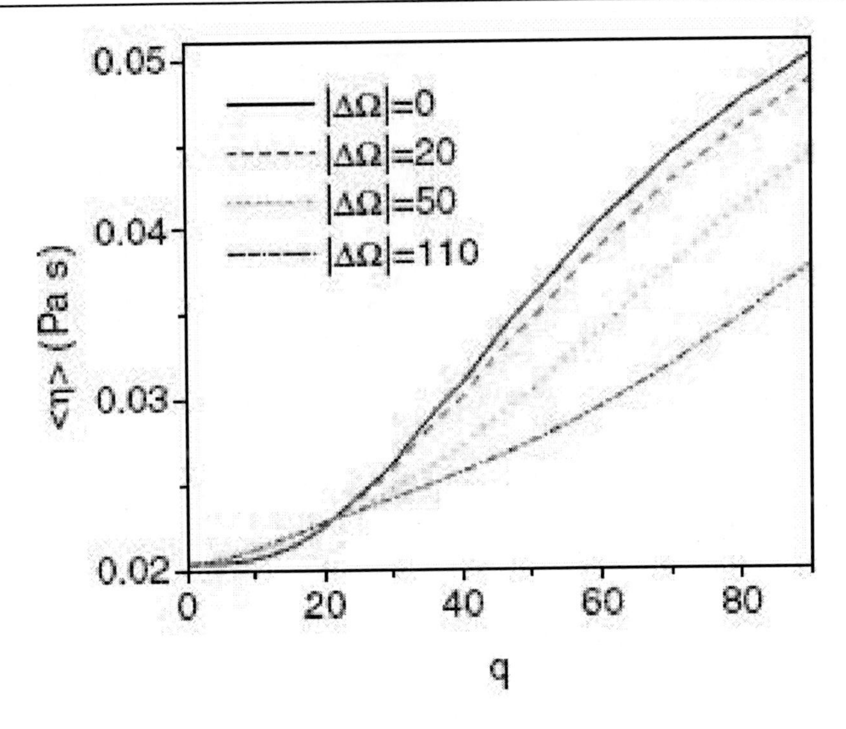

Figure 17. Averaged apparent viscosity as a function of q. for a) $q = 21$ and b) $q = 21$. The units of $\Delta\Omega$ are (rad/s).

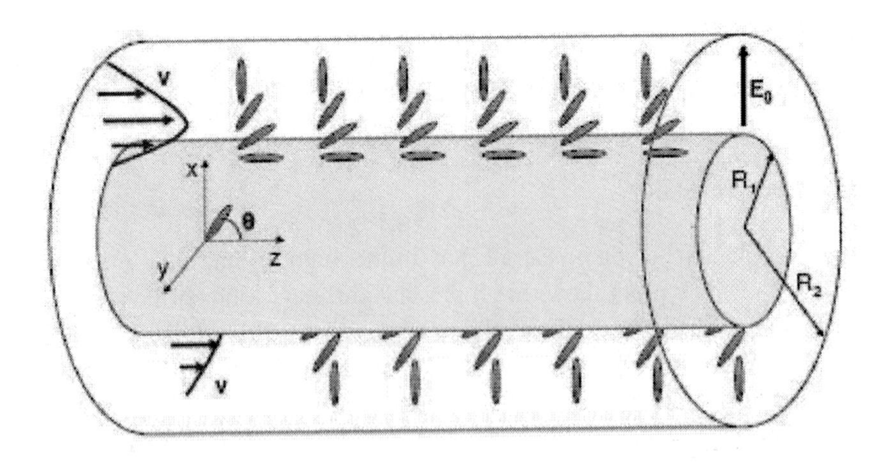

Figure 18. Schematics of a nematic liquid crystal conÞned by two coaxial cylinders and subjected to a radial electric Þfield and a pressure gradient.

pipe and varies with r so that, in a cylindrical coordinate system, the director varies as

$$\hat{\mathbf{n}} = [\sin\theta(r), 0, \cos\theta(r)], \tag{102}$$

and we assume that it satisfies the hybrid hard anchoring conditions

$$\theta(r = R_1) = 0, \quad \theta(r = R_2) = \pi/2, \tag{103}$$

where $\theta(r)$ is the orientational angle defined with respect to the $z-$axis, as shown in Fig. 18. We also assume that the nematic is subjected to a constant pressure drop along the axis of the cylinder that produces a shear flow profile along the region confined by both cylinders given by

$$\mathbf{v} = [0, 0, v_z(r)], \tag{104}$$

which satisfies the non slip boundary conditions

$$v_z(r = R_1) = 0, \;\; v_z(r = R_2) = 0. \tag{105}$$

Both, hard-anchoring and non-slip boundary conditions restrict the validity of our description to moderate values of the external agents acting against the elastic forces at the cylinder's wall.

The elastic free energy of the LC is obtained by integrating Eq. (1) over the cylindrical volume. Then, expressing $\nabla \cdot \hat{\mathbf{n}}$ and $\nabla \times \hat{\mathbf{n}}$ in cylindrical coordinates we obtain the free energy per unit length, $f \equiv \mathcal{F}/L$, with L the length of the cylinders:

$$\begin{aligned} f &= \pi K_1 R_2 \left\{ \int_{R_1/R_2}^{1} \left[\left(\frac{d\theta}{dx}\right)^2 \left(\cos^2\theta + \kappa \sin^2\theta\right) + \frac{\sin^2\theta}{x^2} \right] x dx \right. \\ & \left. -q \int_{R_1/R_2}^{1} (\sin^2\theta + \epsilon_\perp^s/\epsilon_a^s) \frac{dx}{x} \right\}, \end{aligned} \tag{106}$$

where $x = r/R_2$, and q is an important parameter defined as

$$q \equiv \epsilon_a^s \Delta\phi^2 / [K_1 \ln^2(R_2/R_1)], \tag{107}$$

where ϵ_a^s is the low frequency dielectric constant. The parameter q represents the ratio of the electric and elastic energies; for $q << 1$ the influence of the applied field is weak, while for $q >> 1$ the field essentially overcomes the Frank's elastic forces. We should mention that there is no Fredericks transition in this system since we are using the hybrid boundary conditions given by Eq. (103).

$$\begin{aligned} 0 &= \frac{d^2\theta}{dx^2} x^2 \left(\cos^2\theta + \kappa \sin^2\theta\right) + \left(\frac{d\theta}{dx}\right)^2 \frac{x^2}{2} (\kappa - 1) \sin 2\theta + \frac{q-1}{2} \sin 2\theta \\ & + x \frac{d\theta}{dx} \left(\cos^2\theta + \kappa \sin^2\theta\right) - \frac{\Lambda}{\eta(\theta)} x^2 \left[x + \frac{b'}{x} \right] (\cos 2\theta - \cos 2\theta_0), \end{aligned} \tag{108}$$

where $b' = -bR_2/(K_1\Lambda)$ while for v_z we obtain

$$v_z(x) = v_0 \Lambda \left[\int_x^1 \frac{s}{\eta(\theta(s))} ds + b'' \int_x^1 \frac{ds}{s\eta(\theta(s))} \right], \tag{109}$$

with

$$b'' \equiv - \frac{\int_{R_1/R_2}^1 \frac{s}{\eta(\theta(s))} ds}{\int_{R_1/R_2}^1 \frac{ds}{s\eta(\theta(s))}}, \tag{110}$$

and $v_0 \equiv 2K_1/R_2$, $\zeta \equiv z/R_2$.. Here, we have applied the boundary conditions given by Eqs. (105) to obtain the velocity profile Eq. (109). In these equations $\cos 2\theta_0 = -\gamma_1/\gamma_2$ and $\eta(\theta)$ was defined in Eq. (41).

In the previous equations, Λ is a dimensionless parameter defined as

$$\Lambda \equiv \frac{1}{2}\frac{dp}{d\zeta}\frac{R_2^2}{K_1}, \tag{111}$$

and represents the ratio of the hydrodynamic and elastic energies; for $\Lambda << 1$ the influence of the applied pressure gradient is weak, while for $\Lambda >> 1$ the flow essentially overcomes the elastic energy. Note that the influence of the pressure gradient is greatly augmented for a large pipe radius R_2.

Once $\theta\,(x)$ has been determined numerically from Eqs. (108) and (103) it can be inserted into Eq. (109) to obtain numerically $v_z\,(x)$.

In what follows, we will present results for the specific case of 4'-n-pentyl-4-cyanobiphenyl (5CB) liquid crystal. The parameters used were $T_{IN} - T = 10$ °C with $T_{IN} = 35$ °C, $\kappa = 1.316$, $K_1 = 1.2 \times 10^{-11}$ N, $\alpha_1 = -.0060$ Pa-s, $\alpha_2 = -.0812$ Pa-s, $\alpha_3 = -.0036$ Pa-s, $\alpha_4 = .0652$ Pa-s, $\alpha_5 = .0640$ Pa-s, $\alpha_6 = -.0208$ Pa-s, $\gamma_1 = .0777$ Pa-s, $\gamma_2 = -.0848$ Pa-s [?], and $R_1/R_2 = 0.5$.

It is convenient to remark that the working temperature is not near enough from the transition temperature to expect critical fluctuations. However, thermal fluctuations in $\widehat{n}$ are present and observable by the scattering of optical fields. This effects is considerably reduced by the presence of the imposed low frequency electric field which is used to decrease the correlation length in $\widehat{n}$ [?].

6.1. Biased Electrorheological Effect

The viscosity function or apparent viscosity, η, is the ratio between the shear stress and the local strain rate. It is a function of the nematic's director by means of the expression Eq. (41) [?] [?]. Since the orientational angle θ is strongly dependent on the electric field and since η depends on θ, it follows that the behavior of the system is non-Newtonian[?].

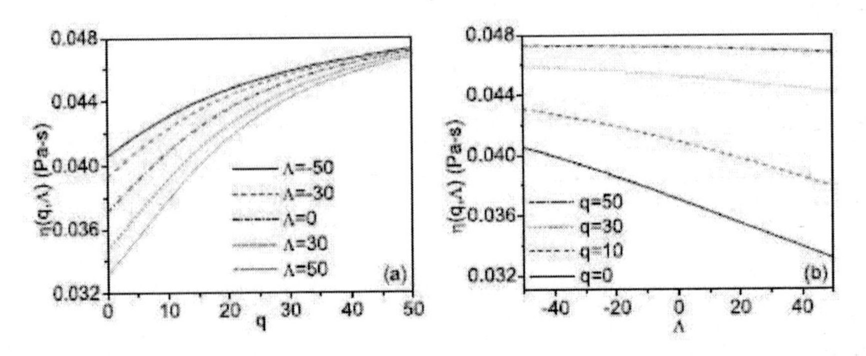

Figure 19. Averaged apparent viscosity as a function of a) the electric field q and b) the pressure gradient Λ.

If we now plot $\overline{\eta}(q,\Lambda)$ as given by Eq. (42) and as a function of q and Λ, we get the results displayed in Fig. 19. Panel (a) shows that $\overline{\eta}$ always increases as a function of the electric field q for any given pressure gradient Λ. Furthermore, for the largest value of q considered ($q = 50$), the value of $\overline{\eta}$ for the largest backward flow ($\Lambda = 50$) increases about 50% with respect to its value in the absence of electric field, whereas it increases only about 20% for the largest forward flow ($\Lambda = -50$). In this sense we can say that electrorheological effects are more evident for backward flow than for forward flow. The reason why the viscosity increases with increasing electric field is that the nematic's director is more aligned with the direction of the field that is perpendicular to the direction of flow. On the other hand, in Fig. 19b, we observe that for a given value of the electric field and for the range of flow considered ($-50 < \Lambda < 50$) the viscosity decreases as Λ increases. This means that for backward flow ($\Lambda > 0$) the viscosity decreases as the magnitude of the flow increases whereas for forward flow ($\Lambda < 0$) the viscosity increases as the magnitude of the flow increases. Therefore, we have flow thinning in one direction and flow thickening in the other. This directional response is due to the fact that the initial undistorted nematic configuration is asymmetrical. Therefore, for backward flow this configuration is distorted so that nematic's molecules are more paralleled oriented in the direction of the flow decreasing the viscosity while in the forward flow the nematic's molecules adopt a more perpendicular position with respect to the direction of the flow increasing the viscosity. Also, in the forward case most of the mechanical energy is elastically accumulated in deforming the nematic's configuration instead of being used to move the fluid, as compared to the backward case. In this sense the undistorted configuration is playing the role of a biased spring inherent to the liquid, which is stiffer in one direction than in the other.

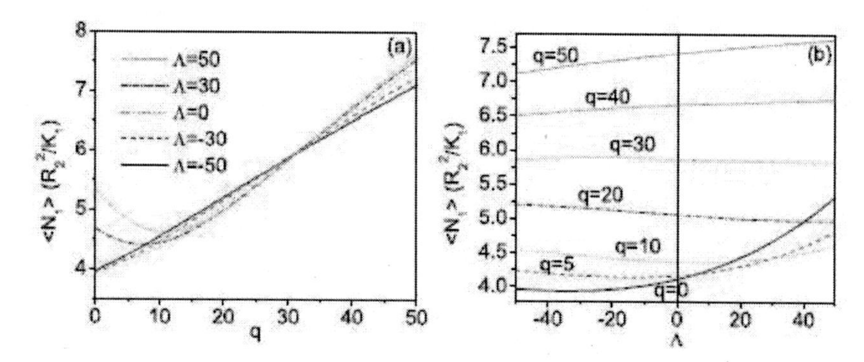

Figure 20. Averaged first normal stress difference as a function of a) q and b)Λ.

6.2. First Normal Stress Difference

Under conditions of shear flow, non-Newtonian fluids usually show positive steady-state first normal stress difference, N_1, over a range of shear rates. Also, N_1 is zero or positive for isotropic fluids at stationary flows over all shear rates. In liquid crystalline solutions, positive normal stress differences are found at low and high shear rates, while negative values occur at intermediate shear rates [?].

On the other hand, Marrucci et al [?] have solved a two dimensional version of the Doi model for nematics [?] in which the molecules are assumed to lie in the plane perpendicular to the vorticity axis, that is, in the plane parallel to both, the direction of the velocity and the direction of the velocity gradient. Despite this simplification, for strong enough shear rates over which N_1 is negative, Doi model is in excellent agreement with observations. This result suggests the possibility that negative first normal stress differences may be produced in a two dimensional flow. Nevertheless, this is not the case for the Leslie-Ericksen approach adopted in this paper as we shall see.

Let us now examine the effects produced by the stresses generated after the reorientation process has taken place by calculating the viscometric functions which relate the shear and normal stress differences. For the geometry under consideration and using the convention of Ref. [?], the first normal stress difference is defined by Eq. (43).Inserting Eqs. (102), (104) (108) and (109) into Eqs. (44) and (43) we get

$$\begin{aligned} N_1 &= \frac{K_1}{R_2^2}\left(\frac{d\theta}{dx}\right)^2\left(\cos^2\theta+\kappa\sin^2\theta\right)+\frac{1}{2R_2}\frac{dv_z(x)}{dx}\left\{\left[\alpha_1(\sin\theta-\cos\theta)+\alpha_5+\alpha_6\right]\right. \\ &\left.\times\cos 2\theta-\lambda\left(\alpha_2+\alpha_3\right)\left[\sin\theta+\cos\theta+\cos 2\theta-\cos 2\theta_0\right]\sin 2\theta\right\}, \end{aligned} \tag{112}$$

which has been expressed in terms of the already calculated quantities $v_z(x)$ and $\theta(x)$.

The integration of the first normal stress difference profile over the cross section of the pipe,renders the averaged first normal stress difference as defined in Eq. (45). This is shown in Figs. 20a and 20b where we have plotted $\overline{N_1}$ as a function of q and Λ. As can be seen from Fig. 20a, $\overline{N_1}$ depends almost linearly on q for backward flow, $\Lambda = -50$, whereas it exhibits a minimum at $q = 10$ for forward flow $\Lambda = 50$. Fig 20b displays clearly how contrasting is the difference between forward and backward flows for small values of the electric field q where a local minimum moves to the right as Λ increases. Also, it is notorious that the degree of the directional dependent non-Newtonian behavior of this confined nematic, can be electrically controlled. We should mention that even though there is a small region near the outer cylinder for which N_1 has small negative values, $\overline{N_1}(q,\Lambda)$ which is a quantity providing the global behavior, is always positive because those regions represent only a small portion of the whole section of the liquid crystal.

It is also useful to calculate the only nonvanishing stress component σ_{rz} from Eq. (44). We find, after doing similar manipulations to those used to obtain Eq. (75), that

$$\begin{aligned} \sigma_{rz} &= \frac{1}{2R_2}\left(\alpha_1\sin^2\theta\cos^2\theta+\alpha_4+\alpha_5\sin^2\theta+\alpha_6\cos^2\theta\right)\frac{dv_z(x)}{dx}+\frac{1}{2}\left(\alpha_2-\alpha_3\right)\times \\ &\sin 2\theta\left[\frac{\lambda}{2}\left(\sin\theta+\cos\theta\right)\frac{dv_z(x)}{dx}+\frac{\Lambda}{g(\theta)}\left[x+\frac{b'}{x}\right]\left(\cos 2\theta-\cos 2\theta_0\right)\right]. \end{aligned} \tag{113}$$

7. Optical Phenomena

In this chapter we present both theoretical and experimental optical studies performed in confined mesophases as planar cells, droplets and optical guides of nematic or smectic liquid crystal.

There are general procedures for describing both acoustic [?] and optics [?] of anisotropic media by using tensor eikonal, however, they do not take advantage of the complete representation given by transverse electric (TE) and transverse magnetic (TM) modes, which considerably simplify the treatment of nonmagnetic media. In the next section, a geometrical formalism to study the ray propagation in a general linear nonmagnetic media is introduced. This formalism uses the TM and TE mode representation that indeed gives a complete representation. More precisely, TM and TE eikonal equations are derived and their ray trajectories are analyzed in the corresponding Lagrangian or Hamiltonian representation. This formalism allows to give a complete description of the propagation of light in systems as the mentioned above.

The study of propagation of light in droplets is interesting due to the electro-optic effect of polymer-dispersed liquid PDLC's with potential applicability in different devices [?] [?]. This effect consists in a change of appearance of the PDLC cell from turbid to transparent, after applying a low frequency electric field to the cell plates. Hence, dispersions of liquid crystal rich droplets in a polymer matrix [?], has brought about a great deal of interest in the study of light propagation in spherical geometries. Some studies in this line are presented in Sec. 7.2.

On the other hand, the study of the hybrid aligned nematic (HAN) cell is extremely important to the liquid crystal (LC) displays industry as the HAN state forms one of the two states in a zenithal bistable device [?] [?]. In a HAN cell, one surface is treated in order to reduce homogeneous alignment of the LC director and on the opposite surface a homeotropically aligning treatment is used. The director profile through the cell changes continuously from one surface to the other. This profile is highly deformed by the application of an electric field and it has been found that this electro-optical effect is more applicable to multicolor LC displays than those of the perpendicular and planar alignments due to their low operating voltage, good color separation, and highly uniform and bright color generation.

Recent studies [?] [?] [?] have applied a geometrical optics formalism to calculate the ray tracing and total internal reflection in HAN cells under the influence of an applied electric field. Moreover, the application of the total internal reflection (TIR) techniques for studying optical properties of nematic cells with an inhomogeneous distribution of the director field are of interest since orientational configurations in hybrid cells may be characterized and identified in terms of the behavior of the reflectivity versus angle of incidence curves [?]. In this chapter such studies giving an insight in propagation of light in HAN cells are presented.

Finally, some theoretically studies of light propagating in smectic bend core liquid crystal fibers are presented, since their optical properties show that they could be useful as optical diodes and frequency filters.

7.1. Optical Limit

In this section the eikonal equations for the TM and TE rays for a general nonmagnetic linear medium are presented, thus, ray trajectories could be analyzed in the corresponding Lagrangian or Hamiltonian representation.

7.1.1. Eikonal Equations

A general time-harmonic field in a nonconducting inhomogeneous and anisotropic medium is considered. Maxwell's equations, in regions free from currents and charges, are given by

$$\nabla \times \mathbf{H} + ik_0\epsilon \cdot \mathbf{E} = 0, \quad \nabla \cdot (\mu \mathbf{H}) = 0 \ , \tag{114}$$

$$\nabla \times \mathbf{E} - ik_0\mu \cdot \mathbf{H} = 0, \quad \nabla \cdot (\epsilon \cdot \mathbf{E}) = 0 \ , \tag{115}$$

where $k_0 = \omega/c$, μ is the magnetic susceptibility and ϵ is the dielectric tensor. Following the usual procedure we assume the following trial form for the fields

$$\mathbf{E} = \mathbf{e}(\mathbf{r})e^{ik_0lW(\mathbf{r})} \quad \mathbf{H} = \mathbf{h}(\mathbf{r})e^{ik_0lW(\mathbf{r})} \ , \tag{116}$$

where $lW(r)$ is the characteristic function of Hamilton which is equal to the difference in optical paths of a ray propagating between two fixed points in the medium, and $\mathbf{e}(\mathbf{r})$, $\mathbf{h}(\mathbf{r})$ are vector functions of the position. Substituting Eq. (116) into Eqs. (114), (115) ad taking the limit of the geometrical optics, $k_0 l \gg 1$, we have

$$\nabla W \times \mathbf{h} + \epsilon \cdot \mathbf{e} = 0 \tag{117}$$

$$\nabla W \times \mathbf{e} - \mu \cdot \mathbf{h} = 0 \tag{118}$$

$$\nabla W \cdot \epsilon \cdot \mathbf{e} = 0 \tag{119}$$

and

$$\nabla W \cdot \mathbf{h} = 0. \tag{120}$$

Thus, the anisotropic nature of the medium expressed by the tensor ϵ, makes W sensible for the polarization of the fields $\mathbf{e}$ and $\mathbf{h}$ as happens for anisotropic elastic media [**?**].

Here, we use an approach, that is exclusive for electromagnetic systems and consists of taking advantage of the complete representation given by the sets of TE and TM modes [**?**], where the only electric or magnetic component, respectively, is transverse. In the case of the TE modes whose only magnetic component is $\mathbf{h}$. Substituting $\mathbf{h}$ from equation (118) into (117), yields

$$(\nabla W_{TE} \times \hat{\mathbf{e}})^2 = \mu\epsilon_{ee} \tag{121}$$

where $\hat{\mathbf{e}} = \mathbf{e}/e$, $e = |\mathbf{e}|$ and $\epsilon_{ee} = \hat{\mathbf{e}} \cdot \epsilon \cdot \hat{\mathbf{e}}$. Then, for the TM modes, equation (117) is solved for $\mathbf{e}$ and the solutions is inserted in equation (118), to obtain

$$(\nabla W_{TM} \times \hat{\mathbf{h}}) \cdot \epsilon^{-1} \cdot (\nabla W_{TM} \times \hat{\mathbf{h}}) = \mu \tag{122}$$

where $\hat{\mathbf{h}} = \mathbf{h}/h$, $h = |\mathbf{h}|$ and ϵ^{-1} is the inverse tensor. Equations (121) and (122) are the so-called eikonal or Hamilton-Jacobi equations corresponding to the TE and TM modes. We can simplify both these equations by expressing then in terms of an orthogonal coordinate system $\{q_1, q_2, q_3\}$. If q_1 is the coordinate along $\hat{\mathbf{h}}$ and $\hat{\mathbf{e}}$, Eqs. (121) and (122) can be rewritten as

$$\frac{1}{h_2^2}\left(\frac{\partial W_{TE}}{\partial q_2}\right)^2 + \frac{1}{h_3^2}\left(\frac{\partial W_{TE}}{\partial q_3}\right)^2 = \mu\epsilon_{11} \tag{123}$$

and

$$\frac{\epsilon_{22}^{-1}}{h_3^2}\left(\frac{\partial W_{TM}}{\partial q_3}\right)^2 - 2\frac{\epsilon_{23}^{-1}}{h_2 h_3}\left(\frac{\partial W_{TM}}{\partial q_2}\right)\left(\frac{\partial W_{TM}}{\partial q_3}\right) + \frac{\epsilon_{33}^{-1}}{h_2^2}\left(\frac{\partial W_{TM}}{\partial q_2}\right)^2 = \mu \tag{124}$$

where h_i, ($i = 2, 3$) are the scale factors and ϵ_{ij}^{-1} are the elements of ϵ^{-1}. Notice that, Eq. (123) show that the TE rays propagate as in an isotropic medium with and index refraction $n^2 = \mu\epsilon_{11}$; which is a consequence of having a unique electric component. In contrast , equation (124) has a different structure which implies a distinct behavior for the TM rays. Similar and more complicated eikonal equations have been found in the context of geometrical acoustics where there exists three independent polarizations, one quasi-longitudinal and two quasi-transverse, that in the general case, cannot be decoupled [?].

7.1.2. Hamiltonian and Lagrange Representation

We could rewrite Eqs. (123) and (124) in the Hamiltonian representation by rewriting them in terms of the variables $p_i = \frac{\partial W}{\partial q_i}$, $i = 1, 2, 3$, which are known as the ray components. It leads to

$$\frac{p_2^2}{h_2^2} + \frac{p_3^2}{h_3^2} = \mu\epsilon_{11} \tag{125}$$

and

$$\frac{\epsilon_{33}^{-1}}{h_2^2}p_2^2 - 2\frac{\epsilon_{23}^{-1}}{h_2 h_3}p_2 p_3 + \frac{\epsilon_{22}^{-1}}{h_3^2}p_3^2 = \mu. \tag{126}$$

It is worth pointing out that from the Hamiltonian $\mathcal{H} = \mu$, we can calculate the four first order differential equations known as Hamiltonian's equations of which three are independent, since equation (126) allows to write one in terms of the others. We can determine the ray trajectories of a particular system from these equations and their boundary conditions. In this section we shall review the Lagrangian formulation of the TM modes which provide us with a useful insight into the understanding of the behavior of its ray trajectories. Hamilton's equation yields

$$(\dot{q}_2 \dot{q}_3) = 2\begin{bmatrix} \frac{\epsilon_{33}^{-1}}{h_2^2} & -\frac{\epsilon_{23}^{-1}}{h_2 h_3} \\ -\frac{\epsilon_{23}^{-1}}{h_2 h_3} & \frac{\epsilon_{22}^{-1}}{h_3^2} \end{bmatrix}\begin{pmatrix} p_2 \\ p_3 \end{pmatrix}.$$

Thus, the Lagrangian $\mathcal{L}$ associated to $\mathcal{H}$ is obtained by using the Legendre transform. It leads to

$$\mathcal{L} = \epsilon_{33}^{-1}h_2^2\dot{q}_2^2 - 2\epsilon_{23}^{-1}h_2 h_3 \dot{q}_2\dot{q}_3 + \epsilon_{22}^{-1}h_3^2\dot{q}_3^2 \tag{127}$$

where $\dot{q}_i = dq_i/d\tau$ are the components of a tangent vector to the trajectory. A ray whose Lagrangian is given by equation (127) does not experience a position dependent refraction index, instead it has an anisotropic kinetic energy in a curved space. Nevertheless, for the case of a uniaxial liquid crystal [?] where $\epsilon_{ij} = \epsilon_\perp \delta_{ij} + \epsilon_a n_i n_j$, we can interpret the system as the 2D projection in a plane described by q_2 and q_3, of a 3D ray propagation whose degrees of freedom are q_2, q_3 and z and which is confined by a constriction of the form $dz = f_2 dq_2 + f_3 dq_3$, where f_2 and f_3 are two functions of q_2 and q_3. Thus, the Lagrangian of the system could be written in the following way:

$$\epsilon_\perp \mathcal{L} = \dot{z}^2 + h_2^2 \dot{q}_2^2 + h_3^2 \dot{q}_3^2 - \lambda(\dot{z} - f_2(q_1, q_3)\dot{q}_2 - f_3(q_2, q_3)\dot{q}_3) \tag{128}$$

where λ is the Lagrange multiplier. Here, f_2 and f_3 are determined by substituting z in terms of q_2 and q_3 given by the constriction and by comparing its terms with the ones of Eq. (127). This leads to the explicit expression for the constriction, given by

$$dz = -\sqrt{\frac{\epsilon_a}{\epsilon_\|}}(n_3 h_2 dq_2 - n_2 h_3 dq_3). \tag{129}$$

By applying the Lagrange equation to the variables λ and z, we find the next expression for λ

$$\lambda = \lambda_0 - 2\sqrt{\frac{\epsilon_a}{\epsilon_\perp}}\vec{v}_\perp \cdot \hat{n}_\perp \tag{130}$$

where λ_0 is an integration constant and $\vec{v}_\perp = h_2 \dot{q}_2 \hat{e}_2 + h_3 \dot{q}_3 \hat{e}_3$ is the velocity component in the $q_1 - q_2$ plane, and $\hat{n}_\perp = n_1 \hat{e}_1 - n_2 \hat{e}_2$.

Due to that the constriction given by Eq. (129) depends on the configuration angle ψ, in general the term dz is not an exact differential $\oint dz \neq 0$, and as a consequence it is not always feasible to find a surface of the form $z = z(q_1, q_2)$, which means that Eq. (129) is a nonholomic constriction [?]. However, Pfaffian theory [?] establishes that for a differential with two independent variables, it is always possible to find and integration factor $\eta(q_1, q_2)$, such that the ratio $d\Gamma = dz/\eta(q_1, q_2)$ is an exact differential $\oint d\Gamma = 0$, then, there exists a surface of the form $\Gamma = \Gamma(q_1, q_2)$. Hence, in terms of the new variable Γ which makes the constriction holonomic, the Lagrangian of the particle adopts the following form:

$$\mathcal{L} = \eta(q_1, q_2)\dot{\Gamma}^2 + h_1^2 \dot{q}_1^2 + h_2^2 \dot{q}_2^2 - \lambda^*(\Gamma - \Gamma(q_1, q_2)), \tag{131}$$

It is worth mentioning that in terms of the new variable Γ, $\eta(q_1, a_2)$ plays the role of its scale factor. The surface can be constructed graphically, by rewriting equation (129) as

$$\vec{v} \cdot \left(\Gamma \hat{z} + \sqrt{\frac{\epsilon_a}{\epsilon_\perp}}\hat{n}_\perp\right) = 0 \tag{132}$$

where $\vec{v} = \dot{\Gamma}\hat{z} + h_1 \dot{q}_1 \hat{e}_1 + h_2 \dot{q}_2 \hat{e}_2$ is the tangent vector to the ray. and $\mu \hat{z} + \sqrt{\epsilon_a/\epsilon_\perp}\hat{n}_\perp$ is the normal vector to the surface.

7.2. Light Propagation in Nematic Droplets

The size of spherical droplets in polymer-dispersed liquid crystals (PDLC's) is usually uniform but can vary between 0.1 and 10 μm. The nematic configuration within droplets depends on surface anchoring and elastic constants, and is responsible for the refractive and birefringent properties of the droplets. The refractive effects are expected to be important in birefringent and inhomogeneous media like liquid crystals even for moderate dielectric anisotropy values, and are able to curve the trajectory of light beams as well as induce local changes of phase instead of global changes which can develop in richer diffraction patterns.

It is possible to study the refractive effect by analyzing the dynamics of transverse modes which propagate in a nematic droplet, in the limit of geometrical optics. Thus, the ray trajectories for both the radial and bipolar configurations, as well as some other physical parameters involved may be calculated and analyzed.

7.2.1. Bipolar Configuration

We consider a spherical nematic droplet of radius R with hard-anchoring bipolar boundary conditions [?]. Fig. 21 shows the nematic director and transverse magnetic modes in the droplet, where the origin of the coordinate system is at the center of the droplet, and for which the bipolar axis is parallel to the z axis. The nematic director can be expressed as

$$\hat{n} = \sin\psi(r,\theta)\hat{e}_r + \cos\psi(r,\theta)\hat{e}_\theta \tag{133}$$

where θ is the polar angle of the spherical coordinates, the bipolar boundary condition is given by $\hat{n}(r = R, \theta) = \hat{e}_\theta$, where $\hat{e}_\theta$ is the unit vector in the direction of increasing θ; the angle ψ is measured from $\hat{e}_\theta$ and contained in the plane define by this vector and the unit vector in the direction of increasing r: $\hat{e}_r$, using the azimuth symmetry of this nematic configuration and the boundary condition given by Eq. (133) imposes the condition $\psi(r = R, \theta) = 0$.

For droplets whose radii R are larger than $R > 100$ nm, surface- induced changes in the nematic order parameter can be neglected [?], and the nematic director $\hat{n}$ can be determined by minimizing Frank's free energy Eq. (1), with the equal elastic constant approximation $K = K_{11} = K_{22} = K_{33}, K_{24} = 0$ and the surface energy is defined due to strong boundary conditions, where $\psi(r = R) = 0$, and $\hat{n}(\theta = 0, \pi, \pi/2) = -\hat{k}$. This means that $\hat{n}$ is parallel to $-\hat{k}$ along the z axis ($\theta = 0, \pi$) and on the equator plane ($\theta = \pi/2$), which is the expected behavior for this configuration. Thus, the configuration of the nematic director is numerically calculated [?] using the shooting method in order to satisfies the corresponding boundary conditions, for a $50 - \mu m$ droplet:

$$\hat{n} = \frac{z\rho\hat{e}_\rho + (\rho^2 - \sqrt{z^2+\rho^2})\hat{k}}{\sqrt{\rho^2(\sqrt{z^2+\rho^2} - 1)^2 + z^2}} \tag{134}$$

where $z = r\cos\theta$ and $\rho = r\sin\theta$ are the cylindrical coordinates, and $\hat{e}_\rho$ is the unit vector in the increasing direction of ρ. Thus, the director lines are

$$\frac{d\rho}{dz} = \frac{n_\rho}{n_z} = \frac{z\rho}{\rho^2 - \sqrt{z^2+\rho^2}}. \tag{135}$$

Fig. 22 shows some director lines given by Eq. (135) that are drawn equally separated at the equator plane ($z = 0$).

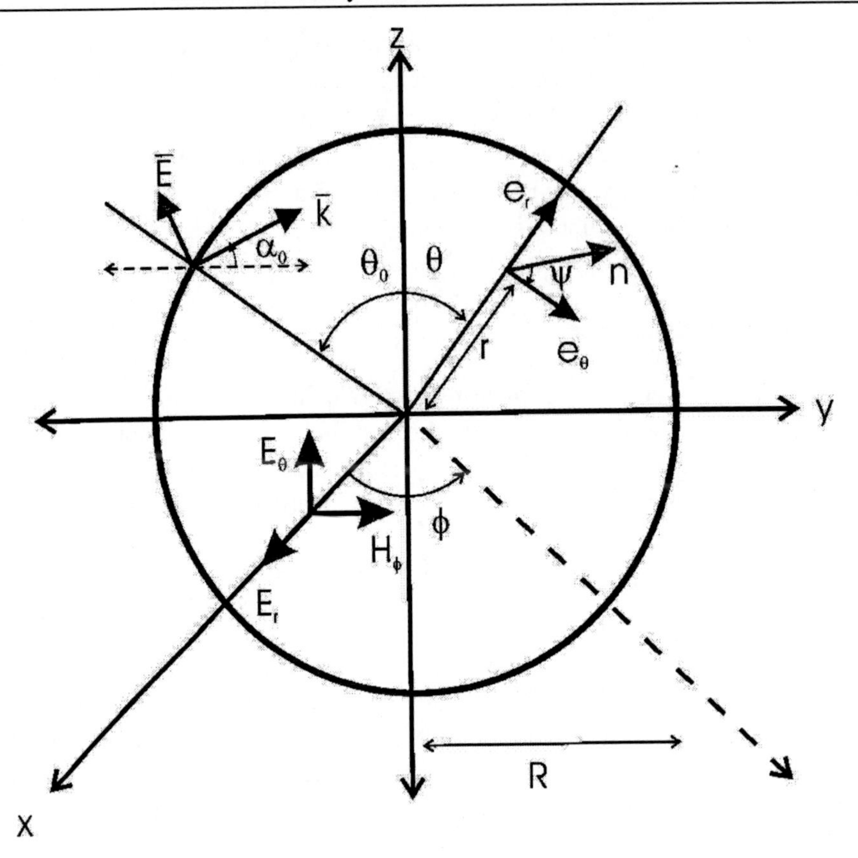

Figure 21. Diagram of a nematic droplet where the nematic director and the transverse magnetic modes are shown [?] Copyright (2010) by The American Physical Society.

7.2.2. Ray Trajectories for Radial Configuration

The radial configuration is represented by Eq. (133) taking $\psi = \pi/2$. The ray trajectories are calculated using the formalism developed in Sec. 7.1. [?], and Eq. (124) that in this case becomes

$$\epsilon_{rr}\left(\frac{\partial W}{\partial x}\right)^2 + 2\frac{\epsilon_{r\theta}}{x}\left(\frac{\partial W}{\partial \theta}\right)\left(\frac{\partial W}{\partial x}\right) + \frac{\epsilon_{\theta\theta}}{x^2}\left(\frac{\partial W}{\partial \theta}\right)^2 = \epsilon_\perp \epsilon_\parallel \tag{136}$$

does not depend explicitly on θ, so that we can separate this variable by using the canonical transformation $W = P(x) - q_0\theta$, where $P(x)$ is a function which depends only on $x \equiv r/R$, and q_0 is the angular momentum which is conserved. Hence W is given by

$$W = -q_0\theta + \left[\epsilon_\perp\left(x^2 - \frac{q_0^2}{\epsilon_\parallel}\right)\right]^{1/2} + q_0\left(\frac{\epsilon_\parallel}{\epsilon_\perp}\right)^{1/2} \arccos\left[\frac{q_0}{x\sqrt{\epsilon_\parallel}}\right] + P_0 \tag{137}$$

According to the Hamilton-Jacobi theory, the ray trajectory is given by $\gamma = \partial W(x, \theta, q_0))/\partial q_0$, where γ is the variable conjugated to q_0, which is invariant in time; that is , an initial condition or constant of motion. Furthermore, it lead to [?]

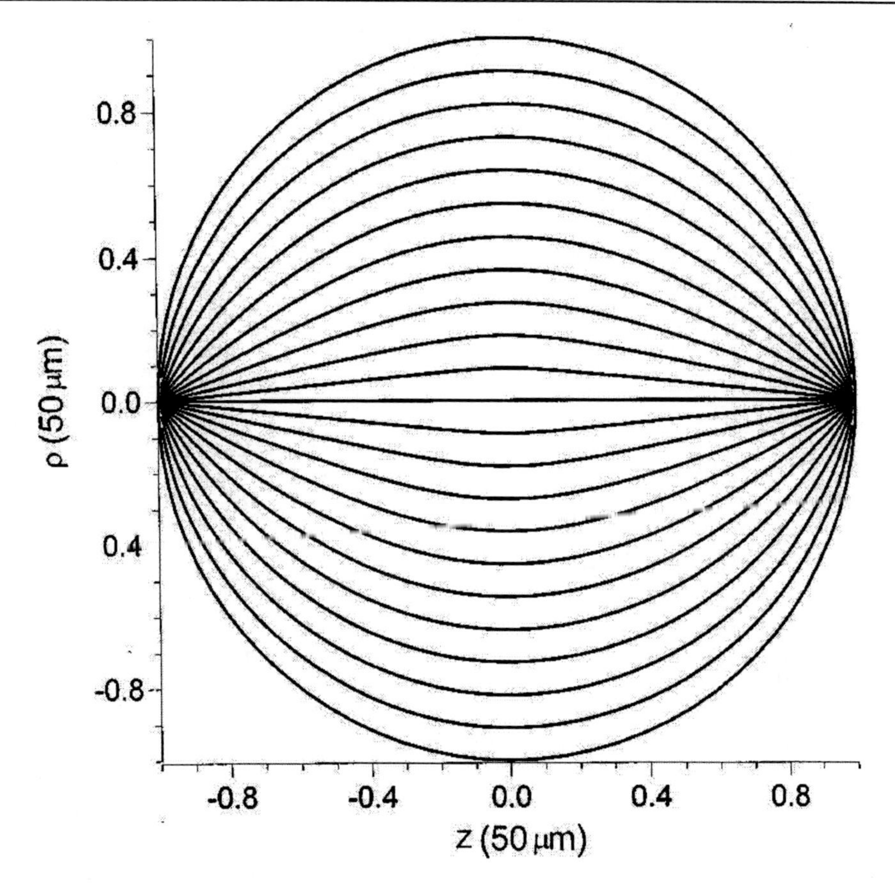

Figure 22. Director lines for various equally separated initial condition at the equator plane [**?**] Copyright (2010) by The American Physical Society.

$$r = \frac{q_0 R / \sqrt{\epsilon_{\|}}}{\cos[\sqrt{\epsilon_{\|}/\epsilon_{\perp}}(\theta - \gamma)]} \tag{138}$$

where, both q_0 and γ can be expressed in terms of the initial θ_0 and α_0 angles at the sphere border: $r = R = 50\ \mu m$ (Fig. 21). Then we can derive [**?**]

$$q_0 = \frac{\sqrt{\epsilon_{\|}}\cos(\alpha_0 + \theta_0)}{\sqrt{(\epsilon_{\perp}/\epsilon_{\|})\sin^2(\alpha_0 + \theta_0) + \cos^2(\alpha_0 + \theta_0)}} \tag{139}$$

$$\gamma = \theta_0 - \left(\frac{\epsilon_{\perp}}{\epsilon_{\|}}\right)^2 \arccos\left[\frac{q_0}{\sqrt{\epsilon_{\|}}}\right], \tag{140}$$

where $\tan\alpha_0 = dz/d\rho$, $z = r\cos\theta$ and $\rho = r\sin\theta$. The form of the trajectories can be inferred from Eq. (139). Thus, when the rays, which start from the right are plotted, initially they are parallel to each other, and are equally separated over the sphere border; for the incident angles, $\alpha_0 = 0$. Here the nematic droplet is considered of radius $R = 50\mu m$, $\epsilon_{\|} = 2.89$ and $\epsilon_{\perp} = 2.25$.

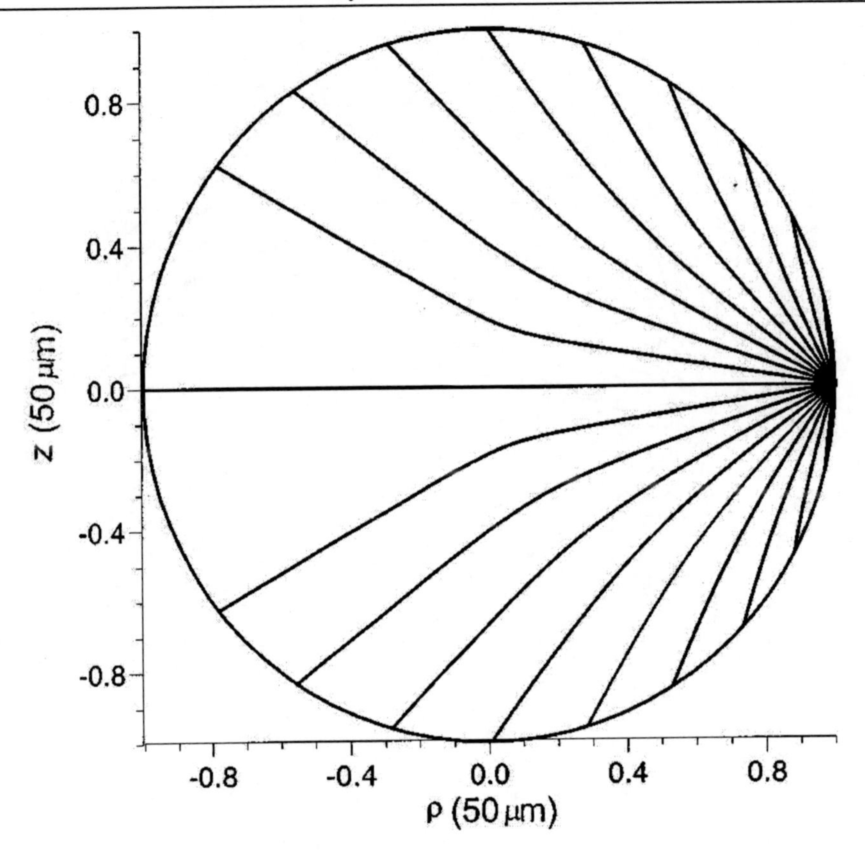

Figure 23. Ray trajectories, for the radial configuration, starting from the right, initially parallel to each other, starting from the same point θ_0 for various equally separated values of α_0 , namely, $\alpha = 0, \pm\pi/20, \pm\pi/10$, ..., and $\pm 9\pi/20$ [?] Copyright (2010) by The American Physical Society. A typical nematic cyanobiphenil for which $\epsilon_\perp = 2.25$ and $\epsilon_\| = 2.89$.

Then, the rays diverge from the equator plane such that the rays nearer to the equator are bent more than the rays which lie further from it. The same behavior is expected for any value of α because of the symmetry of the configuration and we can observe a similar behavior when we plot a set of rays with the same θ_0 and various values of α; the Fig 23 shows clearly the presence of a dark zone induced by the droplet

7.2.3. Ray Trajectories for Bipolar Configuration

In the case of the bipolar configuration, the dielectric tensor components depend on both r and θ, then Eq. (136) can be solved numerically using Charpit's equations [?] [?].

Fig. 24 shows rays which are initially perpendicular to the z axis ($\alpha = 0$), but are deflected by the droplet in such a way that they converge in a region on the equator. On the other hand, when the incidence angles α_0 are larger in increasing order, the deflected ray trajectories converge to a region that lies underneath the equator plane [?]. In Fig. 25 we can see rays that are almost straight lines, for which their deflection angles are small.

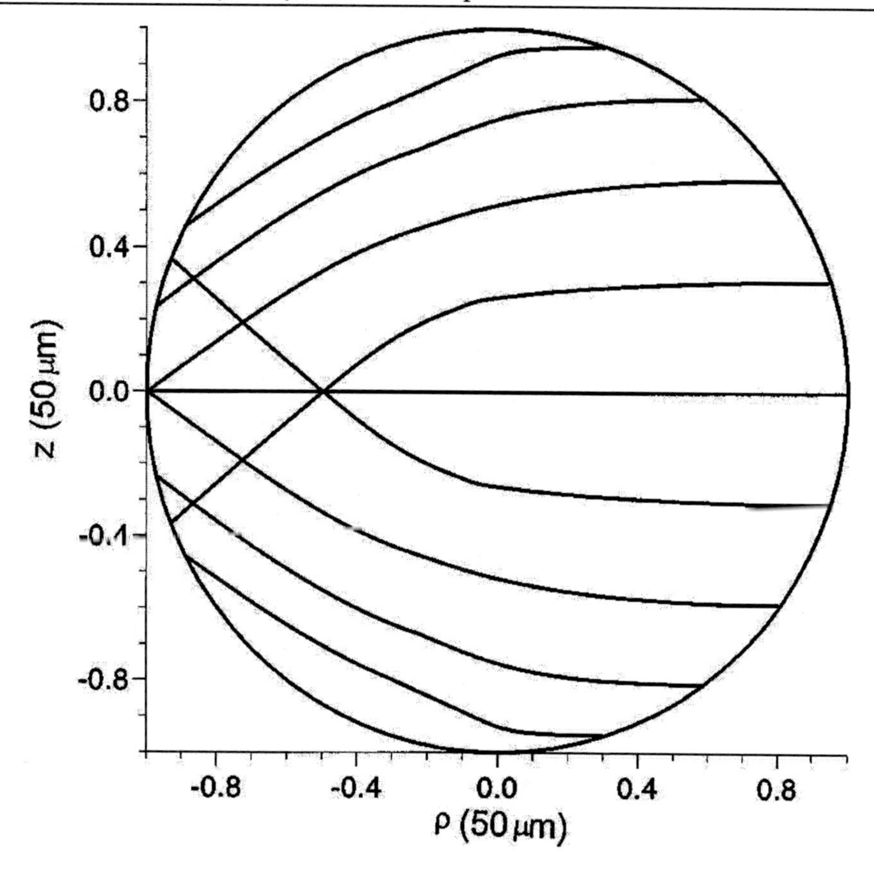

Figure 24. Ray trajectories for the bipolar configuration starting from the right, initially parallel to each other and with equally separated values of θ_0 , namely, $\theta_0 = \pm\pi/8, \pm\pi/4,$ ..., and $\pm\pi/2$; for the incident angle $\alpha_0 = \pi/2$. We use the same substance values as Fig. 23 [?] Copyright (2010) by The American Physical Society.

This means that just in the case when the incident beams are parallel to the bipolar axis, the anomalous-diffraction approach assumed in a previous work [?] is reliable.

Therefore, an initially parallel set of incident rays is scattered by a droplet with the radial configuration, in such a way that there exists a dark zone behind the sphere unreachable for almost every ray. but for the bipolar configuration, for various initially parallel sets of rays, the rays are bend in such a way that they converge in certain regions of the sphere except when the incident beams are parallel to the bipolar axis. It is worthwhile to say that in a geometrical previous study of light propagating in a droplet, reported in Ref. [?], it was found that the rays has an anisotropic kinetic energy as in a curved space, the constraining surfaces were calculated for both a radial and bipolar configuration, and the results are consistent with the presented in this section. In the case of the radial configuration this constriction, due to its curvature, makes that parallel rays diverge from their initial direction such that the central rays are deflected more than the rest of the rays. Moreover, besides the normal incident ray which propagates without being deviated, no one ray is able to reach a solid angle located behind the droplet, i. e., the defocusing effect of the droplet causes the presence of a dark zone behind of the sphere.

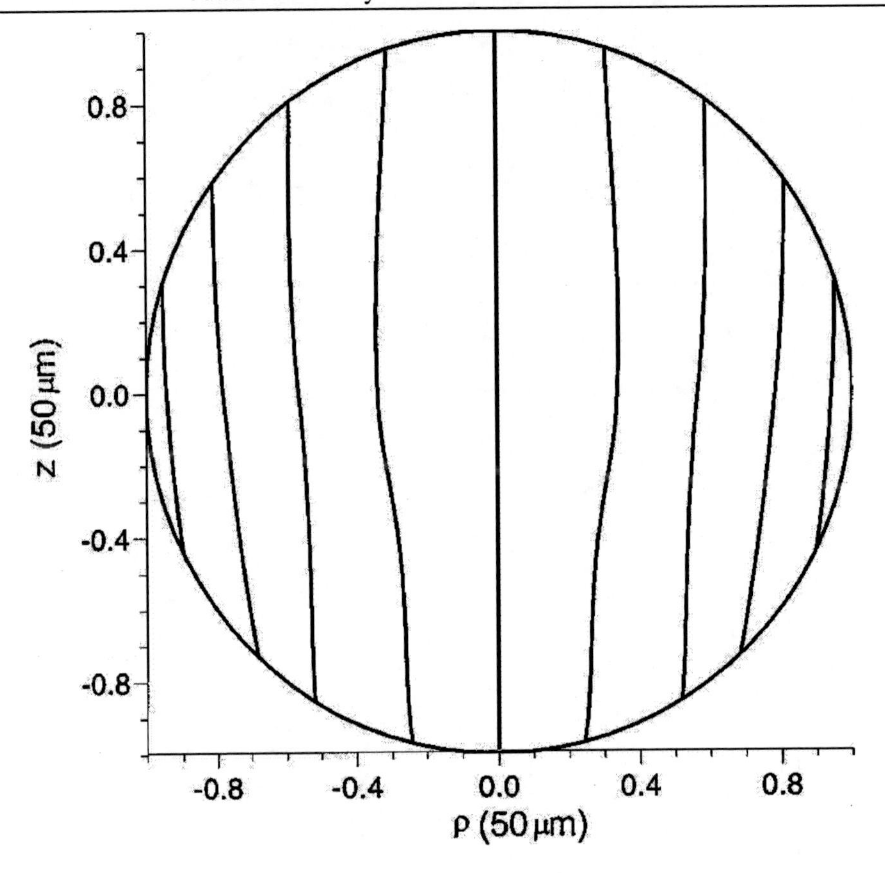

Figure 25. The same as Fig. 24, but for $\alpha_0 = \pi/2$ [?] Copyright (2010) by The American Physical Society.

7.2.4. Electrically Controlled Nematic Droplets

The refractive effects are expected to be important in birefringent and inhomogeneous media like liquid crystals even for moderate dielectric anisotropy values, and are able to curve the trajectory of light beams as well as induce local changes of phase instead of a global one, which can develop richer diffraction patterns. These refractive effects are showed in this section by calculating the ray trajectories corresponding to the transverse electric and magnetic modes when propagating in a nematic droplet in the limit of geometrical optics. The ray trajectories of an initially plane wave when it enters and emerges from a nematic droplet with homeotropic alignment and under the action of an external uniform low frequency field may be calculated, following the formalism of Sec. 7.1., taking into account Frank's energy and the electric energy when the total energy is minimized.

Radial dependence The radial configuration is as described in Sec. 7.2.2. but in this case a external field is acting. The effect of the external field on the nematic is usually opposed to the elastic force effect, particularly at frontiers of the nematic where the molecules anchor the walls. The regions of the frontier of the nematic and in the bulk have different behaviors and can be described simultaneously by using the boundary- layer technique [?] [?]. The solutions for both regions are coupled or matched asymptotically to give rise to a uniform solution valid for all the domain.

Then, it is possible to find the configurational field of the nematic director inside the droplet as [?]

$$\hat{n} = \frac{2(\rho^2+z^2)^h z\rho\hat{e}_\rho + [1+(\rho^2+z^2)^h]z^2 + \rho^2[1-(\rho^2+z^2)^h]\hat{k}}{\sqrt{\rho^2+z^2}\sqrt{z^2[1+(\rho^2+z^2)^h]^2+\rho^2[1-(\rho^2+z^2)^h]^2}} \tag{141}$$

where $z = (r/R)\cos\theta$ and $\rho = (r/R)\sin\theta$ are the cylindrical coordinates and $\hat{e}_\theta$ and $\hat{k}$ are the unit vectors in the ρ direction and z direction, respectively, and h is a dimensionless field defined by $h^2 = \epsilon_a E_0^2 R^2/(4\pi K)$ with E_0 the low frequency field amplitude, ϵ_a the dielectric anisotropy and K is the elastic constant in the equal elastic constant approximation. Thus, the director lines are

$$\frac{d\rho}{dz} = \frac{n_\rho}{n_z} = \frac{2z\rho}{[1+(\rho^2+z^2)^{-h}]z^2+\rho^2[(\rho^2+z^2)^{-h}-1]}. \tag{142}$$

By means of this formula we can exhibit geometrically the nematic configuration, and the description is similar as the electrostatic field lines, i. e., $\hat{n}$ is parallel to the tangent to the curve which touches the point where $\hat{n}$ is to be calculated. Fig. 26 shows, in the insets, some director lines from Eq (142) which are depicted for equidistant angles on the droplet frontier for various values of the dimensionless field h in increasing order. Notice that, as should be expected, the curves are more aligned with the field for larger values of h. Moreover, Eq. (141) confirms the existence of a singular ring located at the equator of the droplet ($\rho = 1$, $z = 0$ for which Eq. (141) gives an undefined director) for strong enough external fields, and strong enough anchoring.

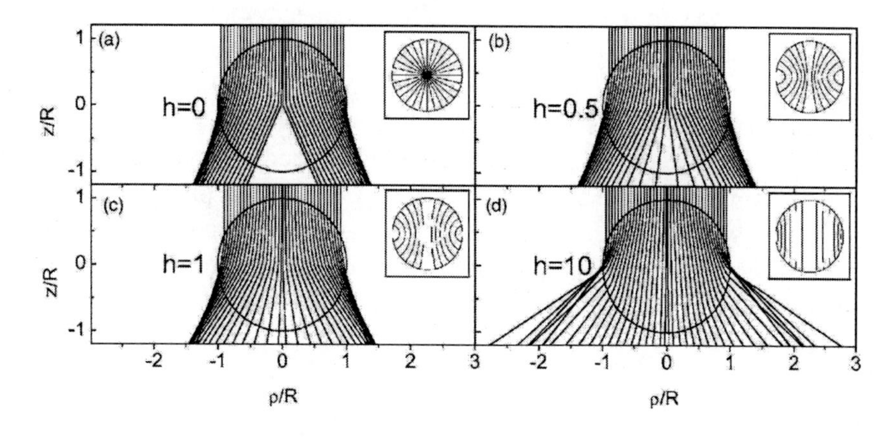

Figure 26. Ray trajectories of a set of initially parallel rays entering a droplet for various values of h. The insets show the director lines acquired by the nematic for the same values of h. The dielectric constants are $\epsilon_{\parallel} = 1.8$ and $\epsilon_{\perp} = 1.3$ which corresponds to $5CB$ at a wavelength of $632nm$ and at a temperature of $25°C$ [?] Copyright (2010) by The American Physical Society.

Light propagation and ray tracing The TM eikonal for the radial configuration can be solved analytically [?] [?]. However, for the configuration in the presence of field, the situation is different since ψ depends on two independent variables r and θ and it is therefore convenient to use the Hamiltonian representation.

The relevant components of the dielectric tensor are given by $\epsilon_{rr} = \epsilon_{\perp} + \epsilon_a \cos^2 \psi(x, \theta)$, $\epsilon_{r\theta} = -\epsilon_a \sin^2 \psi(x, \theta) \cos \psi(x, \theta)$, and $\epsilon_{\theta\theta} = \epsilon_{\perp} + \epsilon_a \sin^2 \psi(x, \theta)$ which are functions of θ and $x = r/R$, Moreover, $p = \partial W/\partial r$ and $q = \partial W/\partial \theta$ are the radial and angular components, and $H = \epsilon_{\perp}\epsilon_{\shortparallel}$ as the Hamiltonian of the system. Thus, Eq. (136) takes the form.

$$\epsilon_{rr}p^2 + 2\frac{\epsilon_{r\theta}}{x}pq + \frac{\epsilon_{\theta\theta}}{x^2}q^2 = H. \tag{143}$$

from which we can obtain the ray trajectories within the droplet by using the Hamilton equations [?].

In Fig. 26 we see the trajectories when the rays of a plane wave front coming from an isotropic medium, of dielectric constant $\epsilon_h = \epsilon_{\shortparallel}$ whose wave vector is parallel to the low frequency electric field, cross a nematic droplet and emerge from it with an emerging angle γ. It can be observed that for different values of h the shadow cone which stands under the droplet when no field is applied vanishes when a field is imposed. Besides, the ray deviations diminish for larger values of h. It is worth to mentioning that the rays crossing the droplet almost by the border are the most deviated because of the hard anchoring condition used to calculate the configuration.

These results allowed to show that a nematic droplet in the absence of a low frequency field deflects the rays crossing the droplet, leaving a shadow below it. In contrast, when the droplet is under the action of a low frequency field, the shadow disappears and the rays suffer a smaller deflection which diminishes for larger low frequency fields. Hence, if a wave front is to traverse a structure having a random arrangement of nematic droplets, the deflection of the rays in the absence of the low frequency field will repeat every time the light crosses each droplet, so that after many droplets the wave front will be divided into pieces, each one propagating in a different direction, therefore, the wave front with defined phase and direction will be scattered and no image will be seen through the matrix with droplets. On the other hand, when the field is applied, a large portion of rays will not be deviated from their initial direction so that they retain their phase and direction after crossing the matrix with droplets, which then appears as a transparent medium. Thus, these results are consistent with observed phenomenology of the electro-optical effect in PDLC cells which scatter light in the absence of applied fields while they become transparent under the action of the field.

7.3. Internal Refraction, Beam Steering and Dispersion in Planar Cells

The propagation of an optical beam in a nematic hybrid cells has been studied both experimentally and theoretically in Ref. [?] and theoretically in Refs. [?] .[?] [?].

In ref [?] the reflectance coefficient R has been measure and analytically calculated as a function of the incidence angle i for the propagation of a linearly polarized Gaussian Beam of finite diameter through a nematic hybrid cell. They propose an analytical model in the optical limit approximation mainly based in the formalism presented above. This model allows to calculate analytically the phase shift, the trajectory and the reflectivity curves of a beam travelling within an extraordinary polarization (P-polarization) inside the cell which shows good agreement with the experimental results.

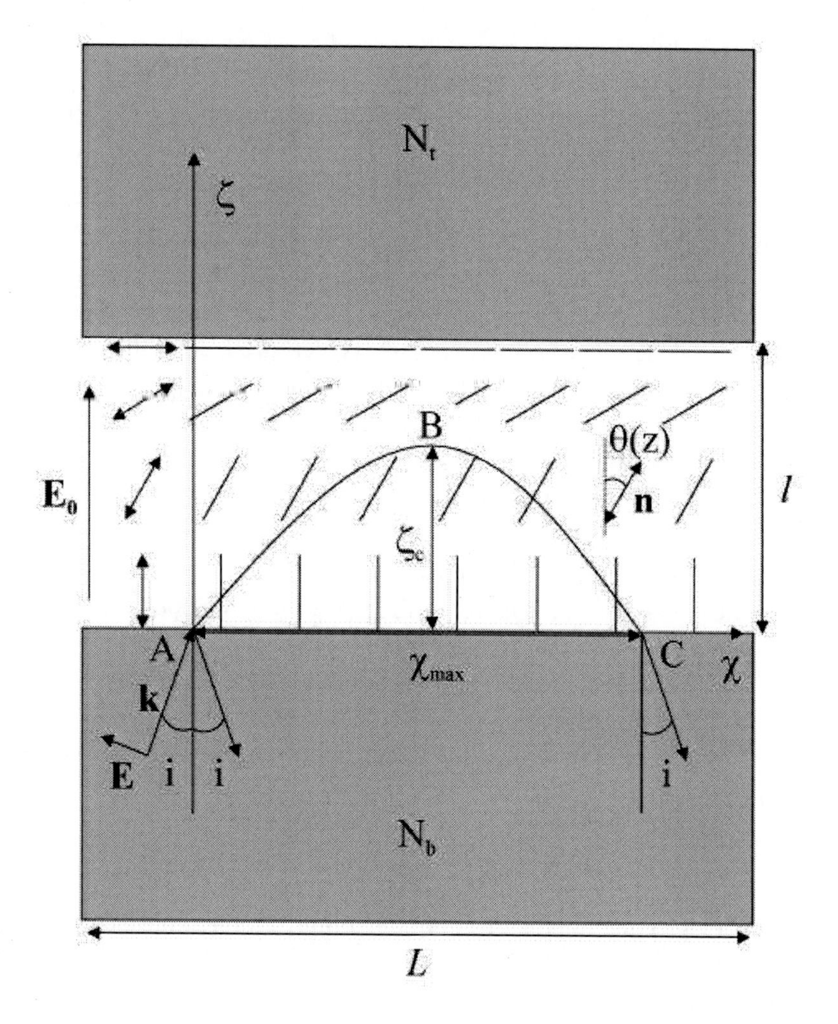

Figure 27. Schematics of the model. A TM-mode is travelling along the hybrid nematic. N_t, $N_b > n_{\|}$, $n_{\perp}$. $\theta(z)$ is the orientational angle of the director [?] Copyright (2010) by The American Physical Society.

The theory was developed considering a pure thermotropic nematic confined between two parallel substrates with refraction indices N_t and N_b, respectively, as depicted in Fig. 27. An obliquely incident laser beam with P polarization (P wave), that is, contained in the incidence plane $x - z$, impinges the nematic with an incident angle i. The intensity of the beam is low enough so that it does not distort the nematic's configuration. The dynamics of this optical field is described by the corresponding Maxwell's equations which contain the dielectric tensor ϵ_{ij}, Eq. (4), and hence depend on θ the orientational angle defined with respect to the z-axis.

The directors initial configuration is spatially homogeneous along the $x - y$ plane and varies with z so that at the boundaries the director

$$\widehat{n} = (\sin\theta(z), 0, \cos\theta(z)) \tag{144}$$

satisfies the hybrid boundary conditions

$$\theta(z=0) = 0, \tag{145}$$

$$\theta(z=l) = \pi/2. \tag{146}$$

where $\theta(z)$ is the reorientational angle defined with respect to the z axis.

By a proper treatment of the boundaries that confine the liquid crystal it is possible to achieve a configuration in which the orientation of the molecules changes continuously from one substrate to other. The cell thickness, l, is measured along the z-axis, transverse to the cell plates, this spatial dimension is small compared with the others. The equilibrium orientational configurations of the director's field are specified by minimizing the total Helmholtz free energy functional [?] along with the electromagnetic energy. Thus, we can obtain a differential equation of the reorientational angle dependent of the electric field and an dimensionless parameter $q \equiv \frac{\epsilon_a E_0^2 l^2}{8\pi K}$ that denotes the ratio between the electromagnetic energy and the elastic energy densities that measures the coupling between the optical field and the nematic and accounts for the effect of the presence of the optical field on the elastic response.

Hence, vanishing values of q represent a linear regime, whereas higher values represent nonlinear effects. In the case where $q \ll 1$, the field does not produce any reorientational effect on the cell and result in the following linear profile for the steady-state orientational configuration [?]:

$$\theta(\zeta) = \pi/2\zeta. \tag{147}$$

The Euler-Lagrange equation that result of the minimization of the energy show that the reorientation dynamics is coupled with the optical field dynamics that is described by the corresponding Maxwell's equations. In the optical limit, taking into account the Hamilton's characteristic function the trajectories of the light inside the cell could be determined [?]. This characteristic function represents the optical path between two fixed points inside the cell.

The behavior of the ray trajectories clearly depends on the incident angle i because the ray component in the x direction $p = k_x/k_0$, may be also given as $p = N_b \sin i$. In Ref. [?] it is reported that there are two relevant regimes for i, the first one corresponds to $i < i_c$, where all the rays always reach the top substrate and part of the ray is transmitted to the top plate. In this case the critical angle $\theta_c = \pi/2$. On the other hand, the second one corresponds to $i > i_c$, when the ray does not reach the top substrate and it is reflected back to the inside of the cell, see Fig. 28. For this case θ_c is not longer constant and it is described by

$$\theta_c = \arccos\sqrt{\frac{p^2 - \epsilon_\perp}{\epsilon_a}} \tag{148}$$

where p is the ray component, $\epsilon_\perp$ is the dielectric constant, perpendicular to the director and ϵ_a is the dielectric anisotropy.

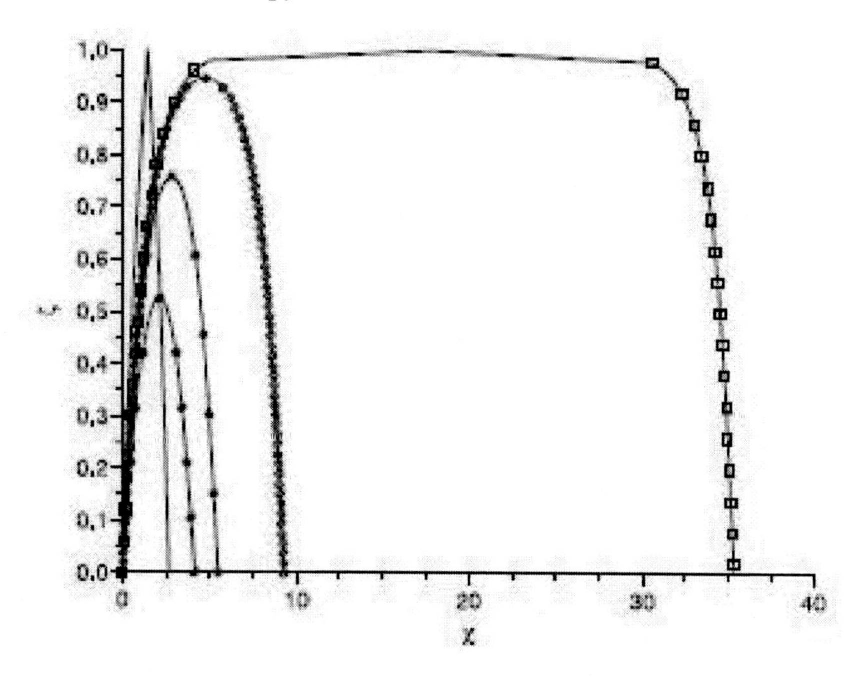

Figure 28. Beam trajectories. (—) $i = 45^o$, $(-\square - \square-)$ $i = i_c$, (−o−o−) $i = i_c + 0.1$, (—) $i = i_c + 2^o$, $(- * - * -)$ $i = i_c + 0.7^o$, $(-\blacklozenge - \blacklozenge-)$ $i_c = 56.3^o$. ζ is the dimensionless coordinate in the direction of propagation and χ is the transversal coordinate. $n_e = 1.735$, $n_o = 1.51$, $\lambda = 632.28$ nm, $N = 1.81$.

Note that for the interval $0^o \leq i \leq i_c \equiv 56.3^o$, the ray always reaches the top substrate, but for i close to but above i_c, the (dimensionless) range of a bending ray χ_{max}, may be as large as 35 times l. If i further increases, the ray penetration ζ_c decreases, as well as its range. It is found that when the value $i_{c2} = 75^o$ is reached, the beams no longer penetrate the liquid crystal cell and a ray is reflected back to the lower substrate. In Fig. 28, the caustic, ζ_c, of the ray trajectory for $i \sim 56.3^o$, is very close to the upper substrate ($\zeta = 1$). For a slightly larger incident angle, $i \sim 56.4^o$, the caustic has a height of $\zeta \sim 0.95$ which corresponds to a distance from the top substrate of about $\sim 10\lambda$. This distance can be quantified more precisely by using the expression for the critical angle θ_c, Eq. (147) and the configurational equation (148).

Following the well-established methods developed in paraxial optics [?], it is possible to extend the model of ray propagation in a nematic hybrid cell, to consider a Gaussian beam instead of an incident plane wave. Then, it is assume that the incident beam is Gaussian with a beam waist ω_0 and angular divergence κ. If the reflectance, R, is measured in a plane Σ normal to the reflected beams, neglected difraction effects, the total reflected electric field can be calculated following the standar procedure for multiple beam reflections in the optical limit in terms of the Fresnel coefficients and the phase shift between adjacent rays [?]. However, in this case all the reflected beams will have a Gaussian profile $\exp(-r^2/\omega_0^2)$.Taking a system of axis in Σ with the origin in the center of the first reflection, the total reflected field, E_R, in this plane is given by

$$E_R = r_{12}\exp(-t^2r^2)+(1-r_{12}^2)\times\sum_{n=1}^{N_R(i)}\left[(-r_{12})^{n-1}(r_{23})^n\exp(jn\Gamma)\times\exp(-t^2(r-nd)^2)\right],$$

where Γ denotes the phase shift between the ray propagating in the liquid crystal and the one reflected back to the lower substrate, the parameter t measures the ratio between the cell thickness and the beam waist

$$t = \frac{l}{\omega_0}, \tag{149}$$

r_{12}, r_{23} are the Fresnel coefficients for the nematic-substrate interface and are given explicitly by

$$r_{12} = \frac{n_o n_e \cos i - N_b\sqrt{n_e^2 - N_b^2\sin^2 i}}{n_o n_e \cos i + N_b\sqrt{n_e^2 - N_b^2\sin^2 i}}, \tag{150}$$

$$r_{23} = \frac{N_t^2\sqrt{n_o^2 - N_b^2\sin^2 i} - n_o n_e\sqrt{N_t^2 - N_b^2\sin^2 i}}{N_t^2\sqrt{n_o^2 + N_b^2\sin^2 i} - n_o n_e\sqrt{N_t^2 - N_b^2\sin^2 i}}, \tag{151}$$

where r denotes the radial coordinate in Σ, and the upper limit N_R, in the summation is the total number of reflections in the cell.

The effects of tunneling through a thin film by means of an evanescent wave are relevant only for films with a thickness of a few wavelengths [**?**]. In our case the effects of a leakage by means of an evanescent wave are localized in the range of incident angles closer to i_c, i. e., $i \in [i_c, i_c + 0.1]$. A study of effects of tunneling [**?**] shows that it can be significant only in the very narrow interval $i_c + 0.1$ for $l = 100\ \mu m$, although, if $l = 10\ \mu m$ the interval may increase by one order of magnitude. Thus, this leakage effect will not alter the general form of the reflectivity curves and only represents a very local correction to these curves.

Both the total energy of the reflected electric field and the total energy that enter into the photodetector are calculated in Ref [**?**]. In this comparative study [**?**] (see Fig. 29) it is found that the reflectance curve shows clearly the different behaviors for each of the angle intervals (a) $i < i_c$ and (b) $i > i_c$. For the first interval we have the usual reflectance behavior for a thin film, i.e., the reflectance coefficient remains small for smaller angles, until i takes values closer to i_c. For $i \gtrsim i_c$ the ray trajectories range can be as large as $35\ l$, (see Fig. 28); the rays travel almost parallel to the top substrate in such way that the separation between the caustics, ζ_c, and this substrate is only of a few wavelengths. As mentioned above, this allows for the possibility of having a tunneling effect. The leakage due to this effect may be considerable for a thin cell since for $i \gtrsim i_c$, the reflectivity coefficient of the bottom substrate-LC, r_{12}, is small and, as a consequence, the transmittance at the bottom-LC interphase is higher, so most of the beam's energy crosses the cell and escapes through the top substrate. For a thicker cell, $l \gg \lambda$, this effect is restricted to a range of a few millidegrees around the incident angle. In the case reported in Ref. [**?**], due to the resolution

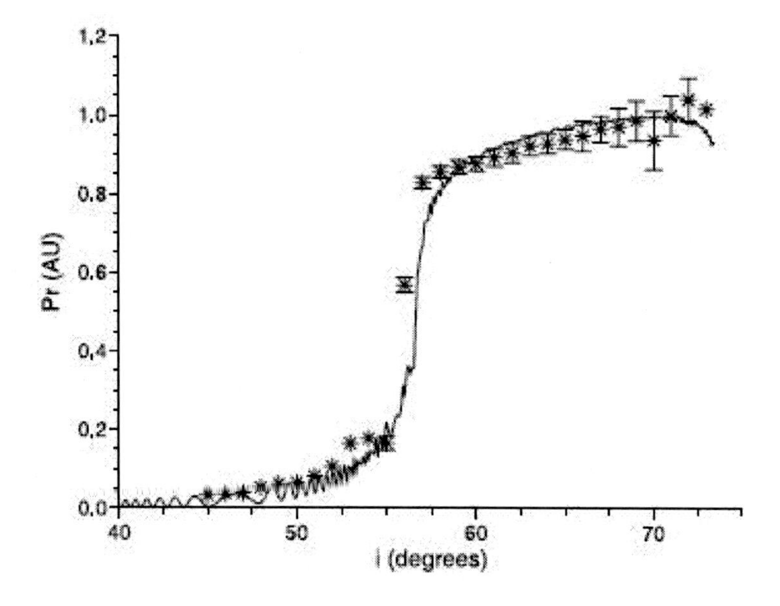

Figure 29. Comparison between experimental and theoretical reflected curves. The continuous line is the theoretical model and ♦ are the experimental points. The size of the bars denotes the size of the experimental error.

of the apparatus these details were practically impossible to observe. However, the model predicts that by reducing the thickness of the cell the effect can be appreciable for a large range of i. It is worthwhile to mention that this tunneling effect allows for the possibility to use this effect in the design of electro-optic switches or variable beam splitters.

Thus, it was shown that an optical beam may propagate in a planar-homeotropic hybrid liquid crystal cell for angles larger than the critical one and to account for its effects in the reflectance curves [**?**].

It is important to say that the beam size effects, reflected in the parameter $t = l/\omega_0$, are considerable when the media are inhomogeneous and the bending trajectories exist. The values of this parameter modify the character of the reflectance curves and make an important difference with respect to the assumption of an incident plane wave, in which the reflected waves are interfering in the whole space. This fact is important because the assumption of incident wave plane fails for reflectivity measurement in inhomogeneous media, and although Berreman's method is well established, it is not suitable when it is applied to the case of bending trajectories, because this method only considers plane waves as incident and reflected fields.

7.3.1. Electrically Controlled Nematic Hybrid Cells

Total internal reflection The local orientational state of a liquid crystal can be modified by the application of low frequency fields. Therefore the local optical axis and its associated refractive index gradient can be modulated by the low frequency electric field. Then, it is possible to electrically control the penetration length, the path, and the range of the beam traveling inside a liquid crystal cell. This phenomenon can be used as the base for a beam steering device or a multiplexor device. In this Sec. we show the possibility to electrically control the range and the depth of penetration of a linearly polarized beam incident on a nematic hybrid cell when total internal reflection conditions are satisfied. The system under study is a pure thermotropic nematic confined between two parallel substrates as the shown in Fig. 27, where a low frequency uniform electric field E_0, parallel to the z axis, is applied. Thus, the equilibrium orientational configurations of the director's field are find out by minimizing the total Helmholtz free energy functional, considering a uniaxial medium for which the dielectric tensor ϵ_{ij} has the general form of Eq. (4), with $\widehat{n} = \widehat{n}[\theta(z)]$

The stationary configuration is then obtained by the corresponding Euler-Lagrange equation which reads

$$\frac{d^2\theta}{d\zeta^2} - q\sin 2\theta(\zeta) = 0. \tag{152}$$

where ζ is a dimensionless variable $\zeta \equiv z/l$ and $q \equiv \frac{\widetilde{\epsilon_a} V^2}{8\pi K}$ is the ratio between the electric energy and the elastic energy densities, therefore it measures the coupling between the electric field and the nematic. Here $\widetilde{\epsilon_a}$ is the low frequency dielectric anisotropy and $V = E_0 l$ is the applied voltage. Thus, we can obtain the trajectory

$$\upsilon = -\chi q + \int_0^\zeta d\eta \frac{\epsilon_{xz} \mp p\sqrt{\epsilon_\perp \epsilon_\shortparallel}/\sqrt{\epsilon_{zz} - p^2}}{\epsilon_{zz}}. \tag{153}$$

In this equation $p = N_b \sin i$, υ is a constant that is determined by the incident point of the beam on the cell and $\chi = x/l$. The $\mp$ sign in Eq. (153) correspond to a ray traveling with its wave vector in the $\pm z$ direction, thus, this equation gives the caustic or ray penetration of the beam. The orientational configurations are shown in Fig. 30 where it is shown that as increasing the intensity of the electric field, the director tends to align to the z axis and it does not grow linearly with ζ as in the case of zero electric field.

The range of a bending ray χ_{max} (see Fig. 27) may be calculated from Eq. (153). with $\theta = 0$.In Fig. 31 we see the range as a function of q (for the second regime when i-$i_c > 0$) for different values of the (shifted) angle of incidence $i - i_c$. It presents a complex behavior: for i close to i_c the range decreases but as we increase the value of i, the range may increase and then decrease (see, for example, the case $i - i_c = 7$ in Fig. 31) while for larger values of i the range increases with q

This peculiar dependence can be explained in terms of the coherence length of the field: $l_c = l/\sqrt{q}$ [?]. When $i \sim i_c$ the range is really large, furthermore, it diverges for $i = i_c$. Now, by increasing the field, the region where the refraction index is varying continuously is compressed to a smaller region located above the top substrate whose thickness is given by $l - l_c$. Thus the curved trajectory is reduced to a smaller region while in the rest of the nematic, the beam propagates following straight lines because there the nematic is almost uniformly aligned with E_0. This compression therefore reduces the range because the straight portions of the trajectory almost do not contribute to this range since for $i \sim i_c$ its component along x is small. If otherwise i is farther from i_c, the range and the penetration length ζc are smaller than those for $i \sim i_c$. Enlarging q leads also to a reduction of the continuously varying index region which now is smaller than that of the former case. However,

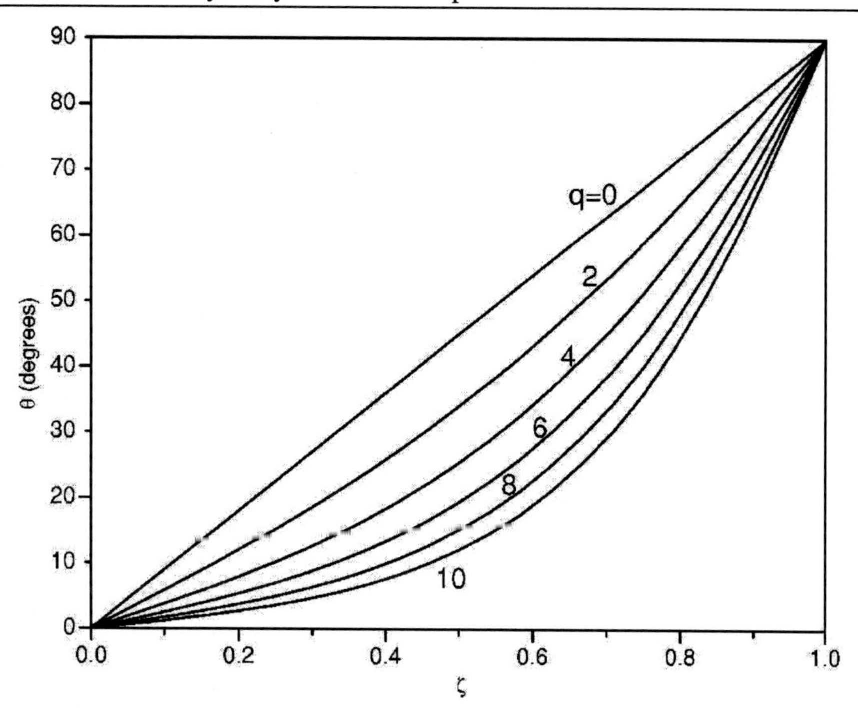

Figure 30. Orientational angle of the director $\theta(z)$ as a function of z, for different values of $q = 0, 2, 4, 6, 8$, and 10 [?] Copyright (2010) by The American Physical Society.

the straight line portions of the trajectory are enlarged by increasing q contributing more to the range because its direction is nearer to x [?]. Thus, the range and the penetration length can be modified by varying the applied voltage. For instance, for a typical nematic $\widetilde{\epsilon_a} = 13$ and $K = 10^{-11}$, then, for $q = 10$ the applied rms voltage is $V \simeq 1.2\ V$. This yields for a cell with a thickness of 100 μm and for $i = i_c + 16^o \simeq 69^o$ a range of $x_{max} \simeq 1$ mm and the corresponding penetration length of $\zeta_c \simeq 50\ mm$. Hence, this effect could be used for the design of optical devices. For example, by combining this hybrid cell with a hemisphere, the output beam can be steered by a refraction process at the hemisphere-air interface since the secondary reflection will leave the hybrid cell far from the center of the hemisphere. It should be remarked that it is expected that most of the energy will leave at the secondary reflection since the incident angle is far from the critical angle between the homeotropic layer and the substrate. Another possible application consists in using this nematic cell as a multiplexor for switching among various optical fibers. Because the beam's range can be electrically controlled, the cell can be used to locate the outgoing beam in various cells' positions where some optical fibers were previously coupled. It should be remarked that the beam's output angle does not change by varying the applied voltage, simplifying considerably the optical coupling procedure with the exit fibers.

Light propagation and transmission The dynamics of the optical field is described by

the corresponding Maxwell's equations and the procedure to solve them within the optical limit gives, in the case of normal incidence, the following expressions for the trajectory

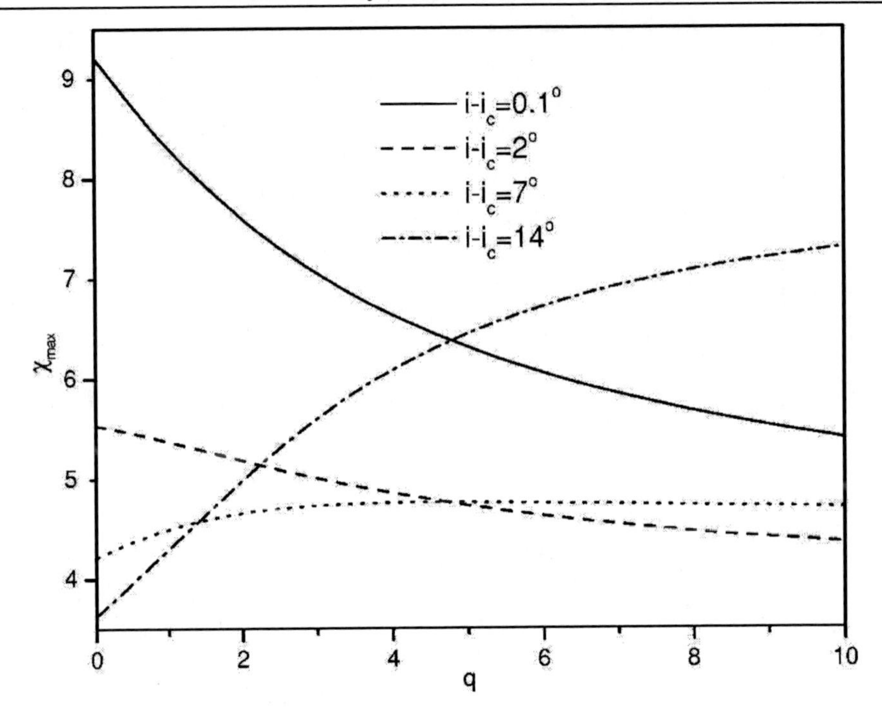

Figure 31. Dimensionless range of a bending ray χ_{max} as a function of q for different values of the (shifted) angle of incidence $i - i_c$ [?] Copyright (2010) by The American Physical Society.

$$\chi(\zeta) = \begin{cases} \int_0^\zeta d\zeta \epsilon_{xz}/\epsilon_{zz}, & TM\ mode \\ 0, & TE\ mode \end{cases} \tag{154}$$

in this equation, $\chi = x/l$ is a dimensionless variable. In the TE mode case, the incident beam is not refracted and continues propagating through the line of incidence without deviation. Strictly speaking the trajectories given by Eq. (??) are valid only for an incident plane wave. However, they will describe correctly the trajectories of beams with Gaussian width as long as the set of rays that are part of the incident beam are parallel to each other. This is true if the diffraction length, defined as $l_r = (k_0\omega_0^2)/2$, with ω_0 the waist of the beam, is large compared to the separation between the beam source and the cell.

The trajectory depends on the wavelength of the incident beam through the dielectric functions $\epsilon_\| \equiv n_\|^2$ and $\epsilon_\perp \equiv n_\perp^2$, which in general are functions of the wavelength. The refractive index $n_\|$ is known as extraordinary refractive index and $n_\perp$ is the ordinary refractive index. For 5CB at $T = 25.1^0$ C, the dependence of the refractive indices on the wavelength is obtained by fitting a three-band model to experimental values [?] [?]. In Fig. 32 we show the trajectories of the beams calculated from Eq. (??), for (a) $q = 0$ and (b) $q = 10$. We notice both the dispersion effect due to the wavelength dependence of the refractive indices and that these trajectories are modified by changing the applied electric field. Note, however, that the horizontal shift of the beam, $\chi_{max} = \chi(\zeta = 1)$, is very small.

In Fig. 33, it is plotted the transmittance on the axis of incidence, $\rho = 0$ ($\rho = r/l$) of a TM mode for the three primary colors as a function of the applied voltage V. We see an

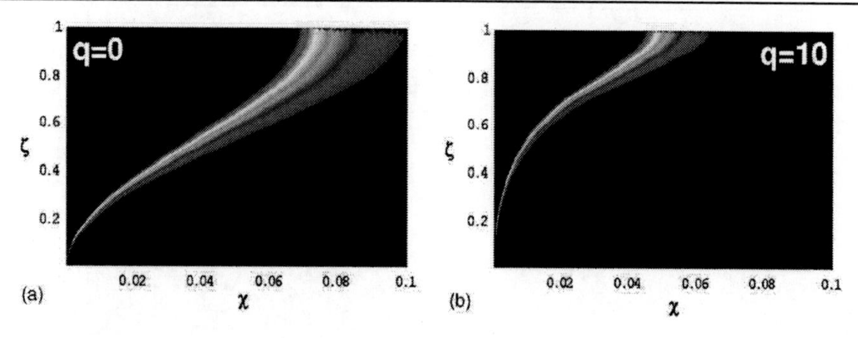

Figure 32. Trajectories of the light beams with TM polarization. (a) $q = 0$ and (b) $q = 10$. Refractive indices of the plates: $N_b = N_t = 1.81$. The colors correspond to wavelengths in the range of $380 < \lambda < 780nm$.

oscillatory behavior consistent with experimental results [?]. On the other hand, since the TE mode is not affected by the reorientation of the nematic produced by the electric field, the transmittance for this mode will not show oscillations. Also, notice that if the applied voltage is very large then the nematic's director will be aligned with the field in the vertical direction. In this case, the transmittance of the TM mode will coincide with that of the TE mode.

Figure 33. Three-color characteristic transmission for a HAN cell located between crossed polarizers as indicated in the text as a function of the applied voltage. Wavelength of colors $\lambda(R) = 632.7$ nm, $\lambda(G) = 554.6$ nm, and $\lambda(B) = 452.0\ nm$. Cell thickness: $l = 8\ \mu m$. Refractive indexes of the plates: $N_b = N_t = 1.81$. Waist of the beam: $\omega_0 = 2$ mm. We used $\tilde{\epsilon_a} \approx 13$ and $K = 10^{-11}$ N, appropriate values for 5CB Ref. [?].

The propagation of a normally incident light beam in the cell depends on the state of polarization of the beam. For the TM mode, the trajectory slightly shifts horizontally and this shift is decreased either by applying an electric field or by increasing the wavelength of the light beam. In contrast, the TE mode propagates without deviation for any light wavelength and applied field. This gives as a result that for a cell between crossed polarizers, the transmittance curves show an oscillatory behavior which is different for different wavelengths. Since only the TM mode is affected by the electric field, this mode is the responsible of the resulting oscillations, which are in good qualitative agreement with the experiments of Ref. [?].

Dispersion effects Furthermore, the cell could be consider to be subjected to a low-frequency electric field applied perpendicular to it, and taking into consideration the dependence of the refractive index of the liquid crystal on the wavelength of the optical beam [?]. A low intensity incident light beam with P-polarization (P-wave) impinges the nematic with an incident angle i. The Euler-Lagrange equation is solved along with the Maxwell equations [?].

The results show that the range and the penetration length of the trajectory of the light beam depends on the color of the beam and that these parameters can be controlled by varying the intensity of the applied electric field, see Fig. 34.

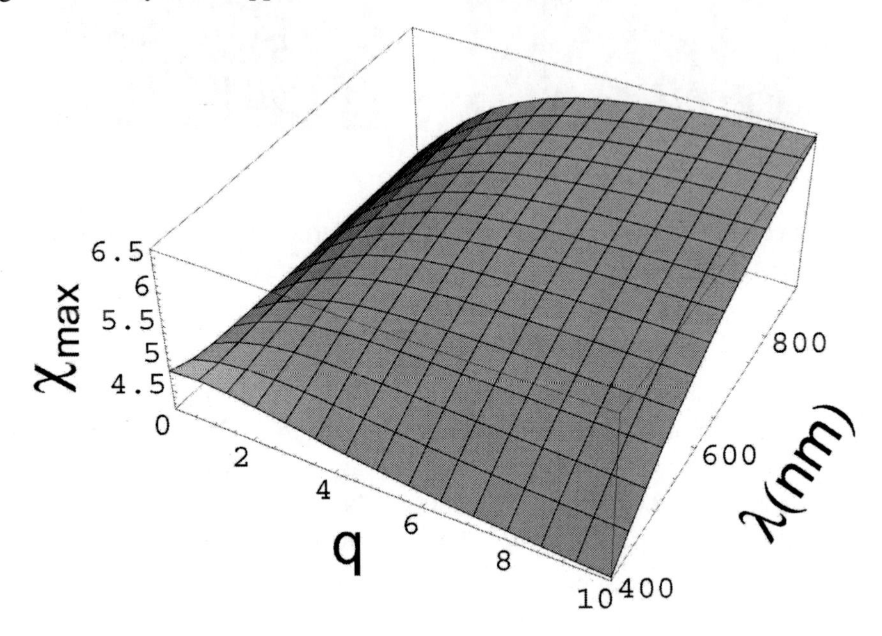

Figure 34. Dimensionless range of a bending ray χ_{max} as a function of q and λ for an angle of incidence $i = 64.54^0$.

In this case the reflectivity curves for incident beams with a Gaussian profile show a clear influence of both the electric field and the color of the beam, see Fig. 35. the reflectivity increases for low incidence angles and decreases for large incidences angles as the value of the electric field increase. Also, as the wavelength of the beam increase the angles of incidence for which there are total internal reflection in the cell shift towards smaller values while keeping the overall shape of the reflectivity curve almost unchanged.

These results indicate that this physical system could be useful for the design of field-controlled wavelength selection devices or for multiplexor applications because all the outgoing rays with different wavelength are parallel to each other [?] [?].

7.4. Smectic Bend Core Guide

Liquid crystals are complex fluids with a wide range of properties: elastic, viscous, electrical, etc. Liquid-crystalline polymers are viscoelastic and can easily formed fiber, just as conventional polymers. Calamitic (bar-like) liquid crystals usually form thin films like membranes. In this sense, it was an outstanding fact to find recently that bend-shape molecules, specifically the phases B_7 and B_2, may form stable fibers instead of films [?]. This feature of self formation of fibers may be useful to design and construct optical fibers and artificial muscles taking into account that smectics, as other thermotropic liquid crystals, are susceptible to external fields, thermal and mechanical perturbations.

Smectic liquid crystals have peculiar properties as an spontaneous electrical polarization which manifests by their anti and ferroelectricity and spontaneous breaking of chiral sym-

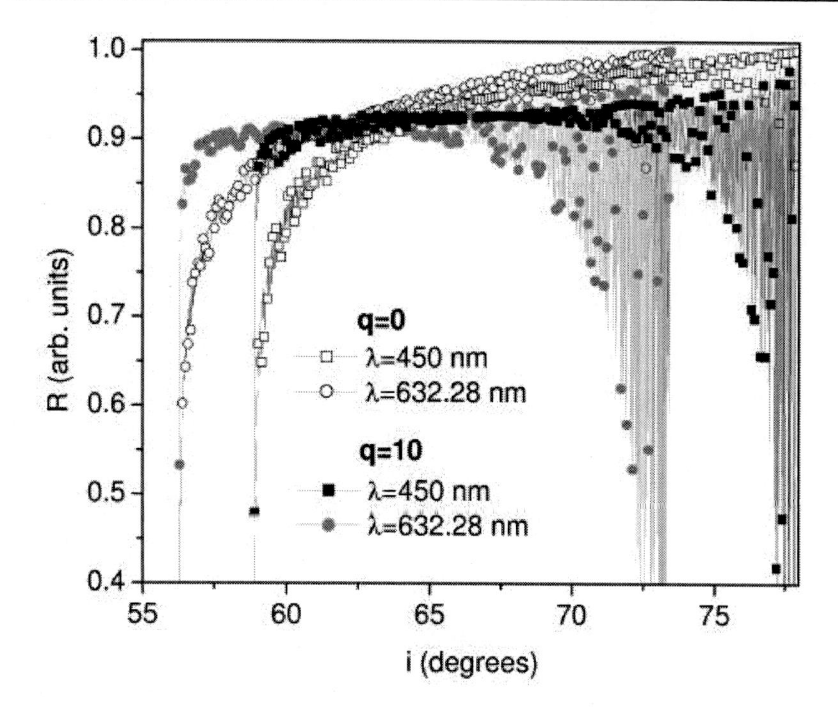

Figure 35. R vs i curve for different values of the electric field q and wavelength λ. $t = 0.1$.

metry in one dimensional phases (smectic phases), nevertheless are formed by molecules which are not chiral intrinsically[**?**].

Thus, the electromagnetic response of these fibers is of interest for its potential technological applications. We begin by considering low intensity beams such that they do not distort the smectic configuration when they propagate through the fiber. In the next sections the optical field within a cylindrical fiber of smectic B liquid crystal core surrounded by an infinite, homogeneous, isotropic cladding is investigated.

7.4.1. Optical Limit for Cylindrical Waveguides

We consider a nonmagnetic optical fiber of smectic liquid crystal consisting of banana-shaped molecules core confined within a very long cylindrical region of radius R, surrounded by an infinite homogeneous isotropic cladding with dielectric constant ϵ_0 (See Fig. 36). The fiber is formed by a series of concentric cylindrical smectic shells where the director, $\widehat{n}$, make an angle ϕ with the normal to the shells and is given by

$$\widehat{n} = cos\phi\widehat{e}_r + sen\phi\widehat{e}_\theta, \tag{155}$$

where $\widehat{e}_r$ and $\widehat{e}_\theta$ are unit vectors in the cylindrical coordinates (r, θ, z). We can define the director's lines in analogy with the electric field lines as the lines parallel to the field at every point, that is

$$\frac{dy}{dx} = \frac{n_y}{n_x}. \tag{156}$$

Then writing the components of Eq. (**??**) in Cartesian coordinates and inserting them into the last equation we get after integration, the following expression.

$$r = Ae^{-\frac{\theta}{\tan[\phi]}}, \tag{157}$$

where A is an integration constant. Notice that this expression defines a plane spiral departing from the origin.

We shall assume that the optical field has low electromagnetic intensity, thus the director field is not distorted by the optical field. Under these circumstances the dielectric tensor of the smectic can be approximated by an uniaxial tensor

$$\overleftrightarrow{\epsilon} = \epsilon_{\perp}\mathbf{I} + \epsilon_a \widehat{n} \cdot \widehat{n} \tag{158}$$

where $\epsilon_{\perp}$ is the dielectric permittivity perpendicular to the long axis of the molecule and ϵ_a is the dielectric anisotropy, considering that the dielectric anisotropy in the second optical axis $\epsilon_b \ll \epsilon_a$, the smectic is biaxial, but we investigate an uniaxial approximation. Thus $\overleftrightarrow{\epsilon}$ can be expressed in terms of the tilt angle as:

$$\overleftrightarrow{\epsilon} = \begin{pmatrix} \epsilon_{11} & \epsilon_{12} & 0 \\ \epsilon_{12} & \epsilon_{22} & 0 \\ 0 & 0 & \epsilon_{33} \end{pmatrix} = \begin{pmatrix} \epsilon_{\perp} + \epsilon_a \cos^2\phi & \epsilon_a \cos\phi \sin\phi & 0 \\ \epsilon_a \cos\phi \sin\phi & \epsilon_{\perp} + \epsilon_a \sin^2\phi & 0 \\ 0 & 0 & \epsilon_{\perp} \end{pmatrix}. \tag{159}$$

Here we have assumed that the tilt angle is uniform for all the fiber. Notice that the dielectric tensor exhibits explicitly the anisotropy which is controlled by ϵ_a. Since the medium is nonmagnetic its magnetic susceptibility tensor is $\overleftrightarrow{\mu} = \delta_{ij}$. Following the usual procedure we can obtain from Maxwell's equations in the frequency space.

If the wavelength of the light is smaller than the typical length of the system l, we can use the optical limit $k_0 l >> 1$ to simplify the equation governing the electromagnetic field. For isotropic media, this procedure leads to a scalar eikonal equation [**?**]. In contrast, for anisotropic media whose transverse electric modes are decoupled from the electric field, the optical field can be described with only two eikonal equations [**?**]: one for the transverse electric modes and the other for the transverse magnetic ones. Nevertheless, for the general anisotropic case Barkovskii [**?**] has proposed a generalization of the well known trial function used to find the eikonal equation,

$$E_j = B_i e^{ik_0 l \psi_{ij}} \tag{160}$$

where ψ_{ij} are the components of the eikonal tensor and B_i are complex amplitudes which may be functions of the coordinates. Here, we have used the convention of sum over repeated indexes. After inserting this expression into the wave equation for the induced magnetic field H from Maxwell equations and assuming the optical limit, Barkovskii obtained the following generalized eikonal equation

$$e_{afc}\epsilon_{cb}^{-1} e_{bgd} \frac{\partial \psi_{dh}}{\partial r_g} \frac{\partial \psi_{hl}}{\partial r_f} + \mu_{al} = 0 \tag{161}$$

where e_{afc} is the Levi-Civita pseudotensor.

From Eq. (**??**) we can see that the dielectric tensor couples the r and θ components of the involved field. Thus, an eikonal tensor of the same structure can be considered [**?**], that is ,

(a)

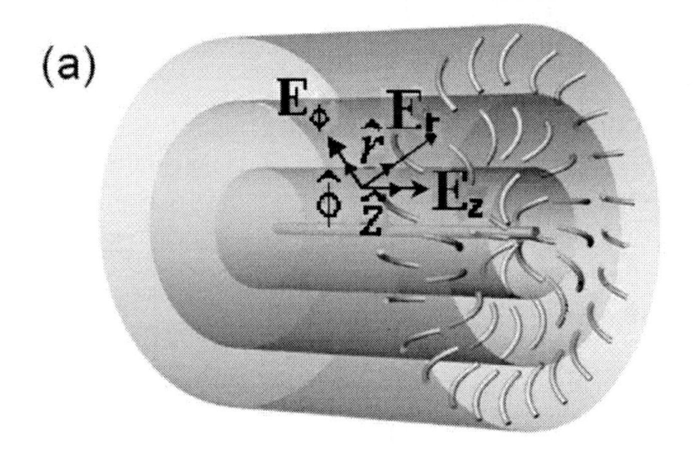

(b)

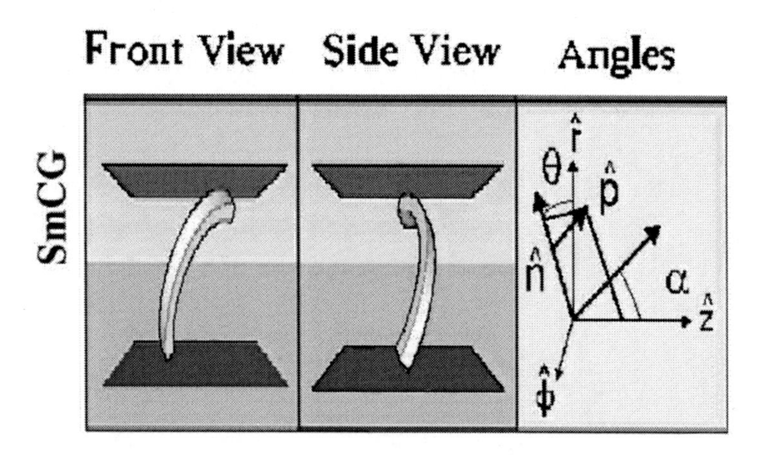

Figure 36. a) Schematic representation of a fiber filled with liquid crystal smectic B_7. b) Molecular orientation of $\hat{n}$ and the polarization director $\hat{p}$ with respect to the smectic layer. The orientation is determined by the tilt angle θ an polarization angle α.

$$\overleftrightarrow{\Psi} = \begin{pmatrix} \psi_{11} & \psi_{12} & 0 \\ \psi_{21} & \psi_{22} & 0 \\ 0 & 0 & \psi_{33} \end{pmatrix}. \tag{162}$$

Then, expanding Eq.(**??**) for the E_x, E_y and E_z components we have explicitly

$$E_x = B_1 e^{i\psi_{11}} + B_2 e^{i\psi_{12}}, \tag{163}$$

$$E_y = B_3 e^{i\psi_{21}} + B_4 e^{i\psi_{22}}, \tag{164}$$

$$E_z = B_5 e^{i\psi_{33}}. \quad (165)$$

For a positively anisotropic smectic the electric field tends to be aligned to the smectic director $\widehat{n}$ or director lines because the projection of $\overleftrightarrow{\epsilon}$ is maximum in that direction. As a consequence a coupling between the optical field and the orientational configuration of the smectic occurs in such way that the polarization of the optical field rotates following $\widehat{n}$. It is worth noting that the fact that $\widehat{n}$ is constrained to the $x-y$ plane causes that only E_x and E_y are couple, whereas E_z remains always perpendicular to $\widehat{n}$ and therefore decouples from it.

If we write explicitly a rotation of the transversal components of the electrical field (E_x and E_y), we arrive at

$$\mathbf{E} = \begin{pmatrix} A_1 \cos\alpha(\mathbf{r}) e^{iw(\mathbf{r})} - A_2 \sin\alpha(\mathbf{r}) e^{iw(\mathbf{r})} \\ A_1 \sin\alpha(\mathbf{r}) e^{iw(\mathbf{r})} + A_2 \cos\alpha(\mathbf{r}) e^{iw(\mathbf{r})} \\ A_3 e^{iw(\mathbf{r})} \end{pmatrix} \quad (166)$$

where $\alpha(\mathbf{r})$ and $w(\mathbf{r})$ denote the position dependent polarization and phase of the electric field. However, this expression for $\mathbf{E}$ should be consistent with Eqs. (**??**)-(**??**). Then expressing the trigonometric functions in terms of complex exponentials, equating amplitudes and phases of both expressions, we get the following relations

$$w_t(\mathbf{r}) + \alpha(\mathbf{r}) = \psi_{11} = \psi_{21}, \quad (167)$$

$$w_t(\mathbf{r}) - \alpha(\mathbf{r}) = \psi_{12} = \psi_{22}, \quad (168)$$

$$w_l(\mathbf{r}) = \psi_{33}, \quad (169)$$

$$B_1 = \frac{A_1 + iA_2}{2} = B_2^*, \quad (170)$$

$$B_3 = \frac{A_2 - iA_1}{2} = B_4^*, \quad (171)$$

$$A_3 = B_3. \quad (172)$$

It is interesting to point out that Eqs. (**??**), (**??**) and (**??**) not only express the elements of the eikonal tensor Eq. (**??**) in terms of the wave phases w_t and w_l of the transverse and longitudinal components of the electric fields, but also involve the polarization angle α. In this sense the geometrical optical limit is able to describe changes in the polarization of the electric field such as those involved in optical activity phenomena.

The eikonal equations can be solved within the smectic B liquid crystal core by taking into account the coupling between the director and the electric field. This allow us to find expressions for the polarization, optical paths of the electromagnetic fields. By applying Hamilton equations to the Hamiltonian of the system for the TM modes, the parametric ray trajectories are calculated.

It is convenient to remark that since the coefficients of these equations are only functions of r, by applying a convenient canonical transformation we can reduce this system of partial differential equations to a simpler system of ordinary differential equation. If w_t does not depend on z and θ, they are cyclic coordinates and their corresponding conjugated variables p_z and p_θ, are conserved quantities. As a consequence, it is useful to use the canonical transformation

$$w_t = f(r) + p_z^t z + p_\theta^t \theta, \tag{173}$$

$$w_l = v(r) + p_z^l z + p_\theta^l \theta, \tag{174}$$

where $f(r)$ and $v(r)$ are only functions of r. Evidently, the ray components are related with w_t and w_l by the usual expressions

$$p_z^t = \frac{\partial w_t}{\partial z}, \; p_\theta = \frac{\partial w_t}{\partial 0}, \tag{175}$$

$$p_z^l = \frac{\partial w_l}{\partial z}, \; p_\theta = \frac{\partial w_l}{\partial \theta}. \tag{176}$$

Similarly, we can express α in the form

$$\alpha = g(r) + P_z z + P_\theta \theta, \tag{177}$$

where P_z and P_θ are constants and $v(r)$ is only a function of r. Nevertheless, P_z and P_θ are no longer ray components, but instead the ratio of change of the polarization angle with respect z and θ. If we insert Eqs. (**??**), (**??**) and (**??**) into Eqs. that come from (**??**), (**??**) and (**??**) we can obtain three ordinary differential equations for $f(r)$, $g(r)$ and $v(r)$. After substituting back these expressions into Eqs. (**??**), (**??**) and (**??**), we find the solution for $f(r)$, $g(r)$ and $v(r)$

To calculate the ray trajectories we write and solve Hamilton´s equation for these functions expressed in terms of their natural variables, to find parametric equations of their ray trajectories.

$$H_t = \left(\epsilon_{22}^{-1} - \epsilon_{12}^{-1}\right) p_z^{t2} - \frac{1}{r}\epsilon_{33}^{-1} p_r^t p_\theta^t + \epsilon_{33}^{-1} \frac{p_\theta^{t2}}{r^2}, \tag{178}$$

$$H_l = \epsilon_{22}^{-1} p_r^{l2} - \epsilon_{11}^{-1} \frac{1}{r^2} p_\theta^{l2} - 2\epsilon_{12}^{-1} \frac{p_\theta^l}{r}, \tag{179}$$

where $p_z^t = df/dr$ and $p_z^l = dg/dr$. Hence, using one of the Hamilton's equation for w_t, we obtain

$$\frac{d}{d\tau}\begin{pmatrix} r \\ \theta \\ z \end{pmatrix} = \begin{pmatrix} 0 & -\frac{\epsilon_{33}^{-1}}{r} & 0 \\ -\frac{\epsilon_{33}^{-1}}{r} & \frac{2\epsilon_{33}^{-1}}{r^2} & 0 \\ 0 & 0 & 2(\epsilon_{22}^{-1} - \epsilon_{12}^{-1}) \end{pmatrix} \cdot \begin{pmatrix} p_r^t \\ p_\theta^t \\ p_z^t \end{pmatrix}. \tag{180}$$

We observe that the first and second equations of this system can be integrated straightforwardly for $r(\tau)$ and $z(\tau)$ since p_θ^t and p_z^t are conserved quantities. However, for solving the second equation we need to insert $p_r^t = df/dr$ from $f(r)$ and then integrate the resulting expression. The above procedure allows us to write the following parametric equations for the three dimensional ray trajectory

$$r(\tau) = \sqrt{2}\sqrt{C_1 - \frac{p_\theta^t \tau}{\epsilon_\perp}}, \tag{181}$$

$$\begin{aligned}\theta(\tau) \quad = \quad & \frac{1}{2p_\theta^{t2}\epsilon_\perp(\epsilon_a+\epsilon_\perp)}\left[2\left\{(1+C_2)p_\theta^{t2} + 2p_\theta^t\tau - \frac{C_1}{2}\epsilon_\perp\right\}\epsilon_\perp(\epsilon_a+\epsilon_\perp)+\right.\\ & p_z^{t2}(\epsilon_a(1+\cos 2\phi+\sin 2\phi)+2\epsilon_\perp)\\ & \left.(C_1\epsilon_\perp - 4p_\theta^t\tau) - p_\theta^{t2}\epsilon_\perp(\epsilon_a+\epsilon_\perp)\ln\left(C_1 - \frac{4p_\theta^t\tau}{\epsilon_\perp}\right)\right],\end{aligned} \tag{182}$$

$$z(\tau) = C_3 - \frac{4\tau p_z^t}{\epsilon_a(\epsilon_a+\epsilon_\perp)}\left[(1+\cos 2\phi+\sin 2\phi)\,\epsilon_a + 2\epsilon_\perp\right], \tag{183}$$

where C_1, C_2, C_3 are constants to be determined by the initial position, while p_z^t and p_θ^t are the ray components which are to be determined by the initial ray direction.

Thus, in Fig. 37 [?] parametric trajectories are shown starting at the same initial position and value of p_θ, different values of the tilt ϕ, and ray component p_z. We observe that for larger values of p_z, the trajectories begin to get closer, but there is no much difference among the curves corresponding to different values of ϕ. On the other hand for smaller values of p_z the trajectories are quite different. For instance, when $\phi = 30^o$, the spiral turns more than for $\phi = 90^o$. In contrast, for different values of the tilt ϕ, and ray component p_θ a similar behavior could be noticed to that shown for small values of p_z [?]. By changing p_θ the modification in the ray trajectories are much smaller than those obtained when p_z is varied. The rotation of all the trajectories and a varying polarization angle α, show evidently the presence of optical activity in the guide even though the smectic forming the guide is not chiral.

The model showed above is the first for studying the propagation of electromagnetic waves within a smectic B cylindrical guide. Under the optical limit it is possible to find three-dimensional ray trajectories within the fiber of banana-shaped molecules' liquid crystal core. Hence, we can see (Fig. 37) that the trajectories are spirals either diverging or converging from the center, described by a paraboloid in the $r - z$ plane. Since all rays rotate while they are propagating through the fiber, and there exists a varying polarization angle α, the fiber exhibits optical activity for a low intensity beam such that the smectic configuration is not distorted.

7.4.2. Exact Solution

It is possible to find the optical fields inside the fiber by establishing the Maxwell's equations in the usual procedure assuming monochromatic fields propagating of frequency ω along the cylinder, with a dielectric tensor for the fiber ϵ that in the uniaxial approximation is given by Eq. **??**, thus, besides the electromagnetic fields, the electric, magnetic field phase [?] and the dispersion relation could be calculated [?].

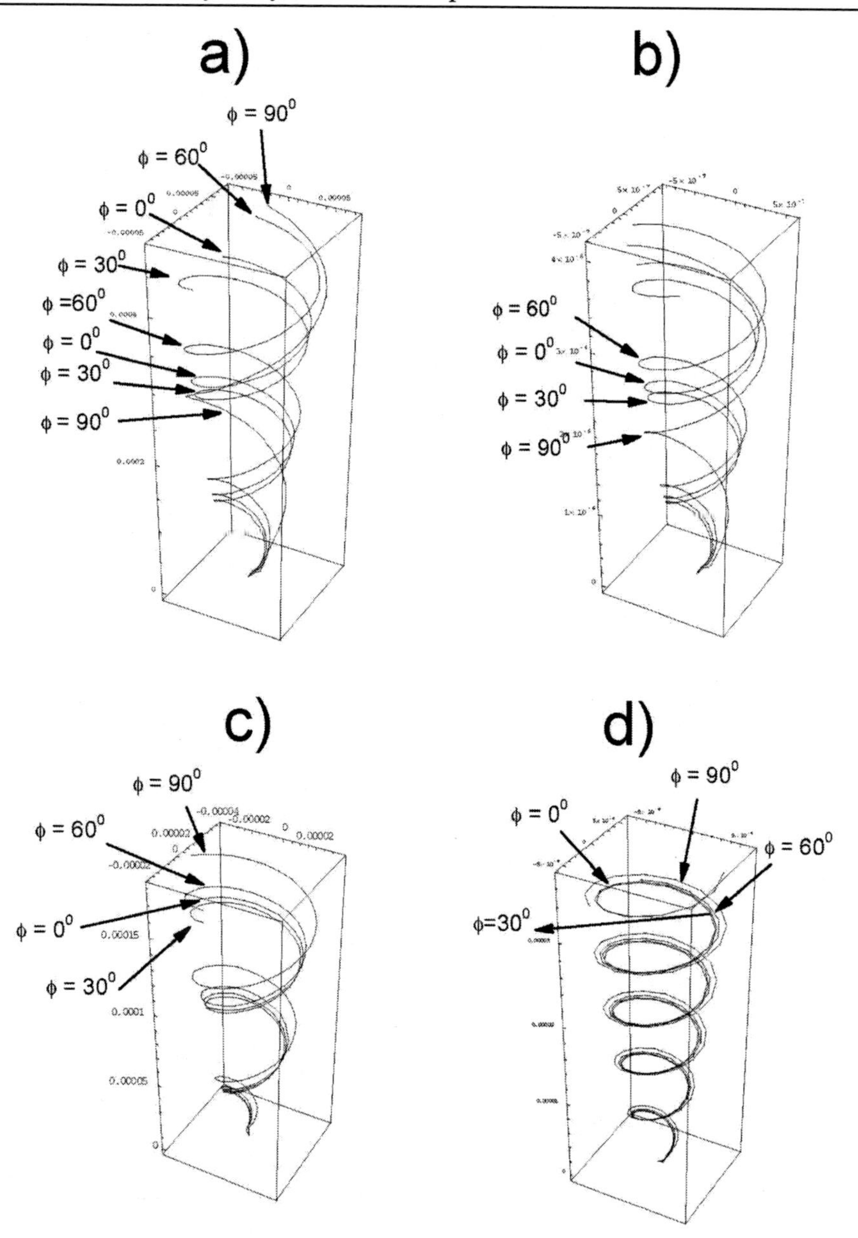

Figure 37. Three dimensional ray trajectories for various values of ϕ as a function of the space coordinates for initial conditions, $r_0 = 10^{-8}cm$, $\theta_0 = 0$, $z_0 = 0$, p_θ =-10^{-4} and a) $p_z = 1$; b) $p_z = 1.05$, c) $p_z = 5$, d) $p_z = 10$.

Here, we present the exact solution considering a biaxial dielectric tensor ϵ given by

$$\bar{\epsilon} = \epsilon_1 \mathbf{I} + (\epsilon_2 - \epsilon_1)\widehat{\mathbf{n}}\widehat{\mathbf{n}} + (\epsilon_3 - \epsilon_1)\hat{\mathbf{p}}\hat{\mathbf{p}} \tag{184}$$

where $\mathbf{I}$ is the identity matrix, ϵ_1, ϵ_2 and ϵ_3 are the dielectric permittivities along the molecular plane normal $\widehat{\mathbf{m}} = \widehat{\mathbf{n}} \times \hat{\mathbf{p}}$; the long axis of the molecule $\widehat{\mathbf{n}}$ (director); and the polarization director $\hat{\mathbf{p}} = \sin\theta \sin\alpha \widehat{\mathbf{e}}_r - \cos\theta \sin\alpha \widehat{\mathbf{e}}_\phi + \cos\alpha \hat{\mathbf{z}}$, where $\widehat{\mathbf{e}}_r$, $\widehat{\mathbf{e}}_\phi$ and $\hat{\mathbf{z}}$ are unit vectors in the cylindrical coordinates (r, ϕ, z).

Following the usual procedure to describe a waveguide, we assume monochromatic fields propagating of frequency ω, along the cylinder axis of the form

$$\mathbf{E} = (E_r, E_\phi, E_z)\,(x) Exp[i(k_0\beta z - \omega t)]. \tag{185}$$

where $x = r/R$, $k_0 = \omega/c$ and β is the propagation parameter. After substitution of this expression in Maxwell's equations, we can write them in the matrix form $(H + iV)\mathbf{\Psi} = \beta^2\mathbf{\Psi}$. where $\mathbf{\Psi} = (E_r, E_\phi, F_z)$ and

$$\mathbf{H} = \begin{bmatrix} \epsilon_{rr} & \epsilon_{r\phi} & D \\ \epsilon_{r\phi} & \epsilon_{\phi\phi} - DD^T & 0 \\ \epsilon_{rr} D^T & \epsilon_{r\phi} D^T & \epsilon_{zz} + \beta^2 \end{bmatrix}, \mathbf{V} = \begin{bmatrix} 0 & 0 & -\epsilon_{rz}/\beta \\ 0 & 0 & -\epsilon_{\phi z}/\beta \\ \beta\epsilon_{rz} & \beta\epsilon_{\phi z} & \epsilon_{rz} D^T/\beta \end{bmatrix}. \tag{186}$$

Here $F_z = i\beta E_z$, $D = (1/k_0 R)(d/dx)$ and $D^\top = (-1/k_0 R x)(d/dx)x$ are mutual Hermitian adjoint operators under a cylindrical norm $< \phi\,|D|\,\psi >= \int \phi D\psi x dx = \int \psi D^\top \phi x dx$, and ϵ_{ij} are the cylindrical components of ϵ, for instance $\epsilon_{rr} = \epsilon_1 + (\epsilon_2 - \epsilon_1)\sin^2\theta + (\epsilon_3 - \epsilon_1)\sin^2\alpha\sin^2\theta$. A direct comparison of these equations with those corresponding to a nematic fiber adopting a escaped configuration [?], displays that for that case the modes are separated into transverse magnetic TM and electric TE modes even under the presence of low frequency electric fields and thermal effects [?], whereas in this case they remain fully coupled. It is worthwhile to remark that the parameters coupling the TE and TM modes are precisely the dielectric anisotropies: $\epsilon_2 - \epsilon_1$ and $\epsilon_3 - \epsilon_1$ contained in the non diagonal elements of ϵ_{ij}.

The electromagnetic fields satisfy the boundary conditions:

$$E_\phi/(dE_\phi/dx)|_{x=1} = E_r/F_z|_{x=1} = K_1(k_0\sqrt{\beta^2 - 1})/(\sqrt{\beta^2 - 1}K_0(k_0\sqrt{\beta^2 - 1})), \tag{187}$$

where $K_0(u)$ and $K_1(u)$ are the modified Neumann functions of order 0 and 1 [?]. These equations account for the continuity of the tangential electromagnetic components E_ϕ, E_z, H_z and the continuity of the normal component of the electric displacement field D_r, at the border of the liquid crystal region.

In the uniaxial limit for which $V = 0$, the system can be solved analytically and its general solution for E_ϕ is given by

$$E_\phi = c_1 J_1(k_0 R\gamma_1 x) + c_2 J_1\,(k_0 R\gamma_2 x) \tag{188}$$

where c_1 and c_2 are arbitrary constants to be determined by the boundary conditions. Here $J_1(u)$ is the Bessel function of order 1 and $\gamma_1 = \sqrt{\epsilon_1 - \beta^2}$ and $\gamma_2 = \sqrt{\epsilon_1(\epsilon_2 - \beta^2)/\epsilon_{rr}}$, where the corresponding eigenvalues β are real numbers so that all the modes are propagating. Furthermore, the fact that the $z-$components of the electric and magnetic fields are pure imaginary quantities makes that the only nonvanishing component of the Poynting vector is also S_z. Thus, for the uniaxial limit the modes do no have angular momentum (they are linearly polarized) and there are no energy leaks in the guide.

Upon the application of the boundary conditions and the substitution of the general solution we can obtain a transcendental equation for β in the uniaxial limit which provides us the band structure. Setting $\beta = 1$ in the mentioned equation we find the following cut-off frequencies

$$\omega_{0na} = \frac{c\chi_{0,n}}{R\sqrt{\epsilon_1 - 1}}, \omega_{0nb} = \frac{c\chi_{0,n}\sqrt{\epsilon_{\theta\theta}}}{R\sqrt{\epsilon_1(\epsilon_2 - 1)}}, \tag{189}$$

where $\chi_{0,n}$ denotes the zeros of the Bessel function of order zero. In contrast to the isotropic dielectric waveguide case for which the TM and TE cut-off frequencies are degenerated, here they are not. Indeed, ω_{0nb} depends on the tilt angle whereas ω_{0na} does not.

The biaxial system has complex solutions and eigenvalues β because $\mathbf{V}$ is not a self adjoint matrix. By means of the boundary conditions it is possible to find the electromagnetic modes using the shooting method [**?**].

Figs. 38 show the real β' and imaginary β'' parts of β versus k_0R for both type of modes where we have used the parameter values $\epsilon_1 = 2.25$, $\epsilon_2 = 2.65$, $\epsilon_3 = 2.45$ and $\theta = \alpha = \pi/4$. Fig. 38(a) has the following features: i) the cut-off frequencies for both type of modes are alternated ordered starting by a magnetic mode, ii) in contrast to the axial case [**?**] these curves never cross each other, iii) there exist k_0-intervals for which the branches deform to get near each other. From Fig. 38(b), i) it can be inferred that besides the first mode, these are slowly leaky modes because their small attenuation values β'' allow them to travel thousands of times the fiber's radius before being considerably attenuated, ii) the k_0R-dependences for all the branches but the first one are damped and non-harmonic oscillating, iii) there are three small k_0R-intervals for which β'' take positive values. It is worthwhile to mention that S_r is not vanishing and thus there are energy leaks. Moreover, S_r is negative for the same intervals and modes in which β'' is negative, whereas it is positive for the rest of the values [**?**].

Since we are not interested in receptor antennas but in guiding systems, Fig. 38(b) manifests strong differences between backward and forward propagation. While for forward propagation the positive part of Fig. 38(a) resembles a standard waveguide, for backward propagation there are only three narrow band gaps represented by the curves under the k_0R-axis. In other words, outside these bands the guide will not propagate any signal in backward direction whereas in forward direction will conduct various modes. Thus, the smectic bend core waveguide behaves in similar way as an optical diode or a very restrictive narrow filter in backward direction.

The $z-$ component of the electromagnetic angular momentum M_z of these modes is not null and changes of sign against k_0R two or three times for all the modes (See Fig. 39). This implies that the modes are elliptically polarized and shift various times their handedness versus k_0R.

Therefore, in a bend core liquid crystal guide for the backwards direction, the propagation is restricted to certain narrow frequency band gaps while, for the forward one the fiber behaves similarly to a standard waveguide, and besides the zero order mode, the fiber modes are elliptically polarized slowly-leaky modes, whose attenuation coefficients are damped oscillating functions of frequency. Thus, this guide could be used to construct both, an optical diode and a very sensitive frequency filter.

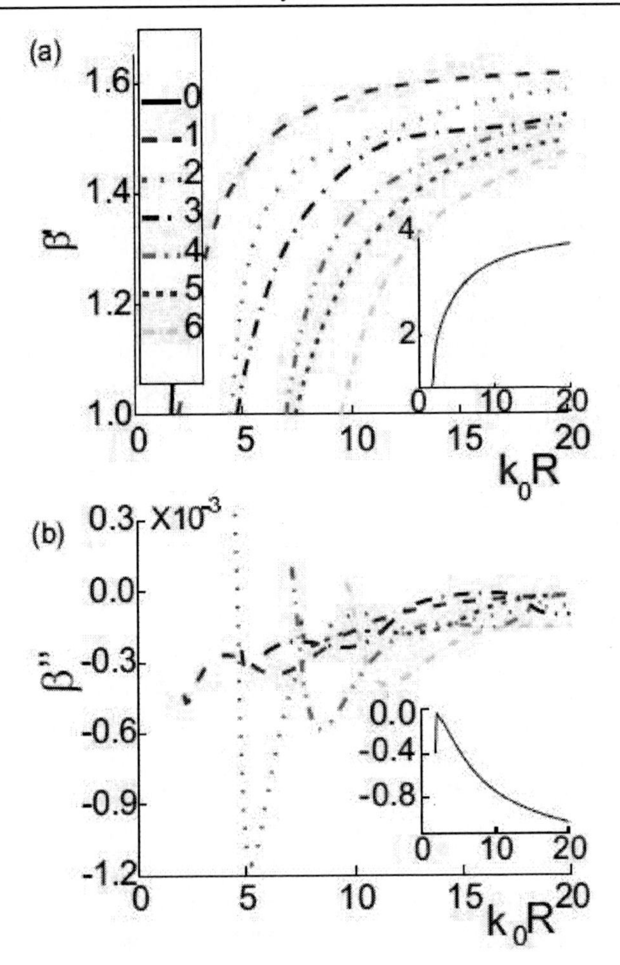

Figure 38. (a) Real β' and (b) imaginary β'' parts of β vs k_0R for both types of modes the parameter values $\epsilon_1 = 2.25$, $\epsilon_2 = 2.65$, $\epsilon_3 = 2.45$,and $\theta = \alpha = \pi/4$.

References

[1] P.G. de Gennes, J. Prost, The Physics of Liquid Crystals, (Clarendon Press, Oxford, UK, 1993)

[2] S. Chandrasekhar, Liquid Crystals (Cambridge University Press, Cambridge, 1992)

[3] F. C. Frank, *Faraday Soc. Discuss.* **25** 19 (1958)

[4] L. P. Kadanoff y P. C. Martin, *Ann. Phys.* **24**, 419 (1963)

[5] P. Hohenberg y P. C. Martin, *Ann. Phys.* **34**, 291 (1965)

[6] I. M. Khalatnikov, *Introduction to the Theory of Superfluidity* (Benjamin, New York, 1965)

[7] N. N. Bogoljubov, *Phys. Abhandl. SU* **6**, 229 (1962)

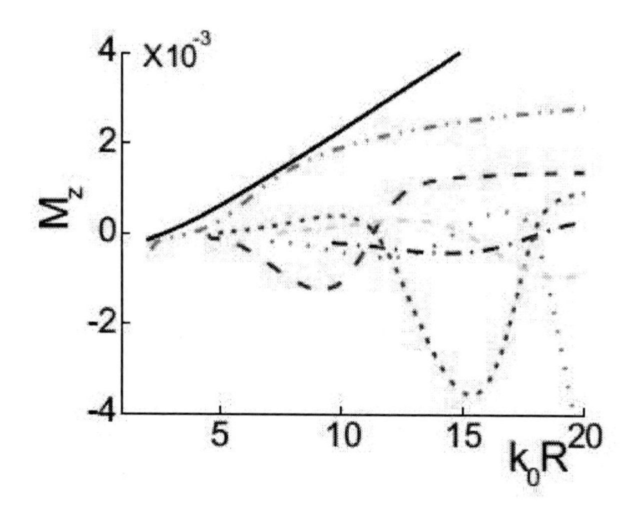

Figure 39. Radial component of the Poynting vector S_r vs k_0R for each mode in Fig. 38.

[8] O. Parodi, *J. Phys. (Fr)* **31**, 581 (1970)

[9] D. Forster, *Hydrodynamic Fluctuations, Broken Symmetry and Correlation Functions* (Benjamin, Reading, 1975)

[10] M. J. Stephen y J. P. Straley, *Rev. Mod. Phys.* **46**, 617 (1974)

[11] H. R. Brand y H. Pleiner, *J. Phys (Paris)* **41**, 553 (1980)

[12] H. R. Brand y H. Pleiner, *Phys. Rev. A* **24**, 1783 (1981); **26**, 1783 (1982); **30**, 1548 (1984)

[13] H. Pleiner, *Liq. Cryst.* **1**, 197 (1986)

[14] H. Pleiner, *Phys. Rev. A* **37**, 3986 (1988)

[15] H. R. Brand y H. Pleiner, *Physica A* **208**, 359 (1994)

[16] L. D. Landau y E. Lifshitz, *Theory of Elasticity* (Pergamon, New York, 1964) 3rd. edition

[17] H. B. Callen, *Thermodynamics and an Introduction to Thermostatistics* (Wiley,New York, 1985) 2a. ediciön

[18] L. M. Blinov y V. G. Chigrinov, *Electrooptic Effects in Liquid Crystal Materials* (Springer, New York, 1994)

[19] G. Ahlers, D. S. Cannell, L. I. Berge y S.Sakurai, *Phys. Rev. E* **49**, 545 (1994)

[20] T. Carlsson, *Mol. Cryst. Liq. Cryst.* **104**, 307-334 (1984).

[21] M. Miesowicz, *Nature* **17**, 261 (1935).

[22] G. Kiss and R. S. Porter, *J. Polym. Sci. Polym. Symp.* **65**, 193 (1978)

[23] G. Marrucci and P. L. Maffettone, *Macromolecules* **22,** 4076 (1989)

[24] M. Doi and S. F. Edwards, *The Theory of Polymer Dynamics* (Oxford University Press, New York, 1986)

[25] R. B. Bird, R. C. Armstrong and O. Hassager, *Dynamic of Polymeric Liquids,* (Wiley, New York, 1977) Vol 1

[26] A. Corella-Madueño and J. Adrián Reyes, *Opt. Commun.* **264**, 148 (2006).

[27] G. P. Crawford, D. W. Allender, and J. W. Doane, *Phys. Rev. A* **45**, 8693 (1992).

[28] P. Halevi, J. A. Reyes-Avendaño, and J. A. Reyes-Cervantes, *Phys. Rev. E* **73**, 040701 (R) (2006).

[29] A. Corella-Madueño, A. Castellanos-Moreno, S. Gutiérrez-Lóopez, and R. A. Rosas and J. Adrian Reyes, *Phys Rev. E* **78** 022701 (2008)

[30] I. K. Yang and A. D. Shine, *J. Rheol.* **36**, 1079-1104 (1992)

[31] R. F. Rodríguez, J.A. Reyes and O. Manero, *J. Chem. Phys.,* **110**(16), 8197-8204 (1999)

[32] I. C. Khoo, *Nonlinear Optics of Liquid Crystals, Progress in Optics, Vol.* **26**, E. Wolf, editor (North Holland, Amsterdam, 1988)

[33] R. F. Rodríguez and J. A. Reyes, *Mol. Cryst. Liq. Cryst.* **282**, 287-296 (1996)

[34] R. F. Rodríguez and J. F. Camacho, *Rev. Mex. Fis.* **44** 1859-1862 (1998)

[35] J. A. Reyes, O. Manero and R. F. Rodríguez, *Rheol. Acta,* **40** 426-443 (2001)

[36] Carlsson T, Sharp K (1981) *Mol. Cryst. Liq. Cryst.* **78**: 157-171

[37] Ian W. Stewart, The Static and Dynamic Continuum Theory of Liquid Crystals Taylor & Francis, London, 2004.

[38] W. H. Press, B. P. Flannery, S. A. Teukolsky, and W. T. Vetterling, *Numerical Recipes: The Art of Scientific Computing* (Cambridge, 1986)

[39] J. Adrian Reyes, A Corella Madueño and Carlos I. Mendoza, *J. Chem. Phys.* **129** 084710 (2008)

[40] Carlos I. Mendoza, A. Corella-Madueño, and J.Adrián Reyes, *Phys rev. E.* **77** 011706 (2008)

[41] *Ray propagation in inhomogeneous anisotropic media,* J. A. Reyes, *J. Phys A Math Gen,* **32**(18) 3409-3418 (1999)

[42] *Ray propagation in nematic droplets*, J. A. Reyes, *Phys. Rev. E* **57** 6700-6705 (1998).

[43] *Ray Tracing and reflectivity measurements in nematic hybrid cells*, J. A. Olivares, R. F. Rodríguez, J. A. Reyes, *Opt. Commun.* **221**(4-6) 223-239 (2003)

[44] *Electrically controlled total internal reflection in nematic hybrid cells,* C. I. Mendoza, J. A. Olivares and J. A. Reyes, *Phys. Rev. E* **70** 062701 (2004)

[45] *Geometrical analysis of the electro-optical effect in nematic droplets*, L. Jiménez*, Carlos I. Mendoza and J. Adrián Reyes, *Phys. Rev. E,* **72** 051705 (2005)

[46] *Dispersion effects on the ray tracing and reflectivity in a nematic cell hybrid cell under an electric field.* Carlos I. Mendoza, R. de la Teja, J. A. Olivares and J. A. Reyes *Rev. Mex Fis.* **52**(5) (2006)

[47] *Light propagation and transmission in hybrid-aligned nematic liquid crystal cells: Geometrical optics calculations*, Carlos I. Mendoza and J. Adrian Reyes, *Appl. Phys. Lett* **89** 091912 (2006). Copyright [2010], American Institute of Physics.

[48] *Guiding of Optical Fields in a Smectic B Liquid Crystal*, Laura O. Palomares, J. Adrián Reyes. *Rev. Mex. Fis. S* **52** (5) 71-79, 2006.

[49] *Band structure for a bend-core liquid crystal fiber* Laura O. Palomares, P. Castro-Garay, and J. Adrian Reyes. *Appl. Phys. Lett.* **94**, 181903 (2009). Copyright [2010], American Institute of Physics.

[50] Born M and Wolf E 1975 *Principles of Optics* (New York: Pergamon)

[51] Barreman D W 1972 *J. Opt. Soc. Am.* **62** 502

[52] Oldano C 1989 *Phys. Rev. A* **40** 6014

[53] Santamato E and Shen Y R 1987 *J. Opt. Soc. Am.* A **4** 356

[54] Kratsov Y A, Naida O N and Fuki A A 1996 *Sov. Phys. - Usp.* **39** 129

[55] Barkovskii L M and Khan Fo T N 1990 *Sov. Phys.-Acoust.* **36** 550

[56] Barkovskii L M and Khan Fo T N 1991 *Opt. Spectrosk.* **70** 34

[57] Naida O N 1977 *Sov. Phys.-Dokl.* **22** 575

[58] Jackson J D 1975 *Classical Electrodynamics* (New York: Wiley)

[59] Fedorov I V 1965 *Theory of Elastic Waves in Crystals* (Moscow: Nauka)

[60] Goldstain H 1998 *Classical Mechanics* " 2nd edn (Reading, MA: Addison-Wesley) section 10.3

[61] Sneddon I N 1957 *Partial Differentia Equations* (New York: McGraw-Hill) section 2.10

[62] D. Riviere, Y. Levy, C. Imbert, *Opt. Commun.* **25** (1978) 206.

[63] Y. Levy, D. Riviere, C. Imbert, M. Boix, *Opt. Commun.* **26** (1978) 225.

[64] D. Riviere, Y. Levy, E. Guyon, *J. Phys. Lett.* **40** (1979) L215.

[65] S. Naemura, *Appl. Phys. Lett.* **33** (1978) 1. [5] D. Riviere, Y. Levy

[66] H. Lin, P. Pallfy-Muhoray, and M. A. Lee, *Mol. Cryst. Liq. Cryst.* **204**, 180 (1991)

[67] J. A. Reyes and R. F. Rodríguez, *Opt. Commun.* **134**, 349 (1997).

[68] J. W. Doane, N. A. Vaz, B.-G. Wu, and S. Zumer, *Appl. Phys. Lett.* **48**, 269 (1986); J. W. Doane, *MRS Bull.* **16**, 22 (1991).

[69] M. Schadt, *Liq. Cryst.* **14**, 73 (1993)

[70] H. S. Kitzerov, *Liq. Cryst.* **16**, 1 (1994).

[71] M. Schadt, *Liq. Cryst.* **14**, 73 (1993).

[72] H. S. Kitzerov, *Liq. Cryst.* **16**, 1 (1994).

[73] I-C. Khoo, *Liquid Crystals: Physical Properties and Nonlinear Optical Phenomena*, (Wiley, New York, 1994).

[74] E. Santamato, B. Daino, M. Romagnoli, M. Settembre, and Y. R. Shen, *Phys. Rev. Lett.* **64**, 1377 (1990); G. Abbate, P. Maddalena, L. Marrucci, L. Saetta, and E. Santamato, *J. Phys. II* **1**, 543 (1991).

[75] D. W. Allender, G. L. Henderson, and D. L. Jhonson, *Phys. Rev. A* **24**, 1086 (1981).

[76] S. Zumer and J. W. Doane, *Phys. Rev. A* **34**, 3373 (1986)

[77] G. Pelzl, S. Diele, A. Jakli, C.H. Lischka, I. Wirth and W. Weissflog, *Liq. Cryst.* **26**, 135 (1999).

[78] G. Liao, S. Stojadinovic, G. Pelz, W. Weisflog, S. Sprunt, A. *Jakli. Phys. Rev E.* **72** 021710, 2005.

[79] Willian H. Press et al, *Numerical Recipes: The Art of Scientific Computing* Third edition, Sec. 17.1, (Cambridge, Cambrige University Press, 2007).

[80] 3S. Matsumoto, M. Kawamoto, and K. Mizunoya, *J. Appl. Phys.* **47**, 3842 (1976).

[81] 1A. Bogi and S. Faetti, *Liq. Cryst.* **28**, 729 (2001); A. V. Zakharov and R. Y. Dong, *Phys. Rev. E* **64**, 031701 (2001).

[82] G. B. Arfken, *Mathematical Methods for Physics*, 3rd ed. (Academic, Ohio, 1985).

[83] H. Schlichting, *Boundary Layer Theory* McGraw-Hill, New York, 1968 .

[84] M. C. Bender and M. C. Arszog, *Advanced Mathematical Methods for Scientists and Engineers* McGraw-Hill, New York, 1978 , Sec. 9.2.

In: Advances in Condensed Matter ... Volume 10
Editors: H. Geelvinck and S. Reynst
ISBN: 978-1-61209-533-2

Chapter 7

DETERMINATION METHODS OF DENSITIES AND ENERGY LEVELS OF IMPURITIES AND DEFECTS AFFECTING MAJORITY-CARRIER CONCENTRATION IN NEXT-GENERATION SEMICONDUCTORS

Hideharu Matsuura*
Osaka Electro-Communication University, Japan

Abstract

Wide bandgap semiconductors such as SiC, GaN, and diamond have a potential for use in high power and high frequency devices, while narrow bandgap semiconductors such as the GaSb family have a potential for near- and mid-infrared laser diodes and photo-detectors for detecting CO_2, CH_4, NO_x, and SO_x. In these next-generation semiconductors, it is essential to precisely determine the densities and energy levels of dopants (donors or acceptors) as well as unintentionally-introduced impurities and defects, which affect the majority-carrier concentrations in semiconductors.

We have developed a graphical peak analysis method called Free Carrier Concentration Spectroscopy (FCCS), which can accurately determine them using the temperature dependence of the majority-carrier concentration without any assumptions regarding dopant species, impurities, and defects. We have determined the densities and energy levels in undoped, N-doped or Al-doped SiC. Moreover, the dependence of the energy level of each dopant species on dopant density has been obtained. From the temperature dependence of the majority-carrier concentration in SiC irradiated by high-energy electrons, the dependence of the density of each dopant or defect on fluence has been determined.

Because wide bandgap semiconductors have a low dielectric constant and a hole effective mass heavier than an electron effective mass, the energy levels of acceptor species are expected to be deep. When x is larger, on the other hand, the donor level of Te in $Al_xGa_{1-x}Sb$ is changed from shallow to deep, just like $Al_xGa_{1-x}As$. In these semiconductors, the density of dopants determined from the temperature dependence of the majority-carrier concentration is much higher than the concentration of dopant

*E-mail address: matsuura@isc.osakac.ac.jp

species determined by secondary ion mass spectroscopy. When the energy level of dopant species is deep, the excited states of the dopant should affect the majority-carrier concentration. Therefore, we have introduced the occupation function of the dopant with deep energy level, which includes the influence of its excited states, instead of the Fermi-Dirac distribution function. Using the occupation function we have proposed, we have investigated semiconductors with deep level dopants by FCCS using the temperature dependence of the majority-carrier concentration.

Recently, high-resistivity or semi-insulating semiconductors have been required to use as substrates of GaN field-effect transistors and as active layers of X-ray detectors capable of operating at room temperature. Because these defects affect the majority-carrier concentrations in high-resistivity or semi-insulating semiconductors and degrade the device performance, it is required to accurately determine the densities and energy levels of defects with deep energy levels. Transient capacitance methods such as Deep Level Transient Spectroscopy cannot apply higher-resistivity semiconductors. Therefore, we have been developing a graphical peak analysis method using an isothermal transient current in the diodc, called Discharge Current Transient Spectroscopy (DCTS). Using DCTS, we have determined the densities, emission rates, and energy levels of deep defects in semi-insulating SiC and thin insulators. Moreover, the density-of-states in high-resistivity amorphous semiconductors are investigated.

PACS 71.55.-i, 71.55.Ht, 72.20.Jv, 73.61.Le

Keywords: Determination of densities and energy levels, Next-generation semiconductor, SiC, GaN, Diamond, InGaSb, AlGaSb, Donor, Acceptor, Deep level.

1. Introduction

In this chapter, we discuss methods to evaluate impurities and defects that affect the majority-carrier concentration in a semiconductor whose resistivity varies from low to extremely high.

The densities and energy levels of traps (i.e., impurities or defects) have usually been evaluated using deep level transient spectroscopy (DLTS) [1]. However, a quantitative relationship between the majority-carrier concentration and the trap densities cannot be obtained using DLTS. The reason for this is that in the DLTS analysis, the following approximation is assumed.

$$\begin{aligned} C(t) &= C(\infty)\sqrt{1-\frac{N_{\mathrm{T}}}{N_{\mathrm{dopant}}+N_{\mathrm{T}}}\exp\left(-\frac{t}{\tau_{\mathrm{T}}}\right)} \\ &\simeq C(\infty)\left[1-\frac{N_{\mathrm{T}}}{2\left(N_{\mathrm{dopant}}+N_{\mathrm{T}}\right)}\exp\left(-\frac{t}{\tau_{\mathrm{T}}}\right)\right] \end{aligned} \tag{1}$$

when

$$\frac{N_{\mathrm{T}}}{N_{\mathrm{dopant}}+N_{\mathrm{T}}} \ll 1 , \tag{2}$$

where $C(t)$ is the transient capacitance after removal of the filling pulse and return to the steady-state reverse bias voltage (V_{R}) in the DLTS measurement sequence, $C(\infty)$ is the steady-state capacitance at V_{R}, N_{dopant} is the dopant density (i.e., donor or acceptor density), N_{T} is the trap density, and τ_{T} is the time constant corresponding to the trap. Based on Eq. (2), DLTS can determine the density and energy level of the trap only when N_{T} is much lower than N_{dopant}, indicating that the trap determined by DLTS barely affects the majority-carrier concentration. If the densities and energy levels of traps can be determined using the experimental majority-carrier concentration, the relationship between the majority-carrier concentration and the trap densities can be investigated directly. Therefore, we discuss a unique method to characterize them from the temperature dependence of the majority-carrier concentration obtained by Hall-effect measurements.

Transient capacitance methods such as DLTS, furthermore, cannot be applied to high-resistivity semiconductors, because the measured capacitance of a diode is determined by the thickness of the diode, not by the depletion region of the junction due to its long dielectric relaxation time. For example, the resistivity (ρ) of high-purity SiC is strongly affected by intrinsic defects located in its midgap, which capture the majority carriers. Therefore, it is necessary to investigate the nature of the deep levels in high-purity SiC. To characterize the intrinsic defects in SiC, however, N-doped low-resistivity SiC irradiated by electrons has been intensively investigated by DLTS [2, 3].

Thermally stimulated current (TSC) [4] is suitable for characterizing traps in high-resistivity or semi-insulating semiconductors. The TSC signal, $I_{\mathrm{TSC}}(T)$, is theoretically given by

$$I_{\mathrm{TSC}}(T) = qA\sum_{i=1} N_{\mathrm{T}i}\nu_{\mathrm{T}i}\exp\left[-\frac{\Delta E_{\mathrm{T}i}}{kT} - \frac{\nu_{\mathrm{T}i}}{\beta}\int_{T_0}^{T}\exp\left(-\frac{\Delta E_{\mathrm{T}i}}{kT}\right)\mathrm{d}T\right] + I_{\mathrm{ssl}}(T), \quad (3)$$

where q is the electron charge, k is the Boltzmann constant, β is the heating rate, $N_{\mathrm{T}i}$ is the density for an ith trap, $\Delta E_{\mathrm{T}i}$ is the activation energy for the ith trap, $\nu_{\mathrm{T}i}$ is the attempt-to-escape frequency for the ith trap, T_0 is the temperature at which heating is started, $I_{\mathrm{ssl}}(T)$ is the steady-state leakage current at a measurement temperature (T), and A is the electrode area. It should be noted that Eq. (3) is available only in thermionic emission processes. Parameters for fitting a curve to the experimental $I_{\mathrm{TSC}}(T)$ are the sets of $N_{\mathrm{T}i}$, $\Delta E_{\mathrm{T}i}$ and $\nu_{\mathrm{T}i}$. However, it is difficult to analyze experimental TSC data when traps with close emission rates are included. Moreover, because the effect of pyroelectric currents and the temperature dependence of the steady-state leakage current must be considered in the TSC analysis, an isothermal measurement is more suitable for characterizing traps than TSC is. Therefore, we discuss a unique method to characterize them using isothermal transient current measurements.

2. Temperature Dependence of Majority-Carrier Concentration

The temperature dependence of the majority-carrier concentration has a lot of information on impurities and defects. In n-type semiconductors, for example, the temperature dependence of the electron concentration, $n(T)$, can be expressed as [5]

$$n(T) = \sum_{i=1} N_{\mathrm{D}i}\left[1 - f_{\mathrm{FD}}(E_{\mathrm{D}i})\right] - \sum_{j=1} N_{\mathrm{TE}j} f_{\mathrm{FD}}(E_{\mathrm{TE}j}) - N_{\mathrm{A}}, \quad (4)$$

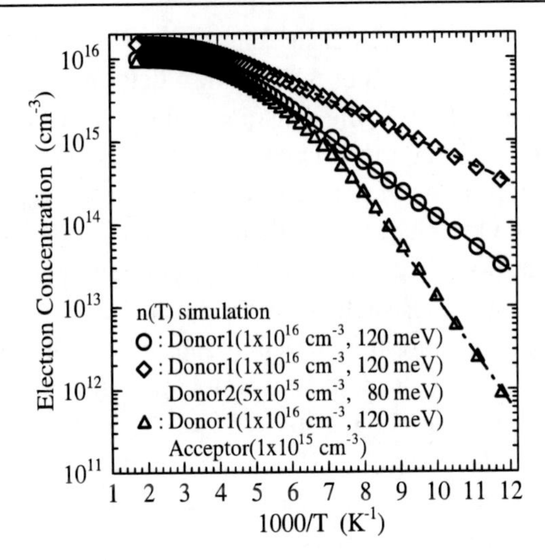

Figure 1. Simulations of temperature-dependent electron concentration for three cases of one donor, two donors, and one donor and one acceptor.

in the temperature range in which a temperature-dependent hole concentration, $p(T)$, is much less than $n(T)$, where $N_{\mathrm{D}i}$ and $E_{\mathrm{D}i}$ are the density and energy level of an ith donor, $N_{\mathrm{TE}j}$ and $E_{\mathrm{TE}j}$ are the density and energy level of a jth electron trap that is negative when it captures an electron from the conduction band, N_{A} is the acceptor density, $f_{\mathrm{FD}}(E_{\mathrm{D}i})$ and $f_{\mathrm{FD}}(E_{\mathrm{TE}j})$ are the Fermi-Dirac distribution function for donors and electron traps, given by [5]

$$f_{\mathrm{FD}}(E) = \frac{1}{1 + \frac{1}{2}\exp\left(\frac{E - E_{\mathrm{F}}(T)}{kT}\right)}, \tag{5}$$

and $E_{\mathrm{F}}(T)$ is the Fermi level at T. From $n(T)$, therefore, we can determine the densities and energy levels of donors and electron traps.

The temperature dependencies of the majority-carrier concentration and mobility can usually be obtained by Hall-effect measurements. When a sample is an n-type epilayer on a substrate, for example, the substrate should be a p-type or semi-insulating semiconductor.

Let us evaluate donors using the $n(T)$ simulations for 4H-SiC in the cases of one donor ($\bigcirc$), two donors ($\diamond$), and one donor and one acceptor ($\triangle$) in Fig. 1. Here, the density and energy level are 1×10^{16} cm^{-3} and $E_{\mathrm{C}} - 0.12$ eV for Donor1, and 5×10^{15} cm^{-3} and $E_{\mathrm{C}} - 0.08$ eV for Donor2, respectively, and the acceptor density is 1×10^{15} cm^{-3}.

In the case of one donor, which is described in university text books, $n(T)$ can be expressed by the following two equations:

$$n(T) = N_{\mathrm{D}}\left[1 - f_{\mathrm{FD}}(E_{\mathrm{D}})\right] \tag{6}$$

and

$$n(T) = N_{\mathrm{C}}(T)\exp\left(-\frac{E_{\mathrm{C}} - E_{\mathrm{F}}(T)}{kT}\right), \tag{7}$$

where

$$N_C(T) = N_{C0} k^{3/2} T^{3/2}, \tag{8}$$

$$N_{C0} = 2 \left(\frac{2\pi m_e^*}{h^2} \right)^{3/2} M_C, \tag{9}$$

N_D is the donor density, E_C is the conduction band maximum, m_e^* is the electron effective mass, M_C is the number of equivalent minima in the conduction band, and h is the Planck's constant. At low temperatures where

$$N_D \gg \frac{N_C(T)}{8} \exp\left(-\frac{E_C - E_D}{kT} \right), \tag{10}$$

the following equation is derived using Eqs. (6) and (7).

$$n(T) \cong \sqrt{\frac{N_C(T) N_D}{2}} \exp\left(-\frac{E_C - E_D}{2kT} \right) \tag{11}$$

$$= T^{3/4} \left(\frac{2\pi m_e^* k}{h^2} \right)^{3/4} \sqrt{M_C} \sqrt{N_D} \exp\left(-\frac{E_C - E_D}{2kT} \right). \tag{12}$$

Therefore, the donor level ($E_C - E_D$) is determined from the slope at low temperatures in Fig. 1. Moreover, N_D is equal to the saturation value in Fig. 1.

From each $n(T)$ at low temperatures in Fig. 1, the straight solid, broken and dotted-dashed lines can be drawn for the cases of one donor, two donors, and one donor and one acceptor, respectively. We can obtain the correct results that the density and energy level of the donor are determined to be 1×10^{16} cm^{-3} and 0.12 eV for $\bigcirc$, respectively. On the other hand, the density and energy level of the donor are determined to be 1.5×10^{16} cm^{-3} and 0.08 eV for $\diamond$, and 0.9×10^{16} cm^{-3} and 0.25 eV for $\triangle$, which are incorrect results. Therefore, this analysis can be applied only when we can confirm that the semiconductor includes only one type of donor species.

In the case of one donor and one acceptor, the following equation is derived using Eqs. (6) and (7).

$$\frac{1}{T^{3/2}} \cdot \frac{n(T)[n(T) + N_A]}{(N_D - N_A) - n(T)} = \left(\frac{2\pi m_e^* k}{h^2} \right)^{3/2} M_C \exp\left(-\frac{E_C - E_D}{kT} \right). \tag{13}$$

We define y as

$$y = \frac{1}{T^{3/2}} \cdot \frac{n(T)[n(T) + N_A]}{(N_D - N_A) - n(T)}, \tag{14}$$

and we try to search the values of N_A and ($N_D - N_A$) that make the relationship between y and $1/T$ straight. From the slope of the straight line, the donor level can be determined using Eq. (13).

In the case of more than two types of donor species, the curve-fitting method is applied. Using

$$n(T) = \sum_{i=1}^{n} N_{Di} \left[1 - f_{FD}(E_{Di}) \right] - N_A, \tag{15}$$

and Eq. (7), the $n(T)$ can be simulated. By a least-squares fit of the $n(T)$ simulation to the experimental $n(T)$, the fitting parameters of n sets of $N_{\mathrm{D}i}$ and $E_{\mathrm{D}i}$ as well as N_{A} are obtained. Prior to the analysis, the number of donor species and the initial values of the parameters should be assumed. The assumptions, however, strongly affect the final results.

In the differential Hall-effect spectroscopy (DHES) [6, 7],

$$-kT\frac{\mathrm{d}n(T)}{\mathrm{d}\Delta E_{\mathrm{F}}(T)} = \sum_{i=1}^{n} N_{\mathrm{D}i}\frac{2\exp\left(\dfrac{\Delta E_{\mathrm{D}i}-\Delta E_{\mathrm{F}}(T)}{kT}\right)}{\left[1+2\exp\left(\dfrac{\Delta E_{\mathrm{D}i}-\Delta E_{\mathrm{F}}(T)}{kT}\right)\right]^2} \times\left[1+\frac{\Delta E_{\mathrm{D}i}-\Delta E_{\mathrm{F}}(T)}{kT}\cdot\frac{\partial(kT)}{\partial\Delta E_{\mathrm{F}}(T)}\right] \tag{16}$$

where $\Delta E_{\mathrm{D}i} = E_{\mathrm{C}} - E_{\mathrm{D}i}$ and $\Delta E_{\mathrm{F}}(T) = E_{\mathrm{C}} - E_{\mathrm{F}}(T)$. When

$$\frac{\Delta E_{\mathrm{D}i}-\Delta E_{\mathrm{F}}(T)}{kT}\cdot\frac{\partial(kT)}{\partial\Delta E_{\mathrm{F}}(T)} \ll 1, \tag{17}$$

Eq. (16) can be rewritten approximately as

$$-kT\frac{\mathrm{d}n(T)}{\mathrm{d}\Delta E_{\mathrm{F}}(T)} \cong \sum_{i=1}^{n} N_{\mathrm{D}i}\frac{2\exp\left(\dfrac{\Delta E_{\mathrm{D}i}-\Delta E_{\mathrm{F}}(T)}{kT}\right)}{\left[1+2\exp\left(\dfrac{\Delta E_{\mathrm{D}i}-\Delta E_{\mathrm{F}}(T)}{kT}\right)\right]^2}. \tag{18}$$

When we define y as

$$y = -kT\frac{\mathrm{d}n(T)}{\mathrm{d}\Delta E_{\mathrm{F}}(T)}, \tag{19}$$

the peak of the y - $\Delta E_{\mathrm{F}}(T)$ characteristics occurs at

$$\Delta E_{\mathrm{F}}(T_{\mathrm{peak}i}) = \Delta E_{\mathrm{D}i} + kT_{\mathrm{peak}i}\ln 2. \tag{20}$$

However, the differential of the experimental data results in an increase in the observational errors.

As a consequence, we adopt the following criteria for an evaluation function that can determine the densities and energy levels of donors and electron traps from the experimental $n(T)$;

1. not to require any assumptions regarding donor species and electron traps before analysis,
2. not to differentiate the experimental data,
3. to have a peak value at the temperature corresponding to the energy level of each donor or electron trap,
4. to be able to verify the values determined.

3. Free Carrier Concentration Spectroscopy

3.1. Basic Concept

Transient capacitance methods, such as DLTS [1] and isothermal capacitance transient spectroscopy (ICTS) [8], can uniquely determine the densities and energy levels of traps in semiconductors, because each peak in the signal corresponds one-to-one to a trap. For example, the ICTS signal is defined as $S(t) \equiv t\mathrm{d}C(t)^2/\mathrm{d}t$. Since $S(t)$ is theoretically described as the sum of $N_{\mathrm{T}i}e_{\mathrm{T}i}t\exp(-e_{\mathrm{T}i}t)$, it has a peak value of $N_{\mathrm{T}i}\exp(-1)$ at a peak time of $t_{\mathrm{peak}i} = 1/e_{\mathrm{T}i}$. Here, $e_{\mathrm{T}i}$ is the emission rate of an ith trap. Therefore, the function $N_{\mathrm{T}i}e_{\mathrm{T}i}t\exp(-e_{\mathrm{T}i}t)$ plays an important role in the ICTS analysis.

To analyze the experimental $n(T)$, we introduced the function theoretically described as the sum of $N_{\mathrm{D}i}\exp(-\Delta E_{\mathrm{D}i}/kT)/kT$ [9, 10]. The function $N_{\mathrm{D}i}\exp(-\Delta E_{\mathrm{D}i}/kT)/kT$ has a peak at $T_{\mathrm{peak}i} = \Delta E_{\mathrm{D}i}/k$. If we can introduce a function in which the peak appears at $T_{\mathrm{peak}i} = (\Delta E_{\mathrm{D}i} - E_{\mathrm{ref}})/k$, we can shift the peak temperature to the measurement temperature range by changing the parameter E_{ref}. This indicates that we can determine $N_{\mathrm{D}i}$ and $\Delta E_{\mathrm{D}i}$ in a wide range of donor levels in a limited measurement temperature range. Therefore, the function to be evaluated should be approximately described as the sum of $N_{\mathrm{D}i}\exp[-(\Delta E_{\mathrm{D}i} - E_{\mathrm{ref}})/kT]/kT$. It should be noted that $N_{\mathrm{D}i}$ and $\Delta E_{\mathrm{D}i}$ determined by this method are independent of E_{ref}.

3.2. Theoretical Consideration

3.2.1. n-type Semiconductors

From Eqs. (4) and (7), a favorable function to determine $N_{\mathrm{D}i}$, $\Delta E_{\mathrm{D}i}$, $N_{\mathrm{TE}j}$, $\Delta E_{\mathrm{TE}j}$ and N_{A} can be defined as

$$H(T, E_{\mathrm{ref}}) \equiv \frac{n(T)^2}{(kT)^{5/2}}\exp\left(\frac{E_{\mathrm{ref}}}{kT}\right). \tag{21}$$

Substituting Eq. (4) for one of the $n(T)$ in Eq. (21) and substituting Eq. (7) for the other $n(T)$ in Eq. (21) yield

$$\begin{aligned} H(T, E_{\mathrm{ref}}) &= \sum_{i=1} \frac{N_{\mathrm{D}i}}{kT}\exp\left(-\frac{\Delta E_{\mathrm{D}i} - E_{\mathrm{ref}}}{kT}\right) I_{\mathrm{n}}(\Delta E_{\mathrm{D}i}) \\ &\quad + \sum_{j=1} \frac{N_{\mathrm{TE}j}}{kT}\exp\left(-\frac{\Delta E_{\mathrm{TE}j} - E_{\mathrm{ref}}}{kT}\right) I_{\mathrm{n}}(\Delta E_{\mathrm{TE}j}) \\ &\quad - \frac{N_{\mathrm{comp}}N_{\mathrm{C0}}}{kT}\exp\left(\frac{E_{\mathrm{ref}} - \Delta E_{\mathrm{F}}(T)}{kT}\right), \end{aligned} \tag{22}$$

where

$$\Delta E_{\mathrm{TE}j} = E_{\mathrm{C}} - E_{\mathrm{TE}j}, \tag{23}$$

$$I_{\mathrm{n}}(\Delta E) = \frac{N_{\mathrm{C0}}}{2 + \exp\left(\dfrac{\Delta E_{\mathrm{F}}(T) - \Delta E}{kT}\right)}, \tag{24}$$

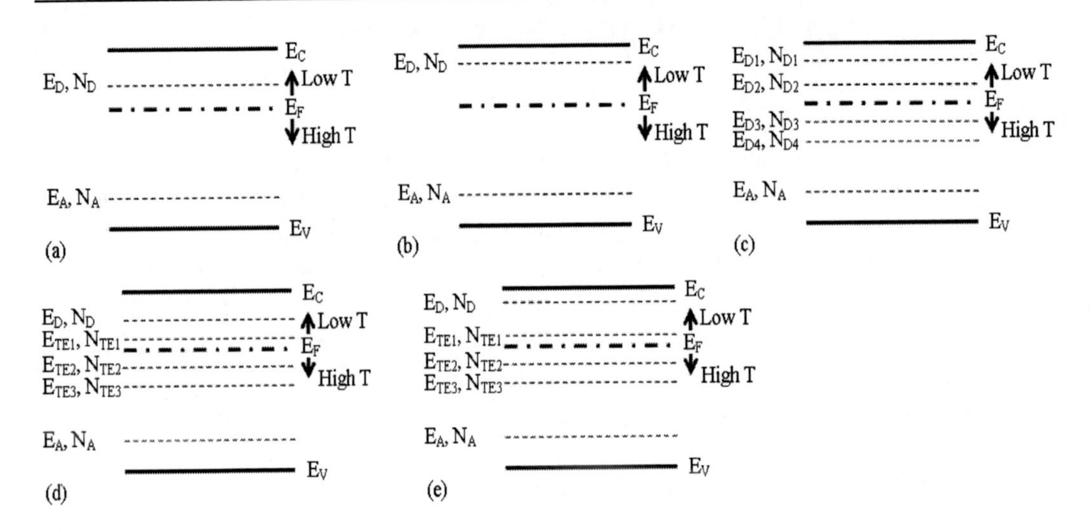

Figure 2. Five cases. The Fermi level moves from LowT to HighT as the measurement temperature rises.

and $N_{\rm comp}$ is the compensating density given by

$$N_{\rm comp} = N_{\rm A} + \sum_{j=1} N_{{\rm TE}j}. \tag{25}$$

The function

$$\frac{N_i}{kT}\exp\left(-\frac{\Delta E_i - E_{\rm ref}}{kT}\right) \tag{26}$$

in Eq. (22) has a peak value of $N_i \exp(-1)/kT_{{\rm peak}i}$ at a peak temperature

$$T_{{\rm peak}i} = \frac{\Delta E_i - E_{\rm ref}}{k}. \tag{27}$$

Although the actual $T_{{\rm peak}i}$ of $H(T, E_{\rm ref})$ is slightly different from the $T_{{\rm peak}i}$ calculated by Eq. (27) due to the temperature dependence of $I(\Delta E_{{\rm D}i})$, we can easily determine the accurate values of ($N_{{\rm D}i}$ and $\Delta E_{{\rm D}i}$) or ($N_{{\rm TE}i}$ and $\Delta E_{{\rm TE}i}$) from the peak of the experimental $H(T, E_{\rm ref})$, using a personal computer. The Windows application software for FCCS can be freely downloaded at our web site (http://www.osakac.ac.jp/labs/matsuura/). This software can also evaluate them by using the curve-fitting method, DHES, and the other methods mentioned in Section 2.

Let us consider the meaning of $N_{\rm comp}$ using Fig. 2. As the measurement temperature rises, the Fermi level moves from LowT to HighT in Fig. 2. In Fig. 2(a), the Fermi level crosses $E_{\rm D}$ in the temperature range of measurement. So, we can determine $N_{\rm D}$ and $E_{\rm D}$. All the acceptors are negatively ionized. Therefore,

$$N_{\rm comp} = N_{\rm A}, \tag{28}$$

and $N_{\rm comp}$ is positive.

Because the Fermi level does not cross E_{D} in Fig. 2(b), we cannot determine E_{D}. All the donors are positively ionized, and all the acceptors are negatively ionized in the temperature range of measurement. Therefore

$$N_{\mathrm{comp}} = N_{\mathrm{A}} - N_{\mathrm{D}}, \tag{29}$$

and N_{comp} is negative because the conduction type is n-type.

Because the Fermi level crosses E_{D2} and E_{D3} in Fig. 2(c), we can determine N_{D2}, E_{D2}, N_{D3}, and E_{D3}. All the donors with E_{D1} are positively ionized, and all the acceptors are negatively ionized in the temperature range of measurement. Because all the donors with E_{D4} are neutral, they do not affect the electron concentration in the measurement temperature range. Therefore

$$N_{\mathrm{comp}} = N_{\mathrm{A}} - N_{\mathrm{D1}}. \tag{30}$$

Because the Fermi level crosses E_{D}, E_{TE1}, and E_{TE2} in Fig. 2(d), we can evaluate E_{D}, N_{D}, E_{TE1}, N_{TE1}, E_{TE2}, and N_{TE2}. All the electron traps with E_{TE3} are negatively ionized because they capture electrons, and all the acceptors are negatively ionized. Therefore

$$N_{\mathrm{comp}} = N_{\mathrm{A}} + \sum_{i=1}^{3} N_{\mathrm{TE}i} \tag{31}$$

and N_{comp} is positive. Here, electron traps are negative when they capture electrons, while they are neutral when they emit electrons.

Because the Fermi level crosses E_{TE1} and E_{TE2} in Fig. 2(e), we can evaluate E_{TE1}, N_{TE1}, E_{TE2}, and N_{TE2}. All the donors are positively ionized, all the electron traps with E_{TE3} are negatively ionized because they capture electrons, and all the acceptors are negatively ionized in the temperature range of measurement. Therefore

$$N_{\mathrm{comp}} = N_{\mathrm{A}} + \sum_{i=1}^{3} N_{\mathrm{TE}i} - N_{\mathrm{D}}. \tag{32}$$

Using FCCS, the densities and energy levels are determined from $n(T)$ for the case of two donors shown by $\diamond$ in Fig. 1. Figure 3(a) shows $H(T,0)$ calculated by Eq. (21). The peak temperature and value of $H(T,0)$ are 249 K and 1.93×10^{36} $\mathrm{cm^{-6}eV^{-2.5}}$, from which the energy level (ΔE_{D1}) and density (N_{D1}) of the corresponding donor species are determined as 120.1 meV and 1.12×10^{16} $\mathrm{cm^{-3}}$.

In order to investigate another donor species included in this semiconductor, the FCCS signal of $H2(T, E_{\mathrm{ref}})$, in which the influence of the previously determined donor species is removed, is calculated using the following equation. It is clear from Eq. (22) that

$$H2(T, E_{\mathrm{ref}}) = \frac{n(T)^2}{(kT)^{5/2}} \exp\left(\frac{E_{\mathrm{ref}}}{kT}\right) - \frac{N_{\mathrm{D1}}}{kT} \exp\left(-\frac{\Delta E_{\mathrm{D1}} - E_{\mathrm{ref}}}{kT}\right) I(\Delta E_{\mathrm{D1}}) \tag{33}$$

is not influenced by the donor with ΔE_{D1}. Figure 3(b) depicts $H2(T, 0.030754)$. Since a peak appears in this figure, another donor species is included. Using the peak temperature of 150.6 K and the peak value of 6.63×10^{36} $\mathrm{cm^{-6}eV^{-2.5}}$, the donor level ($\Delta E_{\mathrm{D2}}$) and the donor density (N_{D2}) are determined as 81.3 meV and 5.08×10^{15} $\mathrm{cm^{-3}}$. The

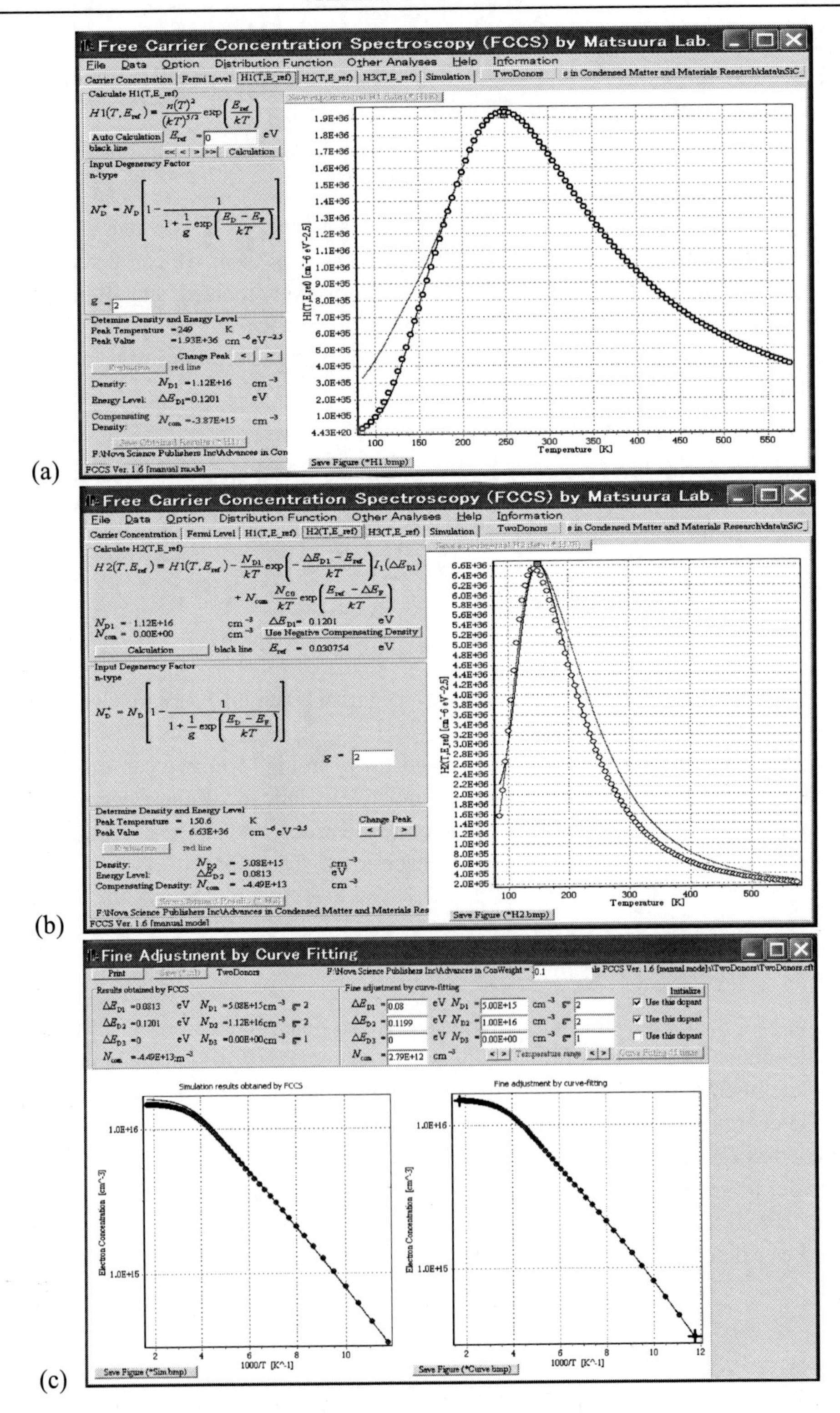

Figure 3. FCCS signal for case of two donors in Fig. 1; (a) $H(T, E_{\text{ref}})$, (b) $H2(T, E_{\text{ref}})$, and (c) curve fitting using values determined by FCCS.

FCCS signal of $H3(T, E_{\text{ref}})$, in which the influences of two donor species previously determined are removed, is calculated. However, $H3(T, E_{\text{ref}})$ is nearly zero, indicating that this semiconductor includes two types of donor species. Since this donor level is considered to be shallowest in the measurement temperature range, N_{comp} is determined to be -4.49×10^{13} cm^{-3} from the value of the signal at the lowest temperature.

Since it is found that two types of donor species are included and the approximate values of ΔE_{D1}, N_{D1}, ΔE_{D2}, N_{D2}, and N_{comp} were estimated, the curve-fitting method can be carried out. Figure 3(c) shows the result by the curve-fitting procedure. By the fine adjustment, the values of ΔE_{D1}, N_{D1}, ΔE_{D2}, N_{D2}, and N_{comp} are finally determined to be 119.9 meV, 1.00×10^{16} cm^{-3}, 80.0 meV, 5.00×10^{15} cm^{-3}, and 2.79×10^{12} cm^{-3}, respectively. The determined donor levels and densities are quite reasonable. Although N_{comp} determined by FCCS is not zero, the value is much less than the donor densities. It is, therefore, elucidated that FCCS can accurately determine the densities and energy levels of donors from $n(T)$.

3.2.2. p-type Semiconductors

From the charge neutrality condition, $p(T)$ can be written as [5]

$$p(T) = \sum_{i=1} N_{\text{A}i} f_{\text{FD}}(\Delta E_{\text{A}i}) - \sum_{j=1} N_{\text{TH}j} \left[1 - f_{\text{FD}}(\Delta E_{\text{TH}j})\right] - N_{\text{D}} \tag{34}$$

in the temperature range in which $n(T)$ is much less than $p(T)$, where $f_{\text{FD}}(\Delta E)$ is the Fermi-Dirac distribution function for acceptors and hole traps, which is given by [5]

$$f_{\text{FD}}(\Delta E) = \frac{1}{1 + 4\exp\left(-\dfrac{\Delta E_{\text{F}}(T) - \Delta E}{kT}\right)}, \tag{35}$$

$N_{\text{A}i}$ and $\Delta E_{\text{A}i}$ are the density and energy level measured from the valence band maximum (E_{V}) of an ith acceptor, respectively, $N_{\text{TH}j}$ and $\Delta E_{\text{TH}j}$ are the density and energy level measured from E_{V} of a jth hole trap, and $\Delta E_{\text{F}}(T)$ is the Fermi level measured from E_{V} at T.

On the other hand, using the effective density of states $N_{\text{V}}(T)$ in the valence band, $p(T)$ is expressed as [5]

$$p(T) = N_{\text{V}}(T) \exp\left(-\frac{\Delta E_{\text{F}}(T)}{kT}\right), \tag{36}$$

where

$$N_{\text{V}}(T) = N_{\text{V0}} k^{3/2} T^{3/2}, \tag{37}$$

$$N_{\text{V0}} = 2\left(\frac{2\pi m_{\text{h}}^*}{h^2}\right)^{3/2}, \tag{38}$$

m_{h}^* is the hole effective mass.

The FCCS signal is defined as

$$H(T, E_{\text{ref}}) \equiv \frac{p(T)^2}{(kT)^{5/2}} \exp\left(\frac{E_{\text{ref}}}{kT}\right). \tag{39}$$

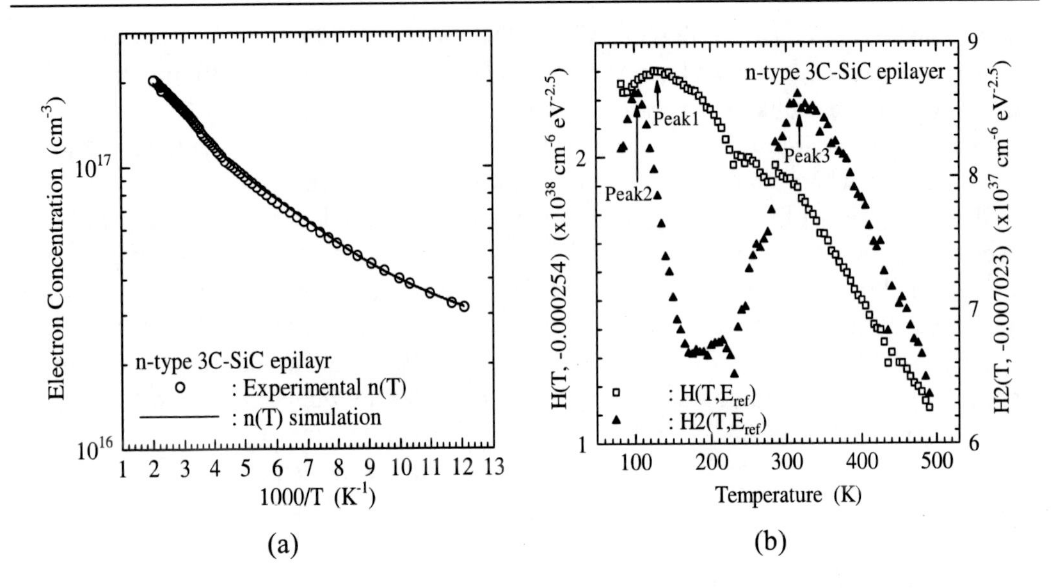

Figure 4. Temperature dependence of electron concentration for undoped 3C-SiC epilayer; (a) experimental and simulated $n(T)$ and (b) FCCS signals of $H(T, E_{\mathrm{ref}})$ and $H2(T, E_{\mathrm{ref}})$.

Substituting Eq. (34) for one of the $p(T)$ in Eq. (39) and substituting Eq. (36) for the other $p(T)$ in Eq. (39) yield

$$
\begin{aligned}
H(T, E_{\mathrm{ref}}) &= \sum_{i=1} \frac{N_{\mathrm{A}i}}{kT} \exp\left(-\frac{\Delta E_{\mathrm{A}i} - E_{\mathrm{ref}}}{kT}\right) I_{\mathrm{p}}(\Delta E_{\mathrm{A}i}) \\
&+ \sum_{j=1} \frac{N_{\mathrm{TH}j}}{kT} \exp\left(-\frac{\Delta E_{\mathrm{TH}j} - E_{\mathrm{ref}}}{kT}\right) I_{\mathrm{p}}(\Delta E_{\mathrm{TH}j}) \\
&- \frac{N_{\mathrm{comp}} N_{\mathrm{V0}}}{kT} \exp\left(\frac{E_{\mathrm{ref}} - \Delta E_{\mathrm{F}}(T)}{kT}\right),
\end{aligned}
\tag{40}
$$

where

$$
I_{\mathrm{p}}(\Delta E) = \frac{N_{\mathrm{V0}}}{4 + \exp\left(\dfrac{\Delta E_{\mathrm{F}}(T) - \Delta E}{kT}\right)}
\tag{41}
$$

and

$$
N_{\mathrm{comp}} = N_{\mathrm{D}} + \sum_{j=1} N_{\mathrm{TH}j}.
\tag{42}
$$

3.3. Experimental Results and Discussion

3.3.1. Determination of Densities and Energy Levels of Dopants

Figure 4(a) shows the experimental $n(T)$ for an undoped n-type 3C-SiC epilayer [11], denoted by ○. Figure 4(b) depicts the FCCS signal of $H(T, -0.000254)$, denoted by □. The peak temperature and value of $H(T, -0.000254)$ are 137.3 K and 2.30×10^{38} cm^{-6}eV$^{-2.5}$,

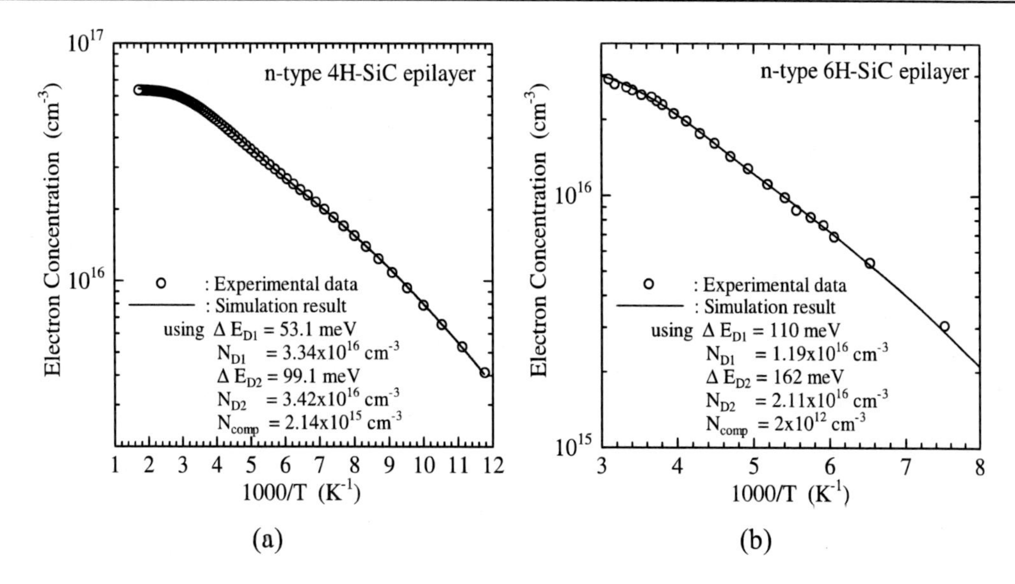

Figure 5. Experimental and simulated $n(T)$ for n-type 4H-SiC (a) and n-type 6H-SiC (b).

respectively, from which the energy level (ΔE_{D1}) and density (N_{D1}) of the corresponding donor species are determined to be 51.0 meV and 7.10×10^{16} cm^{-3}, respectively.

Figure 4(b) also shows $H2(T, -0.007023)$, denoted by ▲. Since two peaks appear, other donor species are included in this epilayer. The low peak temperature and value of $H(T, 0.000254)$ are 98.4 K and 8.54×10^{37} cm^{-6}eV$^{-2.5}$, respectively, and the high peak temperature and value are 315 K and 8.61×10^{37} cm^{-6}eV$^{-2.5}$, respectively. The energy level (ΔE_{D2}) and density (N_{D2}) of the donor species corresponding to the peak at the low temperature are determined as 18.4 meV and 3.84×10^{16} cm^{-3}, respectively, and the energy level (ΔE_{D3}) and density (N_{D3}) of the donor species corresponding to the peak at the high temperature are determined to be 113.9 meV and 1.07×10^{17} cm^{-3}, respectively. However, $H3(T, E_{\mathrm{ref}})$ is nearly zero, indicating that this epilayer includes three types of donor species. Since ΔE_{D2} is shallowest in the measurement temperature range, N_{comp} is determined to be -1.98×10^{15} cm^{-3} from the value of $H2(T, E_{\mathrm{ref}})$ at the lowest temperature.

In order to verify the values obtained by FCCS, the $n(T)$ is simulated using Eqs. (4) and (7). The solid line in Fig. 4(a) represents the $n(T)$ simulation, and is in good agreement with the experimental $n(T)$, indicating that the values determined by FCCS are reliable.

Suzuki *et al.* [12] insisted that the $\sim$ 15 meV donor species came from nonstoichiometric defects in unintentionally doped films. From photoluminescence measurements, Freitas *et al.* [13] and Kaplan *et al.* [14] insisted that the substitutional N donor level was $\sim$ 54 meV. On the other hand, the origin of the donor species with ΔE_{D3} is uncertain.

The open circles in Fig. 5 represent the experimental $n(T)$ and the solid line represents the $n(T)$ simulation with the values by FCCS, for n-type 4H-SiC (a) and 6H-SiC (b) epilayers. The solid lines are in good agreement with the experimental $n(T)$, indicating that the values determined by FCCS are reliable.

According to literature [15, 16], ΔE_{D1} and ΔE_{D2} correspond to the energy levels of

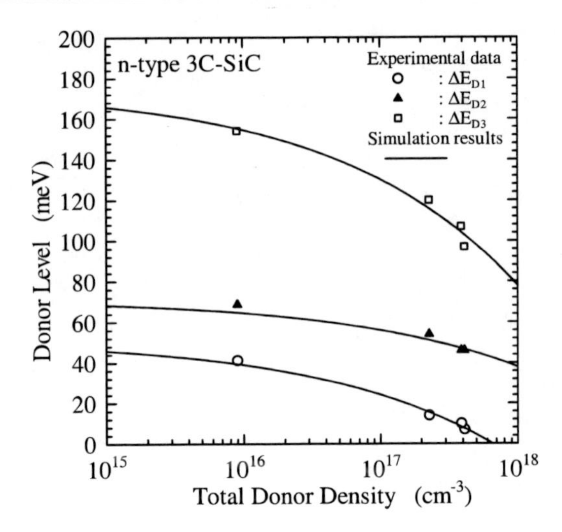

Figure 6. Dependence of each donor level on total donor density in n-type 3C-SiC.

isolated, substitutional N donors at hexagonal and cubic lattice sites in 4H-SiC or 6H-SiC, respectively. Because the ratio of the number of hexagonal lattice sites to the number of cubic lattice sites in 4H-SiC is unit, the probability for N atoms being put into each lattice site is found to be an half, which consists that the experimental result of $N_{\mathrm{D1}} \simeq N_{\mathrm{D2}}$. On the other hand, the ratio of the number of hexagonal lattice sites to the number of cubic lattice sites in 6H-SiC is $1:2$, which consists that the experimental result of $2N_{\mathrm{D1}} \simeq N_{\mathrm{D2}}$.

The acceptor densities and energy levels in undoped GaSb and InGaSb epilayers as well as the donor density and energy level in Te-doped $Al_{0.2}Ga_{0.8}Sb$ were determined [17, 18, 19].

3.3.2. Dopant-Density Dependence

Figure 6 shows the dependence of each donor level on a total donor density ($N_{\mathrm{D,total}} \equiv N_{\mathrm{D1}} + N_{\mathrm{D2}} + N_{\mathrm{D3}}$) [20]. The symbols of ○, ▲ and □ represent ΔE_{D1}, ΔE_{D2} and ΔE_{D3}, respectively. Figure 7(a) and (b) depict the dependence of each donor level in N-doped 4H-SiC epilayers on a total donor density ($N_{\mathrm{D,total}} \equiv N_{\mathrm{D1}} + N_{\mathrm{D2}}$) [21] and the dependence of each acceptor level in Al-doped 4H-SiC epilayers on a total acceptor density ($N_{\mathrm{A,total}} \equiv N_{\mathrm{A1}} + N_{\mathrm{A2}}$) [22], respectively. The symbols of ○ and ▲ in Fig. 7(a) represent ΔE_{D1} and ΔE_{D2}, respectively, and the symbols of ○ and ▲ in Fig. 7(b) represent ΔE_{A1} and ΔE_{A2}, respectively.

An ideal donor level $\Delta E_{\mathrm{D}i}(0)$ is the energy required to emit one electron from the donor site into infinity on E_{C}. However, since an n-type semiconductor is electrically neutral, each positively charged donor is shielded by one electron on E_{C}. This shielding electron is assumed to be located within half ($\overline{r}$) of an average distance ($1/\sqrt[3]{N_{\mathrm{D,total}}}$) of the donors, indicating that the donor level should be lowered by the energy higher than $q/(4\pi\epsilon_{\mathrm{s}}\epsilon_0\overline{r})$ due to Coulomb's attraction. Therefore,

$$\Delta E_{\mathrm{D}i}(N_{\mathrm{D,total}}) = \Delta E_{\mathrm{D}i}(0) - \alpha_{\mathrm{D}i}\sqrt[3]{N_{\mathrm{D,total}}} \tag{43}$$

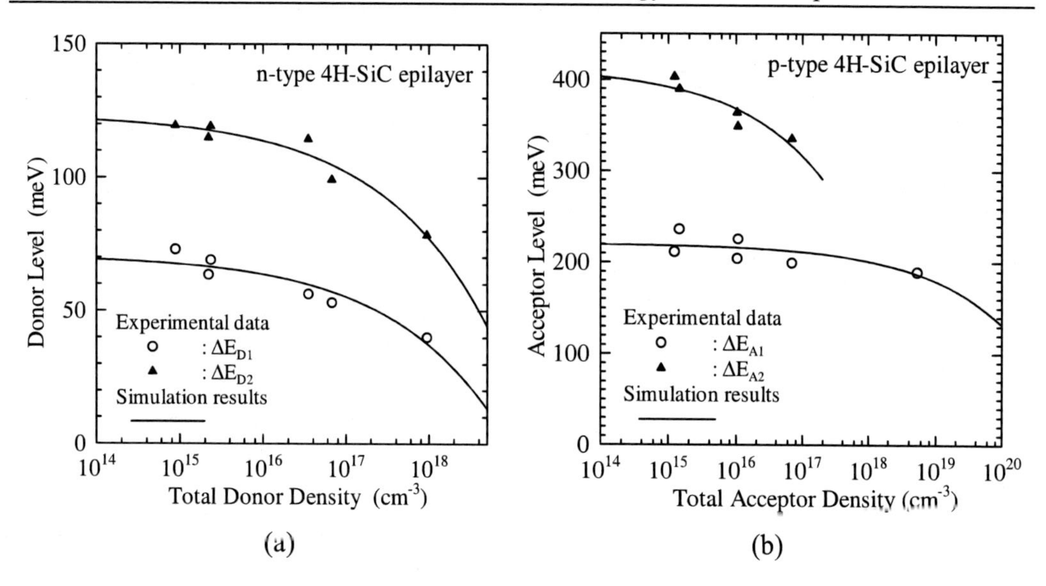

Figure 7. Dependence of each dopant level on total dopant density in 4H-SiC; (a) donors and (b) acceptors.

and

$$\alpha_{\mathrm{D}i} \gtrsim \frac{q}{8\pi\epsilon_{\mathrm{s}}\epsilon_0} = 2.44 \times 10^{-5} \ \mathrm{meV} \cdot \mathrm{cm}. \tag{44}$$

where ϵ_0 is the free space permittivity, and ϵ_{s} is the dielectric constant for SiC. On the other hand,

$$\Delta E_{\mathrm{A}i}(N_{\mathrm{A,total}}) = \Delta E_{\mathrm{A}i}(0) - \alpha_{\mathrm{A}i} \sqrt[3]{N_{\mathrm{A,total}}}. \tag{45}$$

In N-doped n-type 3C-SiC and 4H-SiC, the fitting parameters obtained by a least-squares fit of Eq. (43) to the data in Fig. 6 and Fig. 7(a) are listed in Table 1, respectively, and the $\Delta E_{\mathrm{D}i}(N_{\mathrm{D,total}})$ simulations are denoted by the solid lines in Fig. 6 and Fig. 7(a). Since these $\alpha_{\mathrm{D}i}$ are satisfied with Eq. (44), they are considered to be reasonable. In Al-doped p-type 4H-SiC, the fitting parameters obtained by a least-squares fit of Eq. (45) to the data in Fig. 7(b) are listed in Table 1, and the $\Delta E_{\mathrm{A}i}(N_{\mathrm{A,total}})$ simulations are denoted by the lines in Fig. 7(b). Although these $\alpha_{\mathrm{A}i}$ are a little lower than 2.44×10^{-5} meV $\cdot$ cm, they are considered to be reasonable.

Figure 8(a) and (b) show the temperature dependence of the electron mobility and the hole mobility, respectively. As is clear from Fig. 8(a), the electron mobility at > 250 K can be expressed as

$$\mu_{\mathrm{e}}(T, N_{\mathrm{imp}}) = \mu_{\mathrm{e}}(300, N_{\mathrm{imp}}) \left(\frac{T}{300}\right)^{-\beta_{\mathrm{e}}(N_{\mathrm{imp}})}, \tag{46}$$

where $N_{\mathrm{imp}} = N_{\mathrm{D1}} + N_{\mathrm{D2}} + N_{\mathrm{A}}$. Therefore, $\mu_{\mathrm{e}}(300, N_{\mathrm{imp}})$ and $\beta_{\mathrm{e}}(N_{\mathrm{imp}})$ can be evaluated individually as follows.

$$\beta_{\mathrm{e}}(N_{\mathrm{imp}}) = \beta_{\mathrm{e}}^{\mathrm{min}} + \frac{\beta_{\mathrm{e}}^{\mathrm{max}} - \beta_{\mathrm{e}}^{\mathrm{min}}}{1 + \left(\frac{N_{\mathrm{imp}}}{N_{\mathrm{e}}^{\beta}}\right)^{\gamma_{\mathrm{e}}^{\beta}}}, \tag{47}$$

Table 1. Parameters for donor levels and acceptor in SiC.

Parameters	N-doped 3C-SiC	N-doped 4H-SiC	Al-doped 4H-SiC
$\Delta E_{\rm D1}(0)$ (meV)	51.9	70.9	—
$\alpha_{\rm D1}$ (meV·cm)	5.97×10^{-5}	3.38×10^{-5}	—
$\Delta E_{\rm D2}(0)$ (meV)	71.8	123.7	—
$\alpha_{\rm D2}$ (meV·cm)	3.38×10^{-5}	4.65×10^{-5}	—
$\Delta E_{\rm D3}(0)$ (meV)	176	—	—
$\alpha_{\rm D3}$ (meV·cm)	9.77×10^{-5}	—	—
$\Delta E_{\rm A1}(0)$ (meV)	—	—	220
$\alpha_{\rm A1}$ (meV·cm)	—	—	1.90×10^{-5}
$\Delta E_{\rm A2}(0)$ (meV)	—	—	413
$\alpha_{\rm A2}$ (meV·cm)	—	—	2.07×10^{-5}

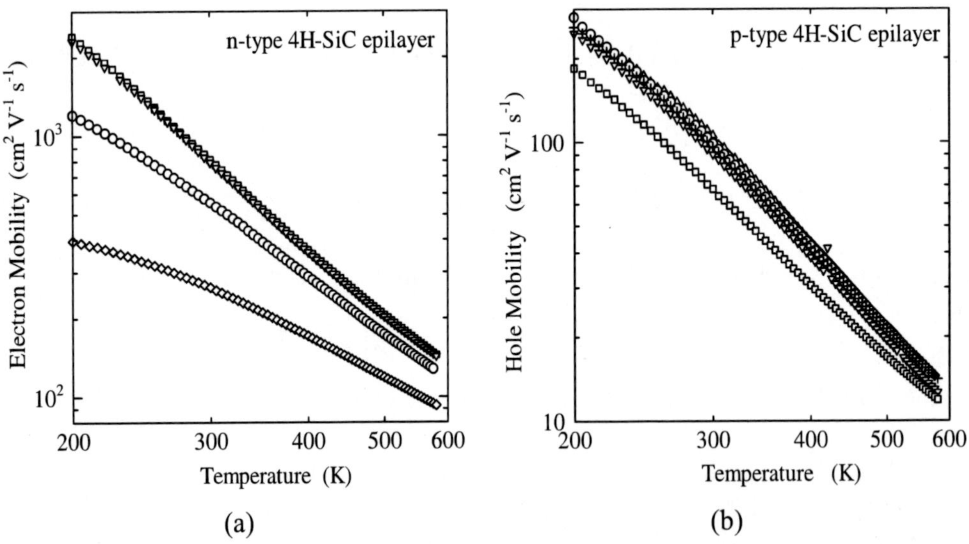

Figure 8. Temperature dependencies of electron mobility for n-type 4H-SiC epilayers (a) and hole mobility for p-type 4H-SiC epilayers (b).

and

$$\mu_{\rm e}(300, N_{\rm imp}) = \mu_{\rm e}^{\rm min}(300) + \frac{\mu_{\rm e}^{\rm max}(300) - \mu_{\rm e}^{\rm min}(300)}{1 + \left(\dfrac{N_{\rm imp}}{N_{\rm e}^{\mu}}\right)^{\gamma_{\rm e}^{\mu}}}, \tag{48}$$

where $\beta_{\rm e}^{\rm min}$, $\beta_{\rm e}^{\rm max}$, $N_{\rm e}^{\beta}$, $\gamma_{\rm e}^{\beta}$, $\mu_{\rm e}^{\rm min}(300)$, $\mu_{\rm e}^{\rm max}(300)$, $N_{\rm e}^{\mu}$, and $\gamma_{\rm e}^{\mu}$ are the fitting parameters. In Fig. 8(b), moreover, the hole mobility at > 250 K can be expressed as

$$\mu_{\rm h}(T, N_{\rm imp}) = \mu_{\rm h}(300, N_{\rm imp}) \left(\frac{T}{300}\right)^{-\beta_{\rm h}(N_{\rm imp})}, \tag{49}$$

where $N_{\mathrm{imp}} = N_{\mathrm{A1}} + N_{\mathrm{A2}} + N_{\mathrm{D}}$. Therefore, $\mu_{\mathrm{h}}(300, N_{\mathrm{imp}})$ and $\beta_{\mathrm{h}}(N_{\mathrm{imp}})$ can be evaluated individually as follows.

$$\beta_{\mathrm{h}}(N_{\mathrm{imp}}) = \beta_{\mathrm{h}}^{\min} + \frac{\beta_{\mathrm{h}}^{\max} - \beta_{\mathrm{h}}^{\min}}{1 + \left(\frac{N_{\mathrm{imp}}}{N_{\mathrm{h}}^{\beta}}\right)^{\gamma_{\mathrm{h}}^{\beta}}}, \tag{50}$$

and

$$\mu_{\mathrm{h}}(300, N_{\mathrm{imp}}) = \mu_{\mathrm{h}}^{\min}(300) + \frac{\mu_{\mathrm{h}}^{\max}(300) - \mu_{\mathrm{h}}^{\min}(300)}{1 + \left(\frac{N_{\mathrm{imp}}}{N_{\mathrm{h}}^{\mu}}\right)^{\gamma_{\mathrm{h}}^{\mu}}}, \tag{51}$$

where $\beta_{\mathrm{h}}^{\min}$, $\beta_{\mathrm{h}}^{\max}$, N_{h}^{β}, $\gamma_{\mathrm{h}}^{\beta}$, $\mu_{\mathrm{h}}^{\min}(300)$, $\mu_{\mathrm{h}}^{\max}(300)$, N_{h}^{μ}, and $\gamma_{\mathrm{h}}^{\mu}$ are the fitting parameters.

Table 2. Parameters for $\beta(N_{\mathrm{imp}})$.

$\beta_{\mathrm{e}}^{\min}$	$\beta_{\mathrm{e}}^{\max}$	N_{e}^{β} (cm^{-3})	$\gamma_{\mathrm{e}}^{\beta}$	$\beta_{\mathrm{h}}^{\min}$	$\beta_{\mathrm{h}}^{\max}$	N_{h}^{β} (cm^{-3})	$\gamma_{\mathrm{h}}^{\beta}$
1.54	2.62	1.14×10^{17}	1.35	2.51	3.04	8.64×10^{17}	0.456

Table 3. Parameters for $\mu(300, N_{\mathrm{imp}})$.

$\mu_{\mathrm{e}}^{\min}$ [cm^2/(V·s)]	$\mu_{\mathrm{e}}^{\max}$ [cm^2/(V·s)]	N_{e}^{μ} (cm^{-3})	$\gamma_{\mathrm{e}}^{\mu}$
0	977	1.17×10^{17}	0.49

$\mu_{\mathrm{h}}^{\min}$ [cm^2/(V·s)]	$\mu_{\mathrm{h}}^{\max}$ [cm^2/(V·s)]	N_{h}^{μ} (cm^{-3})	$\gamma_{\mathrm{h}}^{\mu}$
37.6	106.0	2.97×10^{18}	0.356

From Fig. 8(a), all the parameters in Eqs. (47) and (48) were determined, and are listed in Tables 2 and 3. From Fig. 8(b), on the other hand, all the parameters in Eqs. (50) and (51) were determined, and are listed in Tables 2 and 3. Therefore, the electron and hole mobilities for any N_{imp} at $T > 250$ K can be simulated.

Figure 9(a) shows $p(T)$ for three different Al concentrations [22, 23]. One acceptor level was approximately $E_{\mathrm{V}} + 0.22$ eV, which is assigned to an Al acceptor. Therefore, the density and energy level of this acceptor are represented by N_{Al} and E_{Al}, respectively. The other acceptor level, on the other hand, was approximately $E_{\mathrm{V}} + 0.38$ eV, which has not been assigned. The density and energy level of the deep acceptor are represented by N_{DA} and E_{DA}, respectively. Figure 9(b) shows the relationship between N_{Al} and N_{DA}. From the figure, the following empirical relationship between N_{Al} and N_{DA} was obtained;

$$N_{\mathrm{DA}} = 0.6 \times N_{\mathrm{Al}}. \tag{52}$$

Therefore, it is considered that the deep acceptor is most likely related to Al.

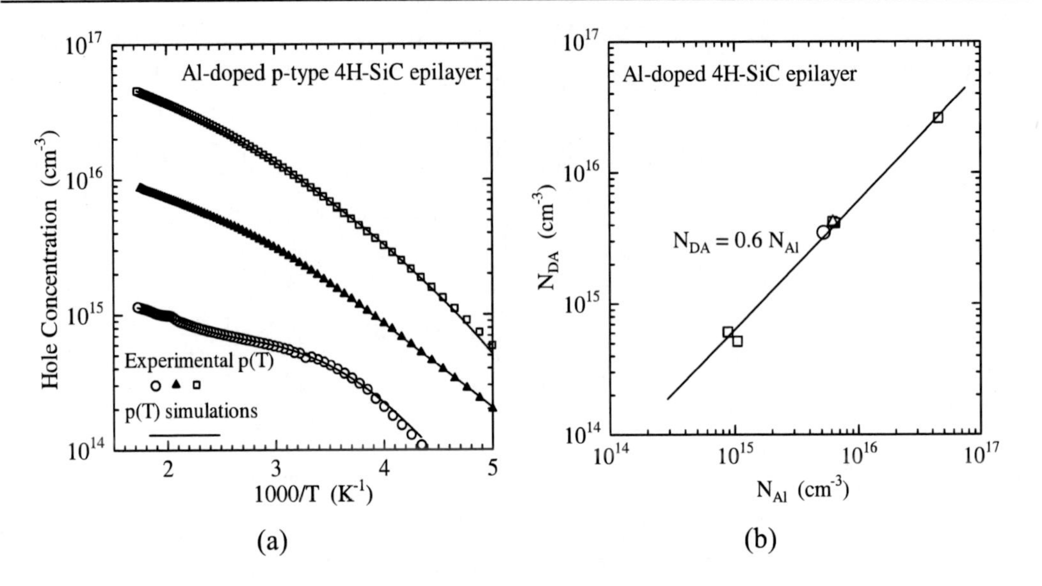

Figure 9. Temperature dependence of hole concentration for three different Al-doping levels (a) and relationship between Al acceptor density and deep acceptor density (b).

3.4. Electron-irradiated 4H-SiC

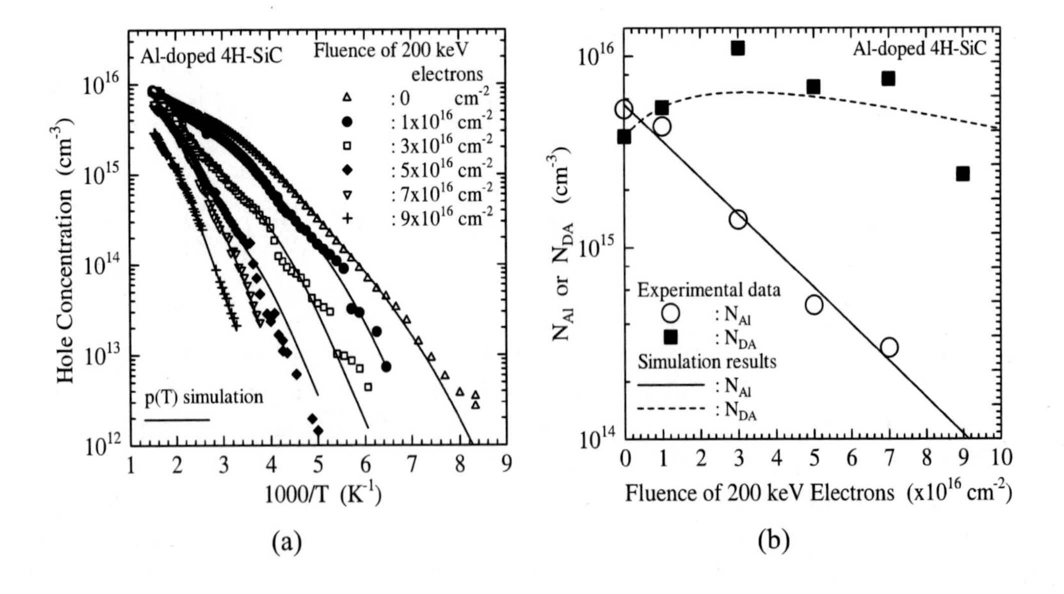

Figure 10. Al-doped p-type 4H-SiC epilayer irradiated by 200 keV electron irradiation; (a) temperature dependencies of hole concentration before and after irradiation and (b) fluence dependencies of Al acceptor and deep acceptor.

Figure 10(a) shows the experimental $p(T)$ in a 10 μm-thick Al-doped p-type 4H-SiC epilayer on n-type 4H-SiC (thickness: 376 μm, with a resistivity of 0.02 Ωcm) for fluences (Φ) of 0 ($\triangle$), 1×10^{16} ($\bullet$), 3×10^{16} ($\square$), 5×10^{16} ($\blacklozenge$), 7×10^{16} ($\bigtriangledown$), and 9×10^{16} cm^{-2} (+), respectively [24]. The $p(T)$ at low temperatures decreased significantly with increasing Φ, whereas the $p(T)$ at high temperatures was changed slightly by the irradiation.

In Fig. 10(a), △ represents $p(T)$ for the unirradiated case, and the solid line is the $p(T)$ simulation with N_{Al} of 5.3×10^{15} cm^{-3}, E_{Al} of $E_{\mathrm{V}} + 0.20$ eV, N_{DA} of 3.7×10^{15} cm^{-3}, E_{DA} of $E_{\mathrm{V}}+0.37$ eV, and N_{comp} of 2×10^{13} cm^{-3}, which were determined by FCCS. Since the $p(T)$ simulation is in good agreement with the experimental $p(T)$ for the unirradiated case, the values obtained by FCCS are reliable.

Figure 10(b) shows the fluence dependencies of N_{Al} and N_{DA}, denoted by ○ and ■, respectively. N_{Al} decreased with increasing Φ, and finally there are no more Al acceptors. On the other hand, N_{DA} initially increased with Φ, and then decreased.

200 keV electron irradiation can replace substitutional C atoms ($\mathrm{C_s}$) [23]. In order to reduce N_{Al} by the 200 keV electron irradiation, the surroundings of the Al acceptor need to be changed. This indicates that the rate of decrease of N_{Al} by the irradiation is proportional to N_{Al}. Consequently, the differential equation that leads to fluence dependence of Al acceptors $N_{\mathrm{Al}}(\Phi)$ is given by

$$\frac{\mathrm{d}N_{\mathrm{Al}}(\Phi)}{\mathrm{d}\Phi} = -\kappa_{\mathrm{Al}} N_{\mathrm{Al}}(\Phi), \tag{53}$$

where κ_{Al} is the removal cross section of Al acceptors for 200 keV electron irradiation. Therefore,

$$N_{\mathrm{Al}}(\Phi) = N_{\mathrm{Al}}(0)\exp(-\kappa_{\mathrm{Al}}\Phi). \tag{54}$$

Figure 10(b) shows $N_{\mathrm{Al}}(\Phi)$ in a semi-logarithmic scale, and the solid line is a straight line obtained by least-squares fitting. Since the straight line is in good agreement with ○ in the semi-logarithmic plots, Eq. (54) is feasible for the fluence dependence of N_{Al}. κ_{Al} is then determined from the slope as 4.4×10^{-17} cm^2.

In Fig. 10(b), at fluences of $\leq 3 \times 10^{16}$ cm^{-2}, N_{DA} increased with increasing Φ, while N_{Al} decreased. Furthermore, the increment of N_{DA} is close to the decrement of N_{Al}. This experimental result may indicate that the 200 keV electron irradiation transforms the Al acceptor into the deep acceptor. In this case, the differential equation describing the fluence dependence of the deep acceptor density $N_{\mathrm{DA}}(\Phi)$ can be expressed as

$$\frac{\mathrm{d}N_{\mathrm{DA}}(\Phi)}{\mathrm{d}\Phi} = -\frac{\mathrm{d}N_{\mathrm{Al}}(\Phi)}{\mathrm{d}\Phi} - \kappa_{\mathrm{DA}} N_{\mathrm{DA}}(\Phi), \tag{55}$$

where κ_{DA} is the removal cross section of the deep acceptors for 200 keV electron irradiation. The broken line in Fig. 10(b) represents the simulated $N_{\mathrm{DA}}(\Phi)$ with κ_{DA} of 1.0×10^{-17} cm^2 using Eq. (55), which shows qualitative agreement with the experimental data, but not quantitative agreement.

Figure 11(a) shows the experimental $n(T)$ before irradiation (○) and after irradiation with 200 keV electrons at Φ of 1×10^{16} cm^{-2} (▲) and 2×10^{16} cm^{-2} (□). From each $n(T)$, two types of donor species were detected and evaluated using FCCS. The energy level of N donors at hexagonal C-sublattice sites (E_{NH}) was $E_{\mathrm{C}} - 70$ meV. The energy level of N donors at cubic C-sublattice sites (E_{NK}) was $E_{\mathrm{C}}-120$ meV. Figure 11(b) shows the fluence dependencies of N_{NH} (○) and N_{NK} (■). The N_{NH} decreased substantially with increasing Φ of 200 keV electrons, whereas N_{NK} decreased only slightly, indicating that N donors at hexagonal C-sublattice sites are less radiation-resistant than N donors at cubic C-sublattice

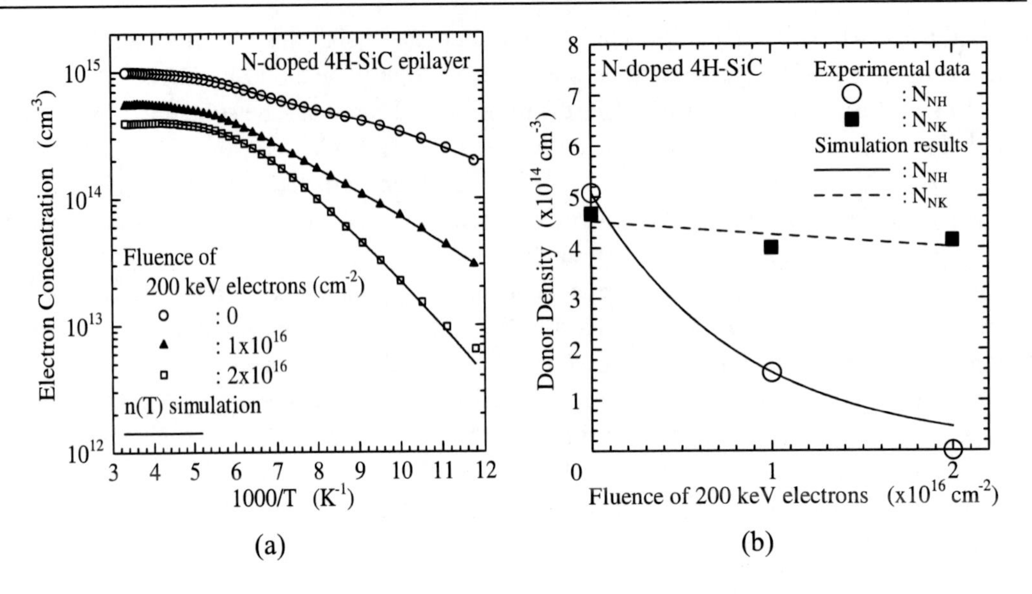

Figure 11. N-doped n-type 4H-SiC epilayer irradiated by 200 keV electron irradiation; (a) temperature dependencies of electron concentration before and after irradiation and (b) fluence dependencies of N donors located at hexagonal and cubic C-sublattice sites.

sites. This finding suggests that 3C-SiC might be the most and 6H-SiC should be the least radiation-resistant of N-doped 3C-SiC, 4H-SiC, and 6H-SiC.

By analogy with Eq. (54), the fluence dependencies of N_{NH} and N_{NK} are expected to be derived from the following differential equations:

$$\frac{\mathrm{d}N_{\mathrm{NH}}}{\mathrm{d}\Phi} = -\kappa_{\mathrm{NH200}} N_{\mathrm{NH}} \tag{56}$$

and

$$\frac{\mathrm{d}N_{\mathrm{NK}}}{\mathrm{d}\Phi} = -\kappa_{\mathrm{NK200}} N_{\mathrm{NK}}, \tag{57}$$

where κ_{NH200} and κ_{NK200} are the removal cross sections for 200 keV electron irradiation of the N donors at hexagonal and cubic C-sublattice sites, respectively. By fitting the curve to the experimental data, the values of κ_{NH200} and κ_{NK200} were determined to be 1.2×10^{-16} and 6.0×10^{-18} cm^2, respectively. The solid and broken curves in Fig. 11(b) represent the simulated fluence dependencies of N_{NH} and N_{NK}, respectively.

3.5. Trap Densities Higher Than Dopant Density

Figure 12(a) shows the $p(T)$ for the 10 Ω cm B-doped FZ-Si irradiated with several fluences of 10 MeV protons [25]. The ◇, △, □, and ○ symbols correspond to the fluences of 0, 1.0×10^{13}, 1.0×10^{14}, and 2.5×10^{14} cm^{-2}, respectively. Figure 12(b) shows the dependencies of B acceptor density and hole-trap densities on proton fluence. The values of ΔE_{TH1}, ΔE_{TH2}, and ΔE_{TH3} were $\sim$ 95 meV, $\sim$ 160 meV, and $\sim$ 300 meV, respectively.

The values of the lowest $\Delta E_{\mathrm{F}}(T)$ in the temperature range of the measurement were 95, 97, 162, and 310 meV for the samples irradiated with fluences of 0, 1.0×10^{13}, 1.0×10^{14},

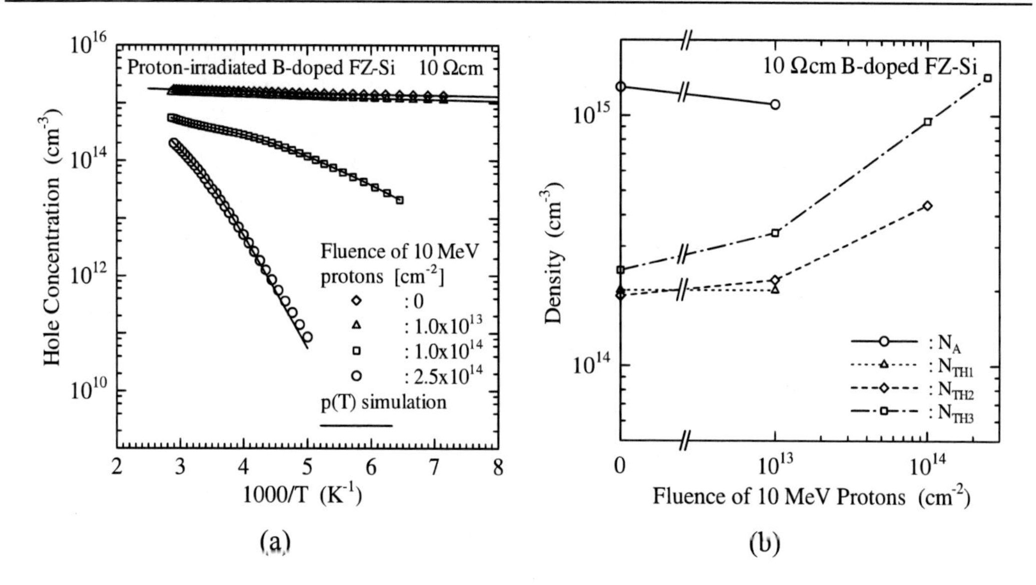

Figure 12. Proton-irradiated Si; (a) temperature dependencies of electron concentration in Si before and after proton irradiation and (b) fluence dependencies of densities of B acceptor and hole traps.

and 2.5×10^{14} cm^{-2}, respectively. Since ΔE_A of B in Si is $\sim$ 45 meV, B acceptors were completely ionized at the lowest measurement temperature, indicating that ΔE_A could not be determined in this measurement temperature range. Because in the case of 1.0×10^{14} cm^{-2} fluence the hole traps with ΔE_{TH1} were not filled with holes at all in the measurement temperature range, the values of ΔE_{TH1} and N_{TH1} could not be evaluated. For the same reason, the values of ΔE_{TH1}, N_{TH1}, ΔE_{TH2}, and N_{TH2} could not be estimated in the sample irradiated with the 2.5×10^{14} cm^{-2} fluence.

The values of N_{TH2} and N_{TH3} increase clearly. Moreover, N_{TH3} of the sample irradiated with 2.5×10^{14} cm^{-2} exceeds N_A of the unirradiated sample. Because the density and energy level of the hole trap with a density higher than the acceptor density can be determined from FCCS, FCCS is superior to DLTS from the viewpoint of the evaluation of traps with high densities.

4. Nondegenerate Heavily-Doped Wide Bandgap Semiconductors

4.1. Problems of Heavily-Doped Case

The excited states of a substitutional dopant in a semiconductor have been theoretically discussed using the hydrogenic model (or the effective mass approximation) [26, 27, 28], and the existence of excited states of the dopant (e.g., B or P) in Si or Ge was experimentally confirmed from infrared absorption measurements at very low temperatures [26]. However, the influence of the excited states on the majority-carrier concentration in Si or Ge was not experimentally confirmed [27, 29], partially because the excited state levels of the dopants in Si or Ge were too shallow and partially because $E_F(T)$ was deeper than the dopant energy level in the temperature range of the measurement. Therefore, by a least squares fit of the charge neutrality equation to the temperature dependence of the majority-carrier concentration experimentally obtained by Hall-effect measurements, the values of dopant density, dopant energy level, and compensating density can be determined using the Fermi-Dirac (FD) distribution function, which does not include the influence of the excited states of the dopant. The FD distribution functions for donors and acceptors are expressed as [5]

$$f_{FD}(E_D) = \frac{1}{1 + \frac{1}{g_{D,FD}} \exp\left(-\frac{E_F(T) - E_D}{kT}\right)} \tag{58}$$

and

$$f_{FD}(E_A) = \frac{1}{1 + g_{A,FD} \exp\left(-\frac{E_F(T) - E_A}{kT}\right)}, \tag{59}$$

respectively, where $g_{D,FD}$ and $g_{A,FD}$ are the degeneracy factors for donors and acceptors, respectively.

Because in p-type wide bandgap semiconductors the experimentally obtained values of E_A have been reported to be deep, $E_F(T)$ is often between E_A and E_V. Furthermore, because the excited state levels of acceptors are as deep as a B acceptor level (i.e., ground-state level) in Si, $E_F(T)$ is close to the excited state levels. The excited states of the acceptor must, therefore, affect $p(T)$.

Using a distribution function including the influence of the excited states of an acceptor derived from the viewpoint of the grand canonical ensemble [27, 30, 31], N_A was determined by fitting the simulation $p(T)$ to the experimental $p(T)$. The N_A determined this way was, however, much higher than the concentration of acceptor atoms determined by secondary ion mass spectroscopy (SIMS). This same situation occurred in the case of the FD distribution function [32, 33].

4.2. Experimental Results

Figure 13 shows $p(T)$ and $E_F(T)$ for the B-doped diamond. Because E_A of B in diamond is approximately $E_V + 0.35$ eV, $E_F(T)$ values were lower than E_A over the measurement temperature range.

N_A and E_A of B, and N_D can be determined using

$$p(T) = N_A F(E_A) - N_D, \tag{60}$$

where $n(T)$ is much less than $p(T)$ over the measurement temperature range, and $F(E_A)$ is the distribution function for acceptors. By replacing $F(E_A)$ with $f_{FD}(E_A)$, the values of N_A, E_A, and N_D were determined by FCCS and are listed in Table 4. In the table, f_{FD}, f_{MC}, and f_{GC} represent the FD distribution function and the distribution functions including the influence of the excited states of a dopant, which are derived from the viewpoint of the microcanonical (MC) and grand canonical (GC) ensembles, respectively, and C is the dopant concentration.

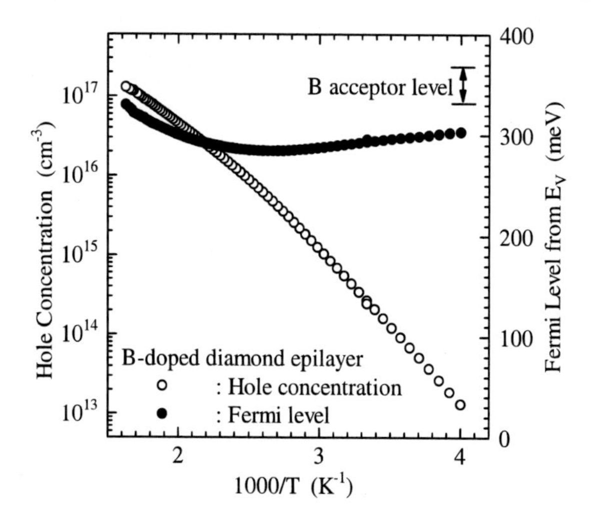

Figure 13. Temperature dependencies of hole concentration and Fermi level for B-doped diamond.

The N_{A} of 9.7×10^{17} cm^{-3} for $f_{\mathrm{FD}}(E_{\mathrm{A}})$ is approximately five times higher than the B concentration (C_{B}) of approximately 2×10^{17} cm^{-3} determined by SIMS. Since N_{A} is the density of B atoms at the lattice sites, N_{A} should be less than or equal to C_{B}. Therefore, $f_{\mathrm{FD}}(E_{\mathrm{A}})$ seems inappropriate to a distribution function for B acceptors in heavily-doped diamond.

Figure 14(a) shows $p(T)$ and $E_{\mathrm{F}}(T)$ for the heavily Al-doped 6H-SiC wafer, while Fig. 14(b) shows $p(T)$ and $E_{\mathrm{F}}(T)$ for the lightly Al-doped 6H-SiC epilayer. $E_{\mathrm{F}}(T)$ values for the heavily-doped 6H-SiC were below E_{A} over the measurement temperature range, whereas $E_{\mathrm{F}}(T)$ values for the lightly-doped 6H-SiC were above E_{A} over almost all the measurement temperature range.

The values of N_{A}, E_{A}, and N_{D} for $f_{\mathrm{FD}}(E_{\mathrm{A}})$ were estimated and are listed in Table 4. The N_{A} obtained for the heavily-doped 6H-SiC is 2.5×10^{19} cm^{-3} that is approximately six times higher than the Al concentration (C_{Al}). On the other hand, the N_{A} obtained for the lightly-doped 6H-SiC is 5.1×10^{15} cm^{-3}, which is nearly equal to the C_{Al}. This suggests that $f_{\mathrm{FD}}(E_{\mathrm{A}})$ can be used only in the lightly doped sample.

As is clear from Table 4, in the Al implanted 4H-SiC layer the N_{A} of 4.9×10^{19} cm^{-3} is approximately five times higher than the C_{Al}, and in the Mg doped GaN epilayer the N_{A} of 8.5×10^{19} cm^{-3} is approximately four times higher than the Mg concentration (C_{Mg}).

4.3. Distribution Function for a Deep Substitutional Dopant

4.3.1. The Number of Configurations of the System

We now consider the microcanonical ensemble. Electrons and holes in semiconductors are fermions, which obey the Pauli exclusion principle. Because of this, in the allowed bands, the multiplicity function $W_{\mathrm{B}}(E_i)$ for $n_{\mathrm{e}}(E_i)$ electrons arranged in $D(E_i)$ states at a given

Table 4. Results for each distribution function.

	B-doped diamond	Heavily Al-doped 6H-SiC	Lightly Al-doped 6H-SiC	Al-implanted 4H-SiC	Mg-doped GaN
f_{MC}					
N_{A} (cm^{-3})	2.8×10^{17}	3.0×10^{18}	4.4×10^{15}	1.2×10^{19}	6.0×10^{18}
E_{A} (eV)	$E_{\mathrm{V}} + 0.32$	$E_{\mathrm{V}} + 0.18$	$E_{\mathrm{V}} + 0.20$	$E_{\mathrm{V}} + 0.18$	$E_{\mathrm{V}} + 0.16$
N_{D} (cm^{-3})	2.0×10^{16}	9.7×10^{16}	3.7×10^{14}	2.2×10^{17}	1.3×10^{17}
f_{FD}					
N_{A} (cm^{-3})	9.7×10^{17}	2.5×10^{19}	5.1×10^{15}	4.9×10^{19}	8.5×10^{19}
E_{A} (eV)	$E_{\mathrm{V}} + 0.34$	$E_{\mathrm{V}} + 0.18$	$E_{\mathrm{V}} + 0.19$	$E_{\mathrm{V}} + 0.16$	$E_{\mathrm{V}} + 0.15$
N_{D} (cm^{-3})	4.0×10^{16}	7.3×10^{17}	8.1×10^{14}	2.5×10^{18}	2.3×10^{18}
f_{GC}					
N_{A} (cm^{-3})	2.1×10^{18}	3.8×10^{20}	5.2×10^{15}	4.9×10^{20}	2.7×10^{20}
E_{A} (eV)	$E_{\mathrm{V}} + 0.38$	$E_{\mathrm{V}} + 0.18$	$E_{\mathrm{V}} + 0.19$	$E_{\mathrm{V}} + 0.17$	$E_{\mathrm{V}} + 0.17$
N_{D} (cm^{-3})	1.4×10^{16}	1.2×10^{19}	8.4×10^{14}	1.7×10^{19}	2.6×10^{18}
Dopant concentration C (cm^{-3})	$\sim 2 \times 10^{17}$	$\sim 4 \times 10^{18}$	$\sim 6 \times 10^{15}$	$\sim 1 \times 10^{19}$	$\sim 2 \times 10^{19}$

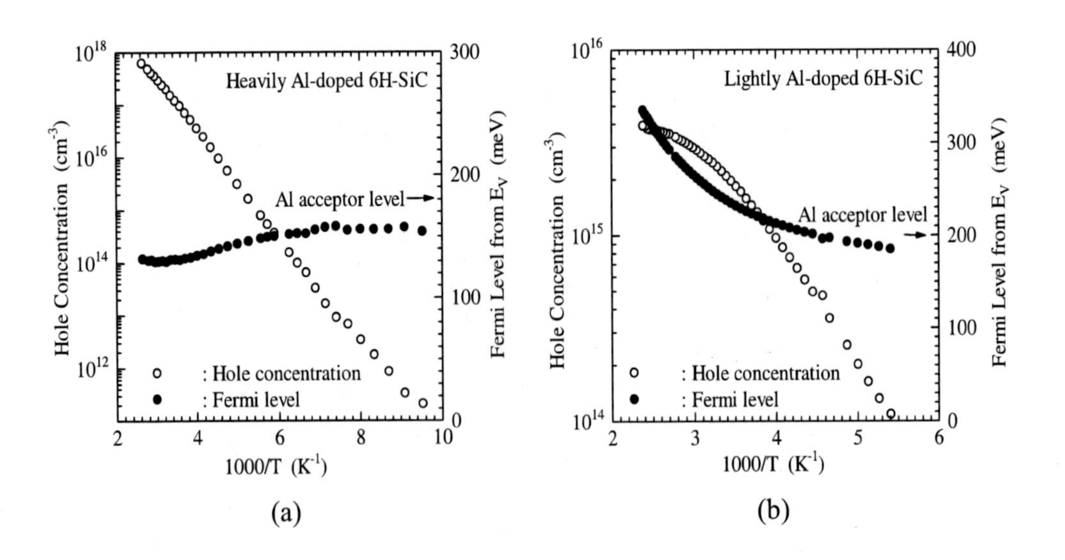

Figure 14. Temperature dependencies of hole concentration and Fermi level; (a) for heavily Al doped 6H-SiC and (b) for lightly Al doped 6H-SiC.

energy (E_i) is expressed as [30]

$$W_{\mathrm{B}}(E_i) = \frac{D(E_i)!}{[D(E_i) - n_{\mathrm{e}}(E_i)]! \cdot n_{\mathrm{e}}(E_i)!}, \tag{61}$$

where $D(E_i)$ is the number of degenerate states per unit volume at E_i and $n_e(E_i)$ is the number of electrons per unit volume at E_i.

In a forbidden band, on the other hand, the multiplicity function for n_D electrons arranged in N_D donors is quite different from Eq. (61), where n_D is the number of electrons bound to donors per unit volume. When spin degeneracy as well as the excited states of the donor is neglected, the multiplicity function (W_{D1}) for the n_D electrons arranged in the N_D donors is given by

$$W_{D1} = \frac{N_D!}{(N_D - n_D)! \cdot n_D!} . \tag{62}$$

Each state, ground or excited, consists of a spin-up state and a spin-down state. When the energy difference between the two states in a magnetic field is denoted by ΔE_{spin}, the multiplicity function (W_{D2}) is given by

$$W_{D2} = \left[1 + \exp\left(-\frac{\Delta E_{spin}}{kT}\right)\right]^{n_D} . \tag{63}$$

When the magnetic field is zero or very weak ($\Delta E_{spin} \cong 0$),

$$W_{D2} \cong 2^{n_D} . \tag{64}$$

In a neutral donor, furthermore, only an electron is located at one state for the ground and excited states of the donor. The multiplicity function (W_{D3}) is given by

$$W_{D3} = \left[g_1 + \sum_{r=2}^{l} g_r \exp\left(-\frac{E_{r,D} - E_D}{kT}\right)\right]^{n_D} , \tag{65}$$

where $E_{r,D}$ is the $(r-1)$th excited state level ($r \geq 2$) of the donor, g_1 is the ground-state degeneracy factor of 1, g_r is the $(r-1)$th excited state degeneracy factor of r^2, and the value of $(l-1)$ is the highest order of the excited states considered here.

Consequently, the multiplicity function (W_D) for the n_D electrons arranged in the N_D donors is expressed as

$$W_D = W_{D1} W_{D2} W_{D3} . \tag{66}$$

Finally, the total number of configurations of the system (W) is obtained from the product of these multiplicities as

$$W = W_D \prod_i W_{Bi} . \tag{67}$$

4.3.2. Thermal Equilibrium Configuration

Thermal equilibrium configuration occurs when the entropy

$$S = k \ln W \tag{68}$$

reaches a maximum value under the following two conservation laws; (1) the total number (n_{total}) of electrons in the system is conserved, that is,

$$n_{total} = n_D + \sum_i n(E_i) = \text{const} \tag{69}$$

and (2) the total energy (E_{total}) of electrons in the system is conserved, i.e.,

$$E_{\mathrm{total}} = \overline{E_{\mathrm{D}}(T)} \cdot n_{\mathrm{D}} + \sum_{i} E_i \cdot n(E_i) = \mathrm{const}, \tag{70}$$

where $\overline{E_{\mathrm{D}}(T)}$ is the average donor level given by

$$\overline{E_{\mathrm{D}}(T)} = E_{\mathrm{D}} + \overline{E_{\mathrm{ex,D}}(T)}, \tag{71}$$

and $\overline{E_{\mathrm{ex,D}}(T)}$ is the ensemble average of the ground and excited state levels of the donor, measured from E_{D}, and is given by

$$\overline{E_{\mathrm{ex,D}}(T)} = \frac{\displaystyle\sum_{r=2}^{l} (E_{r,\mathrm{D}} - E_{\mathrm{D}})\, g_r \exp\left(-\frac{E_{r,\mathrm{D}} - E_{\mathrm{D}}}{kT}\right)}{\displaystyle 1 + \sum_{r=2}^{l} g_r \exp\left(-\frac{E_{r,\mathrm{D}} - E_{\mathrm{D}}}{kT}\right)}. \tag{72}$$

Under these conditions, the distribution functions for electrons including the influence of the excited states of the donor $f_{\mathrm{MC}}(E_{\mathrm{D}})$ can be derived [34].

$$f_{\mathrm{MC}}(E_{\mathrm{D}}) \equiv \frac{n_{\mathrm{D}}}{N_{\mathrm{D}}} = \frac{1}{1 + \dfrac{1}{g_{\mathrm{D,MC}}(T)} \exp\left(-\dfrac{E_{\mathrm{F}}(T) - E_{\mathrm{D}}}{kT}\right)}, \tag{73}$$

where $g_{\mathrm{D,MC}}(T)$ is here called the effective degeneracy factor for donors given by

$$g_{\mathrm{D,MC}}(T) = 2\left[1 + \sum_{r=2}^{l} g_r \exp\left(-\frac{E_{r,\mathrm{D}} - E_{\mathrm{D}}}{kT}\right)\right] \exp\left(-\frac{\overline{E_{\mathrm{ex,D}}(T)}}{kT}\right). \tag{74}$$

4.3.3. The Hydrogenic Donor Case

A neutral donor can be approximately described as a hydrogen atom, that is, a positively ionized donor with an electron in orbit about the ionized donor. In this case, $E_{r,\mathrm{D}}$ is given by [27, 28]

$$E_{r,\mathrm{D}} = E_{\mathrm{C}} - \frac{q^4 m_{\mathrm{e}}^*}{8h^2 \epsilon_{\mathrm{s}}^2 \epsilon_0^2 r^2} = E_{\mathrm{C}} - \frac{\Delta E_{1,\mathrm{D}}}{r^2} \tag{75}$$

and

$$\Delta E_{1,\mathrm{D}} = 13.6 \frac{m_{\mathrm{e}}^*}{m_0} \cdot \frac{1}{\epsilon_{\mathrm{s}}^2} \ (\mathrm{eV}), \tag{76}$$

where m_0 is the free-space electron mass.

Because the Bohr radius (a^*) of the ground state is very small, E_{D} is expressed as

$$E_{\mathrm{D}} = E_{1,\mathrm{D}} - E_{\mathrm{CCC}}, \tag{77}$$

where E_{CCC} is the central cell correction induced due to a strongly localized potential [26]. Therefore, the ground-state level does not obey the hydrogenic model. However, because the wave function extension of the $(r-1)$th excited state is of the order $r^2 a^*$, the excited state levels are expected to follow the hydrogenic model.

4.3.4. Distribution Function for Acceptors

In the case of acceptors, different from those in the conduction band, there are two degenerate valence bands (i.e., a light-hole band and a heavy-hole band), indicating that there is an acceptor state for the light-hole band as well as an acceptor state for the heavy-hole band. When the density of electrons occupied at acceptors ($n_{\rm A}$) is considered, therefore, $f_{\rm MC}(E_{\rm A})$ is derived as [34, 35]

$$f_{\rm MC}(E_{\rm A}) \equiv \frac{n_{\rm A}}{N_{\rm A}} = \frac{1}{1 + g_{\rm A,MC}(T)\exp\left(-\dfrac{E_{\rm F}(T) - E_{\rm A}}{kT}\right)}, \tag{78}$$

where $g_{\rm A,MC}(T)$ is here called the effective degeneracy factor for acceptors given by

$$g_{\rm A,MC}(T) = 4\left[1 + \sum_{r=2}^{l} g_r \exp\left(-\frac{E_{\rm A} - E_{r,\rm A}}{kT}\right)\right]\exp\left(-\frac{\overline{E_{\rm ex,A}(T)}}{kT}\right). \tag{79}$$

Here, $\overline{E_{\rm ex,A}(T)}$ is the ensemble average of the ground and excited state levels of the acceptor, measured from $E_{\rm A}$, and is given by

$$\overline{E_{\rm ex,A}(T)} = \frac{\displaystyle\sum_{r=2}^{l} (E_{\rm A} - E_{r,\rm A})\, g_r \exp\left(-\frac{E_{\rm A} - E_{r,\rm A}}{kT}\right)}{1 + \displaystyle\sum_{r=2}^{l} g_r \exp\left(-\frac{E_{\rm A} - E_{r,\rm A}}{kT}\right)}, \tag{80}$$

$E_{r,\rm A}$ is the $(r-1)$th excited state level of the acceptor, expressed as

$$E_{r,\rm A} = E_{\rm V} + \frac{q^4 m_{\rm h}^*}{8h^2\epsilon_{\rm s}^2\epsilon_0^2 r^2} = E_{\rm V} + \frac{\Delta E_{1,\rm A}}{r^2}, \tag{81}$$

and

$$\Delta E_{1,\rm A} = 13.6\frac{m_{\rm h}^*}{m_0}\cdot\frac{1}{\epsilon_{\rm s}^2}\ \ ({\rm eV}). \tag{82}$$

The average acceptor level $\overline{E_{\rm A}(T)}$ and the acceptor level are expressed as

$$\overline{E_{\rm A}(T)} = E_{\rm A} - \overline{E_{\rm ex,A}(T)} \tag{83}$$

and

$$E_{\rm A} = E_{1,\rm A} + E_{\rm CCC}. \tag{84}$$

4.3.5. Distribution Function Derived from Grand Canonical Ensemble

The grand canonical ensemble is an ensemble of the same subsystems, and particles can transfer from one subsystem into another subsystem, indicating that the number of particles in the subsystem (N) can change. The partition function (Ξ) for the grand canonical ensemble is given by [36]

$$\Xi = \sum_N z^N Z_N, \tag{85}$$

where

$$z = \exp\left(\frac{E_{\rm F}(T)}{kT}\right), \tag{86}$$

Z_N is the partition function for the canonical ensemble for a given N, which is expressed as

$$Z_N = \sum \exp\left(-\frac{\sum_j n_j E_j}{kT}\right). \tag{87}$$

This summation is carried out over all the sets of $\{n_j\}$ under the condition

$$N = \sum_j n_j, \tag{88}$$

where E_j and n_j are the energy level and number of electrons at a jth state in the subsystem, respectively.

A donor is a subsystem. Each donor has one electron or no electron (i.e., $N = 0$ or 1). The electron is located at one state for the ground and excited states (i.e., $n_j = 0$ or 1). Moreover, the number of the spin's states and degenerate excited states should be taken into account. Therefore,

$$Z_0 = 1 \tag{89}$$

and

$$Z_1 = 2\left[\exp\left(-\frac{E_{\rm D}}{kT}\right) + \sum_{r=2}^{l} g_r \exp\left(-\frac{E_{r,{\rm D}}}{kT}\right)\right]. \tag{90}$$

Finally,

$$\begin{aligned}\Xi &= \sum_{N=0}^{1} z^N Z_N = Z_0 + zZ_1 \\ &= 1 + 2\left[\exp\left(-\frac{E_{\rm D} - E_{\rm F}(T)}{kT}\right) + \sum_{r=2}^{l} g_r \exp\left(-\frac{E_{r,{\rm D}} - E_{\rm F}(T)}{kT}\right)\right]. \end{aligned} \tag{91}$$

The mean number $\langle N \rangle$ of electrons in the subsystem in thermal equilibrium is given by

$$\langle N \rangle = \frac{\sum_{N=0}^{1} N z^N Z_N}{\Xi}. \tag{92}$$

Therefore, the distribution function including the influence of the excited states for a donor is derived as

$$f_{\rm GC}(E_{\rm D}) \equiv \langle N \rangle = \frac{1}{1 + \frac{1}{g_{\rm D,GC}(T)} \exp\left(-\frac{E_{\rm F}(T) - E_{\rm D}}{kT}\right)}, \tag{93}$$

where

$$g_{\rm D,GC}(T) = 2\left[1 + \sum_{r=2}^{l} g_r \exp\left(-\frac{E_{r,{\rm D}} - E_{\rm D}}{kT}\right)\right], \tag{94}$$

which coincides with the reported distribution functions [27, 29, 31]. On the other hand, the distribution function for acceptors is derived as

$$f_{\mathrm{GC}}(E_{\mathrm{A}}) = \frac{1}{1 + g_{\mathrm{A,GC}}(T)\exp\left(-\dfrac{E_{\mathrm{F}}(T) - E_{\mathrm{A}}}{kT}\right)}, \tag{95}$$

where

$$g_{\mathrm{A,GC}}(T) = 4\left[1 + \sum_{r=2}^{l} g_r \exp\left(-\frac{E_{\mathrm{A}} - E_{r,\mathrm{A}}}{kT}\right)\right]. \tag{96}$$

4.3.6. Comparison between Three Distribution Functions

If the influence of the excited states for a donor or an acceptor could be ignored (i.e., $l = 1$), $f_{\mathrm{MC}}(E_{\mathrm{D}})$ or $f_{\mathrm{MC}}(E_{\mathrm{A}})$ would coincide with $f_{\mathrm{FD}}(E_{\mathrm{D}})$ or $f_{\mathrm{FD}}(E_{\mathrm{A}})$.

The reason why $f_{\mathrm{GC}}(E_{\mathrm{D}})$ is different from $f_{\mathrm{MC}}(E_{\mathrm{D}})$ is discussed later. Because an electron can be at a higher excited state level at elevated temperatures, the energy of an electron bound to a donor increases with increasing T. Therefore, the average donor level should increase with T, which is consistent with Eqs. (71) and (72).

If, on the other hand, electrons were located at the ground-state level at all temperatures, Eq. (70) could be replaced by

$$E_{\mathrm{total}} = E_{\mathrm{D}} \cdot n_{\mathrm{D}} + \sum_i E_i \cdot n(E_i) = \mathrm{const.} \tag{97}$$

In this case, the distribution function for donors derived from the microcanonical ensemble viewpoint would coincide with $f_{\mathrm{GC}}(E_{\mathrm{D}})$. This suggests that $f_{\mathrm{GC}}(E_{\mathrm{D}})$ is correct only under the assumption that all the electrons bound to donors have E_{D} at all temperatures. In the same way as illustrated for $f_{\mathrm{GC}}(E_{\mathrm{D}})$, if $\overline{E_{\mathrm{A}}(T)}$ could be assumed to be E_{A} (i.e., $E_{\mathrm{ex,A}}(T) = 0$), $f_{\mathrm{MC}}(E_{\mathrm{A}})$ would coincide with $f_{\mathrm{GC}}(E_{\mathrm{A}})$.

4.4. Determination of Reliable Density and Energy Level of Deep Dopant

By FCCS using $f_{\mathrm{MC}}(E_{\mathrm{A}})$, in B-doped diamond, N_{A}, E_{A}, and N_{D} were determined to be 2.82×10^{17} cm^{-3}, $E_{\mathrm{V}} + 0.323$ eV, and 1.95×10^{16} cm^{-3}, respectively. The highest excited state considered here was the sixth excited state (i.e., $l = 7$), under which the best curve fitting was achieved. Because the radius of the sixth excited state is approximately 15 nm and the lattice constant for diamond is 0.356 nm, the number of C atoms in a sphere of radius 15 nm is approximately 2×10^6. Because the C density is 1.8×10^{23} cm^{-3} and the C_{B} is approximately 2×10^{17} cm^{-3}, there is on the other hand, one B atom in approximately 10^6 C atoms. These suggest that the condition (i.e., $l = 7$) is not so bad.

N_{A}, E_{A}, and N_{D} determined by FCCS using $f_{\mathrm{GC}}(E_{\mathrm{A}})$ are also shown in Table 4. In Table 4, all the E_{A} values determined using three distribution functions seems reasonable. However, N_{A} for $f_{\mathrm{GC}}(E_{\mathrm{A}})$ is highest, while N_{A} for $f_{\mathrm{MC}}(E_{\mathrm{A}})$ is lowest. The N_{A} obtained using $f_{\mathrm{MC}}(E_{\mathrm{A}})$ is closest to C_{B}. Therefore, $f_{\mathrm{MC}}(E_{\mathrm{A}})$ is suitable for determining N_{A} from $p(T)$.

Table 4 shows $N_{\rm A}$, $E_{\rm A}$, and $N_{\rm comp}$ for the others determined by FCCS using $f_{\rm MC}(E_{\rm A})$ or $f_{\rm GC}(E_{\rm A})$ As is clear from Table 4, only $f_{\rm MC}(E_{\rm A})$ led to reliable $N_{\rm A}$, $E_{\rm A}$, and $N_{\rm D}$ for heavily-doped cases.

In Te-doped $Al_{0.6}Ga_{0.4}Sb$ where $E_{\rm F}(T)$ was shallower than the donor energy level, moreover, $f_{\rm MC}$ is suitable for determining the density and energy level of the Te donor from the temperature dependence of the electron concentration [19].

Judging from the previous discussions, the distribution function derived from the microcanonical ensemble viewpoint is most appropriate. The reason why a reasonable dopant density can be obtained using Eq. (73) or (78) is discussed from the viewpoint of the effective degeneracy factors. Figure 15 shows $g_{\rm A,FD}$, $g_{\rm A,MC}(T)$, and $g_{\rm A,GC}(T)$ for 6H-SiC,

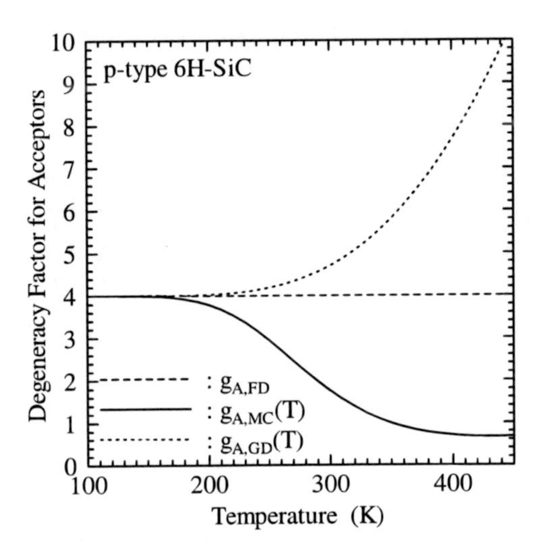

Figure 15. Temperature dependencies of simulated degeneracy factors for acceptors for 6H-SiC.

denoted by broken, solid, and dotted lines, respectively. $g_{\rm A,MC}(T)$ decreases from 4 with increasing T, whereas $g_{\rm A,GC}(T)$ increases. As the effective degeneracy factor for acceptors decreases, the distribution function for acceptors approaches 1 at the same T, indicating that in the case of $f_{\rm MC}(E_{\rm A})$, the ionization efficiency of acceptors is highest at elevated temperatures.

Figure 16(a) depicts the ionized acceptor densities $N_{\rm A}^{-}(T)$ simulated using $N_{\rm A}$ of 3.0×10^{18} cm^{-3}, $E_{\rm A}$ of $E_{\rm V} + 0.18$ eV, $N_{\rm D}$ of 9.7×10^{16} cm^{-3}, and the effective degeneracy factors for acceptors shown in Fig. 15, for $f_{\rm FD}(E_{\rm A})$, $f_{\rm MC}(E_{\rm A})$, and $f_{\rm GC}(E_{\rm A})$, denoted by broken, solid, and dotted lines, respectively. As is clear from the figure, $N_{\rm A}^{-}(T)$ for $f_{\rm MC}(E_{\rm A})$ is highest at elevated temperatures. In the case of $f_{\rm MC}(E_{\rm A})$, therefore, $N_{\rm A}$ required to satisfy the experimentally obtained $p(T)$ is much less than that determined using $f_{\rm FD}(E_{\rm A})$ or $f_{\rm GC}(E_{\rm A})$.

Another interpretation of Fig. 16(a) is as follows. As the temperature increases, the possibility increases that a hole bound to the acceptor is located at a higher excited state level. That is why the acceptor can more easily emit a hole to the valence band at elevated temperatures. This coincides with Eq. (83). Therefore, it is clear that the excited states of

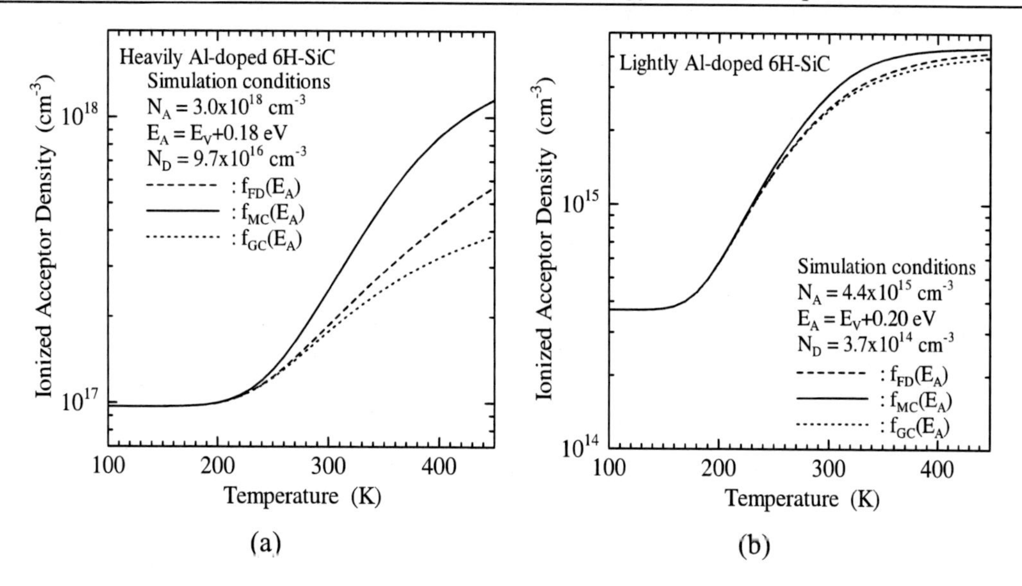

Figure 16. Temperature dependencies of simulated ionized acceptor densities; (a) for heavily Al doped 6H-SiC and (b) for lightly Al doped 6H-SiC.

the acceptor enhance the ionization of the acceptor at elevated temperatures.

Figure 16(b) shows $N_{\mathrm{A}}^{-}(T)$ simulations for the lightly Al-doped 6H-SiC using the effective degeneracy factors for acceptors shown Fig. 15. Here, N_{A}, E_{A}, and N_{D} used in the simulation were 4.4×10^{15} cm^{-3}, $E_{\mathrm{V}} + 0.20$ eV, and 3.7×10^{14} cm^{-3}, respectively. In the figure, the broken, solid, and dotted lines represent the $N_{\mathrm{A}}^{-}(T)$ simulations for $f_{\mathrm{FD}}(E_{\mathrm{A}})$, $f_{\mathrm{MC}}(E_{\mathrm{A}})$, and $f_{\mathrm{GC}}(E_{\mathrm{A}})$, respectively. Although the effective degeneracy factors for acceptors in the lightly-doped case are the same as those in the heavily-doped case, $N_{\mathrm{A}}^{-}(T)$ for $f_{\mathrm{FD}}(E_{\mathrm{A}})$, $f_{\mathrm{MC}}(E_{\mathrm{A}})$, and $f_{\mathrm{GC}}(E_{\mathrm{A}})$ are similar to each other in the lightly-doped case. This is because the effective degeneracy factor for acceptors has little effect on $f_{\mathrm{FD}}(E_{\mathrm{A}})$, $f_{\mathrm{MC}}(E_{\mathrm{A}})$, and $f_{\mathrm{GC}}(E_{\mathrm{A}})$, when $E_{\mathrm{F}}(T)$ is far from the acceptor level. To determine N_{A} from $p(T)$, therefore, $f_{\mathrm{MC}}(E_{\mathrm{A}})$ is the most appropriate among them.

5. Transient Capacitance Method for High-Resistivity Semiconductors

5.1. Dependence of Diode Capacitance on Measurement Frequency

Let us consider the capacitance of a Schottky barrier diode measured at different frequencies. Figure 17(a) and (b) show the energy band diagram and the equivalent circuit of the Schottky barrier diode, respectively. At low frequency ($f \ll 1/2\pi\epsilon_{\mathrm{s}}\epsilon_0\rho$) that is a normal measurement frequency, the equivalent circuit is described as Fig. 17(c), insisting that the capacitance measured is due to the depletion width of the diode.

At high frequency ($f \gg 1/2\pi\epsilon_{\mathrm{s}}\epsilon_0\rho$), on the other hand, the capacitance measured is the geometric capacitance calculated using the semiconductor thickness, because the equivalent circuit can be expressed as Fig. 17(d).

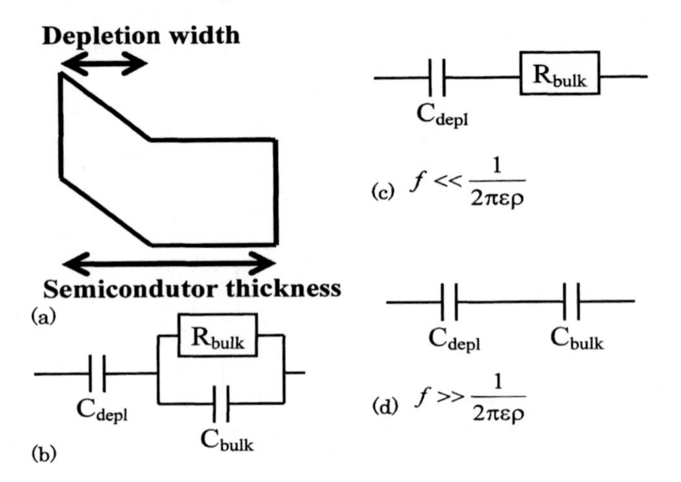

Figure 17. Capacitance measurements for Schottky barrier diode; (a) energy band diagram, (b) equivalent circuit of the diode, (c) equivalent circuit at low measurement frequency, and (d) equivalent circuit at high measurement frequency. Here, ϵ is the permittivity for semiconductor ($\epsilon_s \epsilon_0$).

5.2. Capacitance-Voltage Characteristics

Capacitance–voltage (C–V) and transient capacitance $C(t)$ measurements were carried out in a Schottky barrier diode fabricated using 0.3-mm-thick high-resistivity p-type 6H-SiC at a frequency of 1 MHz in the temperature range between 300 and 600 K, where the electrode area was 3.14 mm^2 [37].

The capacitance of the diode at 300 K was independent of reverse bias as shown by ■ in Fig. 18(a), and was close to the geometric capacitance calculated using the thickness of the 6H-SiC substrate. The ○ in Fig. 18(a) represents the C–V characteristics at 500 K. At $\geq$ 450 K, the capacitance was dependent on reverse bias, indicating that the measured capacitance is determined by the width of the depletion region. Therefore, the capacitance was measured at $\geq$ 450 K. Figure 18(b) shows the C^{-2}–V characteristics of the diode at 500 K. From the slope, the value of N_A was estimated as 9.3×10^{14} cm^{-3}, where N_A includes the acceptor density and hole-trap densities.

5.3. Transient Capacitance

The $C(t)$ was measured after an applied voltage was step-functionally changed from 0 to a reverse bias (V_R) of -20 V. Figure 19 shows the $C(t)$ and ICTS signal $S(t)$ of the diode at 500 K, indicated by broken and solid lines, respectively, where

$$S(t) \equiv t\frac{\mathrm{d}C(t)^2}{\mathrm{d}t}. \tag{98}$$

From the number of peaks of $S(t)$ in Fig. 19, at least three types of hole-trap species could be detected. These hole traps are here referred to as HRH2, HRH3, and HRH4, as shown in Fig. 19. Judging from the ICTS signals in the shorter or longer time range, there are at least

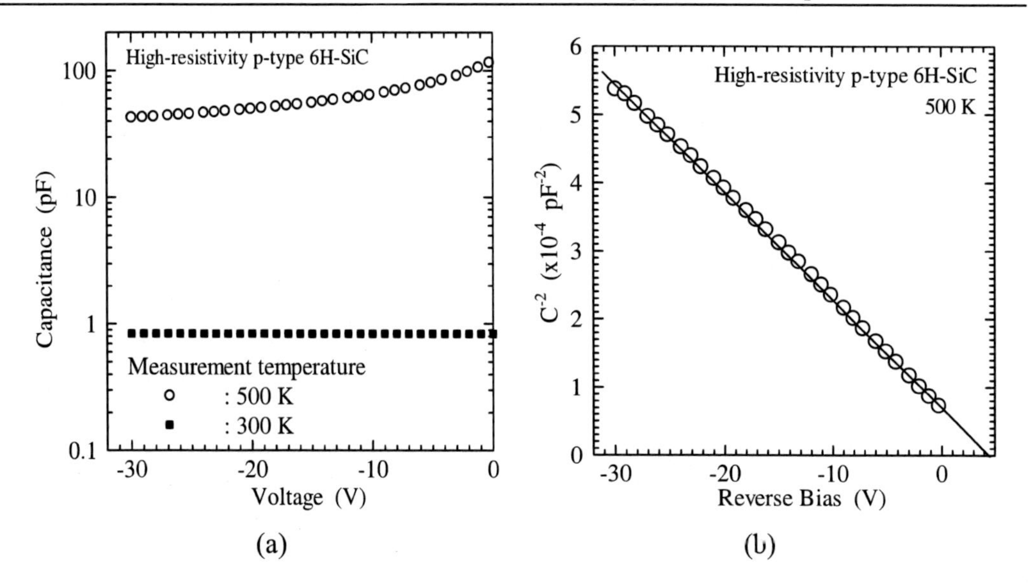

Figure 18. Characteristics of diode; (a) C–V characteristics at 300 K and 500 K and (b) $1/C^{-2}$–V characteristics at 500 K.

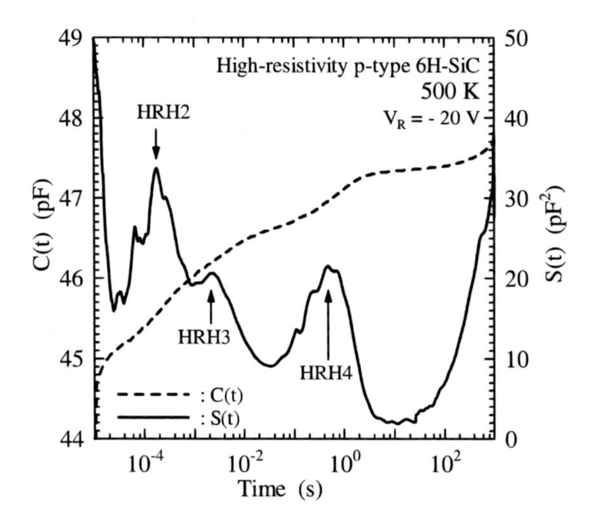

Figure 19. Transient capacitance and ICTS signal at 500 K.

another two unresolved hole traps. Figure 20(a) and (b) show the ICTS signal at 470 and 530 K, respectively, in which these two contributions could be fully resolved and labeled HRH1 and HRH5, respectively. Finally, five types of hole traps could be observed in the temperature range between 450 and 600 K.

The ICTS signal is theoretically expressed as [8]

$$S(t) = \frac{q\epsilon_{\mathrm{s}}\epsilon_0 A^2}{2(V_{\mathrm{D}} - V_{\mathrm{R}})} \sum_i N_{\mathrm{TH}i} e_{\mathrm{TH}i} t \exp\left(-e_{\mathrm{TH}i} t\right), \tag{99}$$

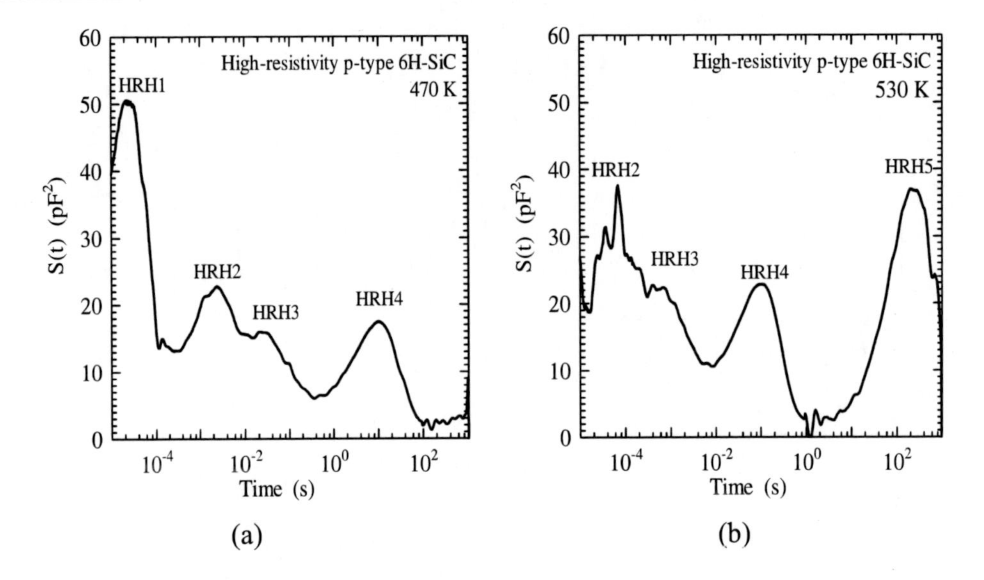

Figure 20. ICTS signal at 470 K (a) and at 530 K (b).

which has a peak at

$$t_{\mathrm{peak}i} = \frac{1}{e_{\mathrm{TH}i}}, \tag{100}$$

where V_{D} is the diffusion potential of the diode, and $e_{\mathrm{TH}i}$ is the emission rate of an ith hole trap, respectively. Moreover, $N_{\mathrm{TH}i}$ is given as [8]

$$N_{\mathrm{TH}i} = \frac{2(V_{\mathrm{D}} - V_{\mathrm{R}})}{q\epsilon_{\mathrm{s}}\epsilon_0 A^2} S(t_{\mathrm{peak}i}) \exp 1. \tag{101}$$

On the other hand, the hole concentration at the ith trap ($p_{\mathrm{TH}i}$) varies as

$$\frac{\mathrm{d}p_{\mathrm{TH}i}}{\mathrm{d}t} = \sigma_{\mathrm{TH}i} v_{\mathrm{th}} p \left(N_{\mathrm{TH}i} - p_{\mathrm{TH}i}\right) - e_{\mathrm{TH}i} p_{\mathrm{TH}i}, \tag{102}$$

where

$$p_{\mathrm{TH}i} = N_{\mathrm{TH}i}\left[1 - f(E_{\mathrm{TH}i})\right], \tag{103}$$

$f(E_{\mathrm{TH}i})$ is the electron occupation probability for the ith hole trap, given as

$$f(E_{\mathrm{TH}i}) = \frac{1}{1 + g_{\mathrm{TH}i} \exp\left(\dfrac{E_{\mathrm{TH}i} - E_{\mathrm{F}}}{kT}\right)}, \tag{104}$$

p is the hole concentration, v_{th} is the thermal velocity of electron, and $\sigma_{\mathrm{TH}i}$ and $g_{\mathrm{TH}i}$ are the cross section and degeneracy factor of the ith hole trap, respectively. In thermal equilibrium ($\mathrm{d}p_{\mathrm{TH}i}/\mathrm{d}t = 0$), $e_{\mathrm{TH}i}$ is derived using Eqs. (102)-(104) as

$$e_{\mathrm{TH}i} = \nu_{\mathrm{TH}i} \exp\left(-\frac{E_{\mathrm{TH}i} - E_{\mathrm{V}}}{kT}\right) \tag{105}$$

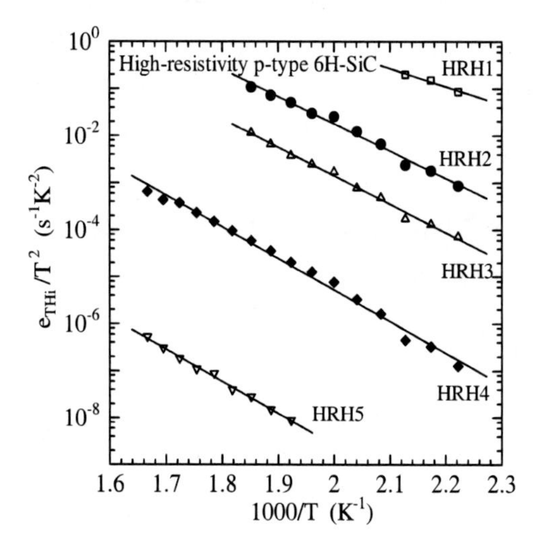

Figure 21. Relationship between $e_{\mathrm{TH}i}/T^2$ and $1/T$.

Table 5. Densities, energy levels and cross sections of hole traps in high-resistivity 6H-SiC determined by ICTS.

Hole-trap species	Density (cm^{-3})	Energy level (eV)	Cross section (cm^2)
HRH1	5.0×10^{13}	$E_\mathrm{V} + 0.76$	1×10^{-14}
HRH2	3.3×10^{13}	$E_\mathrm{V} + 1.15$	2×10^{-12}
HRH3	2.0×10^{13}	$E_\mathrm{V} + 1.20$	5×10^{-13}
HRH4	2.1×10^{13}	$E_\mathrm{V} + 1.33$	4×10^{-14}
HRH5	3.6×10^{13}	$E_\mathrm{V} + 1.35$	4×10^{-17}

and

$$\nu_{\mathrm{TH}i} = \frac{N_\mathrm{V} \sigma_{\mathrm{TH}i} v_\mathrm{th}}{g_{\mathrm{TH}i}}. \tag{106}$$

Since N_V is proportional to $T^{1/2}$ and v_th is proportional to $T^{3/2}$, $\nu_{\mathrm{TH}i}$ is expressed as

$$\nu_{\mathrm{TH}i} = B_i \sigma_{\mathrm{TH}i} T^2 \tag{107}$$

and

$$B_i = \frac{4\pi\sqrt{6\pi} m_\mathrm{h}^* k^2}{g_{\mathrm{TH}i} h^3}. \tag{108}$$

The relationship between $e_{\mathrm{TH}i}/T^2$ and $1/T$ is shown in Fig. 21. In Fig. 21, the optimum straight line fitting to experimental data for each trap could be obtained, from which the values of $E_{\mathrm{TH}i}$ and $\sigma_{\mathrm{TH}i}$ could be determined and listed in Table 5. Here, m_h^* and $g_{\mathrm{TH}i}$ were assumed to be m_0 and 1, respectively. This table also includes $N_{\mathrm{T}i}$ estimated using Eq. (101), which is averaged over the temperature range of the measurement. The sum of the densities of intrinsic defects detected is 1.6×10^{14} cm^{-3}, whereas the sum of the acceptor density and hole-trap densities is approximately 9×10^{14} cm^{-3}. Therefore, the intrinsic defects strongly decrease the majority-carrier concentration in the high-purity 6H-SiC, which makes ρ as high as approximately 10^6 Ω cm.

6. Heterojunction-Monitored Capacitance Method for Amorphous Semiconductors

6.1. Steady-State HMC

The optoelectronic properties of high-resistivity amorphous semiconductors such as undoped hydrogenated amorphous silicon (a-Si:H) are critically linked with the density-of-state (DOS) distribution, $g(E)$, in the mobility gap. Measurement of $g(E)$ and an understanding of the nature of the gap states are, therefore, very important. However, DLTS and ICTS are not feasible to determine the $g(E)$ using a Schottky barrier diode fabricated from high-resistivity amorphous semiconductor, because it is impossible to measure the depletion width of the junction.

In order to measure the depletion width in high-resistivity amorphous semiconductors, an amorphous semiconductor/crystalline semiconductor heterojunction is discussed. The method to determine $g(E)$ in amorphous semiconductors using heterojunctions is referred to as heterojunction-monitored capacitance (HMC) methods [38, 39, 40, 41, 42]. Since the conduction type of undoped a-Si:H is n-type, for example, p-type crystalline Si (c-Si) is selected. Figure 22(a), (b), (c), and (d) show the energy-band diagram, potential, charges in statics, and charges corresponding to AC voltage in the p-type c-Si/undoped a-Si:H/Mg heterojunction, respectively, where Mg forms an ohmic contact with the undoped a-Si:H [38, 43]. In the figure, the gap states as indicated by the black area of (a) are positively charged states, and $\ominus$ represents negatively charged acceptors in the depletion region of p-type c-Si.

The depletion region formed by the heterojunction is considered. When a bias voltage (V) is applied, it produces space-charge layers both in a-Si:H and c-Si. Since this p-type c-Si has only shallow acceptors, the space charge in the c-Si is formed by negatively charged acceptors. However, localized states in a-Si:H distribute within the gap.

Let us discuss the origin of the positive space charge in a-Si:H using Fig. 22. In the neutral region, all the gap states below E_{F} are occupied by electrons, whereas in the depletion region, the gap states above E_{OB2} are devoid of electrons, where E_{OB2} is determined from the condition that the thermal emission rate for electrons is equal to that for holes and given by [39]

$$E_{\mathrm{OB2}} = E_{\mathrm{V}} + E_{\mathrm{g2}} + \frac{kT}{2} \ln\left(\frac{\nu_{\mathrm{h}}}{\nu_{\mathrm{e}}}\right), \tag{109}$$

where ν_{h} and ν_{e} are the attempt-to-escape frequencies for holes and electrons, respectively. Therefore, the gap states in the area painted in black in Fig. 22(a) behave like positive space charges (herein referred to as donor-like states). The density of the donor-like states is constant in the spatial position between 0 and the depletion width in a-Si:H (W_2). This, together with the density of donors (if they exist), gives the effective density of the donor-like states (N_{I}), as shown in Fig. 22(c). Figure 22(b) shows the potential variation with

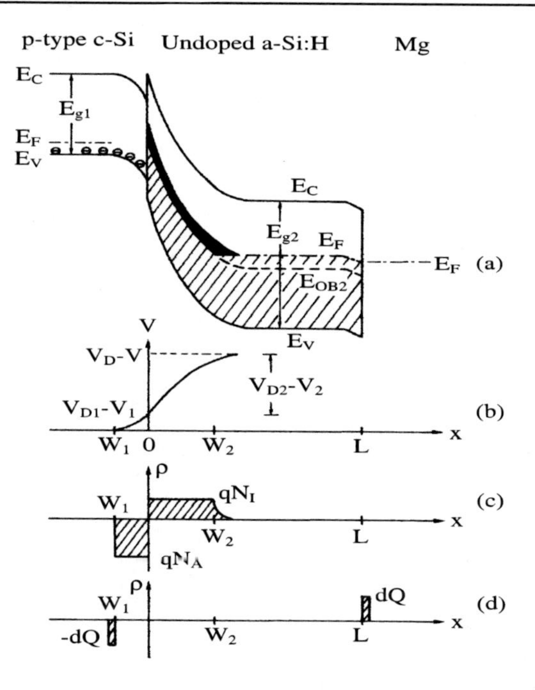

Figure 22. Schematic sketches of undoped a-Si:H/p-type c-Si heterojunction; (a) energy-band diagram, (b) potential, (c) charges in statistic, and (d) charges corresponding to AC voltage.

distance. The depletion widths (W_1 and W_2) are given by

$$qN_{\mathrm{A}}W_1 \cong qN_{\mathrm{I}}W_2 \tag{110}$$

with

$$W_1 = \sqrt{\frac{2\epsilon_{\mathrm{s}1}\epsilon_0\left(V_{\mathrm{D}1} - V_1\right)}{qN_{\mathrm{A}}}} \tag{111}$$

and

$$W_2 \cong \sqrt{\frac{2\epsilon_{\mathrm{s}2}\epsilon_0\left(V_{\mathrm{D}2} - V_2\right)}{qN_{\mathrm{I}}}}. \tag{112}$$

Here, E_{g} is the bandgap and L is the thickness of a-Si:H. The subscripts 1 and 2 refer to c-Si and a-Si:H, respectively.

The capacitance is measured with a small AC voltage at 1 MHz. The resistivity (ρ_1) of c-Si is about 1 Ω·cm so that the dielectric relaxation time ($\epsilon_{\mathrm{s}1}\epsilon_0\rho_1$) becomes 10^{-12} s, indicating that the capacitance C_1 in c-Si is given by

$$C_1 = \frac{\epsilon_{\mathrm{s}1}\epsilon_0 A}{W_1}. \tag{113}$$

On the other hand, the resistivity (ρ_2) of undoped a-Si:H is about 10^9 Ωcm. Then, the dielectric time ($\epsilon_{\mathrm{s}2}\epsilon_0\rho_2$) becomes 10^{-3} s, suggesting that at 1 MHz AC voltage the capaci-

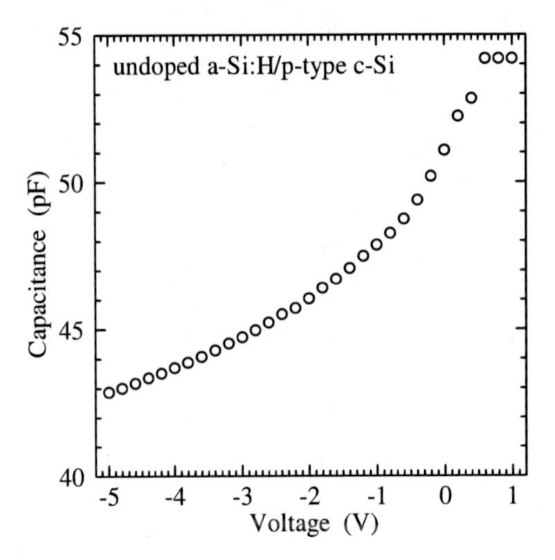

Figure 23. $C - V$ characteristics of undoped a-Si:H/p-type c-Si heterojunction at 1 MHz.

tance C_2 in a-Si:H can simply be given by

$$C_2 = \frac{\epsilon_{s2}\epsilon_0 A}{L}. \tag{114}$$

The measured capacitance C at 1 MHz is expressed as

$$\frac{1}{C} = \frac{1}{C_1} + \frac{1}{C_2}, \tag{115}$$

because the redistribution of charge ($\mathrm{d}Q$) can spatially respond to the 1 MHz AC voltage at W_1 and L, shown in Fig. 22(d). From Eqs. (110)-(112), the following equation is obtained:

$$\frac{V_{D1} - V_1}{V_{D2} - V_2} \cong \frac{N_I \epsilon_{s2}}{N_A \epsilon_{s1}}. \tag{116}$$

Finally, the next equation is derived from Eqs. (111), (113), (115), and (116):

$$\begin{aligned} W_1^2 &= \left[\epsilon_{s1}\epsilon_0 A \left(\frac{1}{C} - \frac{1}{C_2}\right)\right]^2 \\ &\cong \frac{2\epsilon_{s1}\epsilon_{s2}\epsilon_0 N_I (V_D - V)}{q N_A (N_A \epsilon_{s1} + N_I \epsilon_{s2})}. \end{aligned} \tag{117}$$

From the W_1^2–V characteristics, therefore, N_I can be estimated.

Figure 23 shows the high-frequency $C - V$ characteristics of an undoped a-Si:H/p-type c-Si heterojunction. The value of N_A, L, and A are 1.0×10^{16} cm^{-3}, approximately 1.5 μm, and 0.785 mm^2, respectively. Because W_1 is close to 0 at forward biases, the saturated value of 54.2 pF is determined by the thickness of a-Si:H.

Figure 24(a) shows the $C^{-2} - V$ characteristics, and the straight relationship could not be obtained. Figure 24(b) shows the $W_1^2 - V$ characteristics, where $C_2 = 54.2$ pF. Since the

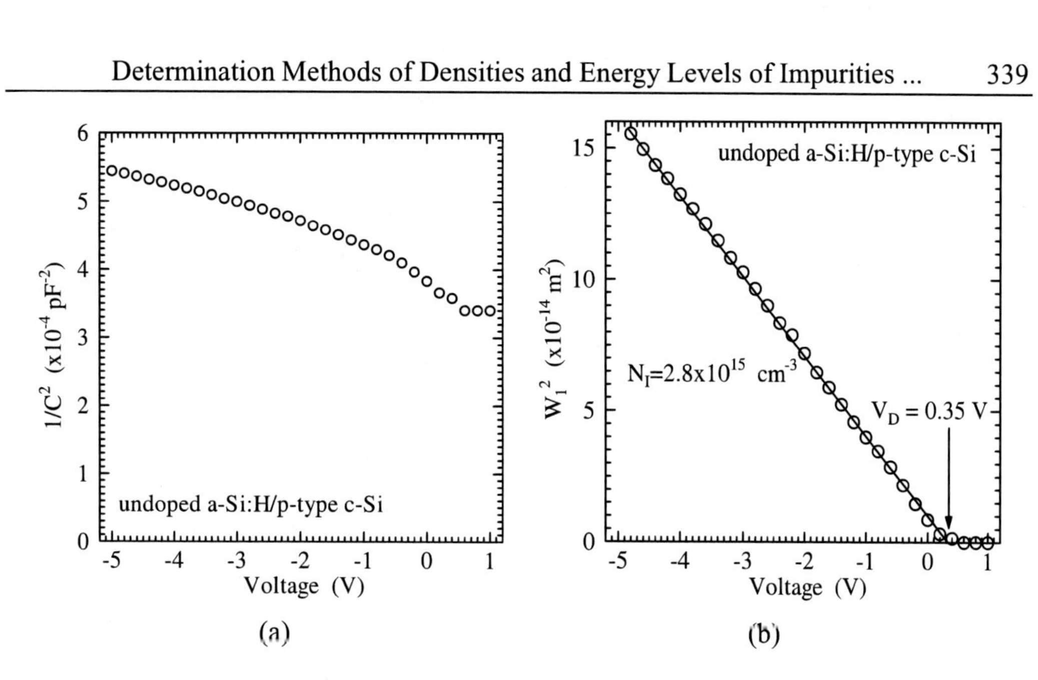

Figure 24. Characteristics of undoped a-Si:H/p-type c-Si heterojunction; (a) $C^{-2} - V$ characteristics and (b) $W_1^2 - V$ characteristics.

data reveal a good linear relationship, N_{I} and V_{D} were determined to be 2.8×10^{15} cm^{-3} and 0.35 V, respectively. The value of W_2 at -5 V is calculated to be approximately 1.4 μm using Eq. (112), indicating that $W_2 < L$. It is noted that W_2 at the maximum reverse bias should be shorter than L.

6.2. Heterojunction-Monitored Capacitance Spectroscopy

In order to determine $g(E)$, the transient HMC is considered after V_{R} is applied to the diode over the zero-bias condition, as shown in Fig. 25. In energy-band diagrams, the gap states as indicated by the hatched area are neutral, and in the depletion region the empty gap states between E_{F} and E_{OB2} behave like positively charged states. This method is referred to as heterojunction-monitored capacitance spectroscopy (HMCS) [39, 41, 42].

Figure 25(a) shows the energy-band diagram and space-charge density at $t < 0$. At $t = 0$, V_{R} is applied across the whole of the amorphous and crystalline components. After the reverse bias has been on for $\epsilon_{\mathrm{s2}}\epsilon_0\rho_2$, the space charge in the vicinity of the heterojunction will redistribute itself in response to the applied potential, as shown in Fig. 25(b). The transient HMC $C_{\mathrm{HM}}(t)$ after $\epsilon_{\mathrm{s2}}\epsilon_0\rho_2$ can be analyzed from Eq. (117), and N_{I} at t can be expressed as

$$N_{\mathrm{I}}(t) = \frac{\epsilon_{\mathrm{s1}} V_{\mathrm{c}}(t) N_{\mathrm{A}}}{\epsilon_{\mathrm{s2}} \left[V_{\mathrm{D}} - V_{\mathrm{R}} - V_{\mathrm{c}}(t) \right]}, \tag{118}$$

with

$$V_{\mathrm{c}}(t) = \frac{q N_{\mathrm{A}} W_1^2(t)}{2 \epsilon_{\mathrm{s1}} \epsilon_0} \tag{119}$$

and

$$W_1(t) = \epsilon_{\mathrm{s1}} \epsilon_0 A \left[\frac{1}{C_{\mathrm{HM}}(t)} - \frac{1}{C_2} \right], \tag{120}$$

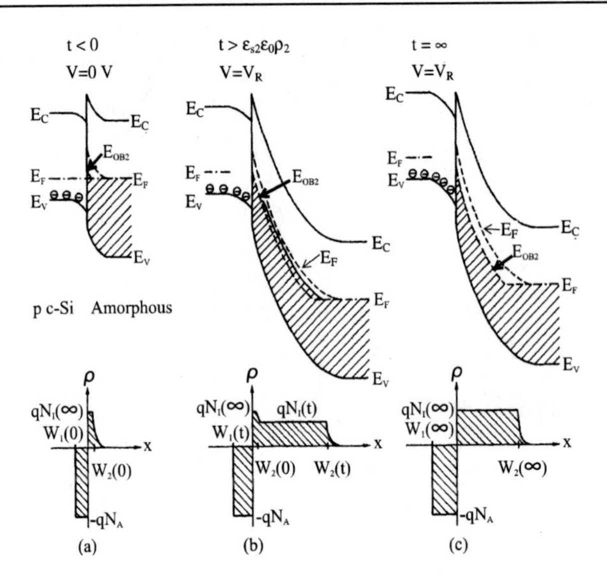

Figure 25. Schematic sketches (energy-band diagram and space-charges density) of undoped a-Si:H/p-type c-Si heterojunction at three different times; (a) $t < 0$, (b) $t > \epsilon_{s2}\epsilon_0\rho_2$, and (c) $t = \infty$.

where $W_1(t)$ is the depletion width at t and $V_c(t)$ is the voltage across the depletion region of c-Si at t. To make this analysis feasible, the absolute value of V_R has to be necessarily much higher than V_D, so that the relationship of $N_I W_2(t) \gg N_I W_2(0)$ is valid and the average value of N_I over the depletion region at t is close to $N_I(t)$. This condition also suggests that interface states do not affect the measurement of HMC.

The function $H(t)$ is defined as

$$H(t) \equiv t\left(\frac{\mathrm{d}N_I(t)}{\mathrm{d}t}\right). \tag{121}$$

On the other hand, $H(t)$ is theoretically derived as

$$\begin{aligned} H(t) &= \int_{E_V}^{E_C} \left[f_{FD}(E') - F_\infty(E')\right] g(E') \left[e_e(E') + e_h(E')t\right] \\ &\quad \times \exp\left\{-\left[e_e(E') + e_h(E')\right]t\right\} \mathrm{d}E' \end{aligned} \tag{122}$$

with

$$f_{FD}(E) = \frac{1}{1 + \exp\left(\dfrac{E - E_F}{kT}\right)}, \tag{123}$$

$$F_\infty(E) = \frac{e_h(E)}{e_e(E) + e_h(E)}, \tag{124}$$

$$e_e(E) = \nu_e \exp\left(-\frac{E_C - E}{kT}\right), \tag{125}$$

and

$$e_{\rm h}(E) = \nu_{\rm h} \exp\left(-\frac{E - E_{\rm V}}{kT}\right). \tag{126}$$

Under the conditions that $e_{\rm e}(E) \gg e_{\rm h}(E)$ and $f_{\rm FD}(E) \sim 1$ (i.e., for the gap states between $E_{\rm F}$ and $E_{\rm OB2}$), Eq. (122) can be approximately expressed as

$$H(t) \cong \int_{E_{\rm V}}^{E_{\rm C}} g(E')e_{\rm e}(E')t \exp\left[-e_{\rm e}(E')t\right] {\rm d}E'. \tag{127}$$

Assuming that the function of $e_{\rm e}(E')t\exp[-e_{\rm e}(E')t]$ behaves as a delta function, $kT\delta(E - E')$, since the integrated value of $e_{\rm e}(E')t\exp[-e_{\rm e}(E')t]$ from $E = 0$ to ∞ using Eq. (125) is kT, we can easily derive the following relationship from Eq. (127):

$$g\left[E(t)\right] \cong \frac{H(t)}{kT} \tag{128}$$

and

$$E_{\rm C} - E(t) = kT \ln\left(\nu_{\rm e} t\right). \tag{129}$$

Using $g(E)$ calculated from Eq. (128) as an initial $g(E)$, we can determine $g(E)$ from which $H(t)$ of Eq. (122) can be obtained to fit the measured $H(t)$.

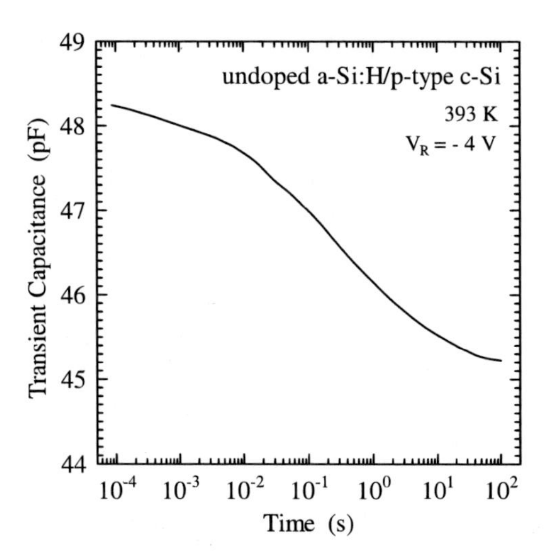

Figure 26. Transient capacitance of undoped a-Si:H/p-type c-Si heterojunction.

Figure 26 shows the change in the capacitance of the heterojunction after $V_{\rm R}$ of -4 V is applied to the diode under the zero-bias condition.

The solid line in Fig. 27 represents $g(E)$ for a-Si:H calculated using $C_{\rm HM}(t)$ in Fig 26, when $\nu_{\rm e} = 10^{13}$ s^{-1}. Here, the value of $\nu_{\rm e}$ can be estimated from the temperature dependence of $H(t)$ [39, 42]. In Fig. 27, the broken and dashed-dotted lines represent $g(E)$ for a-Si$_{1-x}$Ge$_x$:H and a-Si$_{1-x}$C$_x$:H, respectively, where $E_{\rm o}$ is the optical bandgap of amorphous semiconductors.

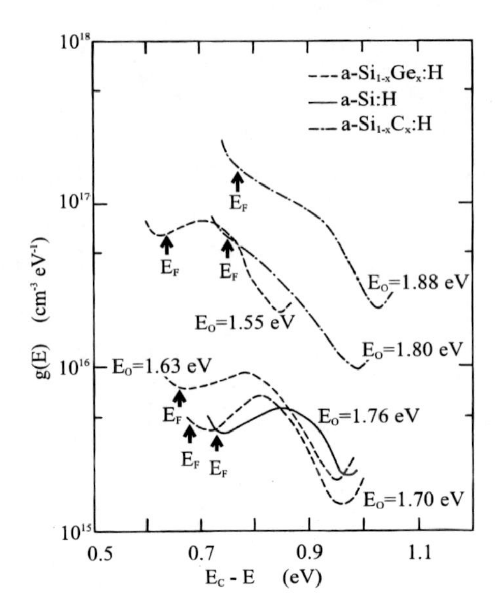

Figure 27. Density of states for a-Si:H, a-Si$_{1-x}$Ge$_x$:H, and a-Si$_{1-x}$C$_x$:H.

7. Discharge Current Transient Spectroscopy

7.1. Basic Concept

Even when the resistivities of materials are too high to measure the change of the depletion width in diodes by capacitance, it is able to measure the transient current in the diodes. Therefore, the transient current due to re-emission (or discharge) of charges from traps is considered.

In graphical peak analysis methods, it is desired that functions to be evaluated have a peak value ($NS_{\mathrm{T}i}$) at a peak discharge time ($t_{\mathrm{peak}i} = 1/e_{\mathrm{T}i}$), where $NS_{\mathrm{T}i}$ is the density per unit area of an ith trap, respectively. One of the functions that satisfy the condition mentioned earlier is expressed as

$$D(t) = \sum_{i=1} NS_{\mathrm{T}i} e_{\mathrm{T}i} t \exp(-e_{\mathrm{T}i} t + 1), \tag{130}$$

where t is the discharge time. Here, when traps are confirmed to be uniformly distributed in the direction of the film thickness, the density per unit volume can be calculated as $NS_{\mathrm{T}i}$ over the film thickness.

When we can introduce a function in which each peak appears at

$$t_{\mathrm{peak}i} = \frac{1}{e_{\mathrm{T}i} + e_{\mathrm{ref}}}, \tag{131}$$

we can shift the peak discharge time to the measurement time range by changing the parameter (e_{ref}). This indicates that we can determine $NS_{\mathrm{T}i}$ and $e_{\mathrm{T}i}$ in a wide emission-rate range, even when the measurement time range is limited. The function that satisfies this condition is

$$D(t, e_{\mathrm{ref}}) = \sum_{i=1} NS_{\mathrm{T}i} e_{\mathrm{T}i} t \exp\left[-\left(e_{\mathrm{T}i} + e_{\mathrm{ref}}\right) t + 1\right]. \quad (132)$$

In this case, each peak value is

$$D(t_{\mathrm{peak}i}, e_{\mathrm{ref}}) = NS_{\mathrm{T}i}\left(1 - e_{\mathrm{ref}} t_{\mathrm{peak}i}\right), \quad (133)$$

at $t_{\mathrm{peak}i}$ expressed as Eq. (131). Therefore, using $t_{\mathrm{peak}i}$ and $D(t_{\mathrm{peak}i}, e_{\mathrm{ref}})$, the values of $NS_{\mathrm{T}i}$ and $e_{\mathrm{T}i}$ can be determined as

$$NS_{\mathrm{T}i} = \frac{D(t_{\mathrm{peak}i}, e_{\mathrm{ref}})}{1 - e_{\mathrm{ref}} t_{\mathrm{peak}i}} \quad (134)$$

and

$$e_{\mathrm{T}i} = \frac{1}{t_{\mathrm{peak}i}} - e_{\mathrm{ref}}, \quad (135)$$

respectively.

In order to obtain Eq. (132), we define the function to be evaluated as

$$D(t, e_{\mathrm{ref}}) \equiv \frac{t}{qA}\left[I_{\mathrm{dis}}(t) - I_{\mathrm{ssl}}(V_{\mathrm{dis}})\right] \exp(-e_{\mathrm{ref}} t + 1), \quad (136)$$

where $I_{\mathrm{ssl}}(V_{\mathrm{dis}})$ is the steady-state leakage current at the discharge voltage (V_{dis}).

In the case of thermionic emission processes, $\Delta E_{\mathrm{T}i}$ can be determined from the temperature dependence of $e_{\mathrm{T}i}$, since $e_{\mathrm{T}i}$ is given by

$$e_{\mathrm{T}i} = \nu_{\mathrm{T}i} \exp\left(-\frac{\Delta E_{\mathrm{T}i}}{kT}\right). \quad (137)$$

When the time dependence of the depolarization of an ith dipole in a dielectric film is given by

$$\exp\left(-\frac{t}{\tau_i}\right), \quad (138)$$

the polarization (P_i) and relaxation time (τ_i) of the i-th dipole can be determined using DCTS. In this case, $e_{\mathrm{T}i}$ and $NS_{\mathrm{T}i}$ in Eq. (132) should be replaced by $1/\tau_i$ and P_i/q, respectively.

7.2. Theoretical Consideration

A capacitor or a diode is considered. When a charge voltage (V_{cha}) is applied to the capacitor in the interval of $-t_{\mathrm{cha}} < t < 0$, a charge current ($I_{\mathrm{cha}}$), which fills traps with charged carriers (electrons or holes), flows through the capacitor. In the case of diodes, V_{cha} is usually 0 V. By $t = 0$, all the traps are assumed to capture charged carriers, indicating that at $t = 0$ the density per unit area, $NS_{\mathrm{T}i}(t)$, of charged carriers captured at the ith trap is expressed as

$$NS_{\mathrm{T}i}(0) = NS_{\mathrm{T}i}. \quad (139)$$

At $t = 0$, the applied voltage is changed from $V_{\rm cha}$ to $V_{\rm dis}$. Since the resistance in the external measurement circuit is very low, the charge due to the geometric capacity disappears in a very short time. At $t > 0$, therefore, the measured $I_{\rm dis}(t)$ arises due to the emission of charged carriers from traps as well as due to $I_{\rm ssl}(V_{\rm dis})$.

Since $NS_{{\rm T}i}(t)$ follows the rate equation

$$\frac{{\rm d}NS_{{\rm T}i}(t)}{{\rm d}t} = -NS_{{\rm T}i}(t)e_{{\rm T}i}, \tag{140}$$

the total charge, $Q(t)$, in the film is expressed as

$$\begin{aligned} Q(t) &= qA\sum_{i=1} NS_{{\rm T}i}(t) \\ &= qA\sum_{i=1} NS_{{\rm T}i}\exp(-e_{{\rm T}i}t). \end{aligned} \tag{141}$$

Because the decrease of $Q(t)$ results in the transient discharge current,

$$I_{\rm dis}(t) = -\frac{{\rm d}Q(t)}{{\rm d}t} + I_{\rm ssl}(V_{\rm dis}). \tag{142}$$

From Eq. (136), $D(t, e_{\rm ref})$ is theoretically expressed as

$$D(t, e_{\rm ref}) = \sum_{i=1} NS_{{\rm T}i}e_{{\rm T}i}t\exp\left[-\left(e_{{\rm T}i} + e_{\rm ref}\right)t + 1\right]. \tag{143}$$

As is clear from Eq. (143), we have obtained a suitable function for the graphical peak analysis method. This analysis is referred to as discharge current transient spectroscopy (DCTS) [44, 45, 46, 47, 48], and can be carried out using software developed in-house (http://www.osakac.ac.jp/labs/matsuura/).

7.3. Shift of Peak Discharge Time to Measurement Time Range

It is demonstrated that $e_{\rm ref}$ in Eq. (136) is a useful parameter when the time range of the measurement is limited. Figure 28(a) shows the transient discharge current, which is simulated assuming the following two kinds of traps. $NS_{\rm T1}$ and $e_{\rm T1}$ for Trap1 are 1×10^{12} cm^{-2} and 0.5 s^{-1}, respectively, $NS_{\rm T2}$ and $e_{\rm T2}$ for Trap2 are 1×10^{12} cm^{-2} and 0.0005 s^{-1}, respectively, and A is 1 mm^2.

Figure 28(b) shows the DCTS signals. The solid line represents the DCTS signal with $e_{\rm ref} = 0$ s^{-1}. In this case, only one peak is detected, and the peak discharge time and peak value are 2.00 s and 1.00×10^{12} cm^{-2}, respectively. From Eqs. (134) and (135), $NS_{\rm T1}$ and $e_{\rm T1}$ are determined to be 1.00×10^{12} cm^{-2} and 5.00×10^{-1} s^{-1}, respectively. Since the other peak is not detected when $e_{\rm ref} = 0$ s^{-1}, however, it is impossible to determine $NS_{\rm T2}$ and $e_{\rm T2}$.

The broken line in Fig. 28(b) represents the DCTS signal with $e_{\rm ref} = 0.001$ s^{-1}. In this case, two peaks appear in the figure, and the two peak discharge times are 2.00 s and 667 s. Using Eqs. (134) and (135), therefore, $NS_{\rm T1}$ and $e_{\rm T1}$ of the trap corresponding to 2.00 s are determined to be 1.00×10^{12} cm^{-2} and 5.00×10^{-1} s^{-1}, respectively, while

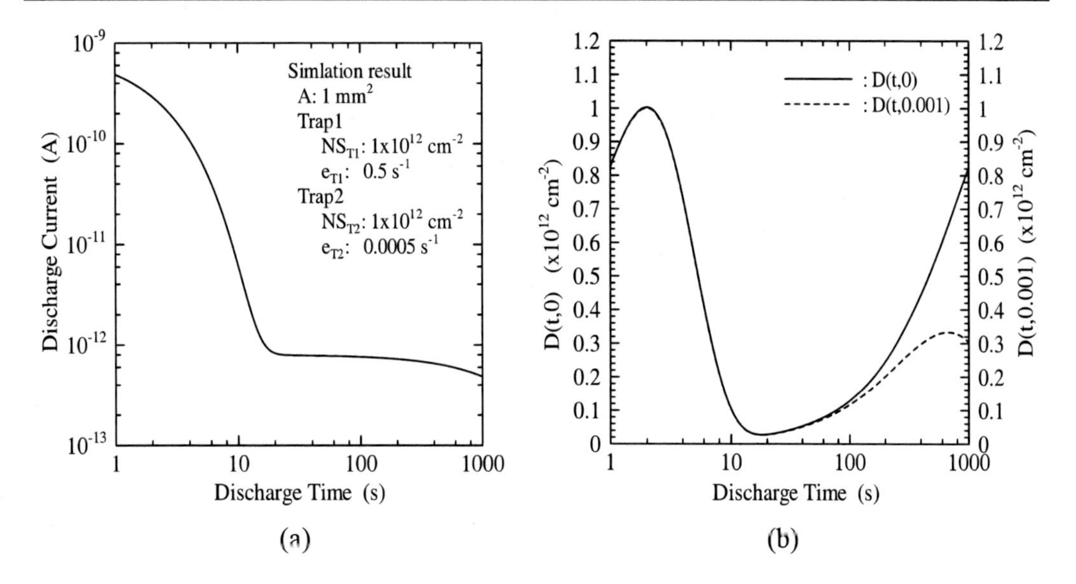

Figure 28. Simulated transient discharge current (a) and DCTS signals (b).

NS_{T2} and e_{T2} of the trap corresponding to 667 s are determined to be 1.00×10^{12} cm^{-2} and 5.00×10^{-4} s^{-1}, respectively. The obtained values are in good agreement with the values using which the transient discharge current was simulated. This indicates that $NS_{\mathrm{T}i}$ and $e_{\mathrm{T}i}$ can be determined in a wide emission-rate range by changing e_{ref}, even when the measurement time range is limited.

7.4. Distinction among Traps with Close Emission Rates

It is demonstrated that e_{ref} in Eq. (136) is also useful when there are traps with close emission rates in the film. Figure 29 shows the transient discharge current, which is simulated assuming the following two kinds of traps. NS_{T1} and e_{T1} for Trap1 are 3×10^{12} cm^{-2} and 0.05 s^{-1}, respectively, NS_{T2} and e_{T2} for Trap2 are 1×10^{12} cm^{-2} and 0.01 s^{-1}, respectively, and A is 1 mm^2. In this case, e_{T2} is very close to e_{T1}.

The solid line in Fig. 30(a) represents the DCTS signal with $e_{\mathrm{ref}} = 0$ s^{-1}. In this case, only one peak is detected, and the peak discharge time and peak value are 22.5 s and 3.47×10^{12} cm^{-2}, respectively. From Eqs. (134) and (135), NS_{T1} and e_{T1} are determined to be 3.47×10^{12} cm^{-2} and 4.44×10^{-2} s^{-1}, respectively. These obtained values are only a little different from NS_{T1} and e_{T1} using which the transient discharge current was simulated. In the figure, the broken and dashed-dotted lines, which correspond to Trap1 and Trap2 respectively, are simulated using the following equation;

$$NS_{\mathrm{T}i} e_{\mathrm{T}i} t \exp\left[-\left(e_{\mathrm{T}i} + e_{\mathrm{ref}}\right) t + 1\right]. \qquad (144)$$

Since it is found that Trap1 mainly affects the DCTS signal with $e_{\mathrm{ref}} = 0$ s^{-1}, it is reasonable that the values determined using the maximum of the DCTS signal are close to NS_{T1} and e_{T1}, respectively.

The solid line in Fig. 30(b) represents the DCTS signal with $e_{\mathrm{ref}} = -0.0089$ s^{-1}. In this case, two peaks appear. One peak value is 4.35×10^{12} cm^{-2} at 29.4 s, and the

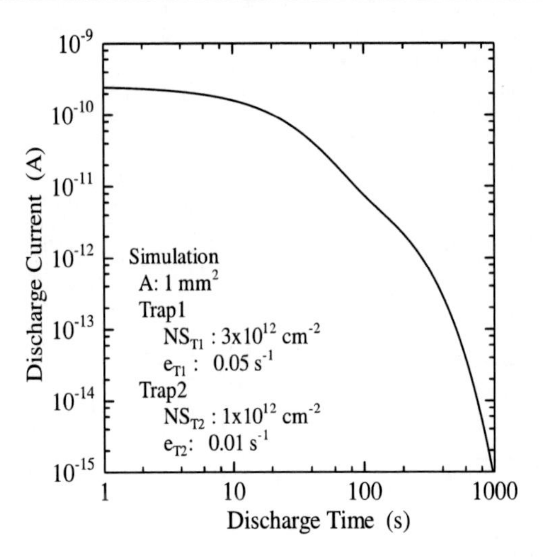

Figure 29. Transient discharge current simulated assuming two kinds of traps.

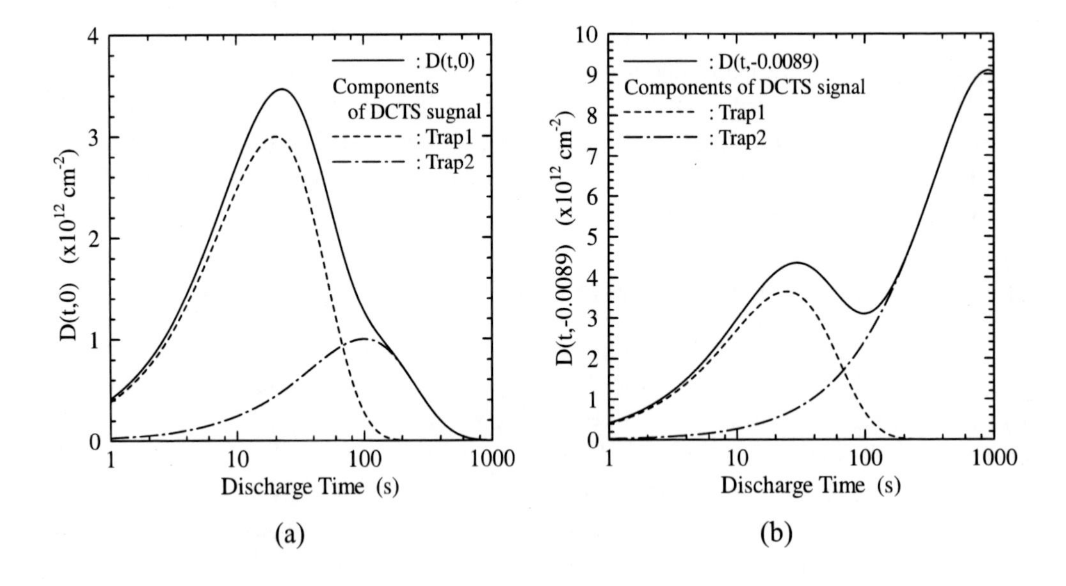

Figure 30. DCTS signals with (a) $e_{\rm ref} = 0$ s^{-1} (solid line) and (b) $e_{\rm ref} = 0.0089$ s^{-1} (solid line). The broken and dashed-dotted lines are simulated using Eq. (144) for Trap1 and Trap2, respectively.

other peak value is 9.09×10^{12} cm^{-2} at 910 s. From Eqs. (134) and (135), $NS_{\rm T1}$ and $e_{\rm T1}$ are determined to be 3.45×10^{12} cm^{-2} and 4.29×10^{-2} s^{-1}, respectively, while $NS_{\rm T2}$ and $e_{\rm T2}$ are determined to be 9.99×10^{11} cm^{-2} and 1.00×10^{-2} s^{-1}, respectively. In the figure, the broken and dashed-dotted lines, which are simulated using Eq. (144), correspond to Trap1 and Trap2, respectively. The peak corresponding to Trap2 is maximum when $e_{\rm ref} = -0.0089$ s^{-1}, although the peak corresponding to Trap1 is maximum when $e_{\rm ref} = 0$ s^{-1}. Therefore, it is natural that the values determined using the maximum of the

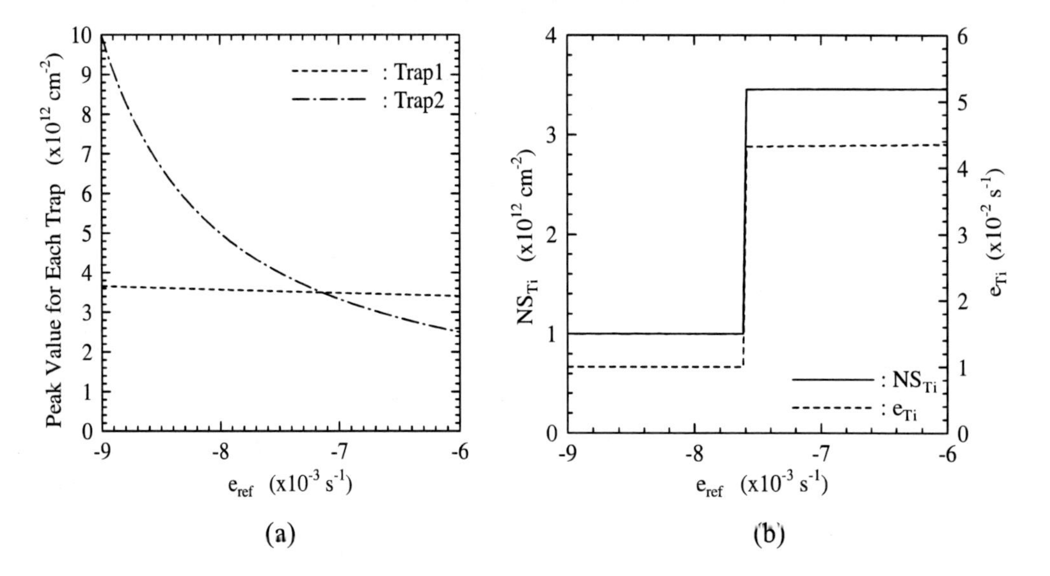

Figure 31. e_{ref} dependence of (a) peak value for Trap1 (broken line) or Trap2 (dashed-dotted line) and (b) $NS_{\text{T}i}$ or $e_{\text{T}i}$ determined from the maximum of the DCTS signal.

DCTS signal with $e_{\text{ref}} = -0.0089\ \text{s}^{-1}$ are close to NS_{T2} and e_{T2}, respectively.

Figure 31(a) shows the e_{ref} dependence of the peak value for Trap1 (broken line) or Trap2 (dashed-dotted line). The peak corresponding to Trap1 is maximum when $e_{\text{ref}} > -0.0071\ \text{s}^{-1}$, while the peak corresponding to Trap2 is maximum when $e_{\text{ref}} < -0.0071\ \text{s}^{-1}$. This suggests that it is possible to distinguish between two traps using the e_{ref} dependence of NS_{T} or e_{T} determined using the maximum of the DCTS signal.

Figure 31(b) shows the e_{ref} dependence of NS_{T} or e_{T} determined from the maximum of the DCTS signal. The two discrete values of e_{T} or NS_{T} clearly appear in the figure. Moreover, the obtained values are close to the values using which the transient discharge current was simulated. Therefore, it is found that DCTS can distinguish among traps with close emission rates by changing e_{ref}.

7.5. DCTS for Semi-insulating Semiconductors

After native oxide layers on a 0.37-mm-thick high-purity semi-insulating on-axis 4H-SiC wafer were removed using HF, Ni electrodes with a radius of 1.25 mm were evaporated in vacuum onto both sides of the samples [47, 48]. Figure 32(a) shows the transient reverse current $I_{\text{dis}}(t)$ at 303 K after a V_{dis} of -100 V is applied to a diode held at thermal equilibrium ($V_{\text{cha}} = 0$ V) for 5 min. In the figure, ○ represents the experimental $I_{\text{dis}}(t)$, and the solid line is calculated by interpolating the experimental $I_{\text{dis}}(t)$ with a cubic smoothing natural spline function.

Figure 32(b) shows $D(t, e_{\text{ref}})$ calculated with an e_{ref} of 0 s^{-1} using Eq. (136). There are two peaks labeled Peak1 and Peak2, and the corresponding traps are here referred to as SI1 and SI2, respectively. From t_{Peak1} and $D(t_{\text{Peak1}}, 0)$, the values of NS_{T1} and e_{T1} for SI1 were determined to be $1.6 \times 10^{11}\ \text{cm}^{-2}$ and $5.5 \times 10^{-2}\ \text{s}^{-1}$, respectively, while from

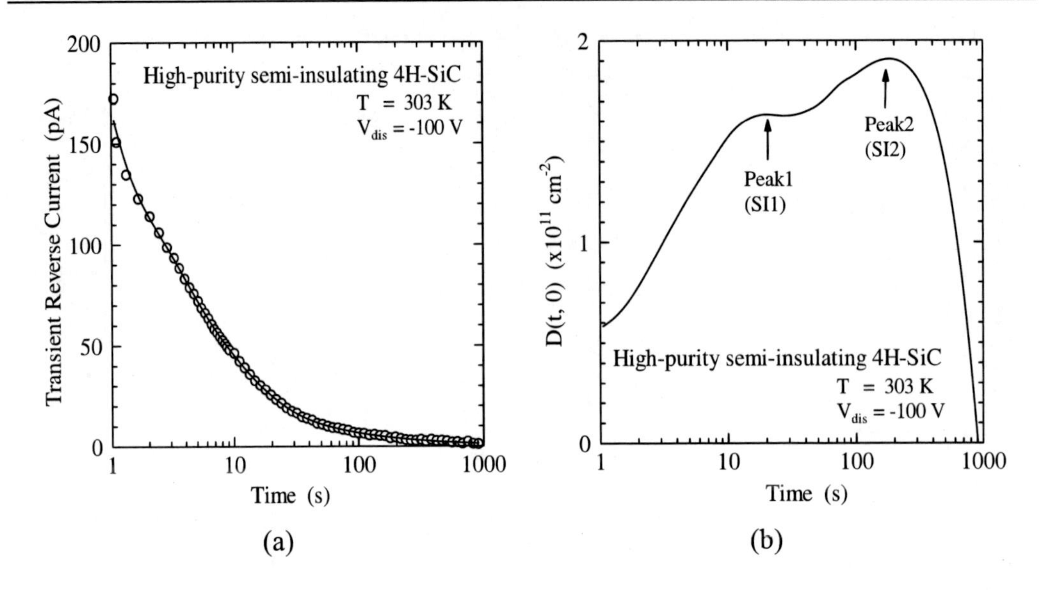

Figure 32. Characteristics of Schottky barrier diode with high-purity semi-insulating 4H-SiC at 303 K; (a) transient reverse current and (b) DCTS signal.

Table 6. Energy levels and cross sections of traps in high-purity semi-insulating 4H-SiC determined by DCTS.

Trap species	Energy level (eV)	Cross section (cm^2)
SI1	0.78	2×10^{-15}
SI2	0.85	2×10^{-15}
SI3	0.90	1×10^{-13}
SI4	1.2	8×10^{-12}
SI5	1.4	9×10^{-10}

t_{Peak2} and $D(t_{\mathrm{Peak2}}, 0)$, NS_{T2} and e_{T2} for SI2 were determined to be 1.9×10^{11} cm^{-2} and 5.5×10^{-3} s^{-1}, respectively.

Finally, five types of traps could be observed in the temperature range between 283 and 373 K. Figure 33 shows the relationship between $e_{\mathrm{T}i}/T^2$ and $1/T$ for five different traps. The optimum straight line fitting to experimental data for each trap could be obtained, from which $E_{\mathrm{T}i}$ and $\sigma_{\mathrm{T}i}$ could be determined using Eqs. (105)-(108), and listed in Table 6.

7.6. DCTS for Crystalline Insulators

Pb(Zr,Ti)O_3 (PZT) thin films were deposited on Pt/SiO_2/(100)Si substrates at 570 °C by metalorganic chemical vapor deposition, using $Pb(C_2H_5)_4$, Zr(O-t-C_4H_9)$_4$ and Ti(O-i-C_3H_7)$_4$ as source precursors [46]. The thickness of the PZT films was 200 nm. The area of the Pt top electrode was 0.79 mm^2. DCTS measurements of Pt/PZT/Pt capacitors were performed at 373 K in an oxidation atmosphere (Ar : $O_2 = 1 : 1$ at 1 atom). The transient

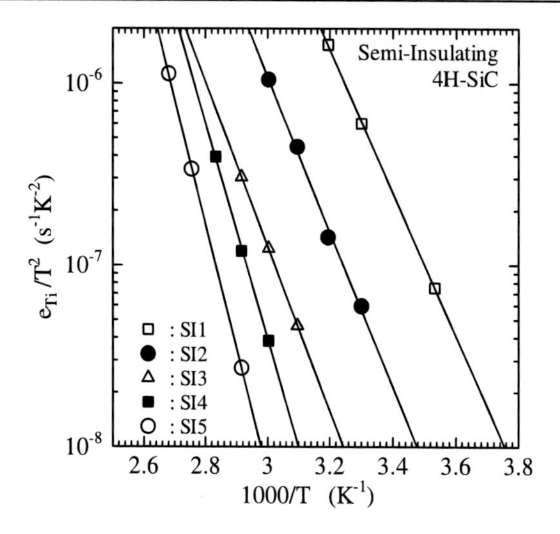

Figure 33. Relationship between $e_{Ti}/T^2 - 1/T$.

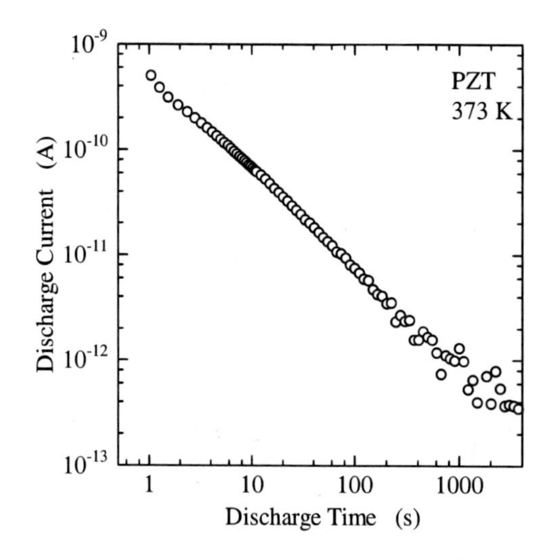

Figure 34. Transient discharge current in Pt/PZT/Pt capacitor.

discharge current was measured at $V_{\text{dis}} = 0$ V, after V_{cha} of 2 V was applied to the capacitor in the interval of 600 s. Figure 34 shows $I_{\text{dis}}(t)$ in the Pt/PZT/Pt capacitor. To search a peak of the DCTS signal precisely, the DCTS signal was calculated by interpolating $I_{\text{dis}}(t)$ with a cubic smoothing natural spline function.

The solid line in Fig. 35(a) represents the DCTS signal with $e_{\text{ref}} = 0\ \text{s}^{-1}$. The maximum discharge time and maximum value were 4.09 s and $1.54 \times 10^{12}\text{cm}^{-2}$, respectively. From Eqs. (134) and (135), NS_{T} and e_{T} were determined to be $1.54 \times 10^{12}\ \text{cm}^{-2}$ and $2.44 \times 10^{-2}\ \text{s}^{-1}$, respectively. In the case that the traps are uniformly distributed in the film, the trap density is estimated to be $7.70 \times 10^{16}\ \text{cm}^{-3}$, because the film thickness was 200 nm. The broken line represents the signal simulated using Eq. (144) with the ob-

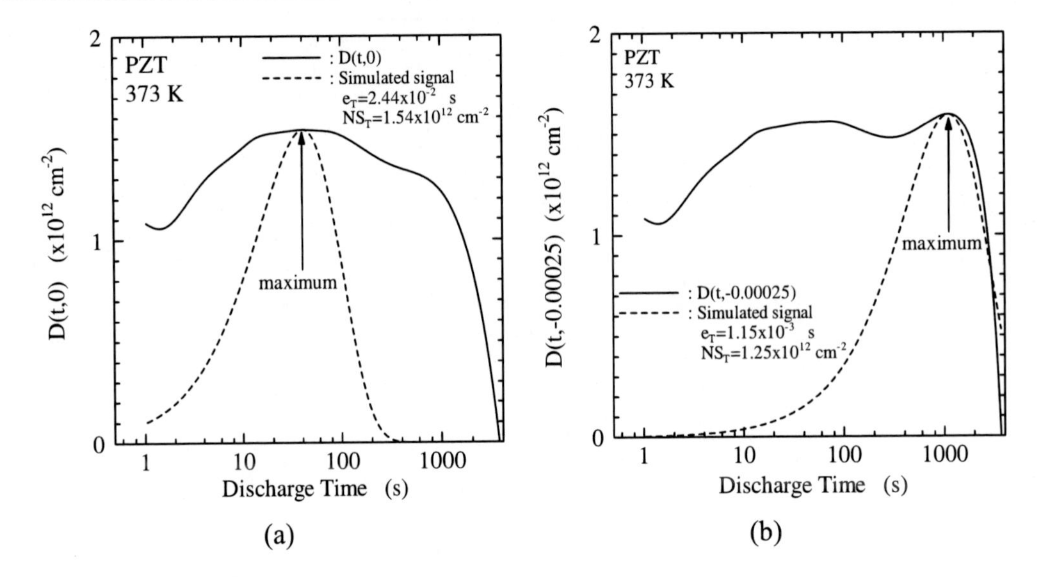

Figure 35. DCTS signal (solid line) with (a) $e_{\mathrm{ref}} = 0\ \mathrm{s}^{-1}$ and (b) $e_{\mathrm{ref}} = -0.00025\ \mathrm{s}^{-1}$, which is calculated by interpolating $I_{\mathrm{dis}}(t)$ with a cubic smoothing natural spline function. The broken line corresponds to the signal simulated using Eq. (144) with NS_{T} and e_{T} determined from the maximum of the DCTS signal.

tained values. Since the solid line is much broader than the broken line, the DCTS signal is considered to be affected by several traps with close emission rates.

The solid line in Fig. 35(b) represents the DCTS signal with $e_{\mathrm{ref}} = -0.00025\,\mathrm{s}^{-1}$. The maximum discharge time and maximum value were 1111 s and $1.60 \times 10^{12}\mathrm{cm}^{-2}$, respectively. From Eqs. (134) and (135), NS_{T} and e_{T} were determined to be $1.25 \times 10^{12}\ \mathrm{cm}^{-2}$ and $1.15 \times 10^{-2}\ \mathrm{s}^{-1}$, respectively.

The broken line represents the signal simulated using Eq. (144) with the obtained values. Since NS_{T} and e_{T} for $e_{\mathrm{ref}} = -0.00025\ \mathrm{s}^{-1}$ are different from those for $e_{\mathrm{ref}} = 0\ \mathrm{s}^{-1}$, at least two kinds of traps with close emission rates are included in the PZT film.

Figure 36(a) shows the e_{ref} dependence of NS_{T} (solid line) or e_{T} (broken line) determined from the maximum of the DCTS signal at $e_{\mathrm{ref}} < 0$. Three discrete values of e_{T} or NS_{T} clearly appear in the figure. Moreover, the e_{ref} range of the constant NS_{T} clearly corresponds one-to-one to the e_{ref} range of the constant e_{T}. Therefore, it is found that DCTS can distinguish among three kinds of traps (Trap1, Trap2, and Trap3) with close emission rates by changing negative e_{ref}. From Fig. 36(a), NS_{T} and e_{T} of each trap can be determined; e_{T1} and NS_{T1} are $1.15 \times 10^{-3}\ \mathrm{s}^{-1}$ and $1.2 \times 10^{12}\ \mathrm{cm}^{-2}$, respectively, and e_{T2} and NS_{T2} are $1.39 \times 10^{-2}\ \mathrm{s}^{-1}$ and $1.53 \times 10^{12}\ \mathrm{cm}^{-2}$, respectively, and e_{T3} and NS_{T3} are $2.3 \times 10^{-2}\ \mathrm{s}^{-1}$ and $1.54 \times 10^{12}\ \mathrm{cm}^{-2}$, respectively.

Figure 36(b) shows the e_{ref} dependence of NS_{T} (solid line) or e_{T} (broken line) determined from the maximum of the DCTS signal at $e_{\mathrm{ref}} \geq 0$. In addition to the three kinds of traps determined in Fig. 36(a), another two kinds of traps (Trap4 and Trap5) are considered to be included in this thin film. e_{T4} and NS_{T4} are approximately $7 \times 10^{-2}\ \mathrm{s}^{-1}$ and $1.5 \times 10^{12}\ \mathrm{cm}^{-2}$, respectively. On the other hand, the emission rate of Trap5 may be

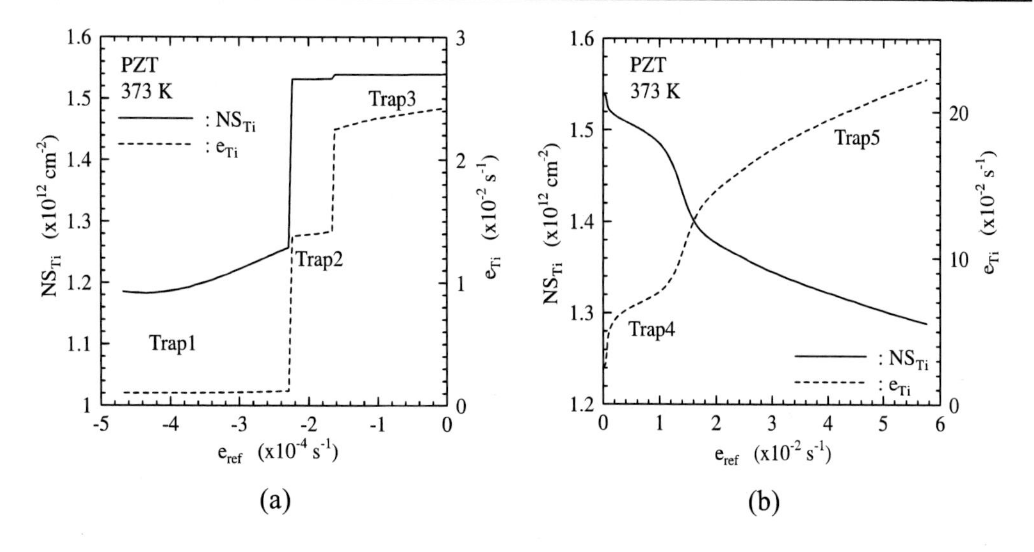

Figure 36. e_{ref} dependence of $NS_{\mathrm{T}i}$ (solid line) or $e_{\mathrm{T}i}$ (broken line) determined from the maximum of the DCTS signal at (a) $e_{\mathrm{ref}} < 0$ and (b) $e_{\mathrm{ref}} \geq 0$.

fluctuated, because both e_{T} and NS_{T} of Trap5 change gradually with e_{ref} in the figure.

7.7. DCTS for Amorphous Insulators

76 nm-thick SiN_x films were deposited on heavily-doped p-type c-Si (p^+ c-Si) at a substrate temperature of 500 °C by direct photo-chemical vapor deposition with a low-pressure mercury lamp using N_2-diluted mixtures of SiH_4 and NH_3 [44, 45]. An Al electrode with 3.14 mm^2 was deposited on SiN_x by vacuum evaporation. Here, p^+ c-Si and Al act as electrodes of a capacitor.

Figure 37(a) shows the I–V characteristics for Capacitors A and B. The leakage current for Capacitor B at low applied biases ($<$ 35 V) was higher than that for Capacitor A. To investigate the reason, $I_{\mathrm{dis}}(t)$ for Capacitors A and B were measured. During the charging time (i.e., in -600 s $< t <$ 0 s), V_{cha} of 50 V was applied to the Al electrode of an Al/SiN_x/p^+ c-Si capacitor. Applying a positive voltage to the Al electrode implies that holes injected from the Al electrode should flow through SiN_x. $I_{\mathrm{dis}}(t)$ was measured at $V_{\mathrm{dis}} = 0$ in $t > 0$ s. Figure 37(b) presents the data of $D(t)$ for Capacitors A and B. $D(t)$ was calculated from the measured $I_{\mathrm{dis}}(t)$ using Eq. (136) with $e_{\mathrm{ref}} = 0\ \mathrm{s}^{-1}$, that is, $D(t) = D(t, 0)$.

When trap levels in the film are continuously distributed in the band gap, Eq. (130) is rewritten as

$$g[E(t)] \cong \frac{D(t)\exp(-1)}{kTL} \tag{145}$$

and

$$E(t) - E_{\mathrm{V}} = kT\ln\left(\nu_{\mathrm{h}} t\right), \tag{146}$$

which are similar to Eqs. (128) and (129). Here, L is the thickness (76 nm) of SiN_x. The $g(E)$ above E_{V} was calculated from Eqs. (145) and (146) and shown in Fig. 37(b), where

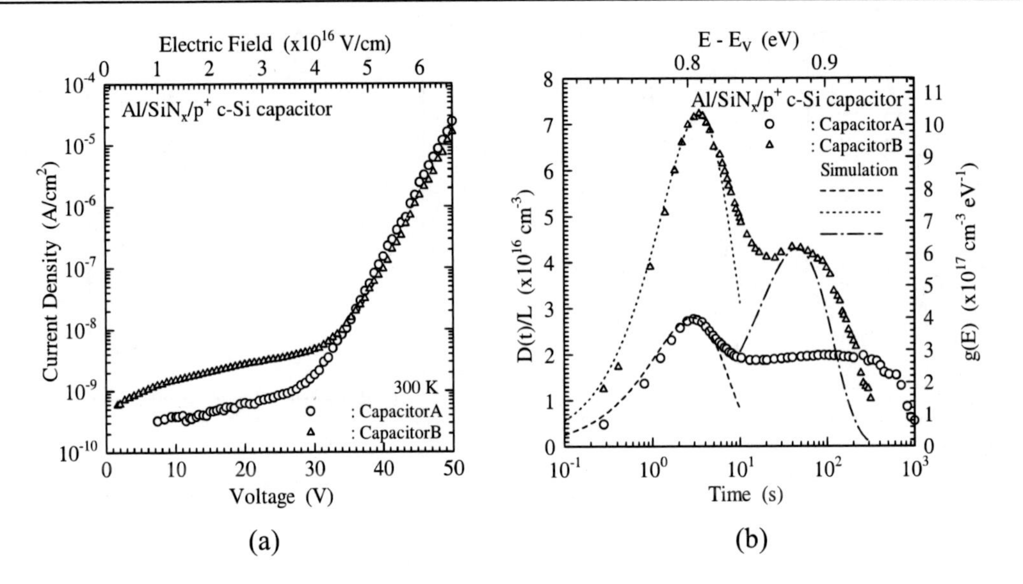

Figure 37. I–V characteristics (a) and DCTS signals (b) for Capacitors A and B. Here, L is 7.6×10^{-6} cm.

it was assumed that $\nu_{\rm h} = 10^{13}$ s^{-1} because $\nu_{\rm h}$ is related to an optical phonon frequency in the film and is equal to or less than the order of 10^{13} s^{-1}. As is clear from Fig.37, the leakage current at < 35 V is correlated with the $g(E)$.

The peak value ($NS_{\rm T1}/L$) and time ($e_{\rm T1}$) of $D(t)$ for Capacitor A were 2.8×10^{16} cm^{-3} and $e_{\rm T1} = 0.35$ s^{-1}, respectively. Assuming that this peak is due to a discrete trap, the component of $D(t)$ corresponding to this peak is calculated using Eq. (144) and shown in the broken line of Fig. 37(b), which is in good agreement with ○. Therefore, one discrete trap with $E_{\rm V} - E$ of ~ 0.81 eV and traps energetically-distributed between 0.85 and 0.96 eV exist in Capacitor A.

From peaks of ($NS_{\rm T1}/L = 7.2 \times 10^{16}$ cm^{-3} and $e_{\rm T1} = 0.29$ s^{-1}) and ($NS_{\rm T2}/L = 4.3 \times 10^{16}$ cm^{-3} and $e_{\rm T2} = 0.022$ s^{-1}) for Capacitor B, on the other hand, the corresponding components represent the dotted and dashed-dotted lines in Fig. 37(b), which are in good agreement with △ at $10^{-1} - 10$ s and $10 - 10^2$ s, respectively. In the SiN$_x$ film of Capacitor B, therefore, two discrete traps with approximately 0.81 and 0.88 eV were included.

8. Conclusion

Several methods to determine densities and energy levels of impurities and defects affecting the majority-carrier concentration in semiconductors whose resistivities vary from low to extremely high have been discussed;

1. Free carrier Concentration Spectroscopy (FCCS) determining densities and energy levels in a semiconductor using the temperature dependence of the majority-carrier concentration without any assumptions regarding impurities and defects,

2. Heterojunction-Monitored Capacitance Spectroscopy (HMCS) determining the

density-of-states in high-resistivity amorphous semiconductors using the transient capacitance of an amorphous/crystalline semiconductor heterojunction,

3. Discharge Current Transient Spectroscopy (DCTS) determining densities and energy levels in a semi-insulating semiconductor or an insulator using the transient current of a diode.

On the other hand, the influence of the excited states of dopants with deep energy levels on the majority-carrier concentration have been discussed, and a unique distribution function for dopants including the effect of the excited states of dopants has been proposed and experimentally tested.

References

[1] Lang, D. V. *J. Appl. Phys.* 1974, vol. 45, 3023-3032.

[2] Danno, K.; Kimoto, T. *Jpn. J. Appl. Phys.* 2006, vol. 45, L285-L287.

[3] Danno, K.; Kimoto, T. *Mater. Sci. Forum* 2007, vol. 556-557, 331-334.

[4] Haering, R. R.; Adams, E. N. *Phys. Rev.* 1960, vol. 117, 451-454.

[5] Sze, S. M. *Physics of Semiconductor Devices* 2nd ed.; Wiley: New York, 1981, chapter 1.

[6] Hoffmann, H. J. *Appl. Phys.* 1979, vol. 19, 307-312.

[7] Hoffmann, H. J.; Nakayama, H.; Nishino, T.; Hamakawa, H. *Appl. Phys. A: Mater. Sci. & Processing* 1984, vol. 33, 47-50.

[8] Okushi, H. *Philos. Mag. B* 1985, vol. 52, 33-57.

[9] Matsuura, H.; Sonoi, K. *Jpn. J. Appl. Phys.* 1996, vol. 35, L555-L557.

[10] Matsuura, H. *Jpn. J. Appl. Phys.* 1997, vol. 36, 3541-3547.

[11] Matsuura, H.; Masuda, Y.; Chen, Y.; Nishino, S. *Jpn. J. Appl. Phys.* 2000, vol. 39, 5069-5075.

[12] Suzuki, A.; Uemoto, A.; Shigeta, M.; Furukawa, K.; Nakajima, S. *Appl. Phys. Lett.* 1986, vol. 49, 450-452.

[13] Freitas, Jr., J. A.; Bishop, S. G.; Nordquist, Jr., P. E. R.; Gipe, M. L. *Appl. Phys. Lett.* 1988, vol. 52, 1695-1697.

[14] Kaplan, R.; Wagner, R. J.; Kim, H. J.; Davis, R. F. *Solid State Commun.* 1985, vol. 55, 67-69.

[15] Ikeda, M.; Matsunami, H.; Tanaka, T. *Phys. Rev. B* 1980, vol. 22, 2842-2854.

[16] Götz, W.; Schöner, A.; Pensl, G.; Suttrop, W.; Choyke, W. J.; Stein, R.; Leibenzeder, S. *J. Appl. Phys.* 1993, vol. 73, 3332-3338.

[17] Matsuura, H.; Morita, K.; Nishikawa, K.; Mizukoshi, T.; Segawa, M.; Susaki, W. *Jpn. J. Appl. Phys.* 2002, vol. 41, 496-500.

[18] Matsuura, H.; Nishikawa, K.; Segawa, M.; Susaki, W. *Jpn. J. Appl. Phys.* 2006, vol. 45, 6373-6375.

[19] Matsuura, H.; Nishikawa, K. *J. Appl. Phys.* 2005, vol. 97, 093711 1-7.

[20] Matsuura, H.; Nagasawa, H.; Yagi, K.; Kawahara, T. *J. Appl. Phys.* 2004, vol. 96, 7346-7351.

[21] Kagamihara, S.; Matsuura, H.; Hatakeyama, T.; Watanabe, T.; Kushibe, M.; Shinohe, T.; Arai, K. *J. Appl. Phys.* 2004, vol. 96, 5601-5606.

[22] Matsuura, H.; Komeda, M.; Kagamihara, S.; Iwata, H.; Ishihara, R.; Hatakeyama, T.; Watanabe, T.; Kojima, K.; Shinohe, T.; Arai, K. *J. Appl. Phys.* 2004, vol. 96, 2708-2715.

[23] Matsuura, H.; Kagamihara, S.; Itoh, Y.; Ohshima, T.; Itoh, H. *Physica B* 2006, vol. 376-377, 342-345.

[24] Matsuura, H.; Minohara, N.; Ohshima, T. *J. Appl. Phys.* 2008, vol. 104, 043702 1-6.

[25] Matsuura, H.; Iwata, H.; Kagamihara, S.; Ishihara, R.; Komeda, M.; Imai, H.; Kikuta, M.; Inoue, Y.; Hisamatsu, T.; Kawakita, S.; Ohshima, T.; Itoh, H. *Jpn. J. Appl. Phys.* 2006, vol. 45, 2648-2655.

[26] Yu, P. Y.; Cardona, M. *Fundamentals of Semiconductors: Physics and Materials Properties* 2nd ed.; Springer, Berlin, 1999; pp. 156 and 160.

[27] Sapoval B.; Hermann, C. *Physics of Semiconductors*; Springer-Verlag, New York, 1993; pp. 73 and 112.

[28] Singh, J. *Semiconductor Devices: An Introduction*; McGraw-Hill, New York, 1994; p. 110.

[29] Smith, R. A. *Semiconductors* 2nd ed.; Cambridge University Press, Cambridge 1978; p. 92.

[30] Brennan, K. F. *The Physics of Semiconductors with Applications to Optoelectronic Devices*; Cambridge University Press, Cambridge, 1999; p. 292.

[31] Ashcroft N. W.; Mermin, N. D. *Solid State Physics*; Holt Rinehart and Winston, Philadelphia, 1976; pp. 581 and 586.

[32] Troffer, T.; Schadt, M.; Frank, T.; Itoh, H.; Pensl, G.; Heindl, J.; Strunk, H. P.; Maier, M. *Phys. Status Solidi A* 1997, vol. 162, 277-298.

[33] Schulze, N.; Gajowski, J.; Semmelroth, K.; Laube, M.; Pensl, G. *Mater. Sci. Forum* 2001, vol. 353-356, 45-48.

[34] Matsuura, H. *New J. Phys.* 2002, vol. 4, 12.1-12.15 [http://www.njp.org/].

[35] Matsuura, H. *J. Appl. Phys.* 2004, vol. 95, 4213-.

[36] Heer, C. V. *Statistical Mechanics, Kinetic Theory, and Stochastic Processes*; Academic Press, New York, 1972; p. 224.

[37] Matsuura, H.; Yanase, H.; Takahashi, M. *Jpn. J. Appl. Phys.* 2008, vol. 47, 7052-7055.

[38] Matsuura, H.; Okuno, T.; Okushi, H.; Tanaka, K. *J. Appl. Phys.* 1984, vol. 55, 1012-1019.

[39] Matsuura, H. *J. Appl. Phys.* 1988, vol. 64. 1964-1973.

[40] Matsuura, H. *Jpn. J. Appl. Phys.* 1988, vol. 27, L513-L515.

[41] Matsuura, H. *Jpn. J. Appl. Phys.* 1988, vol. 27, L516-L518.

[42] Matsuura H.; Okushi, H. In *Amorphous and Microcrystalline Semiconductor Devices II: Materials and Device Physics*; Kanicki, J.; Ed.; Artech House, Boston, 1992; chapter 11.

[43] Matsuura, H.; Okuno, T.; Okushi, H.; Yamasaki, S.; Matsuda, A.; Hata, N.; Oheda, H.; Tanaka, K. *Jpn. J. Appl. Phys.* 1983, vol. 22, L197-L199.

[44] Matsuura, H.; Yoshimoto, M.; Matsunami, H. *Jpn. J. Appl. Phys.* 1995, vol. 34, L185-L187.

[45] Matsuura, H.; Yoshimoto, M.; Matsunami, H. *Jpn. J. Appl. Phys.* 1995, vol. 34, L371-L374.

[46] Matsuura, H.; Hase, T.; Sekimoto, Y.; Uchida, M.; Simizu, M. *J. Appl. Phys.* 2002, vol. 91, 2085-2092.

[47] Matsuura, H.; Kagawa, Y.; Takahashi, M.; Tano, S.; Miyake, T. *Jpn. J. Appl. Phys.* 2009, vol. 48, 056504 1-4.

[48] Matsuura, H.; Takahashi, M.; Kagawa, Y.; Tano, S.; Miyake, T. *Mater. Sci. Forum* 2009, vol. 615-617, 385-388.

In: Advances in Condensed Matter ... Volume 10
Editors: H. Geelvinck and S. Reynst
ISBN: 978-1-61209-533-2

Chapter 8

BASIC THEORY OF FRACTIONAL QUANTUM HALL EFFECT

Orion Ciftja
Department of Physics, Prairie View A&M University,
Prairie View, TX, USA

Abstract

At very low temperatures, a two-dimensional electron system in a perpendicular magnetic field exhibits remarkable quantum phenomena where the fractional quantum Hall effect (FQHE) stands out as one of the most important discoveries in condensed matter physics for the last decades. FQHE represents a unique example of a novel collective quantum liquid state of matter that originates from strong electronic correlations only. A perpendicular magnetic field leads to the creation of massively degenerate discrete quantum states known as Landau levels. The energy quantization associated with such states is the foundation of many new experimental and theoretical advances in condensed matter physics. The filling factor, defined as the ratio of the number of electrons to the degeneracy (number of available states) of each Landau level, represents an important characteristic parameter of the system. Typical liquid FQHE states are more pronounced in the extreme quantum limit of very high perpendicular magnetic field where the lowest Landau level is fractionally filled with electrons. Stabilization of these novel electronic phases happens at special filling factors that generally have odd denominators. Among them, the most robust FQHE states correspond to filling factors $1/3$ and $1/5$ and are well described by Laughlin's theory in terms of trial wave functions. Differently from odd-denominator-filled states in the lowest Landau level, even-denominator-filled states with filling factor $1/2$, $1/4$ and $1/6$ do not show typical FQHE features and behave as isotropic compressible metallic Fermi liquid states. The composite fermion theory for the FQHE shed light on the Fermi-liquid nature of such even-denominator-filled states. On the other hand, Wigner crystallization occurs when the filling factor becomes around or less than $1/7$. In this work, we attempt to give a brief overview of the basic theory of the fractional quantum Hall effect. To this effect and for brevity of treatment, we focus our attention only on states on the lowest Landau level and try to explain in a simple manner some of the key ideas and models used to study such an intriguing phenomenon.

1. Two-Dimensional Motion of a Charged Particle in a Uniform Perpendicular Magnetic Field; Symmetric Gauge

In classical physics we know that a charged particle interacts with a magnetic field via a Lorentz force. This interaction leads to many interesting phenomena suitably described by classical physics. The classical description can be extended to the quantum description by the usual procedure where we identify the Hamiltonian which corresponds to the classical motion and then introduce the corresponding operators. In our case, let us consider the two-dimensional (2D) motion of a particle with mass m and charge $q > 0$ (for simplicity we consider a positive charge). The charged particle is subjected to a uniform perpendicular magnetic field in the positive z-direction:

$$\vec{B} = (0, 0, B) \, . \tag{1}$$

The vector potential characterizing the magnetic field, $\vec{B} = \vec{\nabla} \times \vec{A}$ can be written in a symmetric gauge as:

$$\vec{A} = \frac{B}{2} \, (-y, x, 0) \; . \tag{2}$$

Generally, charged elementary particles also possess spin. Even though spin has no classical analogue, there is a straightforward way to describe the interaction of the spin with a magnetic field. For the time being, we ignore all spin effects and thus focus on the spatial motion. The quantum Hamiltonian which adequately describes the spatial motion of a charged particle in a magnetic field is written as:

$$\hat{H} = \frac{1}{2\,m} \left[\hat{\vec{p}} - q\,\vec{A} \right]^2 \, , \tag{3}$$

where $\hat{\vec{p}} = (\hat{p}_x, \hat{p}_y)$ is the 2D momentum operator. This Hamiltonian can also be written as:

$$\hat{H} = \frac{\hat{p}_x^2 + \hat{p}_y^2}{2\,m} - \frac{\omega_c}{2} \, \hat{L}_z + \frac{m\,\omega_c^2}{8} \left(x^2 + y^2 \right) \, , \tag{4}$$

where

$$\hat{L}_z = x\,\hat{p}_y - y\,\hat{p}_x \, , \tag{5}$$

is the angular momentum operator in the z-direction and

$$\omega_c = \frac{q\,B}{m} > 0 \, , \tag{6}$$

is the cyclotron frequency. The Schrödinger's equation for the above Hamiltonian in a 2D polar system of coordinates becomes:

$$\left[-\frac{\hbar^2}{2\,m} \frac{1}{r} \frac{\partial}{\partial r} \left(r \frac{\partial}{\partial r} \right) + \frac{\hat{L}_z^2}{2\,m\,r^2} - \frac{\omega_c}{2} \, \hat{L}_z + \frac{m\,\omega_c^2}{8} \, r^2 \right] \Psi(r, \varphi) = E\,\Psi(r, \varphi) \, , \tag{7}$$

where $\hbar$ is the reduced Planck's constant, $r^2 = x^2 + y^2$, $\hat{L}_z = -i\hbar \frac{\partial}{\partial \varphi}$ and φ is the polar angle. The z-component angular momentum operator, $\hat{L}_z$ acts only on the angular variable φ and commutes with the Hamiltonian so $\Psi(r, \varphi)$ has to be an eigenfuction of $\hat{L}_z$ as well:

$$\hat{L}_z \, \Psi(r, \varphi) = \hbar\,m_l\,\Psi(r, \varphi) \quad ; \quad m_l = 0, \pm 1, \pm 2, \ldots \tag{8}$$

We know that the eigenfuctions of $\hat{L}_z$ are given by

$$\hat{L}_z\,\Phi_{m_l}(\varphi)=\hbar m_l\,\Phi_{m_l}(\varphi)\quad;\quad \Phi_{m_l}(\varphi)=\frac{e^{i\,m_l\,\varphi}}{\sqrt{2\pi}}\quad;\quad \int_0^{2\pi}d\varphi\;\Phi^*_{m_l}(\varphi)\,\Phi_{m_l'}(\varphi)=\delta_{m_l\,m_l'}\,. \tag{9}$$

Therefore we search the general solution $\Psi(r,\varphi)$ as product of a function of r alone and $\Phi_{m_l}(\varphi)$,

$$\Psi(r,\varphi)=R(r)\,\Phi_{m_l}(\varphi)\,. \tag{10}$$

By substituting Eq.(10) into Eq.(7) we obtain the following differential equation for the radial wave function:

$$\left[-\frac{\hbar^2}{2\,m}\frac{1}{r}\frac{\partial}{\partial r}\left(r\frac{\partial}{\partial r}\right)+\frac{\hbar^2\,m_l^2}{2\,m\,r^2}-\frac{\hbar\,\omega_c}{2}m_l+\frac{m\,\omega_c^2}{8}\,r^2\right]R(r)=E\,R(r)\,. \tag{11}$$

We can rewrite this result as:

$$\left[-\frac{\hbar^2}{2\,m}\left(\frac{\partial^2}{\partial r^2}+\frac{1}{r}\frac{\partial}{\partial r}-\frac{m_l^2}{r^2}\right)+\frac{m\,\omega_c^2}{8}\,r^2\right]R(r)=E'\,R(r)\,, \tag{12}$$

where

$$E'=E+\frac{\hbar\,\omega_c}{2}m_l\,. \tag{13}$$

After some algebra we obtain

$$\left(\frac{\partial^2}{\partial r^2}+\frac{1}{r}\frac{\partial}{\partial r}-\frac{m_l^2}{r^2}+\frac{2\,m\,E'}{\hbar^2}-\frac{r^2}{4\,l_0^4}\right)R(r)=0\,, \tag{14}$$

where

$$l_0=\sqrt{\frac{\hbar}{q\,B}}\,, \tag{15}$$

is the magnetic length. The exact mathematical solution of this problem is rather lengthy and is not trivial. Thus, for the sake of brevity, here we skip the full derivation and simply report the final result for the energy values that is:

$$E=\hbar\,\omega_c\left(n_r+\frac{1}{2}+\frac{|m_l|-m_l}{2}\right)\quad;\quad n_r=0,1,\ldots\quad;\quad m_l=0,\pm1,\ldots\,, \tag{16}$$

where n_r is a non-negative integer radial number. The normalized radial function is calculated to be:

$$R_{n_r m_l}(r)=N_{n_r m_l}\,\exp\left(-\frac{r^2}{4\,l_0^2}\right)\left(\frac{r}{l_0}\right)^{|m_l|}L_{n_r}^{|m_l|}\left(\frac{r^2}{2\,l_0^2}\right)\,, \tag{17}$$

where $L_n^k(x)$ are the associated Laguerre polynomials defined as:

$$L_n^k(x)=(-1)^k\frac{d^k}{dx^k}\left[L_{n+k}(x)\right]\quad;\quad L_n(x)=\frac{e^x}{n!}\frac{d^n}{dx^n}\left(x^n e^{-x}\right)\,. \tag{18}$$

The normalization constant is:

$$N_{n_r m_l} = \sqrt{\frac{n_r!}{l_0^2\, 2^{|m_l|}\,(n_r + |m_l|)!}} \; . \tag{19}$$

One can easily check the normalization condition:

$$\int_0^{\infty} dr\, r\, R_{n_r m_l}(r)^* \, R_{n_r' m_l}(r) = \delta_{n_r n_r'} \; , \tag{20}$$

with help of the following formula:

$$\int_0^{\infty} dx\, e^{-x}\, x^k\, L_n^k(x) L_{n'}^k(x) = \frac{(n+k)!}{n!}\, \delta_{nn'} \; . \tag{21}$$

Thus, the normalized eigenfunction for general n_r and m_l is given by:

$$\Psi_{n_r m_l}(r,\varphi) = R_{n_r m_l}(r)\, \Phi_{m_l}(\varphi) = |n_r m_l\rangle \; . \tag{22}$$

The allowed energy values can also be written as:

$$E = \hbar\, \omega_c \left(n + \frac{1}{2}\right) \; , \tag{23}$$

where

$$n = n_r + \frac{|m_l| - m_l}{2} = 0, 1, \ldots \; , \tag{24}$$

represents the Landau level quantum number. In Table. 1 we display the various combinations of quantum numbers n_r and m_l that generate the Landau energies labeled by the quantum number n. Clearly, each Landau level is highly degenerate. The degeneracy per

Table 1. A display of energy values labeled by n and the corresponding n_r-s and m_l-s that generate such energies

$E/(\hbar\,\omega_c)$	n	n_r	m_l
1/2	0	0	0, 1, 2, 3, ...
3/2	1	0	-1
3/2	1	1	0, 1, 2, 3, ...
5/2	2	0	-2
5/2	2	1	-1
5/2	2	2	0, 1, 2, 3, ...
7/2	3	0	-3
7/2	3	1	-2
7/2	3	2	-1
7/2	3	3	0, 1, 2, 3, ...

unit area is the same in each Landau level, but depends on the magnetic field. In a symmetric gauge, one readily obtains the Landau level degeneracy by counting how many states lie inside a disk of certain area, A. It turns out that the total number of states, N_s, counted this way is:

$$N_s = \frac{A}{2\,\pi\,l_0^2}\,, \tag{25}$$

yielding a degeneracy per unit area:

$$\frac{N_s}{A} = \frac{1}{2\,\pi\,l_0^2}\,. \tag{26}$$

All Landau levels irrespective of the value of their index n, have the same degeneracy which coincides with the value obtained above. The single particle states are especially simple to write in the lowest Landau level (LLL) where $n = 0$ thus $n_r = 0$ and $m_l = 0, 1, \ldots$

$$\Psi_{m_l}(z) = |0\,m_l\rangle = \frac{1}{\sqrt{2\,\pi\,l_0^2\,2^{m_l}\,m_l!}}\left(\frac{z}{l_0}\right)^{m_l}\exp\left(-\frac{|z|^2}{4\,l_0^2}\right)\ ;\ m_l = 0, 1, \ldots\,, \tag{27}$$

where $z = x + i\,y$ is the 2D position coordinate in complex notation.

2. FQHE and the Laughlin States in the Lowest Landau Level

The phenomenon known as the fractional quantum Hall effect (FQHE) represents a novel strongly correlated electronic quantum phase arising from the interplay between low-dimensionality, temperature near absolute zero and strong magnetic field [1, 2, 3, 4]. The experimental discovery of FQHE [5] has stimulated extensive studies on the properties of two-dimensional electron systems (2DES) in a strong perpendicular magnetic field. At low enough temperature and low enough amount of disorder, the FQHE represents the condensation of nearly 2D electrons into an incompressible quantum fluid state formed at some specific filling factors, $\nu = N/N_s$ where N is the number of electrons in the system and N_s is the degeneracy of a Landau level. The most striking experimental signature of this new collective state of matter is the observation of quantized plateaus of the Hall resistance,

$$R_H = \frac{h}{\nu\,e^2} \approx \frac{25812.807\,\Omega}{\nu}\,, \tag{28}$$

where the filling factor ν is a fractional number, h is Planck's constant and e is the magnitude of electron's charge. The strong magnetic field quantizes the electrons's motion on the plane and quenches the kinetic energy of each electron to a discrete set of Landau levels (LLs) separated by the relatively large cyclotron energy, $\hbar\,\omega_c = \hbar\,e\,B/m_e$, where $-e(e > 0)$ is the electron charge and m_e is the (effective) mass of electrons. For simplicity, we assume that spin of the electrons is fully polarized by the strong magnetic field rendering them *effectively spinless*. The uniform electron density of the system can be written as:

$$\rho_0 = \frac{\nu}{2\,\pi\,l_0^2}\,, \tag{29}$$

where $l_0 = \sqrt{\hbar/(e\,B)}$ is the electronic magnetic length. The most robust FQHE states occur in the LLL and correspond to filling factors $\nu = 1/3$ and $\nu = 1/5$. Laughlin's theory [6] describes very well the states with filling factor $\nu = 1/3$ and $\nu = 1/5$ in terms of trial wave functions that belong to the Hilbert space of the LLL [7]. Following Laughlin's discovery, a great deal of subsequent theoretical work has shed light on the incompressible nature of such electronic liquid phases [8, 9, 10, 11, 12]. Other FQHE states at more general LLL filling factors are understood in terms of the composite fermion (CF) theory [13]. However, even-denominator filled states, manifest features that are qualitatively different from the FQHE states of the originating sequence [14, 15, 16, 17].

Since a 2DES in the FQHE regime consists of electrons with negative charge, $q = -e(e > 0)$ it is customary to assume that the direction of the perpendicular magnetic field is $\vec{B} = (0, 0, -B)$. The choice of the negative sign of $\vec{B}$ is a matter of convenience, allowing us to express the LLL wave functions for electrons by the same expression as in Eq.(27) with $z = x + iy$ representing the same complex variable as defined earlier. For filling factors of the form $\nu = 1/m$ $(m = 3, 5)$ the unnormalized trial Laughlin [6] wave function for N electrons can be written as:

$$\Psi_m(z_1, \ldots, z_N) = \prod_{i<j}^{N} (z_i - z_j)^m \prod_{j=1}^{N} \exp\left(-\frac{|z_j|^2}{4\, l_0^2}\right) , \tag{30}$$

where $z_j = x_j + i\, y_j$ is the position of the j-th electron in complex coordinates. This wave function [18] gives an excellent description of the true ground state of the electrons for $m = 3$ and 5. For $m \geq 7$ the electrons tend to form a Wigner crystal [19] consistent with the experimental observation [20] that the FQHE does not occur for filling factors $\nu \leq 1/7$.

For the purpose of our study, we adopt a disk geometry model [21] that considers N electrons of charge $-e(e > 0)$ immersed in a uniform positively charged finite disk of area $\Omega_N = \pi\, R_N^2$ where R_N is the radius of the disk. The density of the system (number of electrons per unit area) or otherwise the uniform density of the background, $\rho_0 = N/\Omega_N$, is constant. Since, the density of the system can also be written as $\rho_0 = \nu/(2\,\pi\, l_0^2)$, for a given filling factor, the radius of disk varies with N as:

$$\frac{R_N}{l_0} = \sqrt{\frac{2\,N}{\nu}} \quad ; \quad N \geq 2 \, . \tag{31}$$

The quantum Hamiltonian of the system,

$$\hat{H} = \hat{K} + \hat{V} \, , \tag{32}$$

consists of kinetic and potential energy operators where

$$\hat{K} = \frac{1}{2\, m_e} \sum_{i=1}^{N} \left[\hat{\vec{p}}_i + e\, \vec{A}_i\right]^2 , \tag{33}$$

is the kinetic energy operator written in a symmetric gauge. The potential energy operator:

$$\hat{V} = \hat{V}_{ee} + \hat{V}_{eb} + \hat{V}_{bb} \, , \tag{34}$$

consists of electron-electron (ee), electron-background (eb) and background-background (bb) interaction potentials written as:

$$\hat{V}_{ee} = \sum_{i<j}^{N} v(\vec{r}_i - \vec{r}_j) \,, \tag{35}$$

$$\hat{V}_{eb} = -\rho_0 \sum_{i=1}^{N} \int_{\Omega_N} d^2r \, v(\vec{r}_i - \vec{r}) \,, \tag{36}$$

and

$$\hat{V}_{bb} = \frac{\rho_0^2}{2} \int_{\Omega_N} d^2r \int_{\Omega_N} d^2r' \, v(\vec{r} - \vec{r}') \,, \tag{37}$$

where $v(\vec{r}_i - \vec{r}_j) = e^2/|\vec{r}_i - \vec{r}_j|$ is the Coulomb interaction potential. In all above expressions, $\vec{r}_i$ (or $\vec{r}_j$) denote electronic 2D position vectors, while $\vec{r}$ and $\vec{r}'$ are background coordinates. Each of the N position variables of electrons, $\{\vec{r}_i\}$ extends all over space ($-\infty$ to $+\infty$). However, background coordinates, $\vec{r}$ (or $\vec{r}'$) are confined within the finite disk, namely $0 \leq |\vec{r}| \leq R_N$ (or $0 \leq |\vec{r}'| \leq R_N$).

Since the Laughlin wave function lies entirely in the LLL, the expectation value of the kinetic energy per electron:

$$\frac{\langle \hat{K} \rangle}{N} = \frac{1}{2} \hbar \, \omega_c \,, \tag{38}$$

is an irrelevant constant. Therefore, the only important contribution in the quantum mechanical Hamiltonian originates from the total potential energy operator.

Given the number, N of electrons in the system we have to calculate $\langle \hat{V}_{ee} \rangle$, $\langle \hat{V}_{eb} \rangle$ and $\langle \hat{V}_{bb} \rangle$ where in a short-hand notation $\langle \hat{O} \rangle = \langle \Psi | \hat{O} | \Psi \rangle / \langle \Psi | \Psi \rangle$ denotes the standard quantum expectation value of an operator $\hat{O}$ relative to wave function, $|\Psi\rangle$. Note that the $\langle \hat{V}_{bb} \rangle$ term does not depend in the form of the wave function and can be calculated analytically. We can write $\langle \hat{V}_{eb} \rangle$ as:

$$\langle \hat{V}_{eb} \rangle = -\rho_0 \int d^2r_1 \, \rho(\vec{r}_1) \int_{\Omega_N} d^2r \; v(\vec{r}_1 - \vec{r}) \,, \tag{39}$$

where $\rho(\vec{r}_1)$ is the single-particle density function:

$$\rho(\vec{r}_1) = N \, \frac{\int d^2r_2 \ldots d^2r_N \, |\Psi(\vec{r}_1, \ldots, \vec{r}_N)|^2}{\int d^2r_1 \ldots d^2r_N \, |\Psi(\vec{r}_1, \ldots, \vec{r}_N)|^2} \,. \tag{40}$$

In a similar fashion one has:

$$\langle \hat{V}_{ee} \rangle = \frac{N \, (N-1)}{2} \, \langle v(\vec{r}_1 - \vec{r}_2) \rangle \,. \tag{41}$$

The ground state interaction energy per particle can be written as:

$$\epsilon = \epsilon_{ee} + \epsilon_{eb} + \epsilon_{bb} \,, \tag{42}$$

where $\epsilon = \langle \hat{V} \rangle / N$, $\epsilon_{ee} = \langle \hat{V}_{ee} \rangle / N$, $\epsilon_{eb} = \langle \hat{V}_{eb} \rangle / N$ and $\epsilon_{bb} = \langle \hat{V}_{bb} \rangle / N$ are, respectively, the total, ee, eb and bb interaction energies per particle. As usual, a charge neutralizing background is required to guarantee the stability of electronic systems with repulsive Coulomb interactions. Thus, the total energy expression should include a positive contribution of the background. Since our interest is to study the state at a given (fixed) magnetic field, this implies that the system should be kept at a fixed density. Stated otherwise, this means that when we increase the number N of electrons in our finite system, the area, Ω_N of the background disk also increases in a way that the density, N/Ω_N remains constant. Thus, background disks with different radii will differ in their contributions to the total energy of the system. For a disk with uniform density, $\rho_0 = \nu/(2\,\pi\,l_0^2)$ and radius R_N we calculate that the background-background interaction potential is:

$$\epsilon_{bb} = \frac{\langle \hat{V}_{bb} \rangle}{N} = \frac{8}{3\,\pi}\sqrt{\frac{\nu\,N}{2}}\frac{e^2}{l_0} \;. \tag{43}$$

Numerous techniques have been employed to calculate the expectation value of the potential energy per particle, $\langle \hat{V} \rangle / N$ [see Eq. (34)] in the Laughlin state. For example, Laughlin [6] initially employed the hypernetted-chain method to estimate the value of this correlation energy. Various Monte Carlo (MC) schemes have also been proposed, all of which are essentially exact in the thermodynamic limit. An excellent description of a standard MC computation of the potential energy and other relevant quantities in a disk geometry is given by Morf and Halperin [22]. Spherical geometries are also used quite often since the convergence to the thermodynamic limit is quicker because boundary effects are eliminated [23].

Although considerable more care is needed in the disk geometry to eliminate boundary effects when extrapolating the finite-N results to the thermodynamic limit (in particular due to the long-range nature of the Coulomb potential), there are cases in which the spherical geometry is either inconvenient, or plainly incompatible with the state under consideration (for example for the study of possible quantum Hall nematic phases [24] for which considerable topological defects would be generated at the poles of the sphere).

In a standard MC approach one considers the calculation of the expectation value of the potential energy operators as given in Eq. (35). In order to compute the expectation value of the electron-background interaction potential one conveniently writes it as

$$\hat{V}_{eb} = \sum_{i=1}^{N} \hat{v}_{eb}(\vec{r}_i) \quad ; \quad \hat{v}_{eb}(\vec{r}_i) = -\rho_0 \int_{\Omega_N} d^2r \, \frac{e^2}{|\vec{r}_i - \vec{r}|} \;, \tag{44}$$

where $\hat{v}_{eb}(\vec{r}_i)$ is the interaction potential of a single electron at position $\vec{r}_i$ with the uniform positive background in the finite disk. Such electron-background interaction potential depends on the ratio r_i/R_N, where $r_i = |\vec{r}_i|$ is electron's distance from the center of the disk and R_N is the radius of the finite disk. It can be expressed as

$$\hat{v}_{eb}(r_i) = -\sqrt{2\nu N}\, F\left(\frac{r_i}{R_N}\right) \frac{e^2}{l_0} \quad ; \quad F(x) = \int_0^{\infty} \frac{dz}{z}\, J_0(x\,z) J_1(z) \;, \tag{45}$$

where $J_n(z)$ are Bessel functions of order n. Note that the above expression simply represents the general result for the electrostatic energy between an arbitrary negative charge, $-q_0$ ($q_0 > 0$) and a finite 2D disk uniformly charged with positive charge Q and radius R:

$$V(r) = -q_0\, V_0\, F(r,R) \quad ; \quad F(r,R) = \int_0^{\infty} \frac{dz}{z}\, J_0\left(\frac{r}{R}\, z\right) J_1(z) \;, \tag{46}$$

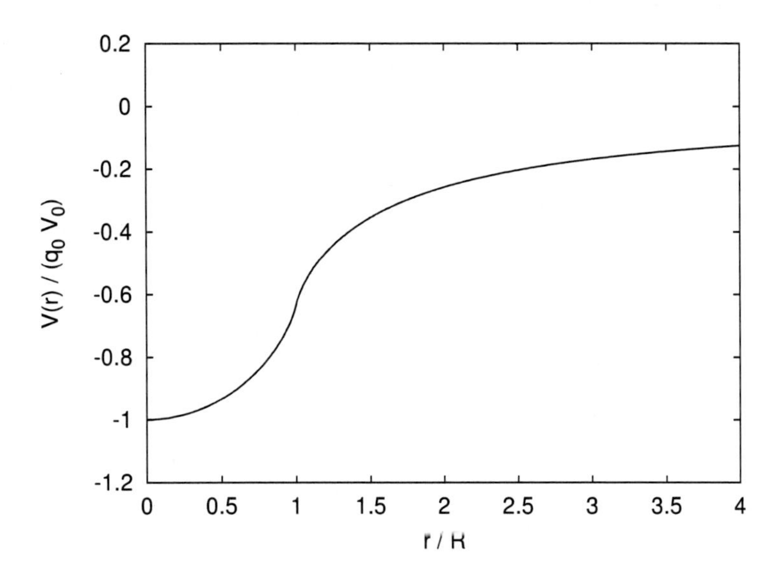

Figure 1. Electrostatic potential energy, $V(r)$ between charge $-q_0$ ($q_0 > 0$) at distance r away from the center of disk and a uniformly charged 2D disk with radius R and positive charge Q. The quantity $V_0 = 2\,k\,Q/R$ is the electrostatic potential at the center of the disk ($r = 0$) and k is the electric Coulomb's constant.

where $F(r, R)$ is a function that depends only on the ratio r/R (given in integral form), $J_n(z)$ are $n-$th order Bessel functions, $r = \sqrt{x^2 + y^2}$ is the distance of the electron from the center of the disk, $V_0 = (2\,k\,Q)/R$ is the electrostatic potential created by the disk at its center ($r = 0$) and $k = 1/(4\,\pi\,\epsilon_0)$ is Coulomb's electric constant. While the integral presentation of $F(r, R)$ is rather convenient, the integration in Eq.(46) can also be carried out analytically resulting in an expression involving complete elliptic integrals of the second kind and hypergeometric functions [25]. Two special values of this function are:

$$F(r = 0, R) = 1 \quad ; \quad F(r = R, R) = \frac{2}{\pi} \,. \tag{47}$$

In Figure 1 we show $V(r)/(q_0 V_0) = -F(r, R)$ as a function of r/R, where r is the distance of the electron from the center of the disk. One immediately notices that the function, $-F(r, R)$ is approximately parabolic for the range $0 \leq r \leq R$ and becomes zero asymptotically at large distances. While most electrons stay within the confines of the neutralizing background, electrons near the edge may spread outside the disk to some extent (although it is extremely unlikely that they will spread to more than a few magnetic lengths from the edge). The expectation value of the electron-background interaction potential during the MC simulation can then be calculated using

$$\frac{\langle \hat{V}_{eb} \rangle}{N} = \frac{1}{N} \left\langle \sum_{i=1}^{N} \hat{v}_{eb}(r_i) \right\rangle \,. \tag{48}$$

Table 2. Correlation energy per particle in the Laughlin state for filling factors $\nu = 1/3$ and $1/5$. These results were obtained after a standard Monte Carlo simulation in a disk geometry. Energies are in units of e^2/l_0

N	$\nu = 1/3$	$\nu = 1/5$
4	-0.38884	-0.32159
16	-0.39766	-0.32328
36	-0.40129	-0.32446
64	-0.40323	-0.32510
100	-0.40445	-0.32550
144	-0.40521	-0.32577
196	-0.40579	-0.32594
400	-0.40675	-0.32624

Finally, the expectation value of $\hat{V}_{ee}$ is accordingly given by

$$\frac{\langle \hat{V}_{ee} \rangle}{N} = \frac{1}{N} \left\langle \sum_{i<j}^{N} \frac{e^2}{|\vec{r}_i - \vec{r}_j|} \right\rangle . \tag{49}$$

In the usual Metropolis MC method [26], the expectation value of an operator can be computed by averaging the value of the operator over numerous configurations $\{\vec{r}_1, \ldots, \vec{r}_N\}$ of the many-body system that obey detailed balance, that is, the probability ratios between pairs of discrete configurations are related by the ratios of the probability distribution for the system. Usually several million configurations are used for each N and the results are extrapolated to the thermodynamic limit by considering a sequence of various increasing N-s.

A MC step (MCS) consists of attempts to move one by one all the electrons of the system by a small distance of order Δ in a random direction. After each attempt (to move the i-th electron from $\vec{r}_i^{\,old}$ to $\vec{r}_i^{\,new}$, the probability ratio between the "new" state and the "old" state is then computed. In the usual Metropolis scheme [26], if this ratio is bigger than a uniformly distributed number in the [0,1] range the attempt is accepted, otherwise it is rejected. The parameter Δ is adjusted so that the acceptance ratio is close to 50%. After attempting to move all electrons (one MCS), the electron configurations are then used to calculate the operator under consideration. Averaging over numerous MCS-s converges gradually (as $1/\sqrt{\text{number of MCS}}$) to the desired expectation value. Normally it is convenient to disregard numerous (several thousand) initial configurations to reach a good "thermalization" before the averaging begins, which significantly reduces the expurious effects of the somewhat arbitrary initial configurations. All the results that we report here were obtained after discarding 100,000 "thermalization" MCS-s and using 2×10^6 MCS-s for averaging purposes. In Table 2 we show the correlation energy per particle for finite systems of N electrons in the Laughlin states $\nu = 1/3$ and $1/5$ obtained using the standard MC method described above. The results are rounded in the last digit.

To get the the bulk (thermodynamic estimate) of the correlation energy per particle one needs to perform a careful extrapolation of the results. We fitted the energies of Table 2 for

$N = 4, 16, 36, 64, 100, 144, 196$ and 400 electrons to a polynomial function as reported in Ref. [22] and obtained:

$$\frac{\langle \hat{V} \rangle_{1/3}}{N} = \left(-0.4094 + \frac{0.0524}{\sqrt{N}} - \frac{0.0225}{N} \right) \frac{e^2}{l_0}, \tag{50}$$

and

$$\frac{\langle \hat{V} \rangle_{1/5}}{N} = \left(-0.3273 + \frac{0.0200}{\sqrt{N}} - \frac{0.0172}{N} \right) \frac{e^2}{l_0}. \tag{51}$$

These interpolation lines are used to estimate the correlation energy per particle in the thermodynamic limit (the first term in each of the parentheses). Our results for thermodynamic limit are similar to those found in Ref.[8] who report the following energies -0.4100 ± 0.0001 and -0.3277 ± 0.0002 (in units of e^2/l_0) for $\nu = 1/3$ and $1/5$, respectively. The estimates in Ref.[8] are derived with the use of the pair correlation function evaluated from MC simulations with up to $N = 256$ electrons and generating as many as 5 million MC configuratons.

2.1. The One-Particle Density

Other physical quantities of interest that may be readily computed are the single-particle density function and the pair distribution function. Given that the Laughlin wave function describes an isotropic liquid state and is rotationally invariant, the single-particle density depends only on the radial distance from the center of the disk. We may compute the single-particle density by counting the number of electrons $N_l(\Delta r)$ found in several 2D shells of width Δr centered around a discrete set of distances to the center $r_l = (l + \frac{1}{2})\Delta r$ ($l = 0, 1, \ldots$):

$$\rho(r_l) \equiv \left\langle \frac{N_l(\Delta r)}{\Omega_l(\Delta r)} \right\rangle, \tag{52}$$

where $\Omega_l(\Delta r) = \pi(\Delta r)^2[(l+1)^2 - l^2]$ is the area of each 2D shell. In the $\Delta r \to 0$ limit the computed quantity corresponds unequivocally to the electron density:

$$\rho(r) = \left\langle \sum_{i=1}^{N} \delta(r - r_i) \right\rangle. \tag{53}$$

The computation of the single-particle density in the Laughlin state, indicates a significant nonuniformity near the boundary (see Figure 2). As the number of electrons increase, a significant portion of the system becomes uniform as expected. Note, however, that the non-uniformity near the edge always persists. This behavior can be used to characterize which electrons are "in the bulk."

2.2. The Pair Distribution Function

Another important quantity related with the trial wave function is the pair distribution function, which corresponds to the conditional probability density to find an electron at a distance r from another electron. For any homogeneous and isotropic liquid with uniform

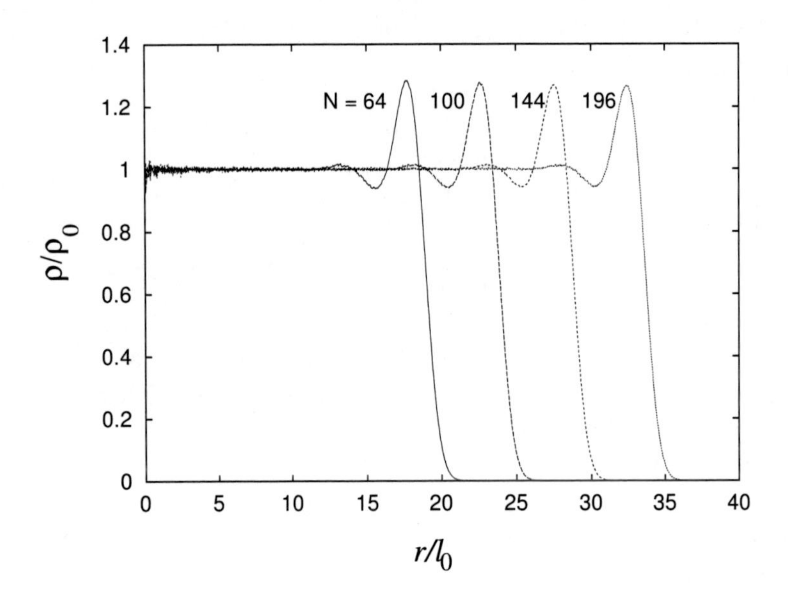

Figure 2. One-body density function, $\rho(r)/\rho_0$, in the Laughlin state $\nu = 1/3$ as a function of the distance r/l_0 from the center of the disk for systems with $N = 64, 100, 144$ and 196 electrons. Note the persistence of an "edge region" of finite width and the development of a "bulk region" for large N. A discretization interval $\Delta r = 0.05\, l_0$ was used in the calculation.

density ρ_0 it is defined as

$$\rho_0\, g(r) = \frac{1}{N}\left\langle \sum_{i=1}^{N}\sum_{j\neq i}^{N} \delta(r - |\vec{r}_i - \vec{r}_j|)\right\rangle . \tag{54}$$

Following the same procedure as before, we discretize in concentric shells around the i-th electron and count the number of electrons $N_l(\Delta r)$ in each shell, which should give $g(r)$ in the $\Delta r \to 0$ limit according to the following equation:

$$\rho_0\, g(r_l) = \frac{1}{N}\frac{1}{\Omega_l(\Delta r)}\left\langle \sum_{i=1}^{N}\sum_{j\neq i}^{N} N_l(\Delta r)\right\rangle . \tag{55}$$

It is evident that electrons near the edges of the system could contribute spuriously to these sums as their "surroundings" are considerably different than those at the bulk. To eliminate as much as possible any boundary effects, it is convenient to consider only (for the "i-electrons" above) the electrons that are within a small circle of radius R_1 around the origin. If N_1 is the average number of electrons that are within this small circle, then the approximation $\hat{g}(r_l)$ for the pair distribution $g(r_l)$ is:

$$\rho_0\, \hat{g}(r_l) = \frac{1}{N_1}\frac{1}{\Omega_l(\Delta r)}\left\langle \sum_{i=1}^{N_1}\sum_{j\neq i}^{N} N_l(\Delta r)\right\rangle , \tag{56}$$

where in this expression one is considering the pairs between any electron i ($i = 1, \ldots, N_1$) lying inside the circle of radius R_1 with all other electrons j ($j = 1, \ldots, i-1, i+1, \ldots, N$) that may lie either inside, or outside that circle. This guarantees that the evaluation of $\hat{g}(r_l)$ involves only pairs, where at least one member lies inside a circle of radius R_1 around the origin, where correlations are believed to be close to those in the bulk of an infinite system.

Figure 3 show plots of the pair distribution function for the states $\nu = 1/3$ and $1/5$ for systems with $N = 4, 16, 36, 64, 100, 144$ and 196 electrons. For our MC simulations we chose $R_1 = 0.25\, R_N$ and a discretization interval $\Delta r = 0.05\, l_0$. Note the gradual decay at large r which reflects the finite size of the system.

The determination of the pair correlation function for a given finite N tends to be quite time-consuming but provides for an alternative way to compute the correlation energy per particle in the thermodynamic limit by using the formula:

$$\frac{\langle \hat{V} \rangle}{N} = \frac{\rho_0}{2} \int_0^{+\infty} d^2r \, \frac{e^2}{r} \left[g(r) - 1 \right] \,, \tag{57}$$

which is valid in the limit of an infinite system. Although the pair distribution function is obtained from a system with a finite number of particles, one can calculate the thermodynamic value of the correlation energy per particle to a very good accuracy by using the slightly modified formula

$$\frac{\langle \hat{V} \rangle}{N} = \frac{\rho_0}{2} \int_0^{R_{cut}} d^2r \, \frac{e^2}{r} \left[\hat{g}(r) - 1 \right] \,, \tag{58}$$

in connjuction with the normalization condition:

$$\rho_0 \int_0^{R_{cut}} d^2r \, \left[\hat{g}(r) - 1 \right] = -1 \,, \tag{59}$$

which defines an upper cuttoff R_{cut}. This approach produces good estimates for the thermodynamic correlation energy per particle as long as $\hat{g}(r)$ is able to reach its asymptotic value.

3. Even-Denominator Filled Fermi Liquid States in the Lowest Landau Level

The phase diagram of a 2DES in a strong perpendicular magnetic field at filling factors $0 < \nu \leq 1$ is intricate with competing liquid and Wigner solid phases. At filling factors, $\nu = 1/3, 2/5, \ldots$ and $\nu = 1/5, 2/9, \ldots$ electrons condense into an incompressible liquid state and the FQHE occurs. It is believed that at even-denominator filling factors, $\nu = 1/2, 1/4$ and $1/6$ the electrons form a compressible Fermi liquid state [27, 28, 29] while for filling factors, $0 < \nu \leq 1/7$, Wigner crystallization occurs [30, 31, 32]. On the theoretical side, the accurate prediction of the critical filling factor, ν_c where the liquid-solid transition occurs is very difficult. For an ideal 2DES system free of any disorder the best available theoretical estimates [19] suggest $\nu_c \cong 1/6.5$. The situation is more complex in the experimental realm. Various measurements [33] on high quality samples, yet with a finite amount of disorder, show the existence of a reentrant solid-like phase around the $\nu = 1/5$ FQHE

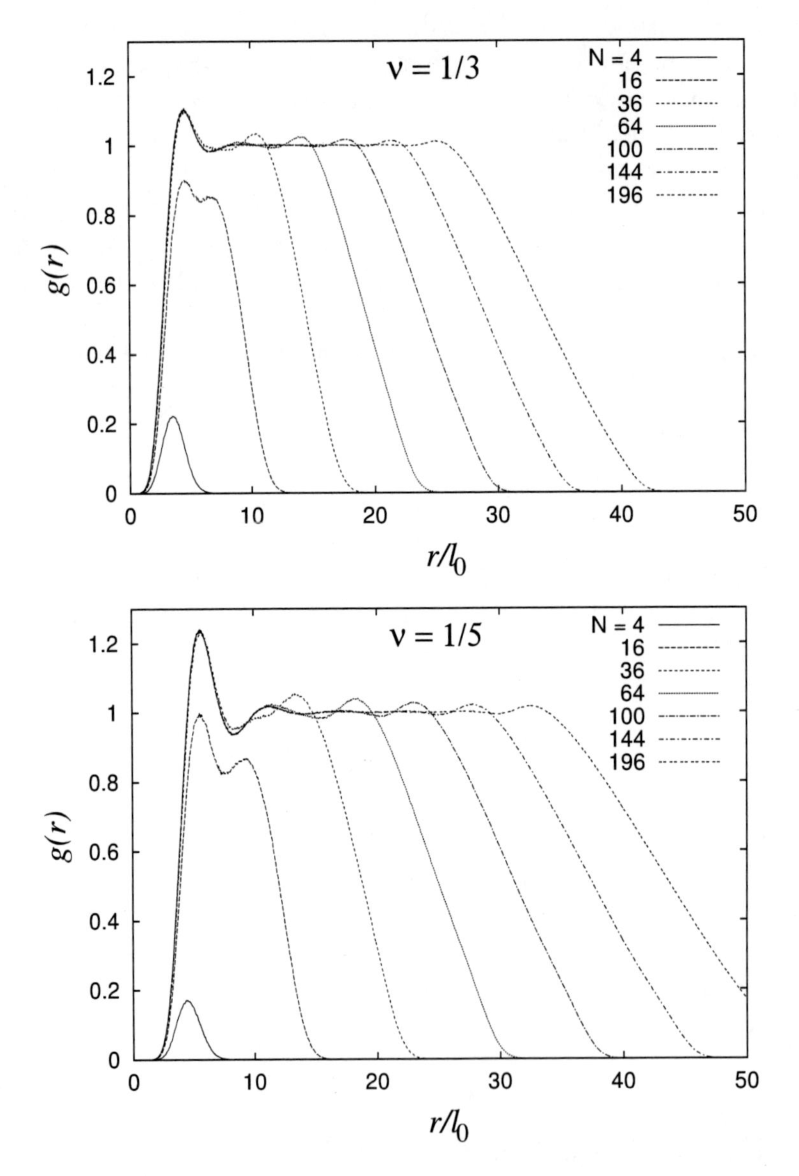

Figure 3. Pair distribution function for the Laughlin states at $\nu = 1/3$ and $1/5$ obtained by a standard Monte Carlo simulation in disk geometry for systems of $N = 4, 16, 36, 64, 100, 144$ and 196 electrons.

state. It is likely that this represents an electronic solid phase pinned to fluctuations of the potential. The majority of theoretical studies which are based on models free of any disorder agree that the critical filling factor, ν_c below which the Wigner crystal forms is close to and probably slightly larger than filling $1/7$ in excellent agreement with the estimate of Lam and Girvin [19].

A recent study [34] based on the exact diagonalization method found strong evidence that the Wigner crystal forms at filling factors, $\nu \leq 1/7$. Such study considered filling factors, $\nu = 1/6, 1/7$, and $1/8$. It was found that for $\nu = 1/8$, as well as $\nu = 1/7$ the translational symmetry (signature of liquid phase) is broken and the system has 2D crystalline order. However, the system continued to have its translational symmetry for the case of filling factor $\nu = 1/6$, suggesting that the value of the critical filling factor, ν_c is between slightly above $1/7$, but below $1/6$. It is now wellknown that the principal FQHE states at $\nu = 1/3$ and $1/5$ are thoroughly explained and very well described by the Laughin wavefunction. On the other hand, other FQHE states at filling factors, $\nu = p/(2\,m\,p{+}1)$ (p, m - integer) are readily understood in terms of the composite fermion (CF) theory [35, 36, 37, 38, 39, 40, 41]. The limit ($p \rightarrow \infty$) of such FQHE states corresponds to even-denominator filled states, $\nu = 1/(2\,m)$ which show Fermi liquid signatures [42]. The basic understanding now is that the low-temperature phase of fully spin-polarized electrons at filling fraction $\nu = 1/(2m)$ is a CF Fermi liquid phase [43, 44, 45].

At $\nu = 1/(2\,m)$ the typical features of a FQHE liquid phase, that is, quantized Hall resistivity, $\rho_{xy} = \nu\,\frac{e^2}{h}$ and vanishing diagonal resistivity, ρ_{xx}, are not observed. Nevertheless, this state shows a broad minimum [27] in diagonal resistivity and exhibits, additionally, anomalous behavior in surface accoustic wave propagation, [28] indicating a different type of correlation. A theory of a compressible Fermi-liquid-like behavior at even-denominator filled states has been proposed by Halperin, Lee, and Read [16]. According to this theory, a 2DES subjected to an external perpendicular magnetic field $\vec{B}$, at even-denominator filling factors, can be transformed to a mathematically equivalent system of fermions interacting with a Chern-Simons (CS) gauge field such that the average effective magnetic field acting on the fermions is zero.

If we assume that $|\Phi(z_1, \ldots, z_N)\rangle$ is a solution of the Schrödinger equation $\hat{H}\,\Phi = E\,\Phi$ then for an even number $q_e = 2, 4, \ldots$, the wavefunction:

$$\Psi(z_1, \ldots, z_N) = \prod_{j<k}^{N} \frac{(z_j - z_k)^{q_e}}{|z_j - z_k|^{q_e}} |\Phi(z_1, \ldots, z_N)\rangle \ , \tag{60}$$

is a solution to the Schrödinger equation $\hat{H}'\,\Psi = E\,\Psi$, with

$$\hat{H}' = \hat{K}' + \hat{V} \ , \tag{61}$$

and

$$\hat{K}' = \frac{1}{2\,m_e} \sum_{j=1}^{N} \left\{ -i\hbar\vec{\nabla}_j + e[\vec{A}(\vec{r}_j) - \vec{a}(\vec{r}_j)] \right\}^2 \ . \tag{62}$$

In the above expression, $\vec{a}(\vec{r})$ represents the CS vector potential:

$$\vec{a}(\vec{r}) = q_e\,\frac{\phi_0}{2\,\pi} \sum_{j=1}^{N} \frac{\vec{z} \times (\vec{r} - \vec{r}_j)}{|\vec{r} - \vec{r}_j|^2} \ , \tag{63}$$

where $\phi_0 = h/e$ is the magnetic field flux quantum and $\vec{z}$ is a unit vector in the z-direction. The CS magnetic field, $\vec{b}(\vec{r})$, associated with the vector potential $\vec{a}(\vec{r})$ is given by:

$$\vec{b}(\vec{r}) = \vec{\nabla} \times \vec{a}(\vec{r}) = q_e\,\phi_0 \sum_{j=1}^{N} \delta(\vec{r} - \vec{r}_j)\,\vec{z} = q_e\,\phi_0\,\rho(\vec{r})\,\vec{z} \ , \tag{64}$$

where $\rho(\vec{r})$ is the local particle density. In other words, the CS transformation can be described as the exact modeling of an electron as a fermion attached to q_e magnetic flux quanta. Assuming a uniform density, the CS flux quanta attached to the fermions are smeared out into a uniform magnetic field of magnitude:

$$\langle b \rangle = q_e \, \phi_0 \, \rho_0 \ , \tag{65}$$

with ρ_0 the average electronic density. At special filling factors $\nu = \phi_0 \rho_0 / B = 1/q_e$ where $q_e = 2, 4 \ldots$, the applied magnetic field precisely cancels the CS magnetic flux. Thus, at the mean field level, the system can be described as fermions in zero magnetic field and should therefore behave as a compressible Fermi-looking liquid state. When ν is away from $1/q_e$, the applied magnetic field and the CS one do not cancel out exactly and, as a result, a residual effective field:

$$B^* = B - q_e \, \phi_0 \, \rho_0 = B \, (1 - q_e \, \nu) \ , \tag{66}$$

is left over. Thus, the mean field system is described as non-interacting system of fermions in a uniform effective field B^*. The effective filling factor for these gauge transformed fermions is, $\nu^* = \phi_0 \, \rho_0 / B^* = p = 1, 2, \ldots$. So, in an effective sense, this model represents the integer quantum Hall effect of the gauge transformed fermions. The "true" filling factor of the electrons, $\nu = \phi_0 \, \rho_0 / B$ is just $\nu = p/(q_e \, p + 1)$ and this is precisely the CF Jain series [13] of FQHE states. In the following we focus our attention on the half-filled LLL state, $\nu = 1/2$, where several related wave functions have been employed to incorporate the physics of CF's on it. These wave functions can be treated as the limit of the series $\nu = p/(2p + 1)$ for $p \to \infty$. The wave function:

$$\Psi^{CS}_{\nu=1/2} = \lim_{p \to \infty} \hat{P}_{LLL} \prod_{j<k}^{N} \frac{(z_j - z_k)^2}{|z_j - z_k|^2} \, |\Phi_p(B^*)\rangle \ , \tag{67}$$

appears as the mean-field solution of the CS theory [46]. The above mean-field CS wave function is unsatisfactory [47, 48] for many reasons. However, these problems can be remedied to a great extent, by throwing away the factor $|z_j - z_k|^2$ in the denominator eventually ending up with:

$$\Psi^{CF}_{\nu=1/2} = \lim_{p \to \infty} \hat{P}_{LLL} \prod_{j<k}^{N} (z_j - z_k)^2 |\Phi_p(B)\rangle \ , \tag{68}$$

which is essentially the CF wave function due to Jain. Note that in this representation, the wave function Φ_p in Eq.(68) must now be evaluated at a magnetic field, B, and not B^*, to ensure the correct filling factor for Ψ^{CF}_{ν} [49] (see also pg. 126 in Ref. [4]). In all the above expressions, $\hat{P}_{LLL}$ is the LLL projection operator and $|\Phi_p(B)\rangle$ is the Slater determinant wave function of p filled Landau levels evaluated at the magnetic field shown in the argument. The two wave functions in Eq.(67) and Eq.(68) have different origins and different short-distance behavior, but they both describe CFs at half filling factor of the LLL.

Let us now go back to the most general case of a filling factor of the form, $\nu = p/(2m \, p + 1)$ (where $q_e = 2 \, m$). In this case, the effective field field felt by the transformed fermions reads $B^* = B \, (1 - 2 \, m \, \nu)$. While not easy to seee through the expression in Eq.(68), note that the Gaussian factors in $|\Phi_p(B)\rangle$ are such that they can be written as:

$$\frac{1}{l_0(B)^2} = \frac{1}{l_0(B^*)^2} + \frac{2\,m\,\nu}{l_0(B)^2}\,, \tag{69}$$

where the magnetic field dependence is explicitely shown in the argument of the magnetic length, $l_0(B)^2 = \hbar/(e\,B)$. This way, one can rewrite the wave function in Eq.(68) for a general even-denominator filling factor $\nu = 1/(2\,m)$ as:

$$\Psi^{CF}_{\nu=1/(2m)} = \lim_{p\to\infty} \Psi^{CF}_{\nu=p/(2mp+1)}\,, \tag{70}$$

where

$$\Psi^{CF}_{\nu=p/(2mp+1)} = \hat{P}_{LLL} \prod_{j<k}^{N} (z_j - z_k)^{2m} \prod_{j=1}^{N} \exp\left[-2m\nu \frac{|z_j|^2}{4l_0(B)^2}\right] \Phi_p(B^*)\,, \tag{71}$$

and $\Phi_p(B^*)$ is the Slater determinant wavefunction of p filled CF Landau levels evaluated at the effective magnetic field, B^*. For the special case of the ground state at $p = 1$ the above CF wavefunction recovers the Laughlin wavefunction which is already known to be a very accurate representation of the exact ground state at $\nu = 1/(2m+1)$.

In the $p \to \infty$ limit, the effective magnetic field, B^*, vanishes. As a result, fermions "see" no net magnetic field, and they form a Fermi sea which does have a uniform density. For such a limit, we recover the following trial wave function for the CF Fermi liquid ground state at $\nu = 1/(2m)$ as written down by Rezayi and Read [50]:

$$\Psi_{Fermi} = \hat{P}_{LLL}\left[det|e^{i\vec{k}_\alpha \vec{r}_i}|\;\Psi_{Bose}\right]\,, \tag{72}$$

where Ψ_{Bose} is the Bose Laughlin wave function [51] for filling factor $\nu = 1/(2m)$:

$$\Psi_{Bose} = \prod_{j<k}^{N} (z_j - z_k)^{2m} \exp\left(-\sum_{j=1}^{N} \frac{|z_j|^2}{4\,l_0(B)^2}\right). \tag{73}$$

The Rezayi-Read (RR) CF Fermi wave function is obtained after the product of the Bose Laughlin wave function with a Slater determinant of plane waves is fully projected into the LLL by means of the projection operator, $\hat{P}_{LLL}$. In the above expressions, N is the number of electrons that occupy the N lowest-lying single-particle plane wave states labeled by the momenta $\{\vec{k}_\alpha\}$ consistent with an ideal 2D spin-polarized Fermi gas and $z_j = x_j + iy_j$ is the position coordinate for the j-th electron in complex notation.

4. Conclusion

Strongly correlated 2DES have always been a prefered target of modern condensed matter physics. The FQHE stands out as a remarkable macroscopic quantum phenomena occuring in 2DES subject to high perpendicular magnetic fields at low temperatures. Low dimensionality combined with a strong perpendicular magnetic field leads to the creation of massively degenerate discrete quantum states known as Landau levels. The energy quantization associated with such states leads to well pronounced quantum behavior and is the foundation of many new experimental and theoretical advances in condensed matter physics. When the magnetic field is very strong, electrons partially fill only the LLL. FQHE originates from electrons forming a strongly correlated incompressible liquid state at special uniform densities that correspond to specific values of the filling factor. In this sense, FQHE represents a unique example of a novel collective quantum liquid state of matter that originates from strong electronic correlations only. The principal liquid FQHE states occur at filling factors $1/3$ and $1/5$. Such states are well described by Laughlin's theory and his trial wave function. Given the importance of such states, in this work we briefly describe the physics of Laughlin states in the LLL. Other odd-denominator FQHE states at filling factors, $\nu = p/(2\,m\,p + 1)$ $(p, m - integer)$ are readily understood in terms of the composite fermion theory. The $p \to \infty$ limit of the above states corresponds to even-denominator filled states, which are belied to be compressible gapless Fermi liquid states qualitatively different from the gapful FQHE liquid states of the originating sequence.

Differently from LLL odd-denominator-filled states, states with filling factor of the form $1/2$, $1/4$ and $1/6$ do not show typical FQHE features and behave as isotropic compressible metallic Fermi liquid states. It turns out that the underlying principle to understand such an unexpected behavior is the Chern-Simons theory which explains how electrons in a magnetic field transform to a mathematically equivalent system of fermions interacting with a Chern-Simons gauge field. For even-denominator filled states in the LLL such transformation happens in a way that the average effective magnetic field acting on the fermions is zero. This is turn explain why these transformed fermions effectively behave as free fermions and can be suitably described within the framework of a Fermi liquid theory.

By its very nature, the field of strongly correlated electronic systems and FQHE involves a large combination of theories, methods and approaches. In this work we try to give only a basic introductory theoretical review of the FQHE states in the LLL. Thus, for brevity of treatment and space limitation, we were forced to leave out of consideration several other aspects of the physics of this very interesting phenomenon. Our objective is not to give a comprehensive review of the vast experimental and theoretical literature published over the last decades. The main emphasis is to clarify only some key physical properties of the FQHE liquid states in the LLL. While not the main scope of this article, we also try to give a brief overview of computational methods applied to the field. To this effect, we briefly explain the implementation of the quantum Monte Carlo simulation method to study systems of electrons in a disk geometry. A lot of work has been done over the last decades and obviously many phenomena shown here have been studied by other authors. While we have tried to give credit to earlier research, certainly and inadvertently we might have overlooked other valuable published work on the topic. Therefore, sincere apologies are extended to those who might feel their work is not suitably represented.

Acknowledgements

This research was supported in part by NSF Grant No. DMR-0804568.

References

[1] *The Quantum Hall effect*, edited by R.E. Prange and S.M. Girvin (Springer Verlag, New York, 1990) ; *The Fractional Quantum Hall Effect*, T. Chakraborty, P. Pietiläinen, (Springer Verlag, New York 1988).

[2] *Composite Fermions*, edited by O. Heinonen, World Scientific, New York (1998).

[3] *Quantum Theory of the Electron Liquid*, G. F. Giuliani and G. Vignale, Cambridge University Press, Cambridge (2005).

[4] *Composite Fermions*, J. K. Jain, Cambridge University Press, New York (2007).

[5] D. C. Tsui, H. L. Stormer, and A. C. Gossard, *Phys. Rev. Lett.* **48**, 1559 (1982).

[6] R. B. Laughlin, *Phys. Rev. Lett.* **50**, 1395 (1983).

[7] O. Ciftja and S. Fantoni, *Europhys. Lett.* **36**, 663 (1996).

[8] D. Levesque, J. J. Weis, and A. H. MacDonald, *Phys. Rev. B.* **30**, R1056 (1984).

[9] S. M. Girvin and T. Jach, *Phys. Rev. B.* **29**, 5617 (1984).

[10] S. M. Girvin, A. H. MacDonald, and P. M. Platzman, *Phys. Rev. B* **33**, 2481 (1986).

[11] G. Fano and F. Ortolani, *Phys. Rev. B.* **37**, 8179 (1988).

[12] R. Morf and N. d'Ambrumenil, *Phys. Rev. Lett.* **74**, 5116 (1995).

[13] J. K. Jain, *Phys. Rev. Lett.* **63**, 199 (1989).

[14] R. K. Kamilla, J. K. Jain, and S. M. Girvin, *Phys. Rev. B.* **56**, 12411 (1997).

[15] O. Ciftja, *Eur. Phys. J. B.* **13**, 671 (2000).

[16] B. I. Halperin, P. A. Lee, and N. Read, *Phys. Rev. B* **47**, 7312 (1993).

[17] O. Ciftja, *Phys. Rev. B.* **59**, 10194 (1999).

[18] R. B. Laughlin, *Phys. Rev. B* **27**, 3383 (1983).

[19] P. K. Lam and S.M. Girvin, *Phys. Rev. B.* **30**, R473 (1984).

[20] E. Mendez, M. Heiblum, L.L. Chang, and L. Esaki, *Phys. Rev. B.* **28**, 4886 (1983).

[21] O. Ciftja and C. Wexler, *Phys. Rev. B.* **67**, 075304 (2003).

[22] R. Morf and B. I. Halperin, *Phys. Rev. B.* **33**, 2221 (1986).

[23] R. Morf and B. I. Halperin, *Z. Phys. B.* **68**, 391 (1987).

[24] C. Wexler and O. Ciftja, *J. Phys.: Condensed Matter* **14**, 3705 (2002); O. Ciftja and C. Wexler, *Phys. Rev. B* **65**, 205307 (2002); *Phys. Rev. B* **65**, 045306 (2002).

[25] O. Ciftja, *J. Computer-Aided Mater. Des.* **14**, 37 (2007).

[26] N. Metropolis, A. W. Rosenbluth, M. N. Rosenbluth, A. M. Teller, and E. Teller, *J. Chem. Phys.* **21**, 1087 (1953).

[27] R. L. Willett, M. A. Paalanen, R. R. Ruel, K. W. West, L. N. Pfeiffer, and D. J. Bishop, *Phys. Rev. Lett.* **65**, 112 (1990).

[28] R. L. Willett, R. R. Ruel, M. A. Paalanen, K. W. West, and L. N. Pfeiffer, *Phys. Rev. B.* **47**, 7344 (1993).

[29] R. L. Willett, R. R. Ruel, K. W. West, and L. N. Pfeiffer, *Phys. Rev. Lett.* **71**, 3846 (1993).

[30] K. Esfarjani and S. T. Chui, *Phys. Rev. B.* **42**, 10758 (1990).

[31] X. Zhu and S. G. Louie, *Phys. Rev. Lett.* **70**, 335 (1993).

[32] X. Zhu and S. G. Louie, *Phys. Rev. B.* **52**, 5863 (1995).

[33] H. W. Jiang, R. L. Willett, H. L. Stormer, D. C. Tsui, L. N. Pfeiffer, and K. W. West, *Phys. Rev. Lett.* **65**, 633 (1990).

[34] K. Yang, F. D. M. Haldane, and E. H. Rezayi, *Phys. Rev. B.* **64**, 081301(R) (2001).

[35] G. Dev and J. K. Jain, *Phys. Rev. B.* **45**, 1223 (1992).

[36] J. K. Jain and R. K. Kamilla, *Int. J. Mod. Phys. B* **11**, 2621 (1997).

[37] R. K. Kamilla and J. K. Jain, *Phys. Rev. B* **55**, 9824 (1997).

[38] R. Morf and N. d'Ambrumenil, *Phys. Rev. Lett.* **74**, 5116 (1995).

[39] K. Park, V. Melik-Alaverdian, N. E. Bonesteel, and J. K. Jain, *Phys. Rev. B.* **58**, R10167 (1998).

[40] O. Ciftja and S. Fantoni, *Phys. Rev. B* **56**, (20) 13290 (1997).

[41] O. Ciftja, S. Fantoni, J. W. Kim, and M. L. Ristig, *J. Low. Temp. Phys.* **108**, 357 (1997).

[42] F. D. M. Haldane, *Phys. Rev. Lett.* **55**, 2095 (1985).

[43] T. Chakraborty, *Phys. Rev. B.* **57**, 8812 (1998).

[44] O. Ciftja and S. Fantoni, *Phys. Rev. B.* **58**, 7898 (1998).

[45] R. K. Kamilla, J. K. Jain, and S. M. Girvin, *Phys. Rev. B* **56**, 12411 (1997).

[46] A. Lopez and E. Fradkin, *Phys. Rev. B* **44**, 5246 (1991).

[47] O. Ciftja and C. Wexler, *Eur. Phys. J. B* **23**, 437 (2001).

[48] O. Ciftja and C. Wexler, *Solid State Comm.* **122**, 401 (2002).

[49] O. Ciftja, *Physica E.* **9**, 226 (2001).

[50] E. Rezayi and N. Read, *Phys. Rev. Lett.* **72**, 900 (1994).

[51] O. Ciftja, *Europhys. Lett.* **74**, 486 (2006).

In: Advances in Condensed Matter ... Volume 10
Editors: H. Geelvinck and S. Reynst

ISBN: 978-1-61209-533-2

Chapter 9

TOPOLOGICAL INSULATORS-TRANSPORT IN CURVED SPACE

D. Schmeltzer
Physics Department, City College of the City University of New York
New York, NY, USA

Abstract

We introduce new methods for investigating propagation of electrons in a multi band system with spin orbit interaction which are time reversal invariant . Therefore Kramer's theorem imposes constraints which give rise to non-trivial Berry connections. As a result chiral zero modes appear at the interface between two domains characterized by parameters which are above or below some critical values. The variation of the parameters is due to disorder, geometry or topological disorder such as dislocations and disclinations . The mechanism might explain the high conductivity coming from the bulk of the topological insulators. We introduced the method of curved geometry and study the effect of dislocations on the Topological Insulators. As a demonstration of the emergent Majorana Fermions we considere a P-wave wire coupled to two metallic rings

I. Introduction: The Berry Connection and Curvature for Particles in Solids

One of the important concepts in Condensed Matter Physics is the idea of Topological order [1, 2, 3, 4, 5, 6, 7, 8, 9, 10, 12, 32]. This concept has been confirmed experimentally by the $3D$ class of Topological Insulators (TI) Bi_2Se_3, Bi_2Te_3 and $Bi_{1-x}Sb_x$ (characterized by a single Dirac cone which lies in a gap)[1, 15, 19, 18]. At the boundary of the $3D$ Topological insulator one obtains a $2D$ surface which an odd number of chiral excitations. The two-dimensional $CdTe/HgTe/CdTe$ quantum wells behave as a $2D$ TI when the exceeds 6.3 nm pushing the s type Γ_6 band to be below the p type Γ_8 [30]. The boundary of the $2D$ surface is described by a $1D$ gap less chiral mode. (It is the single chirality which prohibits back-scattering and gives rise to the quantized spin-Hall effect.) Recently the QSH (Quantum Spin Hall effect) [30] in $CdTe/HgTe/CdTe$ has been discovered [18],[31] and at the edges a spin Hall state dubbed $helical\ liquid$ has been proposed. This spin model is time reversal invariant and therefore obeys Kramer's theorem $T^2 = -1$ (T is the time reversal symmetry $T = -i\sigma^2 K$, K is the conjugation operation and σ^2 is a Pauli spin operator). The helical liquid is given by a one dimensional massless chiral fermion: $right$ fermions with spin up and $left$ fermions with spin $down$. The spin Hall Hamiltonian is a physical realization of the lattice model proposed by [2].

In a variety of Physical problems these properties are described by the topological Chern number, $Ch_n = \int_{T^{D=2n}} Tr[\frac{1}{n!}(\frac{i\mathsf{F}}{2\pi})^n]$.

where $T^{D=2n}$ represents the D dimensional Torus (the First Brillouin Zone (FBZ))in D dimensions and F represents the curvature given in terms of the ground state eigen-spinors .

In a physical language this means that ground state is protected by a gap which is stable to weak perturbations. At the boundary of such a system one finds an odd number of chiral massless fermions. This is revealed by the presence of a single Dirac cone. In the absence of external magnetic fields a well known theorem **the no go theorem** prohibits the existence of a single Dirac cone. The resolution of this paradox is due to Wilson [5] which showed that by adding terms of the form $\sum_{i=1}^{D}[cos(K_i a) - 1]$ to the Dirac Hamiltonian one can choose the parameter of the model such that one has an odd number of chiral fermions on the boundary. This situation has been realized in the class of materials [15, 19] Bi_2Se_3, Bi_2Te_3 and $Bi_{1-x}Sb_x$. Theoretically a number of methods have been proposed for identifying the Chern numbers [3, 4, 5] and the $Chern - Simons$ coefficients which describe the electromagnetic response of the system. The relation between the space dimension D and the index n is $D = 2n$. The first Chern number $n = 1$ exists for $D = 2$ and is non zero for system with broken time reversal symmetry (Quantum Hall effect) [27, 25]. The second Chern number $Ch_{n=2}$ is non zero for $D = 4$ and is not restricted to the case where time reversal symmetry is broken. The time reversal invariants system (TRI) in the presence of the spin orbit interaction can give rise to topological insulators TI. Recently [7] has proposed a method for computing the topological invariants for non interacting electrons in periodic crystals which are time reversal invariant $(TRI)$and obey the inversion symmetry . For system with inversion symmetry the TRI are given in terms of the parity of each pair of the Kramers degenerate occupied bands (at the four (D=2) and eight (D=3) time -reversal and parity invariant points in the Brillouin zone). A general theory based on *particle-hole* symmetry;$time$-$reversal$ symmetry and their product allows to identify 10 symmetry class [14]. However we believe that the calculation of the invariants for systems which do not have parity invariance can be done, directly by computing the zero modes on the boundary. Using the index theorem we can relate the number of zero modes with the Chern numbers.

Recent experiments point to interesting Quantum oscillations in topological insulators [33] evidence of Landau Quantization in TI [34] and Aharonov-Bohm oscillations for cylinders [35]. Some of the experiments demand the use of geometry in curved space [36, 37, 61]. It seems that topological effects such as dislocations and disclinations might play an important role [16, 39].

Zero modes can appear also at the interface between Superconductors and normal metals. For example the Chiral superconductor $(p_x + ip_y)$ in the weak pairing limit is given by a similar hamiltonian as considered for the spin Hall effect. The physical difference is that for the superconductor the spinor is built from *particles* and *holes* and **the negative energy of particles describes the same excitations as positive ones.** As a result, the creation and annihilation operators are given by the **Majorana** fermions [40] built from a linear superposition of a fermion with the **charge conjugated** fermion. The edge states for this case will be given by the one dimensional Ising model which has zero modes with a degenerate Z_2 ground state [41] (Majorana zero mode is trapped in each vortex core which leads to a ground state degeneracy of 2^{n-1} in the presence of $2n$ vortices).

II. The Method for Topological Insulators

In this section we will introduce a set of methods which we propose to use. We will demonstrate our methodology using a simplified version of the spin orbit model studied by [1, 2, 5, 7, 6, 8, 12, 10, 15, 13, 17]. We consider a four dimensional lattice in which we introduce a $defect$ in one of the direction. We propose that for certain parameters a zero mode appears at the interface between domains which have different parameters. When we take into consideration the time reversal points in the Brillouin zone we can introduce an alternative way for determinig the criteria of topological insulator (TI). In section $B2$ we characterize the TI in terms of the Chern numbers. We propose that the different Chern numbers are determined by the symmetry group which connect different different region in the Brillouin zone. In section $B3$ we propose that the Hubbard Stratonovici field which results from the electron-electron interaction can be treated as fourth dimensions giving rise to a TI in three dimensions.

Using the momentum space representation in periodic solids [45, 46, 47, 23] we have constructed a gauge theory in the momentum space which gives rise to $non\text{-}commuting$ [21, 20, 22, 9] coordinates .

For periodic potentials the eigenvalues are characterized by the Bloch eigenfunctions $e^{i\vec{K}\cdot\vec{r}}u_{n,\vec{K}}(\vec{q})$ and eigenvalues $E_n(\vec{K})$ where n is the band index. In the momentum representation the coordinate r^i is represented by $i\partial_{K_i}$. When more than one atom per unit cell are involved we can have band crossing with linear dispersion (in the vicinity of the crossing points). Due to the band crossing, we can not use a single $matrix\ connection$ $< \vec{K}|d|\vec{K} >\equiv< \vec{K}|\partial_{K_i}|\vec{K} > dK^i$ [27] ($|\vec{K} >$ are the eigenvectors) for the entire first Brillouin zone (FBZ) (which is equivalent to the D dimensional torus T^D). Due to the multiple connection [24, 28, 9, 27, 20, 22] one has to introduce $transition$ functions between the different regions in the FBZ. When the number of bands crossing is odd we have an odd number of Dirac cones. For time reversal invariant (TRI) systems (with the the time reversal (TR) operator T) we can apply Kramer's theorem ,$T^2 = -1$ at the special time reversal points $\vec{K} = 0$, $\vec{K}^* = G - \vec{K}^*$ (where G is the reciprocal lattice vector). We find a number of materials which obeys Kramer's theorem with an odd number of Dirac points Bi_2X_3,$(X = Se, Te)$, $Bi_{1-x}Sb_x$ coined Topological Insulators TI. These materials are characterized by strong spin-orbit scattering and have orthorombic crystal structure with the space group D^5_{3d} and have a layer structure with five atomic layers per unit cell. Using a projected atomic $Wannier$ function we can write a projected Hamiltonian in the basis of the six p orbitals $|p_x,\uparrow>$, $|p_y,\uparrow>$, $|p_z,\uparrow>$ and $|p_x,\downarrow>$, $|p_y,\downarrow>$ $|p_z,\downarrow>$ [17]. Using these orbitals we obtain a six band Bloch Hamiltonian which are further projected to only four bands.

II.A. The Effective Model and the Emergent Chiral Modes at the Interface between Two Domains

In order to demonstrate the physics involved we will consider the toy model of two orbital per unit cell :

$$H = i\lambda_{So}\sum_j[(a_j^\dagger a_{j+1}+a_{j+1}^\dagger a_j)-(b_j^\dagger b_{j+1}+b_{j+1}^\dagger b_j)]+M\sum_j(a_j^\dagger b_j+b_j^\dagger a_j)-t\sum_j[(a_j^\dagger b_{j+1}+a_{j+1}^\dagger b_j) \tag{1}$$

The eigenvalues are given by: $\epsilon(\vec{K}) = \pm\sqrt{(2\lambda_{So}sinK)^2+(M-2tcosK)^2}$.

This model has the interesting property that for $M-2t=0$ it develops a $Dirac$ cone at $K=0$ (at $M+2t=0$ the Dirac cone is at $K=\pm\pi$). For $M-2t<0$ have at $K=0$ a gap given by $2|M-2t|$.

Next we extend the one dimensional model to model which has a boundary at $x=0$. We assume that the Hamiltonian parameters obey for $x<0$, $(M-2t)|_{x<0}<0$ and for $x>0$ the parameter are $(M-2t)|_{x>0}>0$. As a result of the fact that the Hamiltonian for $x<0$ is different from the Hamiltonian for $x>0$ gives rise to a mid gap state with zero energy which is to a zero mode.

In order to demonstrate that numbers of zero mode is odd and are chiralic we will we will include a spin orbit interaction . We introduce the basis $|\uparrow>\otimes|1>$, $|\downarrow>\otimes|1>$ $|\uparrow>\otimes|2>|\uparrow>\otimes|2>$. We will use the Pauli matrix τ to describe the band and σ to describe the $spin$. In this basis we can describe the spin orbit interaction by the term:

$$i\lambda_{SO}\sum_{i=1}^{D}\Psi^\dagger(\vec{r})(\sigma^i\otimes\tau_1)\Psi(\vec{r}+a_i)+h.c. \tag{2}$$

The energy difference between the two orbitals is represented by

$$M\Psi^\dagger(\vec{r})(I\otimes\tau_1)\Psi(\vec{r}) \tag{3}$$

The hopping energy between the different orbitals is given by

$$-t\sum_{i=1,2,..D}\Psi^\dagger(\vec{r})(I\otimes\tau_1)\Psi(\vec{r}+a_i)+h.c. \tag{4}$$

This Hamiltonian captures the same physics as used by [7, 12, 13]. The Hamiltonian can be rewritten using the $Dirac\ matrices$ with γ^0 given in the chiral representation . We have: $\gamma^1=\gamma^0(\sigma^1\otimes\tau_3)$,$\gamma^2=\gamma^0(\sigma^2\otimes\tau_3)$, $\gamma^3=\gamma^0(\sigma^3\otimes\tau_3)$ and for the fourth direction we use γ^5. We introduce the notations $\Psi^\dagger(\vec{r})\gamma^0\equiv\overline{\Psi}(\vec{r})$, $H=\int d^Dr\hat{H}(\vec{r})=\sum_{\vec{K}}\hat{H}(\vec{K})$:

$\hat{H}=\sum_{j=1}^{D-1}i\lambda_{SO}(\overline{\Psi}(\vec{r})\gamma^j\Psi(\vec{r}+a_j)-\overline{\Psi}(\vec{r}+a_j)\gamma^j\Psi(\vec{r}))+\lambda_{SO}(\overline{\Psi}(\vec{r})\gamma^5\Psi(\vec{r}+a_4)-\overline{\Psi}(\vec{r}+a_4)\gamma^5\Psi(\vec{r}))+$

$M\overline{\Psi}(\vec{r})\Psi(\vec{r})-t\sum_{j=1}^{D}(\overline{\Psi}(\vec{r}+a_j)\Psi(\vec{r})+\overline{\Psi}(\vec{r})\Psi(\vec{r}+a_j))$

$$\hat{H}(\vec{K}) = \overline{\Psi}(\vec{K})[\lambda_{SO}\sum_{j=1}^{D}\gamma^{j}sin(K_{j}a) + (M - 2t\sum_{j=1}^{D}cos(K_{j}a))]\Psi(\vec{K}) \quad (5)$$

The eigenvalue of this Hamiltonian are given by:
$\epsilon(\vec{K}) \equiv \pm\sqrt{\sum_{j=1}^{D}(2\lambda_{SO}Sin(K_{j}a))^{2} + [M - 2t\sum_{j=1}^{D}cos(K_{j}a)]^{2}}$ where D is the spatial dimension. For $\vec{K} \approx 0$ the system has a gap given by $M - 2tD$ which vanishes at the *critical* value $\frac{M}{2tD} = 1$ and for $\vec{K} \approx [\pi,\pi,\pi,\pi]$ the gap occurs at $M + 2tD$. In order to investigate the effect of the interface we will assume that in the D direction (which we take discrete to be j) we have a defect region with the parameters $M(j)$,$t(j)$ and $\lambda_{SO}(j)$. For the rest of $D-1$ dimensions we have a perfect lattice which obeys periodic boundary conditions. We note that the eigenfunction for such a problem takes the form $\Omega_{\vec{\hat{K}}}(\vec{\hat{r}},j) = e^{i\vec{\hat{K}}\cdot\vec{\hat{r}}}U_{s}(j)$ where $U_{s}(j)$ is a four component spinor. $\vec{\hat{K}}$ and $\vec{\hat{r}}$ are the momentum and coordinates in $D-1 = 3$ dimensions. The spinor $U_{s}(j)$ obeys the *zero* eigenvalue equation:

$$[\sum_{i=1}^{3}\gamma^{i}sin(K_{i}a) + M(j) - 2t(j)\sum_{i=1}^{3}cos(K_{i}a)]U_{s}(j) + [t(j)\gamma^{5} + \lambda_{SO}(j)]U_{s}(j+1) -$$
$$[t(j-1)\gamma^{5} - \lambda_{SO}(j-1)]U_{s}(j-1) = 0 \quad (6)$$

We consider a situation where $M(j) = 2t(j)\sum_{i=1}^{3}cos(K_{i}a) + \epsilon(j)$ where $\epsilon(j) = \delta_{j,-1}\epsilon_{0} - \delta_{j,1}\epsilon_{0}$. To simplify the problem we take $t(j)$ and $\lambda_{SO}(j)$ to be constant. The spinor $U_{s}(j)$ obeys $\gamma^{5}U_{\pm}(j) = \pm U_{\pm}(j)$. Since the zero mode must be normalized we expect that only one of the spinors $U_{\pm}(j)$ will satisfy the zero mode equation . Therefore we have a chiral massless fermion at the interface. This condition can be used for any time reversal points in the Brillouin zone. For $D = 3$ the 8 time reversal points are given $\Gamma_{l=n_{1},n_{2},n_{3}} = \pi(n_{1},n_{2},n_{3})$ where $n_{i} = 0,1$,$i = 1,2,3$ and the gaps occur at $\vec{K} = \pi[n_{1},n_{2},n_{3}]$. Using the the critical condition $M = 2t\sum_{i=1}^{3}cos(K_{i}a)$ we find that for a fixed value of $\frac{M}{2t}$ the chiral modes will occur at $\vec{K} = \pi[n_{1},n_{2},n_{3}]$ only if $\frac{M}{2t}$ obeys $\frac{M}{2t} < [cos(\pi\cdot n_{1}) + cos(\pi\cdot n_{2})cos(\pi\cdot n_{3})]$. Following [7] we identify the value $\delta_{i(n_{1},n_{2},n_{3})} = -1$ with the points where $\frac{M}{2t} < [cos(\pi\cdot n_{1}) + cos(\pi\cdot n_{2}) + cos(\pi\cdot n_{3})]$ and $\delta_{i(n_{1},n_{2},n_{3})} = 1$ with the points where $\frac{M}{2t} > [cos(\pi\cdot n_{1}) + cos(\pi\cdot n_{2}) + cos(\pi\cdot n_{3})]$. This allows us to make contact with the condition for a strong topological insulator :$(-1)^{\upsilon_{0}} = \prod_{i=1}^{8}\delta_{i(n_{1},n_{2},n_{3})}$, when the product is -1 we have a strong TI. Our condition is applicable to any interface between two domains, all we need to consider is the equation $M(j) - 2t(j)[cos(\pi\cdot n_{1}) + cos(\pi\cdot n_{2}) + cos(\pi\cdot n_{3})]$.

The application of these results to $D = 3$ will allow to treat the $D-1 = 2$ boundary as a two dimensional surface which gives rise to a Helical metal [32]. (The particles ($\epsilon(\vec{K}) > 0$) have an opposite chirality in comparison with the antiparticles (the holes). For a TI in $D = 2$ dimensions the boundary corresponds to one dimensional chiral fermions where spin up fermions propagate in one direction and spin down electron propagate in the opposite direction [31].

II.B. The Formulation of the TI in Terms of the Chern Numbers

In order to compute the response of a TI to an external Electromagnetic EM field we need to know the non-zero Chern numbers which are the coeficients of the the Chern-Simons action obtained after integrating the Fermions. For the remaining part we will consider the TI situation with the inverted band condition $M - 2tD < 0$ at $\vec{K} \approx 0$. Using the eigen-spinor we will obtain the transformed coordinate representation, since this can not be done with a single transformation (due to singular points on the torus) we find that the coordinates do not commute and therefore the system has a non zero curvature.

The spectrum of the TI has electron like spinors $u(\vec{K}, s)$ s=1,2 (positive energy) with eigenvalue $\epsilon(\vec{K}) > 0$ and the antiparticles (hole like) spinors $v(\vec{K}, s)$ $s = 1, 2$ for $\epsilon(\vec{K}) < 0$. This allows to introduce the $electron$ like operator $c(\vec{K}, s)$ and the $anti - particle$ (hole) operator $b(\vec{K}, s)$ which annihilate the fully occupied ground state $|G_0>$, $c(\vec{K}, s)|G_0 >= 0$,$b(\vec{K}, s)|G_0 >= 0$. The four component spinor operator $\Psi(\vec{r}) = [\Psi(\vec{r})_{1,\uparrow}, \Psi(\vec{r})_{1,\downarrow}, \Psi(\vec{r})_{1,\downarrow}, \Psi(\vec{r})_{1,\uparrow}]^T$ has the momentum expansion:

$$\Psi(\vec{r}) = \int_{T^D} \sum_{s=1,2} [c(\vec{K}, s)u(\vec{K}, s)e^{i\vec{K}\cdot\vec{r}} + b^\dagger(-\vec{K}, s)v(-\vec{K}, s)e^{i\vec{K}\cdot\vec{r}}] \tag{7}$$

The ground state eigenspinor is given by $|v(\vec{K}, s)>$. Following [9, 21] we find that the transformation $|\vec{K}, \sigma =\uparrow, \downarrow>\rightarrow |v(\vec{K}, s)>$ replaces the coordinate representation $r^i = i\partial_{K_i} \rightarrow R^i = i\partial_{K_i} + A^-(\vec{K})_i$ where $A^-(\vec{K})$ is the connection matrix [9]:

$$A^-(\vec{K}) = \sum_{s=1,2} \sum_{s'=1,2} < v(\vec{K}, s)|dv(\vec{K}, s') > |v(\vec{K}, s) >< v(\vec{K}, s')| =$$

$$\sum_{i=1}^{D} \sum_{s=1,2} \sum_{s'=1,2} < v(\vec{K}, s)|\partial_{K^i} v(\vec{K}, s') > |v(\vec{K}, s) >< v(\vec{K}, s')|dK^i =$$

$$\sum_{i=1}^{D} [A_i^{-,(0)}(\vec{K})I + \sum_{\alpha=1,2,3} \frac{1}{2}\sigma^\alpha A_i^{-,(\alpha)}(\vec{K})]dK^i \tag{8}$$

The connection matrix $A^-(\vec{K})_i$ forms an $SU(2)$ group. The transformed coordinates obey new commutation rules given by the curvature $F_{i,j}$:

$$[R^i(\vec{K}), R^j(\vec{K})] = (-i)F_{i,j} \tag{9}$$

where $F_{i,j} = \sum_{\alpha=1,2,3} \sigma_\alpha F^\alpha_{i,j}$ and

$$F^\alpha_{i,j} = \partial_{K^i} A_j^{-(\alpha)}(\vec{K}) - \partial_{K^j} A_i^{-(\alpha)}(\vec{K}) + \epsilon_{\alpha,\beta,\epsilon} A_i^{-(\alpha)}(\vec{K}) A_j^{-(\beta)}(\vec{K}) \tag{10}$$

Following [26] we define the $n'th$ Chern number using the notation: $Trace[\mathsf{F}] \equiv Tr[\mathsf{F}]$ for the curvature $F_{i,j}(\vec{K})$.

$$Ch_{n=\frac{D}{2}} = \int_{T^{D=2n}} Tr[\frac{1}{n!}(\frac{i\mathsf{F}}{2\pi})^n] \tag{11}$$

Using the Heisenberg equation of motion for the D dimensional TI we find that the Dirac matrices γ are proportional to the velocity operator $\frac{dR^i(\vec{K})}{dt} \equiv \gamma^i$ [52]. In the presence of weak external fields $a_\mu(\vec{r},t)$ we find with the help of the non-commuting coordinates and the relation $\frac{dR^i(\vec{K})}{dt} \equiv \gamma^i$ that the response to the external fields (in agreement with [12] is given by:

$$\delta S_{eff.} = Ch_{n=\frac{D}{2}} \Gamma^D_{C.S.} \tag{12}$$

where

$$\Gamma^D_{C.S.} = \int d^D r \int dt \epsilon^{\mu_1,\mu_2 \ldots \mu_{D+1}} a_{\mu_1} \partial_{\mu_2} a_{\mu_3} \ldots.. a_{\mu_{D+1}} \tag{13}$$

a)The second Ch_2 number for TI

We propose to use the theory of first class constraints [43, 44, 29] to derive the second Chern number for the TI. For $D = 4$ the only non zero Chern number is Ch_2 [26]. The index theorem [26] relates the nummber of chiral fermions to the Chern number :

$$N_+ - N_- = Ch_{n=\frac{D}{2}} \tag{14}$$

which allows us to make the connection between the number of zero modes of opposite chirality to the $Ch_{n=\frac{D}{2}}$. When $M - 2tD < 0$ at $\vec{K} = 0$ one finds on the boundary only a single zero mode. According to the index theorem this will imply that the Chern number must be one. The time reversal points give rise to two connections matrices $A^-(\vec{K})$ and $A^-(-\vec{K})$, which are defined on the upper and lower side of the torus T^D respectively. Each half torus can be mapped to a sphere S^D. Using homotopy we have the map $S^4 \to SU(2)$. According to [26] we obtain $\Pi_4[SU(2)] = Z_2$. We conclude that the Chern number $Ch_{n=2} = Z_2$. To compute the Chern number we will use $A^-(\vec{K})$ for the North sphere (upper torus) and $A^-(-\vec{K})$ for the South sphere (lower part of the torus). Following [26] we have the identity: $Tr[(\mathsf{F})^2] = d(Tr[AdA + \frac{2}{3}A^3])$ $(\mathsf{F})^2$ is *closed* but not *exact* (we can not satisfy the identity with one connection A on the entire torus). Therefore the Ch_2 number will be given by a difference of *two* three form Chern-Simons terms:

$$Tr[(\mathsf{F})^2] = dTr[AdA + \frac{2}{3}A^3] \tag{15}$$

Using the connections $A^-(\vec{K})$ and $A^-(-\vec{K})$ we find:

$$\begin{aligned} Ch_2 &= \int_{T^{D=4}} Tr[\frac{1}{2!}(\frac{i\mathsf{F}}{2\pi})^2] = \int_{S^4-North} Tr[\frac{1}{2!}(\frac{i\mathsf{F}_{\mathsf{North}}}{2\pi})^2] + \int_{S^4-South} Tr[\frac{1}{2!}(\frac{i\mathsf{F}_{\mathsf{South}}}{2\pi})^2] \\ &= \frac{1}{8\pi^2}\int_{S^3}(Tr[(A^{-,North}(\vec{K}))^3] - Tr[(A^{-,South}(\vec{K}))^3]) = \\ &\frac{1}{8\pi^2}\int_{S^3}(Tr[(A^-(\vec{K}))^3] - Tr[(A^-(-\vec{K}))^3] = \frac{1}{24\pi^2}\int_{S^3}[(g^{-1}dg)^3] = Z_2 \end{aligned} \tag{16}$$

The difference between the two connections $A^-(\vec{K}) - A^-(-\vec{K}) = g^{-1}dg$ defines the transition function. At the TRI points [7] the time reversal operator $T = -i\sigma_2 \otimes IK$ defines the matrix elements $< -\vec{K}, s|T|\vec{K}, s' >= g_{s,s'}(\vec{K}) = -g_{s',s}(-\vec{K})$. This relation imposes a constraint on the connection field. The solution of this constraint follows from the $BRST$ Cohomology [29]. The solution of the constraint will be that $g^{-1}dg$ is an element of the Z_2 group therefore $\frac{1}{24\pi^2}\int_{S^3}[(g^{-1}dg)^3]$ takes two values 0 or 1.

b)The first Chern numbers for D=2

Following the theory of the Chern numbers for $D = 2$ the first Chern number is non-zero if time reversal symmetry is violated. Therefore the proposal [30] that the Spin Hall conductance is quantized for the two dimensional heterostructure $CdTe/HgTe/CdTe$ needs clarifications.

b1)- Ch_1 in the absence of time reversal symmetry for D=2

We consider the Hamiltonian:

$$h(\vec{K}) = K^1\sigma_2 - K^2\sigma_2 + M(\vec{K})\sigma_3 \tag{17}$$

When $M(-\vec{K}) = M(\vec{K})$ the time reversal symmetry is broken (the TR operator is given by $T = -i(I \otimes \sigma^2)\mathbf{K}$, where $\mathbf{K}$ is the conjugation operator).

For this case we have a single eigenstate for the ground state $|v(\vec{K}) >$. Using this eigenstate we compute the $connection\ one\ form$ [26] $A^-(\vec{K}) =< v(\vec{K})|dv(\vec{K}) >$. Due to the singularity at $\vec{K} = 0$ [26] we map the upper torus T^2 to the $North$ sphere S^2 defining the north connection $A^{-(North)}(\vec{K})$ and the lower part of the torus T^2 is mapped to the south sphere with the South connection $A^{-(South)}(\vec{K})$. The North eigenvector is related to the South eigenvector through a gauge transformation. When we rotate the eigenstate $|v(\vec{K}) >$ by 2π the state is modified by $e^{i\pi}$ (note the factor $\frac{1}{2}$ which is due to the spin half system). Therefore we have the relation between the north and south state :

$$|v(\vec{K}) >_{North}= e^{i\frac{\varphi}{2}}|v(\vec{K}) >_{South} \tag{18}$$

which gives:

$$\begin{aligned} Ch_1 = \int_{T^{D=2}} Tr[\frac{1}{1!}(\frac{i\mathsf{F}}{2\pi}] = \int_{S_2} Tr[\frac{1}{1!}(\frac{i\mathsf{F}}{2\pi}] = \\ \frac{1}{2\pi}\oint_S i[A^{-(North)} - A^{-(South)}] = \frac{1}{2\pi}\oint_S \frac{1}{2}d\varphi = \frac{1}{2} \end{aligned} \tag{19}$$

This implies that the Hall conductance for the Dirac equation is $\frac{e^2}{2h}$ instead of $\frac{e^2}{h}$. In the presence of a boundary which restricts the system to $x > 0$ the term $M(\vec{K})\sigma_3$ will give rise to a $single\ spin\ Polarized$ massless fermion.

b2)-The first Chern number for time reversal system when D=2

Following [30] and the experimental discovery [18, 19] the Hamiltonian for the two dimensional heterostructure $CdTe/HgTe/CdTe$ with the thickness $d_{Q.W.}$ exciding 6.3 nm has inverted bands and is given by :

$$h(\vec{K}) = K^1\tau^2 \otimes I - K^2\tau^1 \otimes \sigma^3 + [M - 2t(cos(K_1a) - 2tcos(K_2a)]\tau^3 \otimes I \quad (20)$$

For $\vec{K} \approx 0$ the condition for TI is $\frac{M}{4t} < 1$. This model has the special feature that for spin *up* and spin down we have two Hamiltonians:

$$h^{\uparrow} = (\vec{K}) = K^1\tau^2 - K^2\tau^1 + [M - 2t(cos(K_1a) - 2tcos(K_2a)]\tau^3 \quad (21)$$

$$h^{\downarrow} = (\vec{K}) = K^1\tau^2 + K^2\tau^1 + [M - 2t(cos(K_1a) - 2tcos(K_2a)]\tau^3 \quad (22)$$

Therefore both Hamiltonians can be diagonalized separately in the band space τ. In each band the time reversal operator (does not include the spin operator) is given by $T_\tau = \mathbf{K}$ and obeys $T_\tau^2 = 1$. Since $h^{\uparrow}$ and $h^{\downarrow}$ (separately) are *not time reversal invariant*. We conclude that in the τ space each Hamiltonian has Ch_1 number denoted by $Ch_1^{\uparrow}$ and $Ch_1^{\downarrow}$. Following the discussion presented above above we have:

$$[R^i(\vec{K}), R^j(\vec{K})] = f_{i,j}^3\tau^3 \otimes \sigma^3 \quad (23)$$

Therefore the Chern number will be an integer for each spin polarization:

$$Ch_1^{\uparrow} = \int_{T^{D=2}} Tr[\frac{1}{1!}(\frac{iF}{2\pi}] = \int_{S_2} Tr[\frac{1}{1!}(\frac{iF}{2\pi}] = \frac{1}{2\pi}\oint_S i[A^{-(North)} - A^{-(South)}] = \frac{1}{2\pi}\oint_S d\phi = 1 \quad (24)$$

Contrary to the previous case $(b1)$ where the transition function between the North and the South pole was $\frac{1}{2}d\phi$, for the present case the transition function will be $d\phi$ (not half)! Repeating the calculation for $Ch_1^{\downarrow}$ we find $Ch_1^{\downarrow} = -1$. Therefore the spin Hall conductance will be given by $2 \cdot e^2 h$ characterized by the first spin Chern number $Ch_1^{spin} = \frac{1}{2}[Ch_1^{\uparrow} - Ch_1^{\downarrow}]$.

II.C. Time Reversal Invariant Topological Insulators in $3+1$ Dimensions Induced by Electron-Electron Interactions Reproduce Effectively the Theory in $4+1$ Dimensions

The external fields are restricted to three dimensions. Using the effective action $\delta S_{eff.} = Ch_2\Gamma_{C.S.}^{D=4}$ we will take the fourth dimension field $a_{\mu=4}(\vec{r}, t)$ arbitrarily. We compactify the fourth dimension $\phi(\vec{x} = x^1, x^2, x^3, t) = \oint dx^4 a_{\mu=4}(\vec{r} = x^1, x^2, x^3, x^4, t)$ [12] and obtain :

$$\delta S_{eff.}^{(3)} = Ch_{n=2} \cdot \phi(\vec{x}, t) \int d^3x \int dt \epsilon^{\mu,\nu,\sigma,\tau} \partial_\mu a_\nu(\vec{x}, t) \partial_\sigma a_\tau(\vec{x}, t) \quad (25)$$

There are only two values $\theta \equiv Ch_{n=2} \cdot \phi(\vec{x}, t)$ which respect the time reversal symmetry, $\theta = \pi$ corresponds to a TI and $\theta = 0$ correspond to a regular insulators.

We introduce a construction which is entirely in three dimensions space and is generated by the electron-electron interactions

The Hubbard-Stratonovich transformation allows the representation :

$$\sum_a \rho_a(\vec{r})\overline{\Psi}(\vec{r})\gamma^a\Psi(\vec{r}) + (\rho_a(\vec{r}))\frac{1}{2\kappa(\vec{r},\vec{r'})}(\rho_a(\vec{r'})) \tag{26}$$

When the channel $\rho_5(\vec{r})\gamma^5$ dominates the two body interaction we can use single variable $q \equiv \rho_5(\vec{r})$. We obtain a four dimensional model build from the three dimensional momentum K_1, K_2, K_3 with the new field (due to interactions) q. This allows us to use the $D = 4$ formulation with the vector $\vec{K} = [K_1, K_2, K_3, q]$. (For an arbitrary interaction we can construct a four dimensional vector in the following way:

$$\vec{K} = [K_1 + \rho_1, K_2 + \rho_2, K_3 + \rho_3, \rho_5] \tag{27}$$

Following the procedure from the previous section we compute the second Chern number:

$$Ch_2 = \frac{1}{32\pi^2}\int_{T^4} d^3K \int_{q_{min.}}^{q_{max.}} dq\epsilon^{i,j,k,l} Trace[F_{i,j}F_{k,j}] \tag{28}$$

Due to the integration measure $(\rho_a(\vec{r}))\frac{1}{2\kappa(\vec{r},\vec{r'})}(\rho_a(\vec{r'}))$ we approximated the range $q = \rho_5(\vec{r})$ to $q_{min.} < q < q_{max.}$.

In order to compute the polarization we must consider the external electromagnetic interactions h^{EM}:

$$h^{EM} = e\Phi(\vec{r},t) + \frac{e}{2}[\frac{d\vec{r}}{dt}\cdot\delta\vec{A}^{(M)}(\vec{r},t) + \delta\vec{A}^{(M)}(\vec{r},t)\cdot\frac{d\vec{r}}{dt}] \tag{29}$$

For a weak electric and magnetic field in the the z direction we introduce $E_3^{ext.}(\vec{Q}) = -\frac{\dot{A}_3^{ext.}(\vec{Q})}{iQ_3}$ and $A_2^{ext.}(\vec{Q}) = \frac{r_1 B_3^{ext.}.(\vec{Q}_2,..)}{iQ_2}$

We substitute the transformed coordinates $r^i \to R^i$ The derivative of the coordinate is related to the commutator $\frac{dR^i}{dt} = ...E_3^{ext.}[R^i, R^3]$,$i = 1, 2$. We substitute the expression for $\frac{dR^i}{dt}$ into the magnetic interaction $\frac{e}{2}[\frac{d\vec{r}}{dt}\cdot\delta\vec{A}^{(M)}(\vec{r},t) + \delta\vec{A}^{(M)}(\vec{r},t)\cdot\frac{d\vec{r}}{dt}]$ using a slowly varying magnetic field, $A_2^{ext.}(\vec{Q}) = \frac{r_1 B_3^{ext.}(\vec{Q}_2,..)}{iQ_2}$. Keeping terms to first order in Q we find that the interaction can be written as the nonlinear polarizability similar to the $axion$ $electrodynamics$ [12, 60, 11].

$$\frac{1}{32\pi^2}\int_{q_{min.}}^{q_{max.}} dq \int_{T^3} d^3K\epsilon^{i,j,k,l} Tr(< [R^i, R^j][R^k, R^l] >)E_3(\vec{Q})B_3(\vec{Q}) \equiv \vartheta\frac{e^2}{2\pi h}\int d^3x \int dt\vec{E}\cdot\vec{B} \tag{30}$$

where the expectation value $< [R^i, R^j][R^k, R^l] >$ is taken with respect to the interaction measure $\sum_a(\rho_a(\vec{r}))\frac{1}{2\kappa(\vec{r},\vec{r'})}(\rho_a(\vec{r'}))$.

III. Transport on the Boundary of a Topological Insulator

The three dimensional T.R.I. materials Bi_2Se_3,Bi_2Te_3 ,Sb_2Te3 support on their two dimensional boundary (surface) chiral helical metals [12] . The chiral excitations are described by a field which transforms as a two component spinor. $\Psi(\vec{r}) = [\Psi(\vec{r})_\uparrow, \Psi(\vec{r})_\downarrow]^T$ .This means that if the vector $\vec{r}$ is rotated by an angle ϕ the transformation $\vec{r} \to \vec{r}'$ generates a spinor transformation $\Psi(\vec{r}) \to \Psi'(\vec{r}') = e^{i\phi\cdot\frac{\sigma_3}{2}}\Psi(\vec{r})$. This implies that when we perform a rotation of $\phi = 2\pi$ the spinor accumulates a Berry phase of π. This effect suggests that in the presence of *disorder* the state $\Psi(\vec{r})$ is scattered to a new state $\Psi'(\vec{r}')$. When the scattered state is returned back to the point $\vec{r}$ it will have an extra Berry phase of π. Since localization effects are determined by the interference between the closed path and a time reversal path, the extra phase of π will give rise to destructive interference, which means *absence* of localization. This idea has not been confirmed experimentally. The main difficulty being the fact that the experiments use non trivial geometries such as cylinders or spheres. The *topology* affects the Dirac equation by adding new Berry phases. The Aharonov -Bohm oscillations performed in Nano-wires Bi_2Se_3 by [35] seems to support the idea that the maximum conductance should occur for a zero or integer flux $\Phi = n\frac{hc}{e}$ (in zero flux we have a Berry phase of π) and the minimum should be at $\Phi = (n+\frac{1}{2})\frac{hc}{e}$ where localization is supposed to take place . The experiment of [35] has been performed in a cylindrical geometry . Following the [37, 61] one finds that the Dirac equation for a cylinder takes the form: $h^{cylinder} = \hbar[(n_z\cdot\vec{\sigma})(-i)\partial_z + (n_\varphi\cdot\vec{\sigma})(-i)\partial_\varphi + \frac{1}{2R_o}i(n_R\cdot\vec{\sigma})]$ where z,φ are cylindrical coordinates and R_0 is the radius of the cylinder. The presence of the extra $\frac{1}{2R_0}i(n_R\cdot\vec{\sigma})]$ term modifies the quantization rules (an extra term of a half is added). This will imply an extra Berry phase. When a flux is pierce the cylinder we obtain gapless fermions which propagate along the z axis only when the flux is $\frac{1}{2}\frac{hc}{e}$ This means that the maximum of the conductance will occur for fluxes $\Phi = (n+\frac{1}{2})\frac{hc}{e}$ and the minimum should be for zero flux or $\Phi = (n)\frac{hc}{e}$.

Applying an external magnetic field to the two dimensional surface reveals that the Landau quantization and the Hall conductance are modified on the boundary of TI. Due to the Dirac form the Hall conductance is given by $\sigma_{x,y}^{surface} = \frac{e^2}{h}(n+\frac{1}{2})$. Recent experiments of scaning tunneling spectroscopy (STS) reveal the existence of a state at $n = 0$ which is not affected by the value of the magnetic field (the Landau levels for $n >\geq 1$ scale with the magnetic field $\sqrt{B}$).

In curved geometries this result changes. For example the Hall conductance for the Dirac equation on a sphere is $\sigma^{sphere}(x,y) = 2\cdot\sigma^{surface}(x,y) = (2n+1)\frac{e^2}{h}$ [36]. This suggests that for a thin sample $\sigma^{thin}(x,y) = \sigma^{top}(x,y) + \sigma^{bottom}(x,y)$ (the measurement might show contribution from the *top* and from the *bottom* surface). This suggests that for a thin sample the Hall conductance will be similar to the Hall conductance on a sphere.

The experimental results for the conductance of the TI $Bi_{1-x}Sb_x$ show interesting oscillation in the presence of a magnetic field as a function of the angle between the axes C_3 and C_2 [33]. It seems that the results are not ruling out the surface conducting contribution , but suggests contribution from the *bulk*. The theory of TI predicts that when a magnetic field perpendicular to the two dimensional surface is applied the conductance $\sigma_{x,x}$ *decreases* since the Berry phase which in the absence of magnetic field is π, will become in a magnetic field $\pi(1 \pm \frac{\Delta}{\sqrt{(\Delta)^2+|K|^2}})$. Therefore backscattering is allowed (The parameter Δ contains two contributions, the normal component of the external magnetic field and the magnetic exchange which is due to the *magnetic impurities*. Both effects can be described by a modification of the Hamiltonian by $\delta H = \Delta \cdot \sigma^3$. It is known that in a system with spin orbit interaction the conductivity also decreases in the presence of a magnetic field [42]. Therefore a quantitative analysis might be needed in order to separate the two effects. If the TI. are to be responsible for the conductivity one needs to understand the surface-bulk contributions. It seems that a variety of defects might play an important role in transport. Some of the recent experiments might hint that the dislocation lines play a crucial role. This experimental results suggests that interaction disorder and Topological defects such as *dislocations* (screw and edge) and *disclinations* might play an important role in transport.

III.A. The Effect of Static Disorder on the Second Chern Number

The transport experiments [33] performed on the mixed crystal $Bi_{1-x}Sb_x$ TI for $0.7 < x < 0.22$ show that a significant contribution to the conductivity comes from the bulk. In the absence of disorder the condition $\frac{M}{2tD} < 1$ guarantees the existence of the TI. The presence of disorder, is included trough the impurity potential: $V_{disorder} = V_0(\vec{r})\gamma^0 + V_1(\vec{r})I$ ($V_0(\vec{r})$ represents the space dependent mass term - the orbital splitting energy) . The mass gap M is normalized by both terms.

a)-The *two mode* crystal

For inhomogeneus disorder we have a *two mode* crystal, namely the crystals splits into *domains* where the mass gap M takes values, $\frac{M_{eff}(A)}{2tD} < 1$ (domain$A$) and $\frac{M_{eff}(C)}{2tD} > 1$ (domain C). At the interface between the domains we will have *chiral* massless fermion propagation. The recent transport experiment [33] hints that a significant contribution to the conductivity comes from the bulk. Therefore the *two mode* proposal might explain the high conductivity coming from the domain walls in the bulk.

b)-The homogeneous *virtual* crystal.

When the system is homogeneusly disordered the value of the mass gap can be replaced by $\sqrt{< M^2 >} = \hat{M}$ +plus Gaussian fluctuations of $M - \hat{M}$. The value $\frac{\hat{M}}{2tD} = 1$ represents the transition point between the TI and a regular insulator. Due to the Gaussian fluctuations we expect the following transitions : The region $\frac{\hat{M}}{2tD} < 1 - \epsilon$ is a TI. The region $1 - \epsilon < \frac{\hat{M}}{2tD} < 1 + \epsilon$ corresponds to a chiral metal which will become a regular insulator for $\frac{\hat{M}}{2tD} > 1 + \epsilon$.

In order to perform a quantitative analysis we will consider the Hamiltonian $H_{disorder} = H + V_{disorder}$. Using the eigen-spinors $|v(\vec{K}, s) >,|(u\vec{K}, s) >$ we compute the transformed disorder matrix $\hat{V}_{s,s'}(\vec{K}, \vec{K'})$ (with the matrix elements $V^{c^\dagger,c}_{s,s'}(\vec{K}, \vec{K'})$, , $V^{b^\dagger,b}_{s,s'}(-\vec{K}, -\vec{K'})$, $V^{c^\dagger,b^\dagger}_{s,s'}(\vec{K}, -\vec{K'})$, $V^{b^\dagger,c^\dagger}_{s,s'}(-\vec{K}, \vec{K'})$ computed with the help of the Fourier transform of the disorder potential and the eigen-spinors).

$$H = \int_{T^{D=4}} \sum_{s=1,2} [\epsilon(\vec{K})(c^{\dagger}(\vec{K},s)c(\vec{K},s) + b^{\dagger}(\vec{K},s)b(\vec{K},s)] \tag{31}$$

$$V_{disorder} = \int_{T^{D=4}} \sum_{s=1,2} \int_{T^{D=4}} \sum_{s'=1,2} (c^{\dagger}(\vec{K},s), b(\vec{K},s))[\hat{V}_{s,s'}(\vec{K},\vec{K'})](c(\vec{K'},s]), b^{\dagger}(\vec{K'},s'))^{T} \tag{32}$$

We will compute the $self\ energy$ and we will extract the effective mass gap M. Perturbation theory can be used since time reversal symmetry guarantees that the matrix elements $\hat{V}_{s,s'}(\vec{K},\vec{K'} = -\vec{K})$ are zero. In the presence of disorder the ground state spinor $|v(\vec{K},s)>$ is replaced by $|\mathbf{v}(\vec{K},s)>$:

$$\begin{aligned} |\mathbf{v}(\vec{K},s)> &= |v(\vec{K},s)> + \sum_{s'=1,2}\sum_{\vec{P}} \frac{|V(\vec{K},\vec{P})v^{*}(\vec{K},s)v(\vec{P},s')|^{2}}{\epsilon(\vec{K})-\epsilon(\vec{P})}|v(\vec{P},s')> + \\ &\sum_{s'=1,2}\sum_{\vec{P}} \frac{|V(\vec{K},\vec{P})v^{*}(\vec{K},s)u(\vec{P},s')|^{2}}{\epsilon(\vec{K})-\epsilon(\vec{P})}|u(\vec{P},s')> \equiv \\ &|v(\vec{K},s)> + \sum_{s'=1,2}\sum_{\vec{P}} B^{(s,s')}(\vec{K},\vec{P'})|v(\vec{P'},s')> \\ &+ \sum_{s'=1,2}\sum_{\vec{P}} C^{(s,s')}(\vec{K},\vec{P'})|u(\vec{P'},s')> \end{aligned} \tag{33}$$

In order to test the possible phase transition for the Chern number, we propose to compute the new curvature $\mathbf{F}$ which replaces F. For this purpose we introduce the coordinate $X^{i} = i\partial_{K^{i}}$ and compute the new connection: $\mathbf{A}_{i}^{-}$.

$$\begin{aligned} \mathbf{A}_{i}^{-} &=< \mathbf{v}(\vec{K},s_1)|X^{i}|(\mathbf{v}(\vec{K},s_2) >=< v(\vec{K},s_1)|X^{i}|v(\vec{K},s_2 > + \\ &\sum_{s'=1,2}\sum_{s''=1,2}\sum_{\vec{P}}(B^{(s_1,s')}(\vec{K},\vec{P}))^{*}X^{i}B^{(s_2,s'')}(\vec{K},\vec{P})) \\ &+ \sum_{s'=1,2}\sum_{s''=1,2}\sum_{\vec{P}}(C^{(s_1,s')}(\vec{K},\vec{P}))^{*}X^{i}C^{(s_2,s'')}(\vec{K},\vec{P})... \end{aligned} \tag{34}$$

Using this representation we obtain the curvature matrix: $\mathbf{F}_{i,j} = [X^{j}, \mathbf{A}_{i}^{-}]$.
Using the new curvature $\mathbf{F}_{i,j}$ we obtain the second Chern number:

$$Ch_2 = \int_{T^{D=4}} Tr < [\frac{1}{2!}(\frac{i\mathbf{F}}{2\pi})^{2}] >_{average-disorder} \tag{35}$$

In order to answer the question if the elastic disorder destroys the TI we will perform analytical and numerical calculation. In order to modify the Chern number by disorder the representation for the new eigenvector $|\mathbf{v}(\vec{K},s)>$ must be divergent! When the series converges we find that $[X^j,\mathbf{A}_i^-]=[X^j,A^-(\vec{K})_i]$ and the second Chern number is unchanged.

In order to determine the transport properties we will construct an effective non-linear sigma model using the Keldish formalism [48].

III.B. The TI with Disorder Using the Keldish Method

In order to compute the conductivity we need to construct an effective non-linear sigma model using the Keldish formalism. Following [48] we introduce two fermionic fields which reside on the forward and backward contour. We replace the *four* component spinor $\Psi(\vec{r},\omega)$ by a *eight* component spinor ,$\mathbf{\Psi}(\vec{r},w)=[\Psi(\vec{r},\omega),\Psi^\dagger(\vec{r},-\omega)]^T$. Using this formalism we represent the disorder Hamiltonian:

$$H_{disorder}=\int d^Dr\int d\omega V_{disorder}(\vec{r})(\mathbf{\Psi}^\dagger(\vec{r},\omega)\Sigma_3\mathbf{\Psi}(\vec{r},\omega) \tag{36}$$

Performing the average over disorder we can introduce a set of Hubbard Stratonovich fields [58, 50] $Q_j(\vec{r},(\omega,\omega')$ which act as gauge fields in the space $\Sigma_3\otimes\gamma^j$ (Σ_3 acts on the frequency space). Using the Hamiltonian given in section IIA (the TI in D dimensions) + disorder we have:

$$\begin{aligned}
H=&\int d^Dr\int d\omega\int d\omega'[\sum_{j=1,..D}\mathbf{\Psi}^\dagger(\vec{r},\omega)(\Sigma_3\otimes\gamma^j)((-i)\hat{\partial}_j\\
&+Q_j(\vec{r},\omega,\omega')\mathbf{\Psi}(\vec{r},\omega')+\mathbf{\Psi}^\dagger(\vec{r},\omega)(\Sigma_3\otimes\gamma^0)((M+Q_0(\vec{r},\omega,\omega')\mathbf{\Psi}(\vec{r},\omega')\\
&-t\sum_{j=1,..D}\mathbf{\Psi}^\dagger(\vec{r},\omega')(\Sigma_3\otimes\gamma^j)\hat{\nabla}_j^2\Psi(\vec{r},\omega')]
\end{aligned} \tag{37}$$

where $\hat{\nabla}_j^2\Psi(\vec{r},\omega)\equiv[\Psi(\vec{r}+a_j,\omega)+\Psi(\vec{r}-a_j,\omega)-2\Psi(\vec{r}-a_j,\omega)]$ and the Gaussian gauge fields are controlled by: $H_Q=\int d^Dr\int d\omega\int d\omega'[\sum_{j=1,5}\frac{1}{2V^j}(Q_j((\vec{r},(\omega,\omega'))^2]$.

This Hamiltonian will be used to construct the effective action for transport. Integrating the fermions we obtain the effective action in terms of the gauge fields Q_ν. The effective action is a function of the curvature $F_{\nu,\mu}(\omega,\omega')=\partial_\nu Q_\mu(\vec{r},\omega,\omega')-\partial_\mu Q_\nu(\vec{r},\omega,\omega')$ and Chern-Simon term.

Using the *Renormalization Group* we will establish the phase diagram: In the metallic phase the Maxwell term vanishes . If the coefficient of the chern number obeys,$Ch_{n=2}\neq 0$ we have a chiralic metal. When $Ch_{n=2}=0$ we have a regular metal. When the Maxwell term is non zero we have either a regular insulator $Ch_{n=2}=0$ or a T.I for $Ch_{n=2}\neq 0$.

III.C. Interaction and Disorder

For the one dimensional boundary TR symmetry forces the backscattering term to vanish. Since *umklapp* is allowed [31] the presence of the Coulomb interactions for $D=1$ drives the system to the localized phase. We have found that the Luttinger parameters decreases below $\frac{1}{4}$ [51, 52] causing the chiral edge excitations to localize.

The two dimensional boundary of the TI gives rise to the $D = 2$ chiral metal. [17] described by the Hamiltonian:

$$h(\vec{K}) = v(K^2\sigma_1 - K^1\sigma_2) + \lambda[(K^1 + iK^2)^3 + (K^1 - iK^2)^3]\sigma_3 \tag{38}$$

Using the time reversal operator $T = -i\sigma^2\mathbf{K}$ we observe that the $h(\vec{K})$ is time reversal invariant, $h(-\vec{K}) = Th(\vec{K})T^{-1}$. This Hamiltonian has two eigen -spinors $|u(\vec{K}) >$ for particles and $|v(\vec{K}) >$ for antiparticles which have opposite chirality. We introduce the two component spinor : $\Psi(\vec{r}) = [\Psi(\vec{r})_\uparrow, \Psi(\vec{r})_\downarrow]^T$. Using the ground state $|G_0 >$ we introduce the *particles* and *anti-particles* operators: $c(\vec{K})|G_0 >= 0$,$b(\vec{K})|G_0 >= 0$. For this case we observe that the spinor operator is given by the chiral Weyl form: $\Psi(\vec{r}) = \int_{T^2}[c(\vec{K})u(\vec{K})e^{i\vec{K}\cdot\vec{r}} + b^\dagger(-\vec{K})v(-\vec{K})e^{i\vec{K}\cdot\vec{r}}]$.

The effect of interaction might favor the formation of a Ferromagnetic state and therefore the system will become a regular insulator which in the presence of disorder can localize. It is therefore important to construct an effective action with disorder and interaction. We will consider first the effect of interactions (at a later stage disorder and interactions will be considered together).

We start with the Helical metal in $2 + 1$ dimensions in the presence of short range repulsion interactions away from half filling:

$$\begin{aligned} H &= \int d^2r(\Psi^\dagger(\vec{r})[\sigma^1(-i\partial_2 - N^1(\vec{r})) - \sigma^2(-i\partial_2 + N^2(\vec{r})) \\ &\quad +\sigma^3 N^3(\vec{r}) + \mu]\Psi(\vec{r}) + \frac{3}{8U}(\vec{N}(\vec{r}) \cdot \vec{N}(\vec{r})) \end{aligned} \tag{39}$$

$\mu > 0$ is the chemical potential which we choose to be in the positive. As a result the low energy excitations of holes and electrons will have the same chirality. The fermion representation will be determined only by the electron like spinor $|u(\vec{k}_F + \vec{q}) >$ and $|u(\vec{k}_F - \vec{q}) >$ where $|\vec{k}_F| = \mu$,

$$\Psi(\vec{r}) = \int d^2q[c(\vec{k}_F + \vec{q})u(\vec{k}_F + \vec{q})e^{i(\vec{k}_F+\vec{q})\cdot\vec{r}} + b^\dagger(\vec{k}_F - \vec{q})v(\vec{k}_F - \vec{q})e^{-i(\vec{k}_F-\vec{q})\cdot\vec{r}}] \tag{40}$$

As a result the electron excitations will depend on the angle on the Fermi surface. At the point $\vec{k}_F = [\vec{k}_{F1}, \vec{k}_{F2}] = [|\vec{k}_F|, 0]$ the excitation will be $\epsilon(\vec{q}) \propto (q_1 + \frac{q_2^2}{2|\vec{k}_F|})$.(Similar behavior will occur at the points $\vec{k}_F = [0, \pm|\vec{k}_F|]$, $\vec{k}_F = [\pm|\vec{k}_F|, 0]$). As result we find from dimensional analysis that the presence of the dispersion q^2 will make short range interaction *marginal* in 2+1 dimensions! For this reason we expect critical Ferromagnetic fluctuations which might stabilize a ferromagnetic state.

III.D. Topological Insulators in Curved Spaces

It seems that in curved spaces caused by geometry or topological disorder such as dislocations and disclinations we can have the following scenario. For a TI we dislocations might create boundaries such that chiral excitations might be bound to the dislocations. On the other hand if we have already massless fermions some of the topological defects might cause localization.

In order to investigate the effect of the Topological Disorder on the TI we will introduce a geometric formulation for curved spaces [54, 55, 57, 55, 61, 36, 37, 56, 37]. We consider an n dimensional manifold where a mapping from the curved space X^a, $a = 1, 2...n$ to the $local\ flat$ space exists, x^μ, $\mu = 1, 2...n$. We introduce the tangent vector [37] $e^a_\mu(\vec{x}) = \frac{\partial X^a(\vec{x})}{\partial x^\mu}$ which satisfies the orthonormality relation $e^a_\mu(\vec{x}) e^b_\mu(\vec{x}) = \delta_{a,b}$ (here we use the convention that we sum over indices which appear twice). Representing the curved coordinate in terms of local flat coordinates gives rise to the metric tensor $e^a_\mu(\vec{x}) e^a_\nu(\vec{x}) = g_{\mu,\nu}(\vec{x})$. We introduce the linear connection which is determined by the Christoffel tensor $\Gamma^\lambda_{\mu,\nu}$ which is defined by:

$$\nabla_{\partial_\mu} \partial_\nu = \Gamma^\lambda_{\mu,\nu} \partial_\lambda \tag{41}$$

Next we introduce the vector field $\vec{V} = V^a \partial_a = V^\mu \partial_\mu$ and the spinor $\Psi(\vec{x})$. The covariant derivative of the vector field V^a is determined by the spin connection $\omega^a_{i,b}$:

$$D_\mu V^a(\vec{x}) = \partial_\mu V^a(\vec{x}) + \omega^\mu_{a,b} V^b \tag{42}$$

Similarly in the basis x^μ the covariant derivative $D_\mu V^\nu(\vec{x})$ is determined by the Christoffel tensor:

$$D_\mu V^\nu(\vec{x}) = \partial_\mu V^\nu(\vec{x}) + \Gamma^\lambda_{\mu,\nu} V^\lambda \tag{43}$$

The spinor connection $\omega^\mu_{a,b}$ determines the covariant derivative $D_\mu V^a(\vec{x})$ defined above and as well determine the covariant derivative of the spinor $\Psi(\vec{x})$ given in term of the gamma matrices γ.

$$D_\mu \Psi(\vec{x}) = (\partial_\mu + \frac{1}{2} \omega^{a,b}_\mu [\gamma^a, \gamma^b]) \Psi(\vec{x}) \tag{44}$$

We first evaluate the Christoffel tensor which is a function of the metric tensor $g_{\mu,\nu}(\vec{x})$.

$$\Gamma^\lambda_{\mu,\nu} = -\frac{1}{2} \sum_{\tau=1,2,..n} g^{\lambda,\tau}(\vec{x}) [\partial_\nu g_{\nu,\tau}(\vec{x}) + \partial_\mu g_{\nu,\tau}(\vec{x}) - \partial_\tau g_{\mu,\nu}(\vec{x})] \tag{45}$$

The relation between the spin connection and the linear connection can be obtained from the fact that the two covariant derivative of the vector $\vec{V}$ are equivalent.

$$D_\mu V^a = e^a_\nu D_\mu V^\nu \tag{46}$$

Since we have the relation $V^a = e^a_\nu V^\nu$ it follows from the last equation

$$D_\mu [e^a_\nu] = D_\mu \partial_\nu e^a = (D_\mu \partial_\nu) e^a + \partial_\nu (D_\mu e^a) = 0 \tag{47}$$

Using the definition of the Christoffel index we find the relation between the spin connection and the linear connection:

$$D_\mu [e^a_\nu] = \partial_\mu e^a_\nu(\vec{x}) - \Gamma^\lambda_{\mu,\nu} e^a_\lambda(\vec{x}) + \omega^a_{\mu,b} e^b_{nu}(\vec{x}) \equiv 0 \tag{48}$$

From this equation we determine the spin connection:

$$\omega_{\mu}^{a,b} = \frac{1}{2}e^{\nu,a}(\partial_{\mu}e_{\nu}^{b} - \partial_{\nu}e_{\mu}^{b}) - \frac{1}{2}e^{\nu,b}(\partial_{\mu}e_{\nu}^{a} - \partial_{\nu}e_{\mu}^{a})$$
$$-\frac{1}{2}e^{\rho,a}e^{\sigma,b}(\partial_{\rho}e_{\sigma,c} - \partial_{\sigma}e_{\rho,c})e_{\mu}^{c} \quad (49)$$

This allows to rewrite the Dirac equation $h = \gamma^{a}(-i\partial_{a})$ as :

$$h^{curved} = e_{a}^{\mu} \cdot \gamma^{a}(-i)(\partial_{\mu} + \frac{1}{8}\omega_{\mu}^{ab}[\gamma_{a}, \gamma_{b}]) \quad (50)$$

For the surface case $n = 2$ the calculations are simplified since any Riemann surface can be brought to conformal form , $ds^{2} = e^{-\beta(\vec{x})}[dx^{2} + dy^{2}]$ where $\beta(\vec{x})$ is the conformal transformation from the flat to curve surface. For this case one obtains [36] a simple form which has been used for a sphere.

a)Bended cylinder pierced by flux

When the cylinder is bended we obtain a modification of the results in comparison with a straight cylinder. For a straight cylinder pierced by the flux $\Phi = \frac{1}{2}\frac{hc}{e}$ gives rise to a chiral metal. When the cylinder is bend the check if the theory of [57, 56] predict the formation of localized bound states and therefore a insulating behavior is induced.

c)The effect of Topological disorder on the Topological Insulators- Dislocations and Disclinations-Computation of the connection.

The presence of *dislocations* and *disclinations* deform substantially the lattice structure and therefore might affect the electronic transport. Following [39] the deformation of the crystal give rise to a coordinate transformation which will affect the Dirac equation. Given a perfect lattice with coordinates x^{i}, lattice deformation will modify the coordinate to $X^{a}(\vec{x}) = x^{a} + e^{a}(\vec{x})$ where $e^{a}(\vec{x})$ is the local lattice deformation. As a result the metric is given by $g_{\mu,\nu}(\vec{x}) = \frac{\partial X^{a}}{\partial x^{\mu}}\frac{\partial X^{a}}{\partial x^{\nu}}$.

Screw dislocations are characterized by the *Burger* vector $\vec{B}$ which is parallel to the dislocation line contrary to the edge dislocation where the Burger vector is perpendicular to the dislocation line . The metric tensor for a screw dislocation is : $ds^{2} = (dz + \frac{B^{z}}{2\pi}d\theta)^{2} + d\rho^{2} + \rho^{2}d\phi^{2}$.

Using the tangent vector we represent the burger vector in term of the *torsion*:

$$\oint dX^{\mu}e_{\mu}^{a} = \int\int dX^{\mu}dX^{\nu}(\partial_{\mu}e_{\nu}^{a} - \partial_{\nu}e_{\mu}^{a}) = -B^{a} \quad (51)$$

The Dirac equation in the presence of *edge* and *screw* dislocations:
$h^{dislocation} = e_{a}^{\mu} \cdot \gamma^{a}[(-i)\partial_{\mu} + \eta_{\mu}(\vec{x}) + \gamma^{5}S_{i}(\vec{x})]$,
where the real vector $\eta_{i}(\vec{x})$ characterizes edge dislocations and $S_{i}(\vec{x})$ is an *axial* vector which describes the screw dislocations.

The axial vector which describes the screw dislocations breaks time reversal symmetry, contrary to the real vector $\eta_{i}(\vec{x})$ which does not destroy the chiral excitations .

A similar analysis shows that a $disclination$ has no torsion but has $curvature$. The metric tensor for a disclination is given by $ds^2 = dz^2 + d\rho^2 + \alpha^2\rho^2 d\phi^2$ The effect of disclinations can be understood as an angle deficit $2\pi \cdot \alpha$,therefore once a rotation of 2π is performed the spinor will rotate by an angle $\pi \cdot \alpha < \pi$. Therefore we obtain a Berry phase of $\pi \cdot \alpha$ which will allow backscattering and will allow localization. (Quantitatively the effect of the line deficit can be expressed as the strength of a vortex.) We believe that the dislocation might play an important role for computing transport properties, contrary to regular disorder some of the dislocations and disclinations can cause localization.

d)The Topological insulator in a curved space induce by Dislocations

Using the coordinate transformation induced by the dislocations we can rewrite the model given in section IIA. The new model will contain the Dirac equation in 4+1 dimensions $curved\ space$ (due to dislocations) plus the non relativistic mass term, $M\overline{\Psi}(\vec{r})\Psi(\vec{r}) - t\sum_{j=1}^{D}(\overline{\Psi}(\vec{r}+a_j)\Psi(\vec{r}) + \overline{\Psi}(\vec{r})\Psi(\vec{r}+a_j))$ which in the continuum will be replaced by the Laplacean $\nabla^2 = \Delta$ which in the curved space will becomc the **Laplace-Beltrami** operator Δ_{LB} [57, 56]. The dislocation is believed to give rise to a $chiral\ zero\ mode$ (along the dislocation line).

e)An explicit demonstration of the method presented for an edge dislocation in two dimensions

We consider the two dimensional chiral Dirac equation in the absence of any dislocation:

$$h = \sigma^1(-i\partial_2) + \sigma^2(i\partial_1) \tag{52}$$

We consider an edge dislocation with the burger vector B^2 (the value of the vector is the lattice constant in the y direction).

The dislocation is given in terms of the $torsion$ tensor $T^{a=2}_{\mu=1,\nu=2}$

$$T^{a=2}_{\mu=1,\nu=2} = \partial_1 e^2_2 - \partial_2 e^2_1 = B^2\delta^2(\vec{x}) \tag{53}$$

Using the solution of the Laplace equation we obtain the tangent components:

$$e^2_1 = \beta\frac{y}{(x^2+y^2)} \qquad e^2_2 = 1 - \beta\frac{x}{(x^2+y^2)} \tag{54}$$

where $\beta = \frac{B^2}{2\pi}$ and $e^1_1 = 1$,$e^1_2 = 0$.

Using the tangent components we obtain the metric tensor $to\ first\ order$ in β:

$$g_{11} = 1, \quad g_{12} = \beta\frac{y}{x^2+y^2}, \quad g_{22} = 1 - \beta\frac{y}{x^2+y^2}, \quad g_{21} = 0 \tag{55}$$

Using the the vectors components e^a_μ and the results obtained in equations $49 - 50$ we compute the transformed Pauli matrices:

$$\sigma^1(x) \approx \sigma^1, \quad \sigma^2(x) \approx \sigma^2 - \beta\frac{y}{x^2+y^2}\sigma^1 \tag{56}$$

The spin connection is obtained from equation 50:

$$\omega^{12}_1 = -\omega^{12}_2 \approx \pi\beta\delta^2(\vec{x}) \tag{57}$$

As a result the chiral Dirac equation in the presence of an edge dislocations becomes:

$$h^{edge-dislocation} = \sigma^1[-i\partial_2 - \frac{1}{8}\pi\beta\delta^2(\vec{x})\sigma^3] + +[\sigma^2 - \beta\frac{y}{x^2+y^2}\sigma^1][i\partial_1 + \frac{1}{8}\pi\beta\delta^2(\vec{x})\sigma^3] \quad (58)$$

IV. Topological Z_2 Realization for the $p_x + ip_y$ Chiral Superconductor Wire Coupled to Metallic Rings in an External Flux

The experimental realization of the $p_x + ip_y$ pairing order parameter (p-wave superconductors) in Sr_2RuO_4 , $^3He - A$ and the $\nu = \frac{5}{2}$ case represents interesting new excitations. In these (weak coupling) $p_x + ip_y$ pairing the excitations are half vortices which are zero mode energy Majorana fermions. We consider a p-wave superconductor confined to a one-dimensional wire. At the edges of the wire $x = 0$ and $x = L$ the pairing order parameter vanishes and two zero modes appear at the edges (x=0 and x=L). Due to the charge conjugation of the Bogoliubov spectrum these zero modes should be Zero mode Majorana Fermions [53]. Mapping the problem to the Ising model one can show that the ground state is a Z_2 doubly degenerate ground state. As a result the single particle excitations are non-local. **As a concrete example we couple the p-wave wire to two rings which are pierced by external fluxes.** In the first stage we will study a spinless model and in the second stage we will consider the effect of spin.

The model for the spinless case is :$H = H_{P-W} + H_1 + H_2 + H_{(P-W,1)} + H_{(P-W,2)}$.

H_{P-W} is the p-wave superconductor with the pairing gap Δ and polarized fermion operator $C(x) \equiv C_{\sigma=\uparrow}(x)$. $H_{P-W} = -t\sum_{x,x'}(C^+(x)C(x')+h.c.)+\lambda\sum_x C^+(x)C(x)-\Delta\sum_{x,x'}[\gamma_{x,x'}C^+(x)C^+(x')+\gamma_{x,x'}C(x')C(x)]$. The time reversal and parity symmetry are broken, $\gamma_{x,x'=x-a} = -\gamma_{x,x'=x+a}$. Following [53] one introduces the Majorana fermions $\eta_1(x)$, $\eta_2(x)$ [53] and finds: $C(x) = \eta_1(x) + i\eta_2(x)$; $C^+(x) = \eta_1(x) - i\eta_2(x)$ and $\eta_1(x) = \frac{1}{2}(C(x) + C^+(x))$ and $\eta_2(x) = \frac{1}{2i}(C(x) - C^+(x))$. The H_{P-W} Hamiltonian has the pairing boundary conditions $\Delta(x = 0) = \Delta(x = L) = 0$ and $\Delta(x) = \Delta_0$ for $0 < x < L$. This model is equivalent to the 1d-Ising model in a transverse field [53] : $H_{P-W} = \frac{i}{2}\int dx\hat{\eta}^T[\sigma_3\partial_x + i\Delta(x)\sigma_2]\hat{\eta}$ where $\hat{\eta}^T = (\eta_1(x), \eta_2(x))$.

The two eigenfunctions are charge conjugated with two bound states at x=0 and x=L.

$\eta_1(x) = \eta_L\Phi^L(x) + ...$ and $\eta_2(x) = \eta_0\Phi^0(x) + ...$

$\Phi^0(x)$ and $\Phi^L(x)$ are the bound states at x=0 and x=L and η_0 and η_L are the zero modes Majorana operators. They obey $(\eta_0)^+ = \eta_0$, $(\eta_0)^2 = \frac{1}{2}$ and $(\eta_L)^+ = \eta_L$, $(\eta_L)^2 = \frac{1}{2}$ .This suggests that the ground state is Z_2 doubly degenerate.

Using the zero modes Majorana Fermions η_0 and η_L we obtain a fermion operator $q = \frac{1}{2}(\eta_L + i\eta_0)$ and $q^\dagger = \frac{1}{2}(\eta_L - i\eta_0)$ which obey anti-commutation relations $[q, q^\dagger]_+ = 1$ which obeys $q^\dagger|0>=|1>$ and $q|1>=|0>$. The p-Wave wire is described by the ground state $|0>$ when the wire contains only pairs. When one electron is added to the wire the state is $|1>$. The energy of this state is given by ϵ . The value of ϵ depends on the length of the wire, when $L \to \infty$ we have $\epsilon \to 0$ [53].

$H_1 + H_2$ represent the Hamiltonian for the two metallic rings each pierced by a flux $\hat{\varphi}_i$,$i = 1, 2$ [22].

$H_i = \int_0^{l_{ring}} dx \left[\frac{\hbar^2}{2m}\psi_i^\dagger(x)(-i\partial_x - \frac{2\pi}{l_{ring}}\hat{\varphi}_i)^2\psi_i(x)\right] i = 1, 2,$

In the long wave approximation the $p - Wave$ Hamiltonian H_{P-W} is replaced by $H_{P-W} = \epsilon q^\dagger q$.

$H_{(P-W,1)} + H_{(P-W,2)}$ represents the coupling between the $p - Wave$ wire to the two rings trough the matrix element g.

$H_{(P-W,1)} + H_{(P-W,2)} = ig[q^\dagger(\psi_1^\dagger + \psi_1 + \psi_2^\dagger - \psi_2) + q(\psi_1^\dagger + \psi_1 - \psi_2^\dagger(0) - \psi_2)]$

Using the Heisenberg equation of motion we substitute $q(t)$ and $q^\dagger(t)$ and reduce the problem to a eigenvalue problem in terms of the rings spinless electron operators $\psi_1(x)$, $\psi_2(x)$. We will work with the Fourier transform in the frequency z representation. We replace the two electron operators by the four component spinor,

$\mathbf{\Psi}(x, z) = [\psi_1(x, z), \psi_2(x, z), \psi_2^\dagger(x, -z), \psi_1^\dagger(x, -z)]^T.$

$$z\begin{pmatrix} \psi_1(x,z) \\ \psi_2(x,z) \\ \psi_2^+(x,-z) \\ \psi_1^+(x,-z) \end{pmatrix} = \tag{59}$$

$$\begin{bmatrix} e_1 & 0 & 0 & 0 \\ 0 & e_2 & 0 & 0 \\ 0 & 0 & -e_2 & 0 \\ 0 & 0 & 0 & -e_1 \end{bmatrix}\begin{pmatrix} \psi_1(x,z) \\ \psi_2(x,z) \\ \psi_2^+(x,-z) \\ \psi_1^+(x,-z) \end{pmatrix} + \frac{2g^2\delta(x)}{\epsilon^2 - z^2}\begin{bmatrix} -z & \epsilon & \epsilon & z \\ \epsilon & -z & -z & -\epsilon \\ \epsilon & -z & -z & -\epsilon \\ z & -\epsilon & -\epsilon & -z \end{bmatrix}\begin{pmatrix} \psi_1(z) \\ \psi_2(z) \\ \psi_2^+(-z) \\ \psi_1^+(-z) \end{pmatrix}.$$

We use the notation: $e_i = (-i\partial_x + \varphi_i)^2$.

A full solution involves a numerical computation for the eigenvalues and eigenvectors. We find that the problem simplifies in two limits $\epsilon >> g$ and $\epsilon << g$.

a)When $\epsilon << g$ the current in each ring will depend only on the flux applied. For equal fluxes we will find that the current and the magnetization are proportional to the applied flux. This case corresponds to the $Andreev$ reflection at each interface.

b)When $\epsilon >> g$ the current through the current and the magnetization will be proportional to the $(flux)^2$. This case corresponds to a $crossed\ Andreev\ reflection$

We believe that measuring the current and the magnetization as a function of the flux will allow an indirect confirmation for Majorana fermions.

V. Conclusion

This article aims to introduce a variety of powerful geometrical methods for condensed matter systems such as connection and curvature. Electrons in solids are characterized by Bloch bands. When the band crosses one obtains non-trivial phases which gives rise to non -commutativity in momentum space characterized by the Chern numbers. The Topological insulators are material with spin-orbit interaction. At special point in the Brillouin zone the model is time reversal invariant . As a result a new type of insulators characterized by the second Chern number is identified. At the boundary of the Topological Insulators an odd number of Chiral states are identified. Those states are robust againt disorder and represent perfect metals.

The idea of protected boundary can be generalized to any interface including disorder or Topological defects such as dislocations and disclinations.

We introduce Majorana fermions and show that they can be detected by measuring the persistent current in metallic rings attached to a P-wave superconductor.

References

[1] B.A.Volkov and O.A. Pankratov "'Twodimensional massless electrons in an inverted contact"' *JETP LETT.* vol.42,179(1985)

[2] F.D.M. Haldane "'Model For Quantum Hall Without Landau Levels "'Phys.Rev.Lett.**61**,2015(1988).

[3] Karl Jansen "'Chiral Fermions,Anomalies and Chern-Simons Currents on the Lattice"' hep-lat/921203.

[4] David B.Kaplan "'A Method For Simulating Chiral Fermions On The Lattice "' hep-lat/920601.

[5] Michael Creutz and Ivan Horwath "'Surface States and Chiral Symmetry On The Lattice"' *Phys.Rev.***50**,2297(1994)

[6] C.L. Kane and E.J. Mele "'Quantum Spin Hall Effect In Graphene "' *Phys.Rev. Lett.* **95** 226801 (2005)

[7] C.L. Kane and E.J. Mele "'Z_2 Topological Order And The Quantum Spin Hall Effect "'*Phys.Rev.Lett.* **95**,146802(2005).

[8] Liang Fu and C.L. Kane "'Topological Insulators With Inversion Symmetry"' cond-mat/0611341

[9] D.Schmeltzer,"'Topological Spin Current Induced By Non-Commuting Coordinates: An application to the Spin-Hall Effect "'*Phys.Rev.B* **73**,165301(2006).

[10] J.E.Moore and L.Balents "'Topological Invariants of Time-Reversal-Invariant Band Structures"' cond-mat/0607314

[11] Andrew M.Essin and J.E.More "'Topological Insulators Beyond The Brillouin Zone Via Chern Parity"' cond-mat/0705.0172.

[12] Xiao-Liang Qi, Taylor Hughes and Shou-Cheng Zhang "'Topological Field Theory Of Time Reversal Invariant Insulators"' *Phys.Rev.B***78**,195424(2008)

[13] Xiao-Liang Qi and Shou-Cheng Zhang "'Topological Insulators and Superconductors"' cond-mat/1008.2026

[14] Andreas P.Schnyder , Shinsei Ryu, Akira Furusaki, Andreas W.W. Ludwig "'Classification Of Topological Insulators and Superconductors In Three Spatial dimensions"' cond-mat/0803.2786.

[15] M.Z.Hasan and C.L. Kane "'Topological Insulators"' cond-mat/1002.3895

[16] Ying Ran et.al "'One-dimensional Topologically Protected Modes in Topological Insulators with Dislocations"' *Nature Physics* vol 5,298 (2009)

[17] Chao-Xing Liu et al, "'Model Hamiltonian for Topological Insulators "' *Physical Review B* **82** 045122 (2010 "'

[18] Konig et. al *J.Phys. Soc. Jpn,* **77** 031007, (2008)

[19] Y.Xia ,D.Qian, D.Hsieh, L.Wray, A. Pal, A.Bansil, D. Grauer, Y.S. Hor,R.J.Cava and M.Z. Hasan "'Discovery Of Large-Gap Topological-Insulator Class With Spi-Polarized Single Dirac-Cone On The Surface"' cond-mat/ 0908.3513.

[20] D.Schmeltzer , "'Quantum Mechanics For Genus g=2 Persistent Current in Coupled Rings"' *J. Phys:Condens Matter* **20** 335205(2008).

[21] D.Schmeltzer ,and Hsuan-Yeh Chang ,"'Rectified voltage induced by a microwave field in a confined two-dimensional electron gas with a mesoscopic static vortex"' *PMC Physics B* 2008, 1:14 ,October 21 (2008).

[22] D.Schmeltzer and A.Saxena "'The wave functions in the presence of constraints-persistent Current in Coupled Rings "'Phys.Rev.B **81** ,195310 (2010).

[23] E.I. Blount "'Formalism of Band Theory"' *Solid State Physics* edited by F.Seitz and D. Turnubul , (Academic, New York ,1962),Vol.13, pages 305-375.

[24] M.V.Berry "'Quantal Phase Factors Accompanying Adiabatic Changes"' ,*Proc.R.Soc.***A392**,45(1984).

[25] D.Thouless,M.Kohmoto,M.Nightingale et M.den.Nijs, "'Quantized Hall Conductance In a Two-Dimensional Periodic Potential"' *Phys.Rev.Lett.***49**, 405 (1982).

[26] M.Nakahara "'Geometry,Topology And Physics "' Taylor Francis Press 2003.

[27] J. Bellisard "'Change of Chern Number at Band Crossings"' cond-mat/0950403.

[28] B.Simon "'Holonomy the Quantum Adiabatic Theorem and Berry Phase "' *Phys.Rev.Lett.***51**,2167 (1983).

[29] J.W. Holten "'Aspect of BRST quantization"' hep-th/0201124 (2002).

[30] B.Andrea Bernevig Taylor L.Hughes, and Shou -Cheng Zhang"' Quantum Spin Hall Effect and Topological Phase Transition In HgTe Quantum Wells"' cond-mat/061139.

[31] Wu.C.B., A. Bernewig and Shou-Cheng Zhang "'The Helical Liquid and The Edge Of Quantum Spin Hall Systems"' *Phys.Rev.Lett.* **96** 106401 (2006).

[32] Xi Dai, Taylor L. Hughes, Xiao-Liang Qi, Zhong Fang and Shou-Cheng Zhang "'Helical Edge and Surface States in HgTe Quantum Wells and Bulk Insulators "' cond-mat/0705.1516.

[33] A.A.Taskin and Yoichi Ando ,"'Quantum Oscillations in a Topological Insulator"' *Phys.Rev.B* **80**,085303 (2009)

[34] Peng Cheng, Canli Song, Tong Zhang, Yanyi Zhang , Yilin Wang, Jin-Feng Jia, Jing Wang,Yayu Wang , "'Landau Quantization of Topological Surface States in Bi_2Se_3"' *Physical Review Letters* **105** 076801 (2010)

[35] H.Peng et.al "'Aharonov-Bohhm interference in Topological Insulator Nanoribbons "' *Nature Materials* ,**9** 225 -229 (2010)

[36] Ayelet Pnueli "'Spinors and Scalars on Riemann Surfaces "'*J.Phys. A: Math Gen.***27** 1345-1352 (1994)

[37] M.B.Green,J.H. Schwartz and E.Witten, "'Some Differential Geometry"' *Superstrings Theory* vol.2 Cambridge Monographs on Mathematical Physics pages $271 - 277$ (1987).

[38] N.D.Birrell and P.C. Davies "'Quantum Fields in Curved Space"' pages $81 - 85$ Cambridge University Press (1982).

[39] Arnold M. Kosevici "'The Crystal"' Wiley -VCH (2005).

[40] Gordon W.Semenoff and Pasquale Sodano "' Teleportation By A Majorana Medium"' cond-mat/0601261.

[41] Daniel Boyanovsky "'Field Theory of The Two Dimensional Ising Model: Conformal Invariance ,Order, And Disorder ,And Bosonization"' *Phys.Rev.B.* **39**, 6744(1989).

[42] S.Hikami, A.I. Larkin, and Y.Nagaoka "'Theory of Localization in the presence of Spin-Orbit interaction"' *Prog.Theor.Phys.* **63** ,707 (1980).

[43] D.Schmeltzer ,"'The Marginal Fermi Liquids - An Exact Derivation Based on Dirac's First Class Constraints Method"', *Nuclear Physics B* **829**447-477 (2010)

[44] D.Schmeltzer "'A Derivation of The Marginal Fermi Liquid From First Class Constraints"' *Physica B:Condensed Matter* Vol. 404, issue 19, 15 October 2009, pages 3085-3088.

[45] J. Zak, "Dynamics Of Electrons In Solids In External Fields"', *Phys. Rev.* **168**,687 (1968).

[46] J. Zak, "Dynamics Of Electrons In Solids In External Fields,II"', **177**,1151 (1969).

[47] J. Zak, "Berry Phase for energy Bands In Solids", *Phys. Rev. Lett.* **62**, 2747 (1989) .

[48] Alex Kamenev and Anton Andreev "'Electron-Electron Interactions in Disorder Metals : Keldish formalism"' *Phys.Rev.B***60**,2218 (1999)

[49] D.Schmeltzer ,"'Geometrical Non-Abelian Bosonization approach for the Two Dimensional Hubbard Model "' *Phys.Rev.B* **54**,10269,1996.

[50] D.Schmeltzer "'Quantum Criticality at the Metal Insulator Transition "' *Phys.Rev.***63**,075105 (2001).

[51] M.Malard et. al "'Interacting Quantum Wires"' *Physica B* **404**, 3155-3158 (2009)

[52] D.Schmeltzer et. al "' A scaling approach for Interacting Quantum wires"' *J.Phys.Condens. Matter* **22**, 095301 (2010)

[53] Alexei Yu Kitaev "'Unpaired Majorana Fermions In Quantum Wires"' cond-mat/ 0010440.

[54] Bryce S. DeWitt "' "'Dynamical Theory In Curved Spaces .I. A Review Of Classical and Quantum Action Principles"' *Review of Modern Physics* Vol.29 ,3,377 (1957).

[55] R.C.T. da Costa "'Quantum Mechanics Of A Constrained Particle"' *Physical Review* **A23**,1982(1981).

[56] P.Exner and P.Seba "'Bound States In Curved Quantum Waveguides"' *Jounal of Math. Phys.***30**,2574(1989.

[57] J.Goldstone and R.L. Jaffe "'Bound States In Twisted Tubes "' *Physical Review* **B45**,14100(1992).

[58] D.Schmeltzer ,"'Geometrical Non-Abelian Bosonization approach for the Two Dimensional Hubbard Model "' *Phys.Rev.B* **54**,10269,1996.

[59] Cenke Xu and J.E.More "'Stability Of The Quantum Spin Hall Effect :Effects Of Interactions ,Disorder,and Z_2 topology "' cond-mat/0508291.

[60] Andrew M. Essin Joel E.Moore and David Vanderbilt "'Magnetoelectric Polarizability and Axion Electrodynamics In Cristalline Insulators"' cond-mat/0810.2998

[61] N.D. Birrell and P.C.W.Davis "'Quantum Fields In Curved Space"' Cambridge University Press 1982.

In: Advances in Condensed Matter ... Volume 10
Editors: H. Geelvinck and S. Reynst

ISBN: 978-1-61209-533-2

Short Communication

AN ITINERANT-ELECTRON METAMAGNETIC SYSTEM AS A SYSTEM WITH TWO-BAND EIGENVALUE SPECTRUM

M. A. Grado-Caffaro and M. Grado-Caffaro*
Scientific Consultants, Madrid, Spain

ABSTRACT

A generic itinerant-electron metamagnetic system at low temperature is treated as a system of two-band eigenvalue spectrum within the Stoner model. In this context, we establish a mathematical expression for the electronic energy of the system in relation to the Fermi levels of the spin-up and spin-down electron bands and in relation to the magnetic susceptibility. In particular, as a relevant part of the electronic energy, we give an expression for the exchange energy between the spin-up and spin-down electrons.

Keywords: Itinerant-electron metamagnetism; Two-band eigenvalue spectrum; Stoner model; Electronic Energy; Exchange energy; Magnetic Susceptibility

1. INTRODUCTION

Itinerant-electron metamagnetism is a very interesting phenomenon which continues attracting the attention of the people really interested in knowing exciting features of magnetism in solids [1,2]. In order to understand the physics underlying the phenomenon in question, a deep understanding of the fundamental principles of itinerant-electron magnetism is needed (see, for instance, ref.[3]). On the other hand, we note that a lot of questions on metamagnetism remain open so that, from the theoretical standpoint, there are many

* Permanent address: M.A. Grado-Caffaro and M. Grado-Caffaro- Scientific Consultants, C/ Julio Palacios 11, 9-B, 28029-Madrid (Spain); www.sapienzastudies.com; e-mail address: ma.grado-caffaro@sapienzastudies.com.

unexplored issues relative to electronic energy, density of states, exchange energy, and susceptibility. Certainly, elaborating consistent theoretical models to get a real advancement on the above issues is highly desirable.

As a matter fact, the present communication presents a new theoretical formulation relative to the electronic energy of a generic itinerant-electron metamagnetic material within the framework of the well-known Stoner theory (see, for example, refs.[4,5]), which is valid for zero absolute temperature. In itinerant-electron metamagnetism, energy is electronic energy plus lattice energy. The electronic energy is considered as volume-independent in the current literature whereas the lattice energy changes with volume (see, for instance, refs.[4,5]). On the other hand, the electronic energy depends, among other quantities, on the exchange energy between spin-up and spin-down electrons (consider the Stoner model). In the following, we shall give a relationship for the aforementioned exchange energy and another one for magnetic susceptibility. To date, explicit quantitative relations for the above issues cannot be found in the literature.

2. THEORY

First of all, we define a system of two-band eigenvalue spectrum in the framework of the non-relativistic time-independent Schrödinger equation as a system which obeys the above equation through the corresponding Hamiltonian operator with ψ_{ij} as wavefunctions and E_{ij} as energy eigenvalues of the above operator, $i = 1,2$ being the band index and $j = 1,2,...$ labelling the states within the bands. We take the ground state of the system as a state such that all the states in the $i = 1$ band are occupied and all the states in the $i = 2$ band are empty while the excited state is such that all the states in the $i = 2$ band are occupied and all the states in the $i = 1$ band are empty.

Within the Stoner theory, the electronic energy at $T \approx 0^\circ K$ of an itinerant-electron metamagnetic system under a magnetic field is given by (see, for example, refs.[4,5]):

$$E_e(W) \approx \int_{-W}^{E_{F1}} Eg(E,W)dE + \int_{-W}^{E_{F2}} Eg(E,W)dE - \frac{1}{4}\chi^2 JH^2 - \mu_B \chi H^2 \qquad (1)$$

where E stands for energy, W is the energy half-bandwidth, g denotes electronic density of states, E_{F1} and E_{F2} are the Fermi levels of the spin-up electrons and spin-down ones respectively, χ is magnetic susceptibility, J is the exchange energy between the spin-up and spin-down electrons, μ_B is the Bohr magneton, and H is the magnitude of the strength of the applied magnetic field. We will denote by $i = 1$ the spin-up electron band while the spin-down electron band will be designated with $i = 2$. The system in question is assumed to be isotropic (χ is a scalar quantity instead of a tensor as in crystalline lattices) so that this system is at least partially disordered, that is, a relatively distorted lattice (here, of course, this does not mean magnetically disordered).

The exchange energy J is approximately proportional to the square of the difference of the Fermi energies of the spin-up and spin-down electron bands and inversely proportional to the energy of the excited state of the system minus the energy of the ground state so we have:

$$J \propto (E_{F1} - E_{F2})^2 \sum_{i,j} \frac{1}{E_{2j} - E_{1i}} \quad (2)$$

From a generalized model related to a Wannier-type formulation in terms of both coordinate and angular momentum operators [6], we can express the susceptibility as:

$$\chi \propto \sum_{i,j} \frac{\langle \psi_{1i} | \hat{L} | \psi_{2j} \rangle \langle \psi_{2j} | \hat{L} | \psi_{1i} \rangle}{E_{2j} - E_{1i}} \quad (3)$$

where $\hat{L}$ is the angular momentum operator.

Then, by inserting (2) and (3) into (1) and knowing $g(E,W)$, calculation of the electronic energy is feasible. On the other hand, one has that $\chi = \chi_1 + \chi_2$ where χ_1 is the contribution to the susceptibility from the spin-up electron band ($i = 1$) and χ_2 is the contribution due to the spin-down electron band ($i = 2$). Consequently, from formula (3) it follows:

$$\chi_1 \propto \sum_{j=1}^{\infty} \frac{\langle \psi_{11} | \hat{L} | \psi_{2j} \rangle \langle \psi_{2j} | \hat{L} | \psi_{11} \rangle}{E_{2j} - E_{11}} \quad (4)$$

$$\chi_2 \propto \sum_{j=1}^{\infty} \frac{\langle \psi_{12} | \hat{L} | \psi_{2j} \rangle \langle \psi_{2j} | \hat{L} | \psi_{12} \rangle}{E_{2j} - E_{12}} \quad (5)$$

CONCLUSION

In the context of the Stoner model, we have established formulas (2) and (3) which represent significantly new results valid for temperatures near $0^{\circ} K$. Certainly, our preceding formulation will serve as a key point of reference for progressing towards the full elucidation of the main aspects related to itinerant-electron metamagnetism (one of these aspects refers to electronic density of states). Really, finding explicit mathematical expressions for J and χ is a useful task. In particular, we want to remark the relevance of formula (2) which gives the exchange energy between spin-up and spin-down electrons.

REFERENCES

[1] T. Goto, K. Fukamichi, H. Yamada: Itinerant electron metamagnetism and peculiar magnetic properties observed in 3d and 5f intermetallics, *Physica* B 300 (2001) 167-185.

[2] T. Goto, H. Katori, T. Aruga Sakakibara, H. Mitamura, K. Fukamichi, K. Murata: Itinerant electron metamagnetism and related phenomena in Co-based intermetallic compounds, *J. Appl. Phys.* 76 (1994) 6682-6687.

[3] J. Kubler: Theory of Itinerant Electron Magnetism (Oxford University Press, Oxford, 2009).

[4] N.H. Duc, D. Givord, C. Lacroix, C. Pinettes: A new approach to itinerant-electron metamagnetism, *Europhys. Lett.* 20 (1992) 47-52.

[5] M.A. Grado-Caffaro, M. Grado-Caffaro: Some considerations on electronic energy of itinerant-electron metamagnetic materials with volume-dependent exchange energy, *Active and Passive Elec. Comp.* 24 (2001) 63-67.

[6] M.A. Grado-Caffaro, M. Grado-Caffaro: A model for quantum systems of two-band eigenvalue spectrum, *Mod. Phys. Lett.* B 18 (2004) 1449-1452.

INDEX

A

B

C

D

E

F

G

H

I

J

K

L

M

N

O

P

Q

R

S